CALCULUS USING *MATHEMATICA*

CALCULUS USING *MATHEMATICA*

K.D. STROYAN

University of Iowa
Iowa City, Iowa

Academic Press, Inc.
A Division of Harcourt Brace and Company

Boston   San Diego   New York
London   Sydney   Tokyo   Toronto

Mathematica is a registered trademark of Wolfram Research, Inc.
Macintosh is a registered trademark of Apple Computer, Inc.
Windows is a trademark of Microsoft Corporation
NeXT is a registered trademark of NeXT Computer, Inc.

ACADEMIC PRESS, INC.
1250 Sixth Avenue, San Diego, CA 92101-4311

United Kingdom Edition published by
ACADEMIC PRESS LIMITED
24–28 Oval Road, London NW1 7DX

ISBN: Macintosh: 0-12-672971-9
       IBM/DOS: 0-12-672972-7
       NeXT: 0-12-672973-5

Printed in the United States of America
93 94 95 96 MV 9 8 7 6 5 4 3 2 1

Contents

Part 1
Differentiation in One Variable

Part 2
Integration in One Variable

Part 6
Infinite Series

Contents of Mathematica NoteBooks
for
Calculus Using *Mathematica*

Chapter 1. Introduction

aMathcaIntro.ma introduces the Mathematica 'front end' or NoteBook Editor with open and closed cells. It also gives a brief tour of the various kinds of calculations that are possible in Mathematica and leads into the work of Chapters 2 and 3.

> Exact Arithmetic
>
> Floating Point (Approximate) Arithmetic
>
> Symbolic Computations
>
> Graphics
>
> Lists

Part I. Differentiation in One Variable

Chapter 2. Using Calculus to Model Epidemics

FirstS-I-R.ma checks the hand calculations done in solving the first S-I-R model. It also provides an introduction to variable assignment and simple editing in Mathematica. It leads into the loop calculation for the second model.

SecondS-I-R.ma does more sophisticated calculations with the S-I-R model by varying the step size and producing graphs of the solutions. It also provides an introduction to some Mathematica programming structures like the Do loop.

EpidemicRoots.ma describes how to calculate limiting values of the S-I-R model. It does this with Mathematica's numerical root finding algorithm, which it introduces.

Chapter 3. Numerics, Symbolics and Graphics in Science

Functions.ma provides an introduction to function notation in Mathematica. Examples of numerical calculations with functions are given as well as symbolic computations. An example of a function that is actually a procedure is also given.

SlideSquash.ma introduces animations. A series of graphs representing parabolas with a parameter varied are generated and combined into a movie. This provides a dynamic representation of translation and expansion and shows how these are represented analytically.

ExpGth.ma (*see also* Chapter 8) is a demonstration of how rapid exponential growth is. Starting from a simple model with algae cells doubling every 6 hours, the NoteBook demonstrates that 1000 algae cells would completely fill Lake Michigan in only 15 days.

LogGth.ma (*see also* Chapter 8), in contrast to the previous NoteBook, demonstrates how slow logarithmic growth is. A computer adding ten billion terms of the harmonic series every second still takes 3.1×10^6 ages of the solar system to get to 100.

Chapter 4. Linearity vs. Local Linearity

Zoom.ma produces an animation of a graph expanding. The section of the graph to be blown up is surrounded by a small box and local coordinate axes are displayed. This is the live geometric version of the main idea of differential calculus: smooth curves appear linear under powerful magnification.

NonDiffble.ma shows Weierstrass's nowhere differentiable function. Not all functions are smooth and this one is 'kinky' at every point.

Chapter 5. Direct Computation of Increments

Differences.ma illustrates the difference quotient limit approaching the derivative function. It is another way to see the main approximation of differential calculus.

SymbolicIncrem.ma calculates symbolic increments using Mathematica.

Microscope1D.ma animates the main idea of differential calculus, namely that small changes in differentiable functions are locally linear. It also shows how to pre-compute the linear functions with Mathematica. In effect, calculus lets us 'see' with one eye in the microscope without opening the other eye to see the whole graph. Rules tell us what we will see in the microscope.

Chapter 6. Symbolic Differentiation

DiffRules.ma defines rules for a function that allow the function to perform symbolic differentiation. The rules are defined in the same order as the rules for differentiation are presented in the text, so that at any point in learning rules, this symbolic differentiator can only do the problems to which those rules apply.

Chapter 7. Basic Applications of Differentiation

Dfdx.ma shows how to use the built-in Mathematica function for differentiation.

Chapter 8. The Natural Logarithm and Exponential

EulerApprox.ma shows the discrete Euler approximations to $dy = y\, dt$ converging to $e^{\wedge}t$. This illustrates the 'official' definition of the natural exponential function.

ExpDeriv.ma (*see also* Mathematical Background) approximates $d(b^{\wedge}t)/dt$ directly to find Euler's $e = 2.71828 \ldots$ as the base that has constant of proportionality 1.

ExpGth.ma (*see also* Chapter 3) is a demonstration of how rapid exponential growth is. Starting from a simple model with algae cells doubling every 6 hours, the NoteBook demonstrates that 1000 algae cells would completely fill Lake Michigan in only 15 days.

Chapter 8 contains the mathematical 'order of infinity' result that says exponentials grow faster than powers.

LogGth.ma (*see also* Chapter 3), in contrast to the previous notebook, demonstrates how slow logarithmic growth is. A computer adding ten billion terms of the harmonic series every second still takes $3.1 \times 10^{\wedge}6$ ages of the solar system to get to 100.

Chapter 8 contains the mathematical 'order of infinity' result that says powers grow faster than logs.

Chapter 9. Graphs and the Derivative

PlanckL.ma solves for Wein's Law of Radiation using the wavelength form of Planck's Law.

PlanckF.ma solves for Wein's Law of Radiation using the frequency form of Planck's Law.

Chapter 10. Velocity, Acceleration and Calculus

Gravity.ma contains data describing the fall of a body in vacuum. By performing operations on this data, the students can derive Galileo's Law that the acceleration of a falling body is constant (in the absence of air resistance). Students are also asked to reject Galileo's first conjecture that speed is proportional to the distance fallen.

AirResistance.ma (*see also* Scientific Projects) contains data on the fall of a body influenced by air resistance. By performing calculations with the data, students are able to calculate the coefficient of air resistance and develop a model for the fall of a body under the effects of air resistance.

Chapter 11. Maxima and Minima in One Variable

SolveEquations.ma Mathematica root finding is used in student-written NoteBooks to solve max-min problems that are intractable by hand, such as the distance from a point to a curve. See also the Geometric Optimization Exercises in the Mathematical Background.

Chapter 12. Discrete Dynamical Systems

FirstDynSys.ma solves difference equations by iterating the initial condition. It also draws graphs and shows the 'cobweb' iteration diagram.

InitialConds.ma solves difference equations with several initial conditions and shows the 'flow' of simultaneous solutions.

Whales.ma (*see also* Science Projects) solves a difference equation of order 9. This NoteBook comes from a report to the International Whaling Commission and is used in a scientific project.

Part 2: Integration in One Variable

Chapter 13. Basic Integration

Sums.ma shows how Mathematica handles the algebra of summation and contains exercises needed to build the theory of integration.

GraphIntAprx.ma shows various graphical approximations to integrals by left and middle rectangles and by trapezoids.

NumIntAprx.ma computes various numerical approximations to integrals.

Chapter 14. Symbolic Integration

SymbolicIntegr.ma shows how to use Mathematica symbolic integration and discusses several non-elementary integrals.

Chapter 15. Applications of Integration

SliceXamples is a folder of NoteBooks containing many 3-dimensional examples of slicing figures into disks, segments, etc.

Part 3: Vector Geometry

Chapter 16. Basic Vector Geometry

Vectors.ma draws vectors.

Chapter 17. Analytical Vector Geometry

Circles.ma is a basic introduction to Mathematica's ParametricPlot[] function, beginning with an animation of the parametric circle.

Cycloid.ma draws cycloids, epicycloids and hypocycloids, also known as spirographs. The idea is to trace the path of a point on a rolling wheel. The technical tools are the parametric circle and vector addition.

EpiCycAnimate.ma animates a wheel of radius 1 rolling around a wheel of radius 3. Contrary to a hasty 'intuitive' solution, the small wheel turns 4 times in one loop around the big wheel.

Chapter 18. Linear Functions and Graphs in Several Variables

Surface Slices builds up a surface plot with the curves of successive slices and shows the result in an animation.

Explicit Plots teaches you how to use the Plot3D[] function and draw the explicit surface graphs of the functions in text Exercise 15.6.

Contour Plots contains the contour graphs of the functions from text Exercise 15.6. You are to compare these with your solution to the NoteBook ExplicitSurfaces.ma. To make things interesting, one of the contour plots is deliberately WRONG. You are to find it.

FlyBy Surfaces The Mathematica function FlyBy[] helps to visualize a surface. It creates an animation that is analogous to getting in a plane and flying by the surface.

Part 4: Differentiation in Several Variables

Chapter 19. Differentiation of Functions of Several Variables

Microscopes in 3-D 'zooms' into the graph of a function of two variables. It does this for the explicit graph, the contour graph, or a density graph of the function.

Total DifflGrfs computes and plots the total differentials of the functions in Exercise 16.8.

Chapter 20. Maxima and Minima in Several Variables

Maxmin gives Mathematica solutions to the various equations on max-min examples from the text and can be used as a template for further multivariable max-min.

Part 5: Differential Equations

Chapter 21. Continuous Dynamical Systems

EulerApprox.ma shows an animation of better and better discrete "Euler" approximations to the solution of a differential equation.

EULER&Exact.ma shows a discrete Euler approximation and the exact solution of a differential equation.

AccDEsoln.ma is an accurate differential equation solver that is functionally similar to the Euler methods studied in detail by students. Without laboring the details, this is a way to get more accurate numerical solutions.

DirField.ma plots direction fields for two dimensional dynamical systems and plots 3-D vectors along the gradient tangent to a surface.

Flow1D shows an animation of a one-dimensional flow and the associated explicit solutions of a differential equation.

Flow2D makes an animated flow associated with a pair of autonomous differential equations. (It is based on a fixed step Runge–Kutta method.)

Xamples

> CowSheepFlow shows a case of competition between two species.
>
> FoxRabbitFlow This flow animation is the classical Lotka–Volterra predtor–prey system, which is studied in more detail in the projects.

Chapter 22. Autonomous Linear Dynamical Systems

UnforcdSpring.ma makes an animation of the oscillation of a spring without external forcing. The motion is given by the differential equation

$$m\,x'' + c\,x' + s\,x = 0.$$

Chapter 23. Equilibria of Continuous Dynamical Systems

Five Cases of Linear Equilibria

The following separate NoteBooks animate the main types of dynamic equilibria.

> NegativeRoots. Negative characteristic roots make the origin an attractor.
>
> OppositeRoots. Opposite sign characteristic roots make the origin attractive in one direction and repulsive in others.
>
> RepeatedRoots. A repeated negative root makes the origin an attractor, but with one invariant line.
>
> ImaginaryRoots. Purely imaginary roots make the origin a neutrally stable equilibrium; perturbations oscillate indefinitely.
>
> ComplexRoots. Complex roots with negative real parts make the origin a spiral attractor.

LocalStability.ma may be used to compute the symbolic criteria for stability of equilibria in nonlinear systems.

Part 6: Infinite Series

Chapter 24. Geometric Series

ConvergSeries.ma illustrates convergence of the Taylor series of some of the most basic functions.

Chapter 25. Power Series

FormalTSeries.ma shows how to use Mathematica's built-in formal Taylor series.

BesselSeries.ma illustrates the Taylor Series approximation of Bessel function.

AbsSeries.ma The absolute value function has a series which can be built up from powers, however, this is not in the form we call "power series." This NoteBook shows an animation of this convergence. The limit of this series is not differentiable at the kink in the absolute value graph.

Chapter 26. The Edge of Convergence

FourierSeries.ma *(see also* Math Background) shows animations of convergence for several Fourier series.

Scientific Projects

co2 notebook.ma *(see* Chapter 2, Linearity vs. Local Linearity) contains data on the percentage of carbon dioxide in the atmosphere measured since 1958. The NoteBook approximates the data by a simple linear function and uses the linear function to make predictions.

This NoteBook shows the danger of using a linear approximation to project too far into the future. The linear approximation of calculus is only 'local.'

FluDataHelp.ma helps you work the project on matching the theoretical predictions of the S-I-R model to the actual 1968 Hong Kong Flu epidemic as described in the Scientific Projects chapter on Epidemiological Applications.

LadderHelp.ma is a NoteBook to help solve the project on the ladder from the Scientific Projects chapter on Applications to Mechanics.

AirResistance.ma contains data on the fall of body influenced by air resistance. By performing calculations with the data, you can calculate the coefficient of air resistance and develop a model for the fall of a body under the effects of air resistance. This is a project in the Scientific Projects chapter on Applications to Mechanics.

BungeeHelp.ma helps you to solve the project on bungee diving from the Scientific Projects chapter on Applications to Mechanics.

OpticsHelp.ma is a Mathematica package to reflect parallel 'rays of light.'

DrugData.ma is a Mathematica NoteBook for the project on fitting data for drug concentration to the biexponential model.

BytieHelp.ma helps you work the project on discrete price adjustment in the model economy the Scientific Projects on Applications to Economics.

WhaleHelp.ma solves a difference equation of order 9. This NoteBook comes from a report to the International Whaling Commission and is used to study the sustainable harvest level for this particular species of whale. See the section on sustained harvest of whales in the Scientific Projects on Applications to Ecology.

Forced Springs

> SpringFriction shows the effect of varying the forcing frequency on a linear oscillator: mass, spring, and shock absorber.

> Resonance shows the effect of varying the forcing frequescy on a linear oscillator. A maximal response is called a resonant frequency.

Jupiter.ma shows how to use Jupiter as a slingshot to send a satellite out of the solar system.

Lissajous.ma helps you compute trajectories of the spring pendulum from the Scientific Projects chapter on Differential Equations from Vector Geometry.

Mathematical Background

ExpDeriv.ma helps you approximate $d(b\char94 t)/dt$ directly to find Euler's $e = 2.71828 \ldots$ as the base that has constant of proportionality 1.

CoolHelp.ma examines some students' data on cooling and compares it with Newton's Law of Cooling. It is the project suggested in the Mathematica Background Section of the Canary Resurrected.

InvFctHelp.ma calculates the inverse of the Sine function directly and is intended to help you find the inverse to the function $y = x\char94 x$ needed in the project in the Mathematical Background chapter on Inverse Functions.

LeastSquares.ma goes with the Mathematical Background Chapter 9 on Least Squares fit of data.

VanderFlow.ma The Van der Pol nonlinear oscillator is mathematically very interesting because it has a single stable limit oscillation, rather than an attracting equilibrium point. This NoteBook illustrates the flow described in The Mathematical Background chapter on Theory of Initial Value Problems.

Fourier Series Animations showing convergence of several Fourier series.

Preface

Calculus is primarily important because it is the language of science. It is profound mathematics and a key to understanding physical science and engineering, but calculus also has an expanding role in economics, ecology, and some of the most quantitative parts of business and social science. Beginning college students should learn calculus, because without calculus they close the door to many scientific and technical careers. This book is intended for beginning students who want to become users of calculus in any one of these areas.

A vast majority of our students are convinced of the importance of calculus by working on problems that they find interesting. The primary goal of this new calculus curriculum is to show students first hand how calculus acts as the language of change and to answer their question, 'What good is it?' We 'show' students this by developing their basic skills and then having them apply those skills to their choice of topics from a wide variety of scientific and mathematical projects. Students themselves answer such questions as,

Why did we eradicate polio by vaccination, but not measles?

They present their solution in the form of term papers. (Three large papers and several small ones per year.) This core text does NOT stand on its own; rather it is one of four parts of our materials:

Core text:	***Calculus using Mathematica***
Science projects:	***Scientific Projects for Calculus using Mathematica***
Computing:	***Mathematica NoteBooks for Calculus using Mathematica***
Math projects:	***Mathematical Background for Calculus using Mathematica***

Computing with *Mathematica* has changed both the topics we treat and the way we present old topics. It allows us to achieve our primary goal of having students work real scientific and mathematical projects within the first semester. A number of topics formerly considered too advanced are a major part of our course. *Mathematica* can numerically solve basic differential equations and create a movie animating its 'flow.' This allows us to treat deep and important applications in a wide variety of areas (ecology, epidemiology, mechanics), while only developing basic skills about traditional exponential functions. The ingredients in studying nonlinear 2-D systems are high school math for describing the law of change and exponentials for local analysis.

Mathematica has accessible 3-D graphics, which it can also animate, so we can study problems in more than one variable in the first year. Most problems in science have more than one variable and several parameters.

Mathematica has a convenient front end editor (called NoteBooks) that helps us keep the 'intellectual overhead' to a minimum. Our intention is to use computing to study deep mathematics and applications, not to let the tail wag the dog. We weave *Mathematica* into the fabric of the course and introduce the technical features gradually.

Mathematica also helps students learn the central mathematics of calculus. Our students learn the basic skills of differentiation and integration, but also learn how to use *Mathematica* to perform very elaborate symbolic and numerical computations. We don't labor some of the esoteric 'techniques of integration,' or complicated differentiations. If students' basic skills are backed up with modern computing, it is not necessary to drill them ad nauseam in order to make them proficient mathematical thinkers and users of calculus. Our students demonstrate this in several major term papers on large projects. In addition, their performance on traditional style tests is very good (though the tests only comprise half of their grade.) Good understanding of the main computations and knowledge of how to use them with help from modern graphic, numeric and symbolic computation focuses our students' efforts on the important issues.

How Much Does it Count?

Our course grades were first derived in the following way, though we are shifting credit away from exams as we go.

Traditional Skill Exams (computers not available)
   Exam #1 -15%
   Exam #2 -15%
   Final Exam - 20%
Daily Homework (about 3/4 traditional drill and word problems) - 15%
Weekly electronic homework (submitted electronically) - 10%
Term paper length projects (almost all use computing) - 25%

Suggested Course Syllabi

Two suggested course syllabi that we have used are included in the *Mathematica* Note-Books. The accelerated syllabus is for students who are very well prepared in high school math. We have used it at Iowa and BYU. The regular syllabus is for students with ordinary preparation for mainstream calculus in college. We have used it at Iowa, UNC and UW-L.

Acknowledgments

Many people contributed ideas and effort toward developing these materials. The National Science Foundation made it all possible.

Undergraduate students at the University of Iowa, The University of Northern Colorado, The University of Wisconsin - LaCrosse, and Brigham Young University have worked hard in response to our taking them seriously. They showed us which applications of calculus they find interesting. Almost all the projects were developed in close collaboration with students in the course or with undergraduate assistants who took the course.

Faculty and graduate students at these schools also contributed a great deal. The faculty leaders were Walter Seaman at Iowa, Steve Leth at UNC, John Unbehaun at UW-L and Gerald Armstrong at BYU. We have been blessed at Iowa with many marvelous graduate assistant teachers. Francisco Alarcon, Randy Wills, Asuman Oktac, Robert Dittmar, Monica Meissen, Srinivas Kavuri, Kathy Radloff, Robert Doucette, Oki Neswan and others helped make the course a success from the start.

Ideas from other new calculus projects also contributed to our developments. Frank Wattenberg at the University of Massachusetts was especially helpful. The philosophy of the Five College Project, which Wattenberg's materials spin off, contributed a lot to our course. Lang Moore at Duke, Elgin Johnston and Jerry Mathews at Iowa State traded ideas for student projects with us. Deborah Hughes-Hallett, Andrew Gleason and David Lomen of the Harvard project shared rough ideas and offered encouragement at many meetings. Jerry Uhl, Horacio Porta and William Davis of the *Calculus & Mathematica* project did likewise. Our calculus projects share a common goal to show students that calculus is important and thus help train the 21st century's scientists, engineers, mathematicians and technical managers. We are all doing this in different ways and at different levels. The successful feature common to our new courses is: more actively involved students.

CHAPTER 1

Introduction

This book (with its software called "*Mathematica* NoteBooks") is a new approach to an old subject: calculus. Calculus means stone or pebble in Latin. Roman numerals were so awkward that ancient Romans carried pebbles to compute their grocery bills (and gamble), and as a result "calculus" also came to be known as a method for computing things. The specific calculus we will study is "differential and integral calculus," formerly called "infinitesimal calculus." The fundamental things this calculus is used to compute are rates of change of continuous variables through differentials and the inverse computation: accumulation of known rates of change through integrals. A great deal has been built on this foundation.

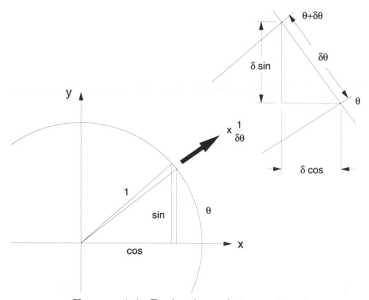

FIGURE 1.1: Derivatives of sine and cosine

Calculus is one of the great achievements of the human intellect. It has served as the language of change in the development of scientific thought for over three centuries. During that time it developed into a coherent polished subject and guided the development of physical science and technology. The contemporary importance of calculus is actually expanding into economics and the social sciences as well as continuing to play a key role in its traditional areas of application. Our first view of calculus in Chapter 2 previews its use in the study of epidemic diseases, where the models can show us such things as the differences in successful vaccination strategies for polio and measles. Similar models have made national health policy changes in screening for gonorrhea and are being used to help combat the spread of AIDS.

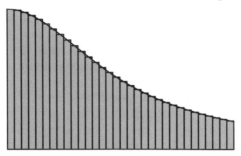

FIGURE 1.2: Arctangent as an Integral

The model in Chapter 2 is a system of differential equations, so it is technically a little ahead of the story. We won't find explicit solutions to systems of differential equations on the second day of this course. The important thing we are trying to bring out in Chapter 2 is that it is relatively easy to describe the spread of a disease like measles by writing formulas that say how it changes. One new thing in our approach to calculus is that you will be expected to use calculus to describe applications that interest you. (We offer many choices, including mathematical ones, and you are encouraged to devise your own.) This will show you the role of calculus as a language – you will have to "speak" it.

The other new thing about our course is our computer laboratory. We will use a scientific software language called *Mathematica* that is both powerful and relatively easy to use. In Chapter 2 we will use *Mathematica* to overcome the technical difficulties in our system of differential equations. By describing epidemic changes in *Mathematica* notation we will immediately get approximate answers to many questions about the spread of S-I-R diseases. Throughout the course we will use *Mathematica* to both help us understand and explore mathematics and to expand the power and applicability of the mathematics we learn. You will leave this course with a good start on learning 21st century scientific computing.

You will also learn all the old things from regular calculus in this course. In particular, you will learn to compute derivatives and integrals. Good high school algebra preparation and effort during the semester are all you need. The rote computations that are often over-drilled in regular calculus take practice, but not very much if you are proficient at high school math. If you make this effort, you will be well prepared for your later science and math courses. *Mathematica* may even help you check some of your rote skills. You will be far better prepared than regular calculus students in understanding the role of calculus in science and mathematics.

We expect you to work hard and stay up-to-date. The computer lab is also staffed for you to get individual help on regular assignments and to work with partners on projects - it's not just for computers.

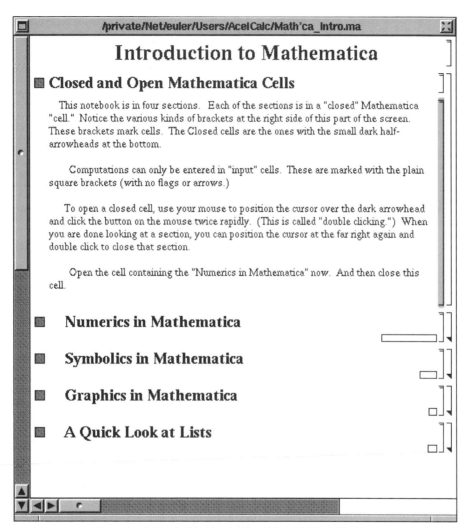

FIGURE 1.3: A Mathematica NoteBook Window

1.1. A *Mathematica* Introduction with aMathcaIntro.ma

The main goal of this chapter is to have you run your first *Mathematica* program. *Mathematica* programs are called "NoteBooks." NoteBooks include a very convenient editor that has "cells" that you can open and close for ease of reading. NoteBooks can include instructions to you, the computation instructions for the computer, places for you to compute, and places to type your own interpretation of the results. *Mathematica* does numerical computations, symbolic computations, and graphs. All these can be combined in a NoteBook.

The NoteBook **aMathcaIntro.ma** uses the features of the editor to give you an introduction to *Mathematica* . When you open this NoteBook on your computer you will see a window that looks something like the one above. (The control buttons may be a little different.)

1.2. *Mathematica* on a NeXT

The following exercises introduce you to a NeXT/*Mathematica* computer network. Separate sections following this one introduce *Mathematica* on Macintosh and on IBM computers. *Mathematica* is the nearly the same on all of these computers, but the details of opening or saving *Mathematica* NoteBooks varies from machine to machine, and even sometimes depends on local network software. Go to your computer lab to complete the first *Mathematica* NoteBook, which is an introduction to *Mathematica*. In the process, you will learn about your particular machine. This may be a little frustrating at first, because you are using *Mathematica* to learn about *Mathematica*, but once you get started, you will see that *Mathematica* NoteBooks are a very helpful and powerful form of scientific computer program. The NoteBooks themselves contain almost all the information about computers that you will need in the course.

You need to learn to "log in" and do "mouse" editing. These things are fussy at first, but easy to do once you get the hang of it.

EXERCISE 1.4. *Log in, change your password, and log out.*

When you first sit down at the NeXT, you will see the following login panel.

FIGURE 1.5: Login Panel

To log in to the network, type your user name in the box marked name. Usually your user name will be the first letters of your last name entered in small letters. After you have entered you user name, press the Return key. If you make a mistake, you can use the Delete key to backspace. You will not have a password at first, so if you press Return again, you will be logged on to the network.

One of the first things you will see after logging in to the network is a column of small pictures on the right side of the screen. We will refer to these pictures as "icons". They all represent applications that are available to run on NeXT. If you position your cursor over any of the icons and click the mouse twice (this is called "double-clicking"), NeXT will start running that application. The icon for *Mathematica* is the left part of the figure below.

FIGURE 1.6: The *Mathematica* Icon and Preferences Icon

The first thing we want you to do after logging on is to change your password. To do this double-click on the Preferences icon shown above (with the date and time). After a few moments you will now see a window that looks like this:

FIGURE 1.7: Password Panel in Preferences

Click the mouse on the lock icon so that it has a light background as shown. You have no password to start, so you can first press return. Then enter your new password and press return. After you have entered and verified your password, a message "Password Changed" will appear. After this, click "Quit" on the Preferences Menu, in the upper left part of the screen. The menu looks like this:

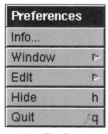

FIGURE 1.8: Preferences Menu

You can experiment with changing other things as well. Clicking on the clock will allow you to personalize your preference icon. Please don't adjust the power button, however; this can cause problems for everyone.

When you have successfully changed your password, you can log out. To do this, double click on the NeXT icon in the upper right corner of the screen. The NeXT icon looks like this:

FIGURE 1.9: Workspace Icon

When you have double clicked on it the Workspace menu will appear in the upper left corner of the screen.

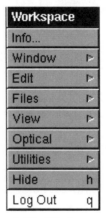

FIGURE 1.10: Workspace Menu

Now simply click on "Log Out." NeXT will ask you to verify that you want to log out. If you are ready to log out click on "OK".

EXERCISE 1.11. *Log on to your account with your new password, run the NoteBook aMathcaIntro.ma, and save your work on your account.*

To open NoteBook **aMathcaIntro.ma**, use the browser. Your browser will look like the following panel. (Your browser will probably be larger.) Ask your instructor for the name of the directory that contains the course NoteBooks and click on it to turn it white. Next, double click on the NoteBook ,**aMathcaIntro.ma**, to turn it white and run it (Single clicking only turns the background white).

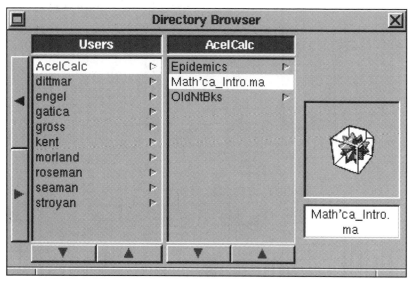

FIGURE 1.12: File Viewer or Browser

After you have done this a window will open on your screen with the *Mathematica* intro-duction. One version of that window is shown above. This window has features common to all windows on the NeXT. The small arrows at the lower left allow you to move in the window by clicking on them. The small X in the upper right corner will close the window if you click on it. You can move windows around the screen by clicking on the title at the top of the window and holding the mouse button down. Then if you move the mouse, you will move the window. If you click and hold down the mouse button on the lower right or lower left corners of a window you can change its size. Before starting the NoteBook you may want to do some experimenting with the window to get some practice manipulating windows.

Read the instructions in the NoteBook and do the exercises in it. Ask a lab monitor for help if you need it. Things may seem difficult at first, but with a little practice you will quickly become proficient. Don't let a fussy computer detail stop you — ask for help.

When you are done with the introductory NoteBook, save it to your account. To do this click on "Window" in the *Mathematica* menu that is in the upper left corner of the screen. The menu looks like this:

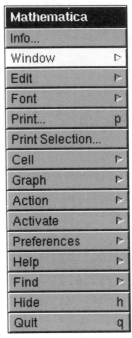

FIGURE 1.13: *Mathematica* Menu

After you have clicked on "Window" a sub-menu will appear off to the side. Click on "Save As" on this menu.

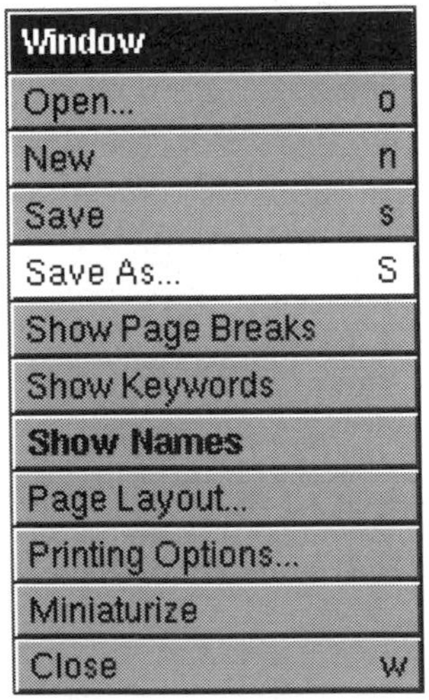

FIGURE 1.14: SaveAs in *Mathematica*

When you do this, the following menu will appear.

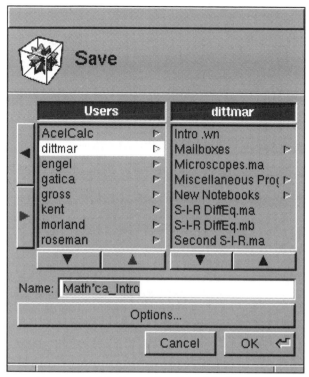

FIGURE 1.15: Save Panel in *Mathematica*

Click on your name to get to your directory, and then enter a name like

HmWk30Aug.ma

for your completed NoteBook. Finally click on "OK", and *Mathematica* will save the file to your directory. If you use the name "HmWk30Aug.ma" (with the due date of your assignment), we can find and grade your work.

When you are done doing the exercises in the introductory NoteBook, make sure you have saved your work and then log out.

1.3. *Mathematica* on a Macintosh

This section outlines the use of *Mathematica* on a Macintosh computer. The programs you will run are called "*Mathematica* NoteBooks." They will be available on one of your disk drives, either on your individual machine or on your network. If your machines are networked, you need to learn about the details of your local network from your instructor. These vary from place to place and your machines may not even be networked.

EXERCISE 1.16. *If your Macintosh computers are networked, log in, change your password, and log out.*

You need to learn to do "mouse" editing. There is a special program 'tour of your Macintosh' that may help get you started. Of course, it is a little frustrating sometimes to use the computer to get an introduction to the computer. Stick with it; once you learn a little "mouse editing" you will find that almost all the computer information you need is contained in the *Mathematica* NoteBooks. You could also ask a friend or lab monitor to help get you into the introductory *Mathematica* NoteBook.

Except for the location of the control buttons, *Mathematica* NoteBooks look and run the same way on a Macintosh, NeXT, or DOS Windows 3.0 machine.

EXERCISE 1.17. *Log on to your account with your new password, run the NoteBook **aMathcaIntro.ma**, and save your work on your account.*

Read the instructions in the NoteBook and do the exercises in it. When you are done with the introductory NoteBook, save it to your account. Your instructor will tell you how you should save your homework. This depends on the kind of network and disk drives that are installed with your system.

1.4. *Mathematica* on DOS Windows (IBM)

This section outlines the use of *Mathematica* on a DOS Windows 3.0 (IBM) machine. The Windows version of *Mathematica* is functionally similar to the one for NeXT or Macintosh. Your DOS Windows machines may be networked or 'stand alone.' Your instructor will tell you which DOS directories contain the course programs called *Mathematica* NoteBooks.

EXERCISE 1.18. *If your DOS Windows system is networked, log in, change your password, and log out.*

You need to learn to drive a "mouse." This is almost a non-verbal motor skill, and it is a little frustrating sometimes to use the computer to learn about the computer. Don't be afraid to ask for help. Once you learn a little "mouse editing" you will find that almost all the computer information you need is contained in the *Mathematica* NoteBooks.

EXERCISE 1.19. *Log on to your account with your new password, run the NoteBook **aMathcaIntro.ma**, and save your work on your account.*

Read the instructions in the NoteBook and do the exercises in it. Ask a lab monitor for help if you need it. When you are done with the introductory NoteBook, save it to your account. Your instructor will tell you how you should save your homework. This depends on the kind of network and disk drives that are installed with your system.

1.5. Free Advice

We know that these first two exercises take a long time and may be a little frustrating for people who are not used to "windows" and "mouse" editing. These initial frustrations will seem very simple in a week or so if you confront them now. DON'T GET BEHIND IN YOUR HOMEWORK.

We encourage you to work with a friend on the computing in this course. Often two heads are better than one, because one of you can concentrate on the machine details while the other thinks about the main mathematical task you are trying to accomplish. As long as you change roles occasionally, you will both learn faster.

1.5.1. Drill Exercises. There are several kinds of problems for you to work in this course. Routine algorithms that just need practice and not much thinking are called drill exercises. Some of these are expanded on in the Mathematical Background book that accompanies this core text. For example, there are proficiency tests there on skills of differentiation and integration. Some instructors like to give similar tests as 'barrier exams.' This means that you have several chances to pass a skill exam like the ones in the background, but that you must get 90% of the problems completely correct. You don't get a grade for the skill test, but can not pass the course unlesss you pass it. The point is that these are necessary skills, but that the skills alone do not mean you understand the material.

1.5.2. Exercises. Regular exercises may require a little more thought about the current material than drill exercises. Calculus is important because the basic skill algorithms have important meanings. The skills themselves are not the whole story and exercises sometimes take a step in the direction of their meaning. Exercises are sometimes 'electronic,' that is, require you to use a *Mathematica* NoteBook. Most instructors assign one large electronic homework assignment per week.

1.5.3. Problems. "Problems" are larger exercises that are meant to help you organize your thinking. They go beyond exercises in that they usually have several parts and ask you to write summary explanations of the combined meaning of the parts. Your written explanations are very important. You must use complete English sentences and complete mathematical thoughts. You will find that it is sometimes difficult to put your ideas in words. You 'understand,' but find it hard to explain the problem. You will find that wrestling with this difficulty improves your understanding and helps you to combine the parts of calculus that you have learned into a coherent understanding.

1.5.4. Projects. "Projects" are larger problems, serious applications of calculus, not fragmented parts of applications or text exercises contrived to come out in whole numbers. Scientific Projects comprise another book accompanying the core text. Mathematical projects are included in the Mathematical Background book. Some of the projects are not very demanding technically and will help you get used to writing technical reports about mathematics. For example, the scientific projects on CO_2 and the Expanding House only use high school math. The mathematical project on computer calculation of derivatives of exponentials or the Canary Resurrected, build on simple ideas from the course.

Part 1

Differentiation in One Variable

CHAPTER 2

Using Calculus to Model Epidemics

This chapter shows you how the description of *changes* in the number of sick people can be used to build an effective model of epidemics. Calculus is the language of change and that language allows us to study change in significant ways that plain English will not allow. Here is an example of the sort of serious scientific question that you will be able to answer with this model: In the U. S., we have eradicated polio and smallpox, yet measles remains a persistent pest despite vigorous vaccination campaigns. Why were we able to eradicate polio, but not measles? You can work out the answer to this question in the Herd Immunity project after you have studied the chapter.

Calculus as the language of change really does give us deep insights. We want you to see an example immediately. The primary goal of our course is to show you that calculus has important things to say about many real problems.

The computer makes this study possible at such an early stage. The change model involves simple, but tedious arithmetic. The formulas for this arithmetic look awfully messy, but you will easily be able to make computations in the short tables contained in exercises. The computer will carry out much longer computations based on the same idea; the electronic exercises show you how to program the appropriate computations. Moreover, *Mathematica* provides us with numerical and graphical results very conveniently, so we can analyse our results and begin to develop insight into the model epidemics.

The messy algebraic formulas describing the arithmetic of our first model not only look awkward, they are awkward. The continuous differential equation model is better because it is simpler and because the tools of calculus that we will develop during the course can be applied to it to answer deeper questions, even about the epidemic model. You might wish to look ahead to Chapter 12 on Discrete Dynamical Systems and compare that to the simpler and more powerful formulas in Chapter 21 Continuous Dynamical Systems. It is true that the differential equations get a little ahead of our story, but they will give you a point of reference throughout the course. (The ideas apply to many applications and in pure mathematics. Later you will have many choices of projects besides epidemics.) Differential equations are only ahead of our story in a technical sense, not conceptually. You will see that the equations say what the various rates of change are, but you will not 'integrate' them symbolically. Concentrate on the idea of how infectious individuals change the number of sick people over time and let the computer do the arithmetic. The moral of the chapter is that these simple rates of change give us important information with the help of *Mathematica*. Seeing this will also raise new questions, but at a higher level. Later, calculus and *Mathematica* together will answer these questions.

2.1. The First Model

An epidemic is an unusually large short term outbreak of a disease. Human epidemics are often spread by contact with infectious people, though sometimes there are "vectors" such as mosquitos or rats and fleas or mice and ticks involved in disease transmission. There are many kinds of contagious diseases such as smallpox, polio, measles and rubella that are easily spread through casual contact. Other diseases such as gonorrhea require more intimate contact. Another important difference between the first group of diseases and gonorrhea is that measles "confers immunity" to someone who recovers from it whereas gonorrhea does not. In other words, once you recover from rubella, you cannot catch it again. This feature offers the possibility of control through vaccination. In this section we will formulate our first model of the spread of a disease like rubella where people are susceptible, then infectious, and finally, recovered and immune. (Rubella is commonly called "German measles." It is usually mild, but may produce severe birth defects if contracted by a pregnant woman during her first trimester.)

A scientific model is necessarily a simplification. We cannot include exact contacts between individuals or even other diseases of the population, diet, the amount of travel, the weather and other factors that have "small" effects on the spread of the disease, because we would not be able to make computations and predictions with such a complicated model. By identifying the "main" effects we can formulate a model that is simple enough to use to compute and predict the overall spread of a disease. That's the good news. The bad news is that our model cannot predict specifics such as when you will contract rubella. Some exercises will ask you to use the model to make predictions, but others will ask you to think about and discuss limitations of the model. The usefulness of a model requires us to make correct scientific judgments about which effects are "small" and which are not. We need to test our models against known data both to measure important parameters and to verify that we have only neglected small effects. Once this is done, we can use the model to make predictions in new situations.

2.1.1. Basic Assumptions. We will make the following assumptions in formulating our model:

(1) **SIR:** Individuals all fit into one of the following categories:
    Susceptible: those who can catch the disease
    Infectious: those who can spread the disease
    Removed: those who are immune and cannot spread the disease.
(2) **The population is large,** but fixed in size and confined to a well-defined region. In the case of rubella, you might imagine the population to be a large public university during the semester when relatively little outside travel takes place.
(3) **The population is well mixed;** ideally, everyone comes in contact with the same fraction of people in each category every day. Again, imagine the multitude of contacts a student makes daily at a large university.

When formulating models it is very important to make your assumptions clear. We have listed them explicitly to help make them clear. The exercises will ask you about the limitations of these assumptions.

FIGURE 2.1: Disease Compartments

The categories of people and the way we have assumed that they move between the categories can be summarized in the graphic "compartments" shown in Figure 2.1.

We want our mathematical model to be able to compute the number of people in each of the compartments at any given time. In mathematical models it is also important to keep a list of variables. Variables are letters that stand for quantities that can change. We must express the main properties of the epidemic in terms of these variables.

2.1.2. Variables for Model 1.

$$
\begin{aligned}
t &= \text{the time in days with t=0 at the start of observation} \\
S &= \text{the number of susceptible individuals} \\
I &= \text{the number of infectious individuals} \\
R &= \text{the number of removed individuals} \\
n &= \text{the total number of people in the population}
\end{aligned}
$$

These variables keep track of the number of individuals in each of the compartments as shown in Figure 2.2. Time t is the independent variable and S, I and R are dependent on time. In other words, they are functions of time, but they are not functions given by explicit formulas, such as $y = x^2$.

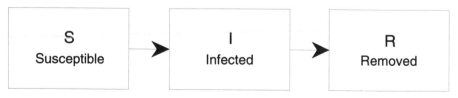

FIGURE 2.2: Compartment Sizes at time t

2.1.3. Derivation of the Equations of Change. Now we want to add a mathematical description of the way individuals move between the susceptible, infectious and removed compartments. This will add two parameters, a and b to our model that are specific to a particular S-I-R disease. Parameters are letters which have a fixed value during the entire problem, but can change from one problem to the next. We begin with the simplest change, how one recovers.

A person with rubella is infectious for about 11 days. The infectious compartment will contain people who have had the disease for one day, two days, and so on up to eleven days. At twelve days an individual moves from the infectious to the removed compartment. Of course, not everyone gets the disease at exactly 8:00 am. Also, as the epidemic spreads more people may be in the one day group than in the eleven day group. Nevertheless, we will assume that the different groups in the infectious compartment are roughly the same size so that we can make a simple statement about recovery: For rubella, the number of people added to the removed compartment tomorrow is $\frac{1}{11}$th of the infectious compartment today,

$$\text{new people in } R\text{-compartment tomorrow} = \frac{1}{11}I$$

A general disease where one recovers and gains immunity has the day's change in the recovered compartment given by,

$$\text{new people in } R\text{-compartment tomorrow} = b\,I$$

where the parameter b controls how quickly people recover, or $1/b$ is the number of days one stays in the infectious compartment. For rubella, $b = 1/11$.

Notice that the units of the right side of the above equation for the day's change in R are 'number of people per day.' We will keep careful account of units.

Mathematical models need to be expressed in terms of variables. The equation above is not written in terms of the variables we listed above. In fact, the left hand side is a phrase. The phrase is clear and we want you to make some computations and think about formulating the equation in terms of the variables. Before you compute that, we need to complete formulation of the equations of change. We can't compute the number of recovered people yet, because we need to know the number of infectious people to use our equation for the change in R.

The next change we will describe is how one spreads the disease. If you are infectious with rubella and you sit through a whole class next to a person who is susceptible, then you will probably spread the disease to that person. If you sit through class next to an immune person, you don't spread the disease. We are supposing that our population is well mixed, so people contact others from each component each day. However, the number of new cases each infectious person causes in a day depends only on contacts with susceptible people. The disease-causing contacts by a single infectious person is given on the average by the number of 'close contacts' per infective each day, a, times the fraction of susceptible people, $a\frac{S}{n}$. This makes the number of new cases of the disease

$$\text{new cases tomorrow} = a\frac{S}{n}I$$

where

$$n = \text{the fixed size of the population}$$

The parameter a is the average number of "close" contacts per infective per day. If such a contact is made with a susceptible person, the disease is transmitted. Some diseases are more contagious than others, so we want a parameter that describes the contagiousness. The problem is: How do we measure a? The parameter a is a combined measure of the level of mixing in the population and the contagiousness of the disease. It turns out that a is measured indirectly with the help of calculus.

Notice the units of a and b

$$a \quad \text{in} \quad \frac{\text{number (of adequate contacts)}}{\text{number (infectives)} \times \text{days (infectious)}}$$

and
$$b \quad \text{in} \quad \frac{1}{\text{days}} \quad \text{or} \quad \frac{1}{b} = \text{number of days infective}$$

So the units of the ratio $c = \frac{a}{b}$

$$\frac{a}{b} \quad \text{in number (of contacts) per infectious person}$$

This ratio, c, is an important number in our model with the intuitive meaning

$$c = \text{average number of close contacts per infective}$$

The parameter c is called the "contact number" because it gives the average number of close contacts each infective has over the whole course of the disease. We know the number of infective days of the disease, $\frac{1}{b}$, from clinical observation. We will see how to measure c from data. Then we compute $a = bc$ from measured quantities.

Since we have assumed that our population has a fixed size n, we have all the information that we need to compute the course of an epidemic. New cases cause the size of S to decrease and the size of I to increase by the same amount. Recoveries cause the size of I to decrease and the size of R to increase by that amount. It is best to write all our equations in terms of increases in the variables S, I, R. We simply take a "negative increase" to mean a decrease. The variable S has a negative increase $-a\frac{S}{n}I$, the variable R has a positive increase bI, while I has an input from the susceptible compartment and an output to removed compartment, $a\frac{S}{n}I - bI$.

FIGURE 2.3: Compartments with Parameters

2.1.4. The First Equations of Change.

$$\text{tomorrow's increase in } S \quad = \quad -a\frac{S}{n}I$$
$$\text{tomorrow's increase in } I \quad = \quad a\frac{S}{n}I - bI$$
$$\text{tomorrow's increase in } R \quad = \quad bI$$

EXERCISE 2.4. *Use the daily changes above to compute 4 days of a rubella epidemic. First, you need the specific parameters, a, b, and c.*

(1) *We know that $\frac{1}{b} = 11$ for rubella. The average contact number is $c = 6.8$. Use these values to compute the parameter a.*

(2) *Suppose a population initially (at $t = 0$) has*

$$S = 10,000$$
$$I = 1,000$$
$$R = 19,000$$

Use the equations of change above to complete a table showing the first four days of the epidemic.

(3) *Make graphs of S versus t, I versus t, and R versus t.*

t	0	1	2	3	4
S	10,000				
I	1,000				
R	19,000				

FIGURE 2.5: Tabular Epidemic

The table you have just filled out is one way to record S, I and R as function of t. To find I at time 3 you simply look in the I row below the 3 in the t row. The mathematical notation for a function is

$$S[t], \qquad I[t], \qquad R[t]$$

The value of $I[3]$ is recorded in the I row below the 3 column of the t row.

When you solved the exercise completing the table you computed $S[1]$, $I[1]$ and $R[1]$ from the change equations and the known values $S[0]$, $I[0]$ and $R[0]$. Next, you used $S[1]$, $I[1]$ and $R[1]$ to compute $S[2]$, $I[2]$ and $R[2]$ and so on. This is a recursive calculation, step 3 requires steps 1 and 2.

EXERCISE 2.6. *Write formulas for (an unknown) $S[t+1]$, $I[t+1]$, and $R[t+1]$ in terms of a (known) $S[t]$, $I[t]$ and $R[t]$. Re-write your formulas so that $S[t+1]$, $I[t+1]$ and $R[t+1]$ are on the left hand side of the equalities and expressions in $S[t]$, $I[t]$, and $R[t]$ are on the right.*

The answer to this Exercise 2.6 is our first formal mathematical epidemic model. It allows us to recursively compute the course of an S-I-R epidemic in steps of one day.

EXERCISE 2.7. *The equations of change are not needed to compute all three compartment sizes. If $S[t]$ and $I[t]$ are known, express $R[t]$ in terms of n, $S[t]$ and $I[t]$. (Hint: What is the total population in terms of these categories?)*

EXERCISE 2.8. *Log onto your computer account and run the Mathematica NoteBook **FirstSIR.ma**. It will help you check the values in your table and begin to learn how to compute with Mathematica .*

EXERCISE 2.9. *Conjectures*
We want you to speculate on the course of different kinds of diseases. Suppose disease "A" is very virulent, that is, it is very easily spread. Sketch the graph of I versus t for such a disease. Now do the same for disease "B" which is not so virulent. Suppose both diseases have an infectious period of 14 days. What is the value of b for both models of these diseases? How does the parameter a for disease A, a_A compare to the parameter a for disease B, a_B? Later in this chapter you should compare your speculations with computed results.

EXERCISE 2.10. *Which parts of our model are invalid for describing a flu epidemic on our campus that begins just before Thanksgiving break? Is our model valid for a flu epidemic that begins just after Thanksgiving break?*

PROBLEM 2.11. *Build a model for a disease that confers a period of immunity for a limited time after recovery, but where you lose immunity after this period. What other basic assumptions do you make? Write a detailed explanation.*

2.2. Shortening the Time Steps

Time units of "days" are natural for rubella since $b = 1/11$ days comes directly from the period of infection. We want to keep these units. However, a whole day of exposure may not be needed to transmit the disease and people may move from the susceptible compartment to the infectious compartment in a shorter period of time. Consider the new infections with daily rate

$$\text{number of new infectious per day} = a\frac{S}{n}I$$

In twelve hours, one half day, there will be

$$\text{new infections in } \frac{1}{2} \text{ day} - a\frac{S}{n}I \cdot \frac{1}{2}$$

and so the number of susceptibles is reduced

$$S\left[\frac{1}{2}\right] = S[0] - \left(a\frac{S[0]}{n}I[0]\right) \cdot \frac{1}{2}$$

In four hours, there will be

$$\text{new infections in 4 hours} = a\frac{S}{n}I \cdot \frac{4}{24}$$

and the reduced susceptibles are given by

$$S\left[\frac{1}{6}\right] = S[0] - \left(a\frac{S[0]}{n}I[0]\right)\frac{1}{6}$$

EXERCISE 2.12.

(1) *Complete the table below where $t = 0, \frac{1}{2}, 1, \frac{3}{2}, 2, \frac{5}{2}, 3, \frac{7}{2}, 4$:*
(2) *Graph S, I and R vs. t.*

t	0	$\frac{1}{2}$	1	$\frac{3}{2}$	2	$\frac{5}{2}$	3	$\frac{7}{2}$	4
S	10,000								
I	1,000								
R	19,000								

FIGURE 2.13: Half-day Table

EXERCISE 2.14.

(1) *How many new infections will there be in one hour? Write formulas for $S[t + \frac{1}{24}]$, $I[t + \frac{1}{24}]$ and $R[t + \frac{1}{24}]$ in terms of $S[t]$, $I[t]$ and $R[t]$ which we assume are known, where we assume t has one of the values $t = \frac{k}{24}$, $k = 0, 1, 2, 3, \ldots, 95$.*

(2) *How many total computations would you need to do to compute $S[4]$, $I[4]$ and $R[4]$ in one hour steps? (Count all additions and multiplications. Time $t = 4$ means 4 days.)*

The computer does not mind doing all the computations required to find $S[4]$, $I[4]$ and $R[4]$. You will use a "Do loop" to write a *Mathematica* program for this keeping the functions S, I and R as *Mathematica* lists.

PROBLEM 2.15. *Run the NoteBook **SecondSIR.ma** from the course software for the following cases of the parameters, a, b, and c. The NoteBook will not only do all the calculations necessary to solve the S-I-R model for small time steps, but will also produce graphs of the results. This will enable you to experiment with the S-I-R model and learn how the parameters affect the behavior of solutions.*

(1) *A rubella epidemic, where $1/b = 11$, $c = 6.8$ and $a = b \cdot c$.*
(2) *Measles, with a very high contact number $c = 15$ and $b = 1/8$.*
(3) *Influenza, with a low contact number of $c = 1.4$ and $b = 1/3$.*

The first part is a simple exercise to learn how to use the Mathematica NoteBook. The details are explained first step-by-step, and then as a complete computation within the Note-Book itself.

The second part of the assignment is to learn how to modify the NoteBook to compute with new parameters. Copy the complete computation in a new input cell and modify the parameters to do the computations for the new diseases.

Finally, the most important part of this problem is to analyse your results. What are the main differences between these three diseases? Which one is highly contagious, which is medium and which least contagious? Which one allows a larger portion of the population to escape infection? How many escape in each case? Write your summary analysis of the comparison between these diseases in a few brief paragraphs. (You may use a text cell in your copy of the NoteBook.)

2.3. The Continuous Variable Model

We will summarize what you should have learned from the text and exercises so far. The summary involves the awfully messy formulas mentioned in the introduction, but these lead to the simpler differential equations and, as you already know, correspond to much simpler looking *Mathematica* statements in **SecondSIR.ma**.

First, it is fairly easy to give formulas for the average daily changes in S, I and R, such as

$$\text{tomorrow's increase in } R = b\, I$$

Second, it is easy to use these equations to recursively tabulate the course of an epidemic as you did in the first exercise. Third, the <u>change</u> in a function of time such as $S[t]$ as time changes from t to $t + \Delta t$ is given by the difference

$$S[t + \Delta t] - S[t]$$

A general form of the solution to Exercise 2.14 may be written in the form

$$\frac{S[t + \Delta t] - S[t]}{\Delta t} = -a\frac{S[t]}{n}I[t]$$

$$\frac{I[t + \Delta t] - I[t]}{\Delta t} = a\frac{S[t]}{n}I[t] - b\,I[t]$$

$$\frac{R[t + \Delta t] - R[t]}{\Delta t} = b\,I[t]$$

where we write both sides of the equations as rates of change. These equations may also be solved for the function at the new time and written

$$S[t + \Delta t] = S[t] - \left(a\frac{S[t]}{n}I[t]\right)\Delta t$$

$$I[t + \Delta t] = I[t] + \left(a\frac{S[t]}{n}I[t] - b\,I[t]\right)\Delta t$$

for $t = 0, \Delta t, 2\Delta t, 3\Delta t, \ldots, k\Delta t, \ldots$ (Computing $R[t] = n - S[t] - I[t]$.) This later form is most useful in *Mathematica*, but looks simpler in **SecondSIR.ma**. The first form is written as a rate of change in units of numbers of people per day,

$$\frac{S[t + \Delta t] - S[t]}{\Delta t} \quad \text{in numbers/day}$$

Recall that we verified that the units of the right hand sides are numbers per day.

The time step Δt (or difference in time, abbreviated by Greek capital delta-t) is arbitrary, so it is not difficult to imagine t and Δt taking arbitrary continuous real values. We also want to let our basic variables take continuous real values so that we can formulate a calculus model.

2.3.1. The Continuous S-I-R Variables. For our continuous epidemic model we choose

$$t = \text{time measured in days continuously from t=0 at the start of the epidemic}$$

$$s = \text{the fraction of the population that is susceptible} = \frac{S}{n}$$

$$i = \text{the fraction of the population that is infectious} = \frac{I}{n}$$

$$r = \text{the fraction of the population that is removed} = \frac{R}{n}$$

where n is the (fixed) size of the total population. Dividing both sides of our previous equations by n, we obtain

$$\left(\frac{S[t + \Delta t]}{n} - \frac{S[t]}{n}\right)\frac{1}{\Delta t} = -a\frac{S[t]}{n}\frac{I[t]}{n}, \text{ etc.,}$$

so in terms of the fraction variables we obtain rate equations

$$\frac{\Delta s}{\Delta t} = \frac{s[t + \Delta t] - s[t]}{\Delta t} = -a\,s[t]\,i[t]$$

$$\frac{\Delta i}{\Delta t} = \frac{i[t + \Delta t] - i[t]}{\Delta t} = a\,s[t]\,i[t] - b\,i[t]$$

where $\Delta s = s[t + \Delta t] - s[t]$ and $\Delta i = i[t + \Delta t] - i[t]$ represent the differences in the susceptible and infectious fractions corresponding to the change in time Δt. We can compute $r[t] = 1 - s[t] - i[t]$. Both sides of the above equations are average rates of change from time t to time $t + \Delta t$.

In calculus, the limit of the rate of change of a continuous function as the time step tends to zero is called the derivative,

$$\frac{ds}{dt} = \lim_{\Delta t \to 0} \frac{\Delta s}{\Delta t} = \lim_{\Delta t \to 0} \frac{s[t + \Delta t] - s[t]}{\Delta t}$$

$$\frac{di}{dt} = \lim_{\Delta t \to 0} \frac{\Delta i}{\Delta t} = \lim_{\Delta t \to 0} \frac{i[t + \Delta t] - i[t]}{\Delta t}$$

Since the right hand side of the above rate equations do not depend on Δt we can interpret the limiting equations as the instantaneous rate of change of the susceptible and infectious fractions. This is the system of

2.3.2. S-I-R Differential Equations.

$$\frac{ds}{dt} = -a\,s\,i$$

(S-I-R DE's) $\qquad \dfrac{di}{dt} = a\,s\,i - b\,i$

$$r = 1 - s - i$$

Methods of calculus will let us compute interesting things about this continuous model, but for now we know that if we let Δt be "small" then we can use *Mathematica* to compute $s[t]$ and $i[t]$ in recursive steps of Δt by specifying a, b, $s[0]$, $i[0]$ and using the *Mathematica* NoteBook **SecondSIR.ma** corresponding to the formulas

$$s[t + \Delta t] = s[t] - (a\,s[t]\,i[t])\Delta t$$
$$i[t + \Delta t] = i[t] + (a\,s[t]\,i[t] - b\,i[t])\Delta t$$

This most useful discrete recursive form is closer to the 'differential' form of the equations

$$ds = -a\,s\,i\,dt$$
$$di = (a\,s\,i - b\,i)\,dt$$

where

$$ds \approx s[t + \Delta t] - s[t] = -a\,s[t]\,i[t]\,\Delta t$$
$$di \approx i[t + \Delta t] - i[t] = (a\,s[t]\,i[t] - b\,i[t])\Delta t$$

The interpretation of the approximation or 'limit' is more difficult for the differential form of the equations, because both sides tend to zero as Δt tends to zero. It is also more useful (for example, in the NoteBook **SecondSIR.ma**.)

EXERCISE 2.16. *What is the meaning of fractional values of S, I, and R in Exercise 2.4 of Section 2.1? Are the fractional variables s, i, and r meaningful in a small population?*

EXERCISE 2.17. *Are the recursive formulas for $s[t + \Delta t]$ exact solutions to the S-I-R differential equations? Run the **SecondSIR.ma** NoteBook and compare solutions with step sizes of $1, \frac{1}{2}, \frac{1}{4}, \frac{1}{8}$ over the time interval $0 \leq t \leq 4$.*

EXERCISE 2.18. *The total population is part of a model with S, I and R, but not with s, i, and r. Explain how we might use the history of a rubella epidemic with one size population to make predictions about a new and different size one.*

EXERCISE 2.19. *Suppose we measure a and b in a college population for a new strain of flu. What limitations would there be in using these parameters in predicting an epidemic in a nursing home?*

PROBLEM 2.20. *Expanding and Declining Epidemics*
We say an epidemic is expanding as long as the fraction of infectious people increases. It is declining when i decreases. See Figure 2.21 for the typical appearance of the graph of i vs t in an epidemic. Where is i increasing on the graph? Where is it decreasing? When are the most people sick at one time? How many are sick at the maximum? At the point where i changes from increasing to decreasing, what can you say about the time rate of change of i, that is, the quantity $\frac{di}{dt}$? How is this related to the size of s, a and b? (Hint: What is the formula for $\frac{di}{dt}$?)

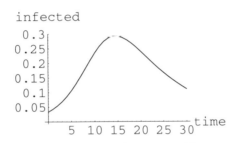

FIGURE 2.21: Infectious vs Time

PROBLEM 2.22. *How many escape?*
1) It seems reasonable that as an epidemic fades out the fraction of infectious people tends to zero,

$$\lim_{t \to \infty} i[t] - 0$$

*Why is this correct? Verify that our model of rubella predicts this by running the **Second-SIR.ma** NoteBook to large values of t.*
2) Does the susceptible fraction tend to zero as an epidemic runs its course,

$$\lim_{t \to \infty} s[t] = 0?$$

*What sorts of intuitive arguments can you make pro or con? Run the **SecondSIR.ma** NoteBook for large values of time and estimate*

$$\lim_{t \to \infty} s[t]$$

from the computer computation. (The graph of s[t] for a four month Rubella epidemic is shown in Figure 2.23. Does s[t] → 0 on that graph?) This is a difficult limit to compute analytically, but it turns out to be a key to a complete understanding of the model.

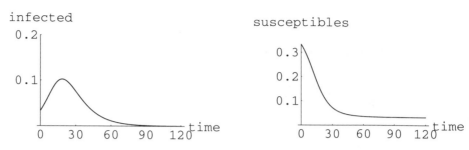

FIGURE 2.23: Four Months of Rubella

PROBLEM 2.24. *Experiment with* $\lim_{t \to \infty} s[t]$ *for rubella, measles and influenza using the **SecondSIR.ma** NoteBook. These parameters are given in Exercise 2.15 above. The long term graphs raise many questions such as: How long does the epidemic last? Is there a formula for* $s[\infty]$*? These are very difficult questions, but the computer will give you practical answers, if not formulas.*

Write a summary of your experiments on how $s[\infty]$ *depends on a, b and* $s[0]$*?*

2.4. Calculus and the S-I-R Differential Equations

Up to this point you should have followed all the mathematical derivations by working the exercises. With the help of the computer you could even answer questions like

$$\text{Does } \lim_{t \to \infty} s[t] \text{ equal zero?}$$

At least approximately. You might wonder, 'Why bother with calculus if you have a computer?' The answer is that you can do more powerful calculations with both calculus and the computer than you can do with either one separately. We want to show you one example, but we do NOT expect you to follow the details in this section until the chapter on continuous dynamical systems.

Using calculus, the S-I-R differential equations

$$\frac{ds}{dt} = -a\,s\,i$$

$$\frac{di}{dt} = a\,s\,i - b\,i$$

show that, no matter what time t,

$$i + s - \frac{1}{c}\operatorname{Log}[s] = k, \qquad \text{for the same constant } k$$

$$i[t] + s[t] - \frac{1}{c}\operatorname{Log}[s[t]] = k$$

Before we show you why the differential equations imply this, here is an example of how to use the result.

EXERCISE 2.25.

(1) *In the rubella epidemic with* $c = 6.8$*,* $s[0] = 1/3$*,* $i[0] = 1/30$*, find* k*.*

(2) *We know from the previous section that $i[t]$ tends to zero as t tends to infinity, so the limiting susceptible fraction also satisfies*

$$0 + s[\infty] - \frac{1}{c} \operatorname{Log}[s[\infty]] = k$$

with the same value of k as you computed in part (1) of the exercise. We could view this as the following question: Find all solutions in s of the equation

$$s - \frac{1}{c} \operatorname{Log}[s] - k = 0$$

*Use the Mathematica NoteBook **EpidemicRoots.ma** to do this. Notice from the graph that there are two roots. The smaller one is $s[\infty]$.*

The real importance of the invariant (in time) equation for s and i is that it can be used to measure c from epidemic data. We have

$$s[\infty] - \frac{1}{c} \operatorname{Log}[s[\infty]] = i[0] + s[0] - \frac{1}{c} \operatorname{Log}[s[0]]$$

since $i[\infty] = 0$ and both sides equal the same constant k. At the start of an epidemic $i[0]$ is small, so

$$s[\infty] - \frac{1}{c} \operatorname{Log}[s[\infty]] \approx s[0] - \frac{1}{c} \operatorname{Log}[s[0]]$$

and therefore approximately,

$$s[\infty] - s[0] = \frac{1}{c} \left(\operatorname{Log}[s[\infty]] - \operatorname{Log}[s[0]] \right)$$

so

$$c = \frac{\operatorname{Log}[s[\infty]] - \operatorname{Log}[s[0]]}{s[\infty] - s[0]}$$

By measuring the susceptible fraction at the beginning and end of an epidemic, we can compute c and then we can find a from the computation $a = b\,c$. Calculus gave us the exact invariant equation which s and i satisfy at all times. No substitute in recursive computer computations provides this equation. However, once we have this information from calculus, the computer again adds to our knowledge.

PROBLEM 2.26. *Use the Mathematica NoteBook **EpidemicRoots.ma** to experiment on how $s[\infty]$ depends on $s[0]$. Specifically, if most of the population is susceptible at the outbreak of an epidemic, will more or less people contract the disease than if only half of the population is susceptible? (This experiment could also be done with **SecondSIR.ma**, by running to a large time.)*

Write a brief paragraph summarizing your experiments.

Later in the course, we will see an analytical way to understand the experiments of the previous problem more clearly.

EXAMPLE 2.27. *How Calculus finds the Time-Invariant Equation*

Without justifying why this works, we compute the ratio

$$\left(\frac{di}{dt}\right) \Big/ \left(\frac{ds}{dt}\right) = \frac{a\,s\,i - b\,i}{-a\,s\,i}$$

cancel dt and do some algebra

$$\frac{di}{ds} = \frac{b}{a}\frac{1}{s} - 1 = \frac{1}{c\,s} - 1$$

We will learn later that if $y = s + k$ for any constant, then $\frac{dy}{ds} = 1$ and that if $\frac{dg}{ds} = \frac{1}{c} \cdot \frac{1}{s}$ for a constant c, then $g = \frac{1}{c}\mathrm{Log}[s]$, the natural logarithm. These facts of calculus say that

$$di = \frac{1}{c}\frac{1}{s}\,ds - 1\,ds$$

$$\int di = \frac{1}{c}\int \frac{1}{s}\,ds - \int ds$$

$$i = \frac{1}{c}\mathrm{Log}[s] - s + k \qquad \text{for some constant } k$$

This is the important time invariant equation for the epidemic model. We will study this in detail at the end of Chapter 23.

2.5. The Big Picture

PROBLEM 2.28. *Write a paragraph explaining how the language of calculus is used to describe the changes in the susceptible, infectious and removed fractions of a population. In other words, explain what the S-I-R differential equations 'say' in English. Notice how cryptic the calculus is compared to English.*

Write a paragraph summarizing the way the computer computes the course of an epidemic by using the change rules given by the differential equations and an initial value of s, i, and r. Be brief and give only a sample calculation and the main idea, rather than a lot of details. How does the computer interact with calculus to find the whole graph of the epidemic from the starting values and prescription for change due to a particular disease?

Finally, explain why the same model applies to a number of diseases by using different parameters a, b, and c. Do these parameters vary as time changes?

2.6. Projects

Once or twice a semester we want you to explore a topic in detail. There is a separate book of detailed Scientific Projects that go with this book. The following subsections give a brief description of the ones most closely related to this chapter. You can work those projects now or come back to them later.

2.6.1. The New York Flu Epidemic.
This project compares the S-I-R model with actual data of the Hong Kong Flu epidemic in 1968-69. The data we have to work with are "observed excess pneumonia-influenza deaths." In this case our 'removed' class includes people who have died. This is a little gruesome, but it is difficult to find data on actual epidemics unless they are extreme.

2.6.2. Vaccination Strategies for Herd Immunity. This project uses the mathematical model of an SIR disease to find a prediction of how many people in a population must be vaccinated in order to prevent the spread of an epidemic. In this project you use data from around the world to make predictions on successful vaccination strategies. Dreaded diseases like polio and smallpox have virtually been eliminated in the last two generations, yet measles is a persistent pest right here on campus. The mathematical model can shed important light on the differences between these diseases. The mathematics is actually rather easy once you understand the basic concept of decreasing infectives.

2.6.3. Endemic S-I-S Diseases. Some infectious diseases do not confer immunity, such as strep throat, meningitis, or gonorrhea. In these diseases, there is no removed class, only susceptibles and infectives. As in the SIR diseases, we can make a mathematical model for SIS diseases.

An S-I-S disease has the potential to become endemic, that is, approach a non-zero limit in the fractions of susceptible and infectious people. Once you have done some experiments, you will be able to show what happens mathematically and explore health policy implications.

CHAPTER 3

Numerics, Symbolics and Graphics in Science

The goal of this chapter is to review the ideas of *independent* and *dependent* variables and *parameters* in the context of some down-to-earth applications. We want you to develop careful working habits that will help you use calculus. We study these ideas from the three points of view you saw in the *Mathematica* NoteBook **aMathcaIntro.ma**: numerics, symbolics and graphics.

Additional high school review material is contained in the accompanying book on Mathematical Background. That review also includes some beginning work with *Mathematica* to help start your computing in familiar territory.

In the previous chapter we used the real variables

$$
\begin{array}{rcl}
t & = & \text{time in days, measured continuously} \\
s & = & \text{susceptible fraction, a ratio of number of people over number of people} \\
i & = & \text{infected fraction} \\
r & = & \text{removed fraction}
\end{array}
$$

to express instantaneous time rate of changes between the variables in an S-I-R epidemic. The expressions also contained parameters a and b that depend on the particular disease, but do not vary with t, s, i, or r. In that model, t is the independent variable, s, i and r are dependent variables, and a and b are parameters. The quantities a and b are considered fixed for a particular disease, t varies freely and $s = s[t]$, $i = i[t]$, and $r = r[t]$ are functions of t.

The letters t, s, i, and r measured quantities in our measles model that changed with time. Given a specific time such as $t = 4$ (days), we could compute $s[4]$, $i[4]$, and $r[4]$ to any desired accuracy by running the *Mathematica* NoteBook **SecondSIR.ma** with small enough step size dt and enough steps to add up to 4. The values of s, i, and r "depend" on the value of t, whereas t is "independent," we can simply specify it. In mathematical terms, s is a function of t, but unlike most functions from high school, the "rule" to find s from t is not a formula in algebra or trig. Functions given by differential equations such as $s[t]$ and $i[t]$ are important in science and mathematics, but so are functions given by explicit formulas. You are already familiar with many of these.

31

The word "review" may bother you if you are used to slightly different terminology. Part of the goal of the chapter is to develop common terminology about topics that you learned in at least a similar form in high school.

THE QUESTIONS:

(1) What are dependent vs independent variables or functions? How did you study functions and variables in high school?
(2) What is the difference between a variable and a parameter? What are some examples from high school?
(3) How do variables and parameters arise in explicit formulas of science and mathematics? (Or, "What good are they?")
(4) What are the numerical, symbolic and graphical aspects of functions with parameters?

3.1. Functions from Formulas

The area of a rectangle is a function of its length and width

$$A = \text{area in sq. in.}$$
$$l = \text{length in inches}$$
$$w = \text{width in inches}$$

$$A = l \cdot w$$

In this case area is a function of two independent variables

$$A = A[l, w] = l \cdot w$$

where the "domains" or permissible input values of l and w satisfy

$$l \geq 0 \quad \text{and} \quad w \geq 0.$$

A is a function of l and w in that once their values are specified, then A is determined. For example, if $l = 2\frac{1}{2}$ and $w = \frac{2}{3}$, then $A = \frac{5}{3}$.

The area of a square is a function of the length of its side.

$$A = \text{area in sq. cm.}$$
$$s = \text{length in cm.}$$

$$A = A[s] = s^2$$

with domain

$$s \geq 0$$

A is a function of a single independent variable. The formula $A = s^2$ makes perfectly good mathematical sense when s is negative but the geometrical meaning as length does not. Geometrically, the function $A = A[s]$ has a domain restricted to $s \geq 0$.

3.1.1. Format for Homework. In homework problems involving applications we want you to

(1) Explicitly list your variables with units and sketch a figure if appropriate.
(2) Translate the information stated in the problem into formulas in your variables. (If this translation is difficult, you may not have chosen the best variables.) Often it is helpful to balance units on both sides of an equation.
(3) Formulate the question in terms of your variables and solve.
(4) Explicitly interpret your solution.

Often it is helpful in solving problems to re-state the question in terms of your variables. We will say more about this as the exercises become more difficult.

EXAMPLE 3.1. *Express the circumference of a circle as a function of its area.*

SOLUTION:
Step1: We use the following variables

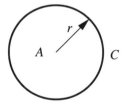

$C =$ the circumference in feet
$A =$ the area in sq. ft.
$r =$ the radius in ft.

Step 2: We know the formulas for circumference and area of a circle:

$$C = 2\pi r \quad \text{and} \quad A = \pi r^2, \quad \text{for } r \geq 0$$

This translation step just translates implicit knowledge of circles.
Step 3: The question asks us to find circumference as a function of area as the independent variable, find: $C = C[A]$

Notice that we have introduced an extra variable, the radius r, to help us solve the problem.

Begin by solving the area equation for r:

$$r^2 = \frac{A}{\pi}$$

$$r = \sqrt{\frac{A}{\pi}}, \quad \text{for } A \geq 0$$

Next, substitute this equation for r in the circumference equation:

$$C \;=\; 2\pi\sqrt{\frac{A}{\pi}} = 2\sqrt{\pi A}$$

This gives the solution:

$$C = C[A] = 2\sqrt{\pi A}, \quad \text{for } A \geq 0$$

Step 4: The interpretation of the solution is simply that this formula gives circumference as a function of area.

EXERCISE 3.2. *Find the perimeter of a square as a function of its area.*

EXERCISE 3.3. *Find the surface area of a sphere as a function of its volume.*

EXERCISE 3.4. *You stand in the middle of railroad tracks with a train approaching you at 100 ft/sec. The train is 15 ft. wide and is 1,000 ft. away at elapsed time 0. Express the angle of view subtended by the train as a function of time. (We do not recommend simulation. How long do you have to get off the tracks?)*

3.1.2. SOH-CAH-TOA. You may need to review some simple facts of high school trigonometry such as "SOH-CAH-TOA" and the Pythagorean theorem. In a right triangle with acute angle θ as shown and sides of lengths o opposite θ, a adjacent to θ and hypotenuse h, we have

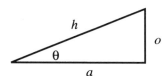
                                   Pythagorean theorem: $h^2 = a^2 + o^2$

$$
\begin{aligned}
\text{Sine} &= \text{opposite/hypotenuse (SOH)}\\
\text{Sin}[\theta] &= \frac{o}{h}\\
\text{Cosine} &= \text{adjacent/hypotenuse (CAH)}\\
\text{Cos}[\theta] &= \frac{a}{h}\\
\text{Tangent} &= \text{opposite/adjacent (TOA)}\\
\text{Tan}[\theta] &= \frac{o}{a}
\end{aligned}
$$

EXERCISE 3.5. *Derive the functional equation*

$$(\text{Cos}[\theta])^2 + (\text{Sin}[\theta])^2 = 1$$

from the Pythagorean theorem and SOH-CAH-TOA (HINT: Divide by h^2).

Trig functions also have associated inverse functions. We may know the ratio o/a, which is the tangent of an angle, and want the angle itself. This is given by the arctangent function which is built into most calculators and *Mathematica* .

$$\theta = \text{ArcTan}[y] \qquad \Leftrightarrow \qquad \text{Tan}[\theta] = y$$

for $-\frac{\pi}{2} < \theta < \frac{\pi}{2}$. For example, if o is fixed at 7.5, then θ as a function of a is

$$\theta = \text{ArcTan}\left[\frac{7.5}{a}\right]$$

This formula was part of the previous speeding train problem (Exercise 3.4).

EXERCISE 3.6. *A plane passes overhead traveling at 600 mph in a straight horizontal line. At elapsed time 0 it is* 5,280 *ft. directly above you. Express the angle from the vertical that you look at the plane as a function of time. (Careful with units.)*

EXERCISE 3.7. *The courtesy light in front of my house is 8 ft. tall and 4 ft. back from the sidewalk. My daughter is 5 ft. tall and walks down the sidewalk. Express the length of the shadow she casts as a function of the distance she is down the sidewalk from the point perpendicular to the lamp.*

EXERCISE 3.8. *Describe the motion of the piston shown in Figure 3.9 when the crankshaft turns 2000 revolutions per minute.*

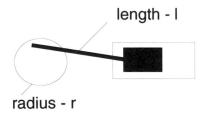

length - l

radius - r

FIGURE 3.9: Crank and Piston

There is additional review of high school trig in the Mathematical Background. You may wish to look at that now, especially to recall the notion of radian measure and the addition formulas for sine and cosine.

3.2. Types of Explicit Functions

Some of the most basic functions defined by explicit formulas that you studied in high school were

Linear Functions:

$$y = m\,x + b \qquad \text{or} \qquad \frac{\Delta y}{\Delta x} = m$$

Polynomials:

$$p = a_0 + a_1 x + \ldots + a_n x^n, \qquad a_j \text{ constants}$$

Power Functions:

$$y = ax^p, \qquad a, p \text{ constants}$$

Trigonometric Functions:

$$y = \text{Sin}[\theta], \text{Cos}[\theta], \text{Tan}[\theta]$$

Inverse Trig Functions:

$$\theta = \text{ArcTan}[y]$$

Exponential Functions:

$$y = a \cdot b^x, \qquad a, b \text{ constants.}$$

especially

The Natural Exponential:

$$y = e^x, \qquad e \approx 2.71828\ldots$$

The Natural Logarithm:

$$x = \text{Log}[y]$$

Note: Natural log is sometimes denoted $\ln(y)$ in high school texts, where $\log(y)$ denotes the base 10 log. We have no use for base 10 logs and use the *Mathematica* name for natural logarithm.

Some of the algebra of these functions is reviewed in the Mathematical Background, but since the main idea of calculus is that 'small changes of smooth functions are approximately linear,' we will give a review of linear functions here. It is important to see how linear functions change.

Before you start Chapter 5, be sure to review the "Laws of Exponents" and "Power Functions" in the "High School Review with *Mathematica*" Chapter of the Mathematical Background.

3.2.1. Linear Functions, Change and Lines. This is a technical algebra summary of 'what makes an explicit equation a line.' In high school you may have memorized various forms of the equation of a line, the point-slope formula, the two point formula, as well as the algebraically simplest slope-intercept formula $y = mx + b$.

From our point of view, the best symbolic way to describe a line is: The ratio of changes in y to the changes in x is constant no matter what pair of points we consider. The change in a variable is a difference, the new value minus the old value, so we abbreviate change by Greek capital delta, and the most important equation of a line says the difference in y over the difference in x is constant:

$$\frac{\Delta y}{\Delta x} = \text{constant}$$

Of course this constant is the slope, so 'a line has constant slope' is essentially the same statement. Geometrically, the constant slope is similar triangles. Algebraically, it is also the statement that the dependent variable changes at a constant 'rate,' when we view the independent variable as time.

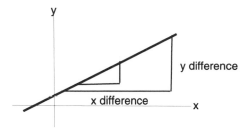

FIGURE 3.10: Similar Triangles

EXAMPLE 3.11. *Point-Slope Formula from Change Formula*

How is this used? If we know a line goes through the point $(x, y) = (-4, 3)$ and has slope $m = -2$, then we take a generic unknown point (x, y) on the line and form the ratio of changes. The change in y moving from 3 to the generic value y is the difference in y, $\Delta y = y - 3$ and the change in x in moving from -4 to the generic value x is the difference in x, $\Delta x = x - (-4) = x + 4$.

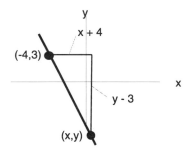

FIGURE 3.12: Changes from $(-4, 3)$ to (x, y)

The ratio of changes is

$$\frac{\Delta y}{\Delta x} = \text{constant}$$

$$\frac{y - 3}{x - (-4)} = \frac{y - 3}{x + 4} = -2$$

That's the equation. We can simplify it algebraically, but it is often best left in this form. The algebraically simplified formula has the form $y = mx + b$, while the difference form shows the point and the slope. You should verify that this is the same as the point slope formula if you memorized that in high school.

EXERCISE 3.13. *Suppose a line goes through the point $(5, -4)$ and has slope -3. What is the x-y-equation of the line? What is the slope-intercept form?*

Suppose a line goes through the fixed (but unknown to you) point (a, b) and has slope m. What is the x-y-equation of the line?

EXAMPLE 3.14. *Two Point Formula from Change Formula*

Here is another example of using change to describe a line. Suppose our line goes through the points $(x, y) = (-2, 3)$ and $(x, y) = (1, 4)$. The two given points tell us the slope,

$$\frac{\Delta y}{\Delta x} = \text{constant}$$

$$\frac{4 - 3}{1 - (-2)} = \frac{4 - 3}{1 + 2} = \frac{1}{3}$$

and the ratio of changes with the general unknown point (x, y) gives us the equation

$$\frac{\Delta y}{\Delta x} = \text{constant}$$

$$\frac{y - 3}{x - (-2)} = \frac{y - 3}{x + 2} = \frac{1}{3}$$

You should verify that this two step procedure is the same as the two point formula if you memorized that in high school.

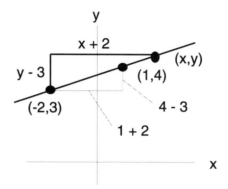

FIGURE 3.15: Two Points Determine a Line

EXERCISE 3.16. *Suppose a line goes through the points $(3, -2)$ and $(-4, 5)$.*
1) Find its slope.
2) Use the slope and the point $(3, -2)$ to find its equation.
3) Use the slope and the point $(-4, 5)$ to find its equation.
4) Compare the two equations by putting both in slope-intercept form.

3.3. Logs and Exponentials

The last major kinds of function that you studied in high school are logarithms and the inverse, exponential functions. We will study these more carefully during the course and there is some additional high school review in the Mathematical Background. This section shows you one application of exponential functions that hinges on the functional identity $b^x \cdot b^y = b^{x+y}$. Functional equations like this are very important. Again, the Mathematical Background Chapter on Functional Identities contains additional information on general identities.

Since the widespread use of inexpensive calculators, base 10 logs and exponentials are no longer of much interest. Natural base or base "e" logs and exponentials are still very important as we shall see throughout the course. The fundamental functions' *Mathematica* names are

$$y = \text{Exp}[x] = e^x, \qquad \text{where } e \approx 2.71828$$

and the inverse

$$x = \text{Log}[y] \quad \Leftrightarrow \quad y = e^x, \, y > 0.$$

Note that some books use $ln[y]$ for the natural logarithm and a few still use $\log[y]$ for base 10 logarithm.

Of course, it still makes perfectly good sense to use other bases, particularly for exponentials. However, calculus becomes much simpler when logs and exponentials are expressed in the "natural" base e. The following example and exercise uses base 2 and the first Exercise in the background review on exponentials asks you to express this base 2 function in terms of base e. The re-expression does not seem advantageous now, but it will be once we have calculus.

Suppose the mold in my basement doubles the number of cells every hour. At midnight there are 56 cells. At 1:00 am there are 2×56 cells. At 2:00 am there are $2 \times (2 \times 56)$ cells and $2 \times 2 \times 56 = 2^2 \times 56$. At 3:00 am there are $2 \times (2^2 \times 56) = 2^3 \times 56$ cells. You can see that at integer hours, $t = 1, 2, 3, \ldots$, there are

$$N \quad = \quad 56 \times 2^t \text{ cells}.$$

or

$$N \quad = \quad N[t] = N_0 2^t$$

Notice the difference between an exponential function

$$y = a \cdot b^x, \qquad a, b \text{ constant}$$

and a power function

$$y = a \cdot x^p, \qquad a, p \text{ constant}$$

EXERCISE 3.17. *Explain the difference between an exponential function and a power function.*

Properties of exponents play an important role in exponential functions. Let's return to the number of mold cells in my basement

$$N = N_0 2^t$$

How many mold cells should there be in $\frac{1}{2}$ hour? If we substitute $t = \frac{1}{2}$, the answer is

$$N\left[\frac{1}{2}\right] = N_0 2^{\frac{1}{2}} = N_0 \sqrt{2}$$

Is this right? If so, how many in the next half hour? We should use the same rule, so

$$N[1] = N\left[\frac{1}{2}\right] \cdot \sqrt{2} = N_0 2^{\frac{1}{2}} \cdot 2^{\frac{1}{2}}$$

Rules of exponents say that this agrees with the integer formula

$$N[1] = N_0 2^1 = N_0 2^{\frac{1}{2} + \frac{1}{2}}$$

PROBLEM 3.18. *Growth of Algae*
*Use the Mathematica NoteBook **ExpGth.ma** for help with these computations.*

   On a warm summer day with plenty of nutrient supplied by runoff, the number of algae cells in a formerly clear pond doubles every 6 hours. This is the time it takes a cell to divide, but naturally, the cells do not all divide simultaneously, so a population of many cells grows almost continuously with time. Suppose you begin with N_0 cells at elapsed time 0. Express the number of cells as a function of time in hours. How many cells are there after 3 hours? What do fractional values of N mean? How many are there after 1day? How many in 1 week? If one cell has mass 1 mg. and we start with 1000 cells, what is the mass of the algae cells after 1 month? What is the approximate mass of all the water in Lake Michigan in these units? Can algae continue to double every six hours?

Use a calculator or the *Mathematica* NoteBook **LogGth.ma** to compute the natural logs in the next exercise.

EXERCISE 3.19. *Logarithmic Growth*
A super duper computer can add ten billion terms of the form $1+1/2+1/3+1/4+1/5+\cdots$ per second with perfect accuracy. The size of the sum $1+1/2+\cdots+1/n$ is approximately $.58+\text{Log}[n]$ (as we shall see next semester). How many centuries would it take this computer to add enough terms to get a sum over 100?

3.4. Chaining Variables or Composition of Functions

3.4.1. Planck's Law. Explicit formulas in science can be complicated. For example, Planck's radiation law can be written

$$I = I[\omega, T] = \frac{a\omega^3}{e^{b\omega/T} - 1}$$

for constants a and b. This expresses the intensity I of radiation at (angular) frequency ω for a body at absolute temperature T. The specific details are not important yet. We want to point out that I is built up using addition, multiplication and division from a polynomial power formula

$$\omega^3$$

and the exponential

$$e^x$$

substituting $\frac{b\omega}{T}$ for x, subtracting 1, multiplying and dividing. When T is fixed, we want to use symbolic formulas of calculus to find

$$\text{Maximum}[I : 0 < \omega < \infty]$$

(The maximum is called Wien's Law of radiation.) Symbolic calculus first gives rules for the basic kinds of functions and then gives rules for functions built up from these. By the end of the semester you will think of Planck's formula as a straightforward combination of simple functions. (The way Planck found the formula is not simple.)

3.4.2. Composition in Function Notation. In Planck's formula we substituted $x = b\omega/T$ into the natural exponential function. Linking variables together in a chain like this is an important symbolic construction. You should also be familiar with the function notation for chains. If

$$y = f(x) = e^x \qquad \text{and} \qquad x = g(\omega) = b\omega/T$$

then the substitution of variables is the function

$$y = f(g(\omega)) = e^{b\omega/T}$$

This notation is useful in some contexts.

EXERCISE 3.20. *These exercises will help you become proficient in manipulating functions, a skill which you will need later on.*
1) Let $h(t) = t^2 + 1$, $q(t) = 7e^t$, $p(t) = \frac{t}{t+3}$. *Write a single expression for the functions:*

a) $h(t) + q(t)$ b) $h(t) \cdot p(t)$ c) $\frac{p(t)}{q(t)}$

2) Let $V(r) = 6r^2$, $A(r) = 2^r + 1$. *Write a single expression for*
a) $V(r+1)$ b) $A(r+3)$ c) $V(r+\Delta r)$
d) $A(r+\Delta r)$ e) $V(A(r))$ f) $A(V(r))$

The *Mathematica* NoteBook **Functions.ma** contains part of the solution to Exercise 3.20. First of all, we define functions in *Mathematica* with the following symbols:

h[t_] := t∧2 + 1 ;
q[t_] := 7 Exp[t] ;
p[t_] := t/(t + 3) ;

The underline _ after the letter t in the function tells *Mathematica* that t is the variable. The := tells *Mathematica* we are defining a function of t_. To check our typing, we enter the computation:

h[t]

and get the output

$$t^2 + 1$$

We may use these functions symbolically. For example, if we enter

p[h[x]]

we get the output

$$\frac{1 + x^2}{4 + x^2}$$

We can substitute symbolic expressions in *Mathematica* functions, too. Try the following:

f[x_] := x∧ 3;
f[x]

Then enter the expression

f[x + dx]

to see the output

$$(dx + x)^3$$

Also enter

Expand[(f[x+dx] - f[x])/dx]

to see the output

$$3x^2 + 3x\,dx + dx^2$$

EXERCISE 3.21. *Let*

$$f(x) = \begin{cases} x + 2 & if\ x \le -1 \\ x & if\ -1 < x \le 3 \\ 4x - 3 & if\ 3 < x \end{cases}$$

Compute $f(-1)$, $f(3)$. Compute other values of $f(x)$ and sketch the graph of $f(x)$. The graph consists of separate "pieces," and $f(x)$ is called a "piecewise-defined" function.
Compute $f(x+1)$ (it will also be a piecewise defined function).

3.5. Graphics and Formulas

The interaction or translation between formulas and graphs and the interpretation of graphs themselves are important parts of this course. Graphs often reveal mathematical results simply and clearly. Here is a simple example.

EXAMPLE 3.22. *A Graphical Minimum*

We want to make a box with a square base and no top. We need the box to hold 100 cubic inches and want to make it out of the least possible amount of cardboard. What dimensions should we use?

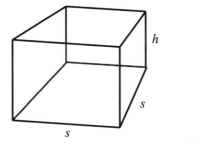

$$A = \frac{400}{s} + s^2$$

The graph of area as a function of the length of the side is:

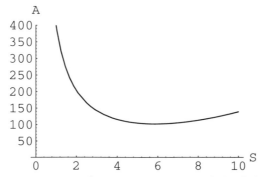

FIGURE 3.23: Area as a Function of the Side

It is clear on this graph that the function decreases as s increases from 0 to about 5.8 and then the function increases as s increases beyond this. Hence the minimum occurs when $s \approx 5.8$ and $h \approx 2.9$.

This example is fine as far as it goes, but we will see that calculus can tell us which parts of graphs of formulas we need to look at. In this specific example, calculus shows that the minimum area occurs when $s = \sqrt[3]{200}$ and the height $h = s/2$, so $A \approx 102.6$ sq. in. The minimizing s and A are readily seen on the graph but values can only be read approximately.

EXERCISE 3.24.
1) Show that the area of a box with an open top and square base of volume 100 cu. in. is given by

$$A = \frac{400}{s} + s^2$$

where s is the length of the edge of the base. The volume of 100 cubic inches can be expressed in terms of height, $100 = hs^2$, so $h = \frac{100}{s^2}$. The area of each side piece is $hs = \frac{100}{s}$, substituting $h = \frac{100}{s^2}$.

2) Sketch the graphs of

$$y = \frac{400}{s} \qquad and \qquad z = s^2$$

*on the same scales for $0 < s < 10$. You may use Mathematica to check your graphs if you wish (see the NoteBook **MathcaIntro.ma**).*

3) Sketch the graph of the sum $y + z$ on the scale of part (2), $A = \frac{400}{s} + s^2$.

4) Locate the side of minimum area as accurately as possible on the graph of A vs s.

Plotting points alone is usually a bad way to sketch graphs because that information alone requires many points to construct a shape and a leap of faith that we've connected the points "correctly". If we only have numerical data, that is all we can do. We need to plot some points even with formulas, but we will learn to use calculus to tell us shape information such as where the graph is increasing or decreasing so that only a few points are required to give qualitatively accurate graphs. Even with the computer, which will moronically plot 1,000 points if you ask it to, we often need to use calculus to decide what range of values contains the important information. Calculus with the computer will give us accurate useful information.

The next exercise shows you a simple reason why points alone are not enough.

EXERCISE 3.25. *Three Points are not Enough*
1) The cartesian pairs $(-2, -2), (0, 0), (2, 2)$ are recorded in the table below the blank graph in the next figure. Plot these points.

2) Show that the graph of $y = x$ contains all three points from part (1) and sketch the graph.

3) Show that the graph of $y = x^3 - 3x$ contains all three points from part (1). Can you sketch it without more points?

4) Find a point on the graph of $y = x$ that is not on the graph of $y = x^3 - 3x$ and plot it.

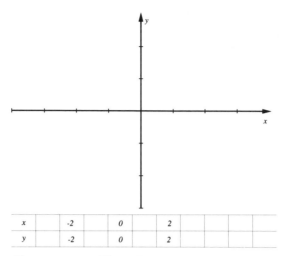

x	-2	0	2			
y	-2	0	2			

FIGURE 3.26: Three Points on Two Graphs

EXERCISE 3.27. *Use Mathematica to plot both functions from the previous exercise on the same graph.*

First start *Mathematica* and then type
y[x_] := x;
z[x_] := x ∧ 3 - 3 ∗ x;
Plot[{ y[x] , z[x] } , {x, -3, 3}, PlotRange -> {-3,3 }];
y[x]
z[x]
and enter the computation.

EXERCISE 3.28. *Find the equations of the lines shown in the graphs below.*

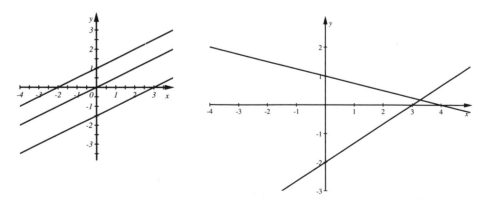

FIGURE 3.29: Three Parallel Lines and Two Intersecting Lines

3.6. Graphs without Formulas

Graphs are primarily good for qualitative information rather than quantitative accuracy. Graphs readily show where quantities are increasing or decreasing, but only give rough approximations to amounts, rates of increase, and the like. Qualitative information can certainly exist without any underlying formulas.

EXERCISE 3.30. *Make a qualitative rough sketch of a graph of the distance traveled as a function of time on the following hypothetical trip. You travel a total of 100 miles in 2 hours. Most of the trip is on rural Interstate highway at the 65 mph speed limit. (What qualitative feature or shape does the graph of distance vs. time have when speed is 65 mph?) You start from your house at rest and gradually increase your speed to 25 mph, then slow down and stop at a stop sign. (What shape is the graph of distance vs. time while you are stopped?) You speed up again to 25, travel a while and enter the Interstate. At the end of the trip you exit and slow to 25, stop at a stop sign and proceed to your final destination.*

3.7. Parameters

The familiar slope-intercept formula for a line is

$$y = mx + b$$

where the letter x stands for the independent variable, y is the dependent variable or $y = y[x]$ is a function of x. The letters m and b are parameters – they are fixed as far as x is concerned.

The S-I-R differential equations

$$\frac{ds}{dt} = -asi$$

$$\frac{di}{dt} = asi - bi$$

contain the parameters a and b. One of the important properties of the S-I-R model is the way that the limiting number of susceptibles $s[\infty]$ depends on the ratio of these parameters $a/b = c$. The parameter b is the reciprocal of the disease-specific infectious period, but varies from one disease to another. Neither a nor b varies with time.

Parameters play an important role in the solution of many scientific problems, where parameters are often measured physical constants. We may find how a maximum oscillation depends on a mass. Specific solutions for known masses do not yield the same scientific insights as the formula with a parameter. So what are parameters mathematically? One book "defined" parameter as 'a variable that is constant.' That "definition" is a contradiction in terms and yet suggests the spirit in which parameters are often used.

In the slope-intercept formula

$$y = mx + b$$

we 'hold m and b fixed' while we vary x. Only varying x gives a single straight line graph. We can also plot a family of lines as we change a parameter. Here are two examples.

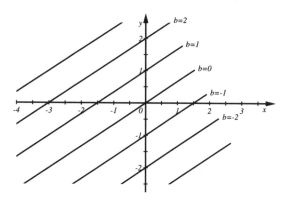

FIGURE 3.31: Variation of the b parameter in $y = \frac{2}{3}x + b$

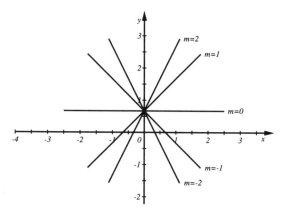

FIGURE 3.32: Variation of the m parameter in $y = mx + \frac{2}{3}$

In short, a parameter is another letter or unknown in our formulas. Roughly speaking, it is called a parameter if it is treated as an unknown constant as far as the independent variables are concerned. We will ask you to work with parameters often. If the extra letters confuse you in a problem, choose a special case or two and work through the problem with specific numbers instead of parameters. Then generalize your work to a letter instead of your specific numbers.

EXERCISE 3.33.
1) What is the formula in terms of parameters for the two solutions of the equation in the variable x,

$$ax^2 + bx + c = 0$$

where a, b and c are parameters? (You learned this in high school.)
2) What are the specific solutions to this equation if a = 2, b = 3 and c = −2?

EXAMPLE 3.34. *Mathematica and Parameters*

We can define functions in *Mathematica* with parameters. For example,
f[x_] := a x∧ 2 + b x + c;
f[x]

gives the output

$$a\,x^2 + b\,x + c$$

In typical scientific problems, we will want to assign values to the parameters. For example we might want to look at various graphs of the $f[x]$ defined above,

a = 1.0;
b = 2.0;
c = 1.0;
Plot[f[x] , { x, -5, 5 }];
Then
a = 1.0;
b = 2.5;
c = 1.0;
Plot[f[x] , { x, -5, 5 }];
Then
a = 1.0;
b = 1.5;
c = 1.0;
Plot[f[x] , { x, -5, 5 }];

EXERCISE 3.35. *Planck's Formula in Mathematica*
Define the intensity i as a function of frequency w in Planck's formula

$$I = \frac{a\omega^3}{e^{b\omega/T} - 1}$$

treating a, b and T as parameters. Use lower case i, because Mathematica uses $I = \sqrt{-1}$. Use w instead of Greek ω.

The next exercise is easy once you understand the geometric meaning of the algebra. This is what we want you to learn (or review). You can check your work on the *Mathematica* NoteBook **SlideSquash.ma**.

PROBLEM 3.36. *Animation of Parameters*
In high school you learned that the graph of every quadratic polynomial is a parabola. A geometrically convenient way to write the parameters is

$$y = \alpha(x - \beta)^2 + \gamma$$

for α, β, γ (unknown) constants.
1) Let $\alpha = \gamma = 1$ and plot the family of curves

$$y = (x - \beta)^2 + 1$$

for $\beta = -1, 0, 1, 2, \ldots$.
2) Let $\beta = \gamma = 1$ and plot the family of curves

$$y = \alpha(x - 1)^2 + 1$$

for $\alpha = -1, 0, 1, 2, \ldots$.
3) Let $\alpha = \beta = 1$ and plot the family of curves

$$y = (x - 1)^2 + \gamma$$

for $\gamma = -1, 0, 2, \ldots$.
4) Can every quadratic of the form

$$p = ax^2 + bx + c$$

also be written in the form

$$y = \alpha(x - \beta)^2 + \gamma$$

and vice versa? Why? What are the restrictions on a and α?
*5) Verify your work using the Mathematica NoteBook **SlideSquash.ma**. The first anima-tion there corresponds to part (1) above, but plots 41 graphs for*

$$\beta = -5.0, -4.75, -4.5, \cdots, 4.75, 5.0$$

The animation uses the graphs to make a computer 'movie' of the graph sliding across the screen. The parts of the NoteBook left for you to work correspond to parts (2) and (3) above.

3.8. Background on Functional Identities

You are familiar with the basic formulas for trig functions. General equations about an unknown function also play an important role in mathematics. In the algae problem 3.18, the basic operation of growth in 3 hours followed by growth for three more hours has to amount to growth in 6 hours. These growths do *not* add, because new individuals in turn reproduce. Formulation of the continuous time model depended on the exponential functional equation $b^s \times b^t = b^{s+t}$. One of the most important examples in the course is a variation on this model which formulates growth as a differential equation.

A differential equation is an equation satisfied by an unknown function. This is analogous to asking which functions satisfy the general identity

$$f(s) \times f(t) = f(s + t)$$

This is a new level of abstraction if you treat the function $f(\cdot)$ as an unknown variable. It is worthwhile abstraction, in fact, the rules of calculus are equations in unknown functions.

General functional equations are studied in the Mathematical Background Chapter on Functional Identities. You will also find the most important identities about trig functions, logs, etc., from high school there, with some review on their use. You will also find a bridge to some important new ideas about treating functions as unknowns.

CHAPTER 4

Linearity vs. Local Linearity

This chapter helps you formulate the main approximation of differential calculus: *small changes in smooth functions are* *approximately* linear.

Linear functions are the easiest functions to work with, nonlinear ones are much more difficult. For example, it is much easier to answer the question: 'For which x is $f[x]$ equal to $g[x]$?' if $f[x]$ and $g[x]$ are of the form $f[x] = 3x+5$ and $g[x] = 2x-1$ than if $f[x] = \text{ArcTan}[x]$ and $g[x] = \text{Log}[x]$.

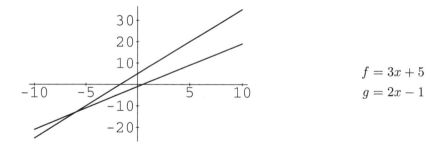

$f = 3x + 5$

$g = 2x - 1$

FIGURE 4.1: A Common Point on Two Linear Graphs

Graphically, these problems are a question of finding the point of intersection.

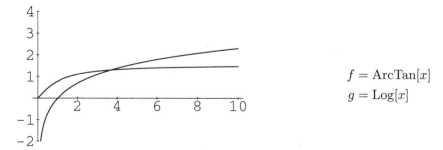

$f = \text{ArcTan}[x]$

$g = \text{Log}[x]$

FIGURE 4.2: A Common Point on Two Nonlinear Graphs

The main idea of differential calculus is that 'small changes in smooth functions are approximately linear.' This chapter helps you formulate a form of this principle in one independent variable. Your formulation of the principle is very important and will be used repeatedly throughout the year. We want you to develop both a graphical and a symbolic understanding of it. We want you to discover it on your own. We will try to help with the discovery, but it really is your job to figure out what 'small changes of a function are approximately linear' means and when the approximation procedure is valid. It will take a while to understand this both intuitively and completely, but it should be a familiar idea by the end of the semester. Next semester we will extend this local approximation to functions of two variables.

The simplest linear function has the form $y = m \cdot x$ in (x, y)-coordinates. The parameter m is the slope of the line that graphs the function. This linear function is increasing if $m > 0$, decreasing if $m < 0$ and horizontal (or constant) if $m = 0$. Calculus tells when a nonlinear function $y = f[x]$ is increasing by computing an approximating linear function, the differential. (The slope of the differential is the derivative.) It is easy to tell if the linear function is increasing and the nonlinear function also increases where the approximation is valid... These words are complicated, but have an easy graphical interpretation. The "local approximation" of differential calculus means that a microscopic view of a smooth function appears to be linear. In other words, calculus can be used to compute what we would see in a powerful microscope focused on the graph of a nonlinear function. If we "see" an increasing linear function, then the nonlinear function is increasing in the range of the microscope.

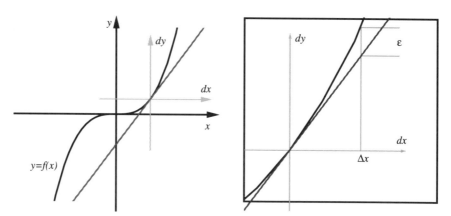

FIGURE 4.3: A Magnified Graph

'Approximation' can mean many things. We take up a perfectly natural kind of approximation in the CO_2 Project in the accompanying separate materials on Scientific Projects. This approximation turns out not to be the kind that is so useful in calculus, but it is worth thinking about a little just for comparison. The next section gives you a clever case for comparison and the last section of the chapter looks at linearity in a new way.

"Local" microscopic approximation is probably not the first kind of approximation you would think to use and it is not always the right one to use. 'When to use it' is part of understanding what it means. Section 4.2 is a self-guided start of a tour of 'local linearity' and the more difficult kind of approximation that makes calculus so successful in the study of *small* changes.

4.1. Linear Approximation of Oxbows

This section shows you one dramatic example of non-local linear approximation. The project on CO_2 in the Scientific Projects shows you a more serious example where long-term linear approximation really is a plausible idea. The local linearity of calculus is only good for small steps; we hope Mark Twain's wit will help you remember this.

The Chapter *Cut-offs and Stephen* of Mark Twain's *Life on the Mississippi* contains the following excerpt about ox-bows on the lower Mississippi. In a flood, the river can jump its banks and cut off one its meandering loops, thereby shortening the river and creating an ox-bow lake.

In the space of one hundred and seventy-six years the Lower Mississippi has shortened itself two hundred and forty-two miles. That is an average of a trifle over a mile and a third per year. Therefore, any calm person, who is not blind or idiotic, can see that in the Old Oolitic Silurian Period, just a million years ago next November, the Lower Mississippi River was upwards of one million three hundred thousand miles long, and stuck out over the Gulf of Mexico like a fishing-rod. And by the same token any person can see that seven hundred and forty-two years from now the Lower Mississippi will be only a mile and three-quarters long, Cairo and New Orleans will have joined their streets together, and be plodding comfortably along under a single mayor and mutual board of aldermen. There is something fascinating about science. One gets such wholesome returns of conjecture out of such a trifling investment of fact.

EXERCISE 4.4. *Linear Oxbows*
How wholesome are Twain's returns? Express the length of the river (in miles) as a function of time in years using the implicit mathematical assumption of Twain's statement. What is this assumption? We are not given the length of the Lower Mississippi at the time of Twain's statement, but we can find it from the information. What is it? What is the moral of this whole exercise in terms of long-range prediction?

4.2. The Algebra of Microscopes

At high enough magnification the graphs $y = x^2$, $y = x^3$, $y = x^{27}$, $y = 1/x$, $y = \sqrt{x}$ appear straight (except at $x = 0$ on $1/x$ and $x \leq 0$ on $\sqrt{x}$.) The *Mathematica* NoteBooks **Microscope.ma** and **Zoom.ma** will let you experiment with these and other functions yourself. The next set of figures shows a 'zoom' at $x = 1$ on the graph of $y = x^2$.

The goal of this section is for you to find a general symbolic formula for the error of deviation from straightness in microscopic views. The general formula will be easy to verify once we have a little theory, but its meaning is very important, so we want you to find the formula for yourself. The formula is confusing because we magnify. The secret is that the actual error must be small compared with the microscope scale, so that what we see in the microscope is small. Before you work any further, think about this question: If you magnified a particle by one million and it then appeared to be 1cm long, how long would it actually be? We need a general formula to answer this kind of question.

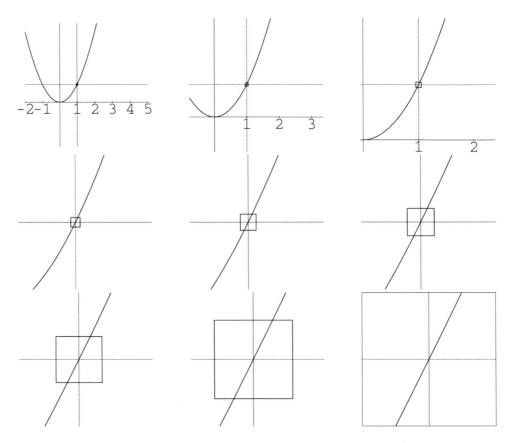

FIGURE 4.5: Successive Magnifications of $y = x^2$

We begin with some numerics to help get the idea. Draw the sketches as best you can on good graph paper so you are connecting numerics, symbolics and graphics. You can use the computers if you wish, but you'll still need to do algebra. Make some more numerical examples of your own if you need help finding the general formula in Exercise4.10.

EXERCISE 4.6. *Your First Magnification*
1) Sketch the graph of $y = x^3$ for $-1.5 \leq x \leq 1.5$, using equal x and y scales.
2) We want to focus our microscope over the x-value $x = 2/3$. What is the corresponding y-value? Draw the local (dx, dy)-axes with its origin at this (x, y)-point, $(x, y) = (2/3, ?)$. The (dx, dy) origin lies at the center of your microscope. We will think of x as fixed and vary dx and dy.
3) The straight line in local coordinates $dy = (4/3)dx$ is tangent to $y = x^3$ at $x = 2/3$. Sketch this as a dotted line on the same graph.
4) Magnify a portion of your graph by 10 and draw the microscopic view with the same scale you used for your original (x, y)-plot. This way $(1/10)$ will appear unit size.

Your result of Exercise 4.6 should look something like the following:

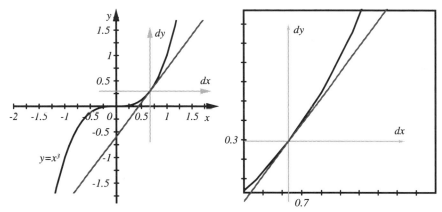

FIGURE 4.7: A Magnification of $y = x^3$

We are interested in the gap between the curve and the straight line in the microscopic view. We will call the amount we measure in the microscopic view using our original scale (but viewed in the microscope) epsilon ε, greek "E" (for error). In this example we want to know: (1) How big is ε in the microscopic view? (2) How big is it in original un-magnified coordinates? Remember the example above. If we magnify by one million and see 1 cm in our original centimeter units when we look in the microscope, we actually have an error of 10^{-6} cm, one one millionth of the apparent error.

EXERCISE 4.8. *Your Next Magnification*
1) *If we see a distance D in the microscope (say $D = 0.3$), how big is the un-magnified distance? Our magnification is $10 = (1/\Delta x)$.*
2) *What are the (x, y)-coordinates of the point where $(dx, dy) = (0, 0)$?*
3) *What are the (x, y)-coordinates of the point on your dotted line $dy = (4/3)dx$ above $dx = \Delta x = 1/10$? (Shown in grey in our figure.)*
4) *The curve $y = x^3$ lies above your dotted line (shown in grey), but in the microscope we only see the dx-dy axes. The x-coordinate of the point $dx = 0.1$ is $0.7666\ldots$, because $dx = 0$ corresponds to $x = 0.6666\ldots$ What is the y-coordinate on $y = x^3$ when $x = 0.7666\ldots$?*
5) *What we see in the microscope as ε on our original scale is the magnified difference between the answer to Part 4 and the answer to Part 3, so its actual size is given by applying the formula from Part 1 to this difference. How much is it?*

You can use *Mathematica* to help check your specific work, but don't forget that understanding the procedure of magnifying is the goal of the section. To plot $f[x] = x^3$ near $2/3$, you could type and enter the program:

```
f[x_ ] := x∧3;
f[x]
m = 10 (*m = Magnification.*)
Plot[ f[x] , { 2/3 - 1/m , 2/3 + 1/m } ];
```

If you also want to plot the tangent line, you need to find the (x, y) equation of the line through $(2/3, f[2/3])$ with slope $4/3$, $l[x] = 4x/3 + ?$, then add the following to your program:

```
l[x_ ] := 4 x/3 + ?;
Plot[ {l[x] , f[x]} , { 2/3 - 1/m , 2/3 + 1/m } ];
```

We want a general formula for the graph of a generic $y = f(x)$ when magnified by an arbitrary $(1/\Delta x)$. Below is a sketch of a general function $y = f(x)$, a pair of local coordinate (dx, dy)-axes centered on the graph over a general point x. A line in local coordinates $dy = mdx$ is shown in grey.

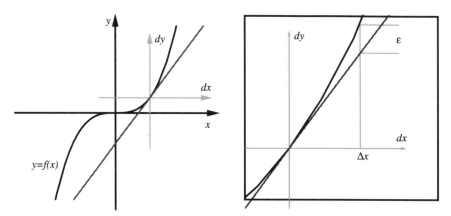

FIGURE 4.9: Magnification of $y = f(x)$

PROBLEM 4.10. *Your General Magnification*
1) If we see a distance D in the microscope how big is the un-magnified distance? Express your answer in terms of Δx and D. (Check your formula when $\Delta x = 1/1,000,000$ and magnification $= 1/\Delta x$.)
2) What is the y-coordinate of the (dx, dy)-origin in terms of the function f and the x center of the (dx, dy)-coordinates?
3) What is the y-coordinate of the point above $dx = \Delta x$ on the line $dy = mdx$?
4) What is the y-coordinate of the point on the graph of $z = f(w)$ above the dx-point $dx = \Delta x$?
5) What we see in the microscope as ε on our original scale is the magnified difference between the answer to Part 4 and the answer to Part 3, so its actual size is _____.
   Write a few brief paragraphs explaining the overall meaning of this problem.

Write the equation

Ans to Exercise 4.10 Part 4 = Ans to Exercise 4.10 Part 3 + Actual error

in terms of f, x, Δx and ε. The condition for local linearity of a graph $y = f(x)$ is that $\varepsilon \approx 0$ when the magnification is large enough, in other words, if $\Delta x \approx 0$. Your equation is the symbolic form of the Increment Principle which we will discuss in the next chapter.
   Linearity in functional notation has a peculiar appearance. This is not difficult, just different. Don't skip the next exercise, because it gives a simple case of the main formula underlying differential calculus. There is more on this topic in the Mathematical Background Chapter on Functional Identities.

PROBLEM 4.11. *Functional Linearity*
We want to consider the linear equation as a function, $f(x) = m\,x + b$ and compare perturbed values of the function symbolically and graphically. Some numerical computations may help

get you started, but you should try to understand the fundamental role of the parameters m and b in general.

1) Choose convenient values of the parameters m and b and make a sketch of the graph of your function $y = f(x)$. Do this for several cases of m and b..

2) Fix one value of x and mark the (x, y) point on your graphs over this x.

3) Choose a perturbation Δx and mark the point above $x + \Delta x$ on your graphs. Your graphs should look something like Figure 4.9.

4) Where does the value $y_1 = f(x)$ appear on your graph?

5) Where does the value $y_2 = f(x + \Delta x)$ appear on your graph?

6) How much do you need to add to the vertical value y_1 to move to the vertical value y_2? Express your answer in terms of m, b, x and Δx. Express your answer in terms of $f(x)$ and $f(x + \Delta x)$.

7) Substitute $x + \Delta x$ into the formula for $f(\cdot)$ and show symbolically that

$$f(x + \Delta x) = f(x) + m \, \Delta x$$

In other words, no matter which x we begin at, the size of the change in y caused by a change of Δx in x is $m \, \Delta x$ and does not depend on b or x. This formula could be expressed

$$f(x + \Delta x) = f(x) + g(\Delta x)$$

where $g(\cdot)$ is the function of only Δx given by $g(\Delta x) = m \, \Delta x$. (Recall that m is a parameter, so does not depend on either x or Δx.)

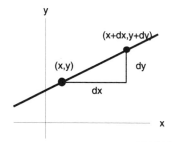

FIGURE 4.12: Linear Changes in x and y

The important formula of the previous exercise is not true for nonlinear functions. Consider $f(x) = a \, x^2$, for example. If I say, 'add 1 to x and tell me how much f changes,' you can only answer in terms of x.

$$\begin{aligned}
f(x + 1) - f(x) &= a \, (x + 1)^2 - a \, x^2 \\
&= a \, [(x + 1)^2 - x^2] \\
&= a \, [x^2 + 2 \, x + 1 - x^2] \\
&= a \, [2 \, x + 1]
\end{aligned}$$

If f were linear, $f(x) = m \, x + b$, the answer would be independent of x.

$$\begin{aligned}
f(x + 1) - f(x) &= [m \, (x + 1) + b] - [m \, x + b] \\
&= m
\end{aligned}$$

EXERCISE 4.13. *Functional Non-Linearity*
Let $f(x) = a\,x^2$. *Compute* $f(x + \Delta x) - f(x)$ *symbolically and show that, in general, we can not write*

$$f(x + \Delta x) = f(x) + g(\Delta x)$$

for a function $g(\cdot)$ *that depends only on* Δx *(but not on* x*).*

4.3. *Mathematica* Increments and Microscopes

The *Mathematica* NoteBooks

Microscope.ma and Zoom.ma

help give you the geometric idea of the Increment Principle. These programs contain "animations" - a computer generated 'movie' of graphs being magnified.

EXERCISE 4.14. *Zooming with Mathematica*
*1) Run the Mathematica NoteBook **Zoom.ma** with its built-in example function,* $y = f[x]$*. Read the instructions in the NoteBook on making Mathematica run an animation. You should see a 'movie' of an expanding graph. Each 'frame' of the movie is expanded 1.5 times. Once the magnification is high enough, the graph appears to be linear.*
2) Focus the microscope at the point $x = 0$ *instead of the built-in* $x = 1$ *and re-Enter the two computation cells.*
*3) Use the Mathematica NoteBook **Zoom.ma** and re-define the function* $f[x_{-}] :=$ *to make animations of microscopic views of some of the functions:*

a) $f[x] = x^2$ b) $f[x] = x^3$ c) $f[x] = \frac{1}{x}$

d) $f[x] = \sqrt{x}$ e) $f[x] = \text{ArcTan}[x]$ f) $f[x] = \text{Log}[x]$

g) $f[x] = \text{Exp}[x] = e^x$ h) $f[x] = \text{Cos}[x]$ i) $f[x] = \text{Sin}[x]$
     Try focusing the microscope over the points $x = 1.5$*,* $x = 0$ *and* $x = -1$ *when all these are possible. (Say why they can't be done if not.)*

4.4. Functions with Kinks and Jumps

Not all functions are smooth or locally linear. This means that when we magnify some functions, their graphs do not become more and more like straight lines. Which functions do and which don't? Symbolic rules of calculus will answer this question easily and it turns out that 'most' functions are smooth so that microscopic views of their graphs do appear linear. This section experiments with some exceptional cases.

EXERCISE 4.15. *Jump Functions*
*Run the Mathematica NoteBook **Zoom.ma** on the functions*

$$j[x] = \frac{\sqrt{x^2 + 2x + 1}}{x + 1} \quad\quad \& \quad\quad j_2[x] = \frac{\sqrt{x^2 - x^4/2}}{x}$$

Focus the microscope at the points $x = +1$ *and* $x = -1$*. Where is this function smooth and locally linear? What happens to this function at* $x =$ *the trouble spot? Why?*

EXERCISE 4.16. *Kink Functions*
*Run the Mathematica NoteBook **Zoom.ma** on the functions*

$$k[x] = \sqrt{x^2 + 2x + 1} \qquad \& \qquad k_2[x] = \sqrt{x^2 - x^4/2}$$

Focus the microscope at the points $x = +1$ and $x = -1$. Where is this function smooth and locally linear?

The main ideas of calculus are based on approximating small changes in the output of a function when a small change is made in the input. We need some notation to indicate a small change. If x_1 and x_2 are nearly equal, we will write

$$x_1 \approx x_2$$

For now, this will just be an intuitive notion, we won't say exactly how close they have to be in order to write $x_1 \approx x_2$.

PROBLEM 4.17. *Continuity*
We say that a function $f[x]$ is continuous when small changes in x only produce small changes in the value of the function

$$x_1 \approx x_2 \Rightarrow f[x_1] \approx f[x_2]$$

Explain in terms of this approximation why the function $j[x]$ in Exercise 4.15 is not continuous, but the function $k[x]$ in Exercise 4.16 is continuous.

The function

$$W[x] = \text{Cos}[x] + \frac{\text{Cos}[3\,x]}{2} + \frac{\text{Cos}[3^2\,x]}{2^2} + \frac{\text{Cos}[3^3\,x]}{2^3} + \cdots + \frac{\text{Cos}[3^k\,x]}{2^k} + \cdots$$

(which is defined by a limiting process, since you can't actually sum to infinity) is continuous, but is not locally linear at any point. In other words, it has a kink at every point on the graph. Weierstrass discovered this function and its graph looks like lots of different size "Ws" all hooked together. It is an old example of a 'fractal.' We want you to graph it and look at it on several scales. When we study series later in the course, we will study it in more detail.

PROBLEM 4.18. *Weierstrass' Wild Wiggles*
Open a new Mathematica window, type and Enter the program:

$$w[x\_] := Sum[Cos[x \; 3 \wedge k]/2 \wedge k, \{k, 1, n\}]$$
$$n = 10$$
$$a = 0$$
$$delta = 1$$
$$Plot[w[x], \{x, a - delta, a + delta\},$$
$$PlotRange- > \{w[a] - delta, w[a] + delta\}, AspectRatio- > 1]$$

Try several scales, "delta". The program above keeps the width of the x-axis and the width of the y-axis on your graph the same. This prevents distortion of slopes.
If your scale is very small, you will need to increase the parameter n in the program. This is because Weierstrass' actual function is the sum to infinity and we only approximate

W[x] by taking the sum of 10 terms. At a smaller scale the error in this approximation could show up. We only need to sum enough terms of Weierstrass' approximation to make our accuracy more than the screen resolution. To see what we mean, run your NoteBook with the a = 0, delta = 1 and n = 3, n = 4, n = 5, · · · .

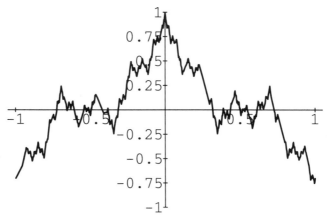

FIGURE 4.19: Weierstrass' Nowhere Differentiable Function

4.5. The Cool Canary - Another Kind of Linearity

It was a cool January night in Iowa, about 0° F outside, but cozy and around 75° F in our snug farm house. Jonnie was still mad at sister Sue for getting a better report card so at 8:00 pm he put her covered canary cage outside. Ten minutes later the canary chirped as he always does at 60°F.

The canary will die when the temperature in the cage reaches freezing. How long do we have to rescue him?

This scenario is a little silly, but we are getting at some serious mathematics. The urgent need of the poor canary should hold your attention. We let T equal the temperature in degrees Fahrenheit and t equal the time in minutes, measured with $t = 0$ at the time Jonnie put the cage outside. Our data are $(t, T) = (0, 75)$ and $(t, T) = (10, 60)$. A linear model of the data looks like the next figure, but is a linear model correct?

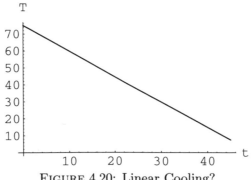

FIGURE 4.20: Linear Cooling?

EXERCISE 4.21. *What is a linear equation whose graph passes through the points* $(0, 75)$ *and* $(10, 60)$? *Use your equation to predict the temperature of the canary at* 6 : 00 *am, ten hours after Jonnie does his vicious deed.*

Would the temperature of the cage go below zero? Of course not. Is the limiting value of the temperature zero? In other words, is the little canary's body temperature zero when Susie finds him in the morning? Could the temperature decline linearly to zero and then abruptly stop declining and stay at zero?

A linear graph of temperature vs. time means that the rate of cooling remains the same for all temperatures. In other words, when the cage is much warmer than the outside air, it only cools as fast as when it has cooled down to almost zero. Does this seem plausible? The next graph shows a nonlinear fit to our same two data points. What happens to the rate of cooling in this case as the temperature decreases toward zero? Does the rate of cooling decrease or increase as the temperature decreases toward zero?

EXERCISE 4.22. *Let* T' *stand for the rate of cooling. What are the units of* T'? *What is the connection between* T *and* T'? *What feature of the graph of* T *vs.* t *represents* T'?

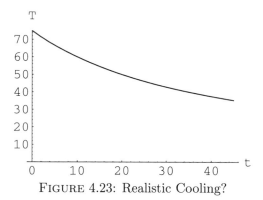

FIGURE 4.23: Realistic Cooling?

There are many explicit graphs with the property that the rate of decrease itself decreases as the dependent variable tends toward zero. One student suggested a variation on $y = \frac{1}{x}$.

EXERCISE 4.24. *Sketch the graph of* $y = \frac{1}{x}$, *except make the variables* T *and* t. *Sketch the family of curves* $T = \frac{k}{t}$ *for various values of the parameter* k. *Note that they all have a 'singularity' at* $t = 0$.

We need to move the graph of $T = \frac{k}{t}$ to the left in order to make it at least a plausible model of the chilly canary. (Certainly the temperature does not tend to infinity as t tends to zero!)

EXERCISE 4.25. *Sketch the family of curves* $T = \frac{k}{t+a}$ *for various values of the parameter* a *and* $k = 1$. *In terms of the parameter* a, *where is the 'singularity' in the equation?*

Now that you see how to stretch and slide the basic graph $T = \frac{1}{t}$, use it to try to make a nonlinear model of the chilly canary.

EXERCISE 4.26. *Find values of the parameters* k *and* a *so that* $T = \frac{k}{t+a}$ *passes through the points* $(0, 75)$ *and* $(10, 60)$. *This is sketched with another nonlinear model in Figure 4.27. What temperature does this model predict for the canary's body at* 6 : 00 *am?*

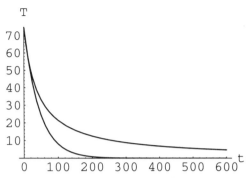

FIGURE 4.27: Two Nonlinear Cooling Models

EXERCISE 4.28. *We expect that when T is big, the rate of cooling is big and that when T is near zero, the rate of cooling is small. Do both of the nonlinear graphs in Figure 4.27 have this feature?*

Guessing explicit formulas for temperature vs. time is a hard way to build a model of the cooling canary. Why not try a family of curves like $T = \frac{k}{(t+a)^2}$ or $T = \frac{k}{\sqrt{t+a}}$? It is better to go right to the underlying question of how the temperature *changes*. The simplest mathematical relationship that says, 'The rate of cooling decreases to zero as T decreases to zero.' is a linear equation between T and T', so we will use linear relationships in new ways once we understand the fundamental 'local' linearity that underlies calculus.

PROBLEM 4.29. *Write an equation that says, "The rate of cooling is proportional to the temperature." Use your equation to sketch a graph of the temperature vs. time that your equation predicts. You only need to know where T starts (75) and your equation for how it changes to make a sketch of its curve. Which of the graphs in Figure 4.27 satisfies this 'law of cooling?'*

You should see that the ideas of the poor freezing canary's plight are simple - at least qualitatively, but it is hard to express the simple ideas in terms of variables. That is because we do not have the proper mathematical *language*. Using the language of change, we formulate the simplest model that says, 'the cage cools slower as temperature approaches zero.' In this language, the last exercise simply asks you for the differential equation:

$$\frac{dT}{dt} = -kT$$

with the initial condition $T = 75$ when $t = 0$. A little more calculus will show us how to find out when to save the canary. Well, at least if our scientific assumptions are correct... If you want to try the physical experiment (a good project), please don't use Susie's parakeet.

CHAPTER 5

Direct Computation of Increments

The main idea of differential calculus is that small changes (or increments) of smooth functions are approximately linear. Geometrically, this means that microscopic views of smooth graphs appear to be linear. This chapter studies direct symbolic calculation of the increments of several basic functions and finds the linear approximation directly.

The computations of the chapter get right at the heart of the approximation of differential calculus. There are three ways to understand this: (1) As intuitive error approximations of views in a microscope. (2) As limits of difference quotient functions. (3) As computations with ideal numbers. The first is the most important, but we will show you how all three approaches work. We just leave the details in terms of rigorous limits or infinitesimals to the Mathematical Background.

We use brute force computation of increments and skillfully collect the 'small' part. Direct computation of increments is harder than applying the differentiation rules of the next chapter, but is needed in order to say what the 'differentiation rules' prove. Differentiation rules are actually theorems that say a local linear approximation is given by the rules of "calculus." The meaning of 'small' can be understood in any of the three ways mentioned above; the main thing is to collect the proper terms in a computation.

We know from the last chapter that some perfectly reasonable functions like $f[x] = \sqrt{x^2 + 2x + 1}$ do not have the local linearity property at some points and some less reasonable ones like Weierstrass' nowhere differentiable function

$$W(x) = \mathrm{Cos}[x] + \frac{\mathrm{Cos}(3x)}{2} + \frac{\mathrm{Cos}(3^2 x)}{2^2} + \cdots + \frac{\mathrm{Cos}(3^n x)}{2^n} + \cdots$$

are continuous, but not locally linear at any point. Calculus gives a simple procedure (in the form of theorems that look like rules of algebra) to find out if a function given by a formula is locally linear. This is both easy to use and powerful mathematics, so we do not want to postpone learning it too long. These rules are the 'little stones' mentioned in the introduction to the book. Of course, the little stones of Chapter 6 are not the whole story, we need to apply what we learn from shuffling the stones to solving interesting problems in projects and later chapters. However, first we want to understand what the approximation means and that is the role of this chapter.

5.1. How Small is Small Enough?

In the last chapter you formulated the deviation of $y = f(x)$ from the straight line $dy = m\,dx$ in a microscopic view. The observed error ε at magnification $1/\Delta x$ satisfies

$$f(x + \Delta x) = f(x) + m \cdot \Delta x + \varepsilon \cdot \Delta x$$

where $\varepsilon \cdot \Delta x$ is the actual error. The number m is the slope of the microscopic straight line $dy = m\,dx$ (in microscope coordinates focused over x). The error ε is the amount of deviation from straightness we see above $x + \Delta x$ at power $1/\Delta x$. With microscopes, we want to let Δx get "small enough" so that ε is below the resolution of our microscope and we cannot detect the difference between the graph and a straight line.

Notice that m may depend on x, but not Δx, since the slope of the curve depends on the point, but not the magnification. As we move the focus point of the microscope, the slope may change, but the graph should always appear straight under the microscope. Since m depends on x and not Δx, it is customary to denote this slope by $f'[x]$ rather than m.

Let's begin with an example, $f(x) = x^3$ and calculate the increment corresponding to a change of Δx

Large Increment of $f(x) = x^3$

$$f(x + \Delta x) = (x + \Delta x)^3 = x^3 + 3x^2\Delta x + 3x\Delta x^2 + \Delta x^3$$
$$f(x + \Delta x) = f(x) + 3x^2\Delta x + (\Delta x[3x + \Delta x])\,\Delta x$$
$$f(x + \Delta x) = f(x) + [\text{term in } x \text{ but not } \Delta x]\Delta x + [\text{observed microscopic error}]\Delta x$$
$$f(x + \Delta x) = f(x) + f'[x] \cdot \Delta x + \varepsilon \cdot \Delta x$$

where $m = f'[x] = 3x^2$ and the observed error is $\varepsilon = (\Delta x[3x + \Delta x])$.

Here is a rough intuitive argument that shows that the error becomes small as Δx becomes small. If x is no more than a thousand, $|x| \leq 1000$, and we want the observed error to be less than one one millionth, $|\varepsilon| < 10^{-6}$, then Δx needs to be small enough so that $|\Delta x|$ times 3001 is less than a millionth, say, $|\Delta x| < 10^{-10}$, because

$$|\Delta x[3x + \Delta x]| < 10^{-10}[3001] < 10^{-6}\frac{3001}{10,000} < \frac{1}{1,000,000}$$

It is clear that as long as x is bounded, we can make the error as small as we please by choosing a small enough Δx. The exact formula for how small Δx needs to be is not so obvious, but it is clear that for any fixed bound on x, the error $\varepsilon \to 0$ as $\Delta x \to 0$ for all $|x| \leq b$.

We summarize this calculation (and the knowledge that ε can be made small by making Δx small) by

$$y = x^3 \qquad \Rightarrow \qquad dy = 3x^2\,dx$$

The intuitive meaning of this formula is that a small difference, dy, in the y-variable, corresponding to a small difference of dx in the x-variable (beginning at a fixed x) is approximately given by the formula shown. This formula, called a differential, omits the error term.

EXERCISE 5.1. $y = x^2 \Rightarrow dy = 2x \, dx$
Let $f(x) = x^2$ and compute $f(x + \Delta x)$, then re-arrange it in the form

$$f(x + \Delta x) = f(x) + [term\ in\ x\ but\ not\ \Delta x]\Delta x + [observed\ microscopic\ error]\Delta x$$
$$f(x + \Delta x) = f(x) + f'[x] \cdot \Delta x + \varepsilon \cdot \Delta x$$

and show that $\varepsilon \to 0$ as $\Delta x \to 0$. "Show" this in any way that you feel is reasonable.
Does x need to be bounded (as in the example above) or can it vary as Δx tends to zero?

5.2. Derivatives as Limits

The derivative in one variable can also be defined by a limit of functions,

$$\lim_{\Delta x \to 0} \frac{f(x + \Delta x) - f(x)}{\Delta x} = f'[x]$$

rather than by using microscopes. This is described in the Mathematical Background Chapter on *Derivatives as Limits*. If you prefer a more conventional introduction to derivatives, you could skip the rest ot this chapter and go to that chapter in the background.

5.3. Small, Medium and Large Numbers

We want to have three different intuitive sizes of numbers, very small, medium size, and very large. Most important, we want to be able to compute with these numbers using the same rules of algebra as in high school and separate the 'small' parts of our computation. In other words, we want to compute ε and 'compute' that it is small.

As a first intuitive approximation, we could think of these scales of numbers in terms of the computer screen. In this case, 'medium' numbers might be numbers in the range -999 to + 999 that name a screen pixel. Numbers closer than one unit could not be distinguished by different screen pixels, so these would be 'tiny' numbers. Moreover, two medium numbers a and b would be indistinguishably close, $a \approx b$, if their difference was a 'tiny' number less than a pixel. Numbers larger in magnitude than 999 are too big for the screen and could be considered 'huge.'

The screen distinction sizes of computer numbers is a good analogy, but there are difficulties with the algebra of screen - size numbers. We want to have ordinary rules of algebra and the following properties of approximate equality:

(1) If p and q are medium, so are $p + q$ and $p \cdot q$.
(2) If ε and δ are tiny, so is $\varepsilon + \delta$, that is, $\varepsilon \approx 0$ and $\delta \approx 0$ implies $\varepsilon + \delta \approx 0$.
(3) If $\delta \approx 0$ and q is medium, then $q \cdot \delta \approx 0$.
(4) $1/0$ is still undefined and $1/x$ is huge only when $x \approx 0$.

You can see that the computer number idea does not quite work, because the approximation rules don't always apply. If $p = 15.37$ and $q = -32.4$, then $p \cdot q = -497.998 \approx -498$, 'medium times medium is medium,' however, if $p = 888$ and $q = 777$, then $p \cdot q$ is no longer screen size... For now, all you should think of is that $\approx$ means 'approximately equals.'

We will extend the 'real' number system that you used in high school to include 'ideal' numbers that obey these simple approximation rules as well as the ordinary rules of algebra and trigonometry. Very small numbers technically are called infinitesimals and what we shall assume that is different from high school is that there are positive infinitesimals.

AXIOM 5.2. *The Infinitesimal Axiom*
The hyperreal numbers contain the ordinary numbers, but also contain nonzero infinitesimal numbers, that is, numbers $\delta \approx 0$ positive, but smaller than all the real positive numbers. Positive infinitesimals satisfy

$$\frac{1}{2} > \frac{1}{3} > \frac{1}{4} > \cdots > \frac{1}{m} > \cdots > \delta > 0$$

for any ordinary natural number $m = 1, 2, 3, \cdots$. We write $a \approx b$ if the hyperreal number $b - a \approx 0$ is infinitesimal.

Two ordinary real numbers, a and b, satisfy $a \approx b$ only if $a = b$, since the ordinary real numbers do not contain infinitesimals. Zero is the only real number that is infinitesimal.

'Ideal' numbers of various kinds are important in many parts of mathematics, for example, the complex numbers add an 'imaginary' $\sqrt{-1}$ to 'real' numbers to simplify algebraic computations. Our new numbers have 'teeny tiny numbers' that will simplify approximation estimates. Direct computations with the ideal numbers produce symbolic approximations equivalent to the function limits described in the Mathematical Background Chapter on Derivatives as Limits, but the rules of Theorem 5.11 give a direct way to compute. Limit theory does not give the answer, but only a way to justify it once you have found it.

If you prefer not to say 'infinitesimal,' just say 'δ is a tiny positive number' and think of $\approx$ as 'close enough for the computations at hand.' The computation rules above are still important intuitively and can be phrased in terms of limits of functions if you wish. The intuitive rules help you find the limit.

The next axiom about the new "hyperreal" numbers says that you can continue to do the algebraic computations you learned in high school. The third axiom says you can continue to do other high school computations - even with trig functions.

AXIOM 5.3. *The Algebra Axiom (Including < rules.)*
Hyperreal numbers obey the same rules of algebra as the real numbers.

The algebra of infinitesimals that you need can be learned by working the exercises in this chapter. We have also given additional drill problems on this topic in the Mathematical Background Chapter on Computation of Limits.

The ideal numbers obey the same rules of algebra as the familiar numbers from high school. We know that $r + \Delta > r$, whenever $\Delta > 0$ is an ordinary positive high school number. Since we want to use the same rules of algebra for the new numbers, we also have new finite numbers given by a high school number r plus an infinitesimal,

$$a = r + \delta > r$$

So the number a is different from r. However, since δ is small, the difference between a and r is small

$$0 < a - r = \delta \approx 0 \qquad \text{or} \qquad a \approx r \quad \text{but} \quad a \neq r$$

Every finite number is near an ordinary number from high school (see Theorem 5.8 below.) The important thing is to learn to compute with approximate equalities.

Functional equations like the addition formulas for sine and cosine or the laws of logs and exponentials are very important. The specific high school identities are reviewed in the Mathematical Background Chapter on High School Review with *Mathematica*. The rigorous form of the next axiom is given in the Mathematical Background Chapter on Functional Identities.

AXIOM 5.4. *The Function Extension Axiom*
Real functions (in any number of variables) have natural extensions to the hyperreal numbers and the extended functions obey the same identities and inequalities as the original functions.

We will make the practical intuitive use of the Function Extension Axiom clearer in the examples and exercises below. This axiom is the key to Robinson's rigorous theory of infinitesimals and it took 300 years to discover. You will see by working with it that it is a perfectly natural idea, as hindsight often reveals.

EXAMPLE 5.5. *The Algebra of Small Quantities*

Let's re-calculate the increment of the basic cubic using the new numbers. Since the rules of algebra are the same, the same basic steps still work, except now we may take x any number and δx an infinitesimal.
Small Increment of $f(x) = x^3$

$$f(x + \delta x) = (x + \delta x)^3 = x^3 + 3x^2 \delta x + 3x\delta x^2 + \delta x^3$$
$$f(x + \delta x) = f(x) + 3x^2 \; \delta x + (\delta x[3x + \delta x]) \; \delta x$$
$$f(x + \delta x) = f(x) + f'[x] \; \delta x + \varepsilon \; \delta x$$

with $f'[x] = 3x^2$ and $\varepsilon = (\delta x[3x + \delta x])$. The intuitive rules above show that $\varepsilon \approx 0$ whenever x is finite. (See Theorem 5.11 and the example following it for the precise rules.)

The idealization of infinitesimals lets us have our cake and eat it too. Since $\delta \neq 0$, we can divide by δ. However, since δ is tiny, $1/\delta$ must be HUGE.

We want to make an official definition of the non-huge numbers:

DEFINITION 5.6. *Finite Numbers*
A hyperreal number x is said to be finite if there is an ordinary natural number $n = 1, 2, 3, \cdots$ so that

$$|x| < n$$

The finite numbers are not just the ordinary real numbers because of the rules of algebra. For example, $\pi - \delta < \pi < \pi + \delta$, but each of these numbers differ by only an infinitesimal, $\pi \approx \pi + \delta \approx \pi - \delta$. Of course, infinitesimals are finite, since $\delta \approx 0$ implies that $|\delta| < 1$. If we plotted the new number line at unit scale, we could only put one dot for all three. However, if we focus a microscope of power $1/\delta$ at π we see three points separated by unit distances.

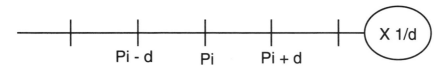

Pi - d   Pi   Pi + d

FIGURE 5.7: Magnification at Pi

The basic fact is that finite numbers only differ from reals by an infinitesimal. (Instructor's note: This is equivalent to Dedekind's Completeness Axiom.)

THEOREM 5.8. *Standard Parts of Finite Numbers*
Every finite hyperreal number x differs from some ordinary real number r by an infinitesimal amount, $x - r \approx 0$ or $x \approx r$. The ordinary real number infinitely near x is called the standard part of x, $r = \text{st}(x)$.

A picture of the ideal number line looks like the ordinary real line at unit scale. We can't draw far enough to get to the infinitely large part and this theorem says each finite number is indistinguishably close to a real number. If we magnify or compress by new number amounts we can see new structure.

You still cannot divide by zero (that violates rules of algebra), but if δ is a positive infinitesimal, we can compute the following:

$$-\delta, \quad \delta^2, \quad \frac{1}{\delta} \qquad \text{What can we say about these quantities?}$$

NEGATIVE INFINITESIMALS:
In ordinary algebra, if $\Delta > 0$, then $-\Delta < 0$, so we can apply this rule to the infinitesimal number δ and conclude that $-\delta < 0$, since $\delta > 0$.

ORDERS OF INFINITESIMALS:
In ordinary algebra, if $0 < \Delta < 1$, then $0 < \Delta^2 < \Delta$, so $0 < \delta^2 < \delta$.

We want you to formulate this more exactly in the next exercise. Just assume δ is very small, but positive. Formulate what you want to draw algebraically. Try some small ordinary numbers as examples, like $\delta = 0.01$. Plot δ at unit scale and place δ^2 accurately on the figure.

EXERCISE 5.9. *Draw the view of the ideal number line when viewed under an infinitesimal microscope of power $1/\delta$. Which number appears unit size? How big does δ^2 appear at this scale? Where do the numbers δ and δ^3 appear on a plot of magnification $1/\delta^2$?*

INFINITELY LARGE NUMBERS
For real numbers if $0 < \Delta < 1/n$ then $n < 1/\Delta$. Since δ is infinitesimal, $0 < \delta < 1/n$ for every natural number $n = 1, 2, 3, \ldots$ Using ordinary rules of algebra, but substituting the infinitesimal δ, we see that $H = 1/\delta > n$ is larger than any natural number n (or is "infinitely large"), that is, $1 < 2 < 3 < \ldots < n < H$, for every natural number n. We can "see" infinitely large numbers by turning the microscope around and looking in the other end.

EXERCISE 5.10. *Backwards microscopes or compression*
Draw the view of the new number line when viewed under an infinitesimal microscope with its magnification reversed to power δ (not $1/\delta$). What size does the infinitely large number H (HUGE) appear to be? What size does the finite (ordinary) number $m = 10^9$ appear to be? Can you draw the number H^2 on the plot?

The new algebraic rules are the ones that tell us when quantities are infinitely close, $a \approx b$. Such rules, of course, do not follow from rules about ordinary high school numbers, but the rules are intuitive and simple. More important, they let us 'calculate limits' directly.

THEOREM 5.11. *Computation Rules for Finite and Infinitesimal Numbers*
(1) *If p and q are finite, so are $p + q$ and pq.*
(2) *If ε and δ are infinitesimal, so is $\varepsilon + \delta$.*
(3) *If $\delta \approx 0$ and q is finite, then $q \cdot \delta \approx 0$. (finite x infsml = infsml)*

(4) $1/0$ *is still undefined and* $1/x$ *is infinitely large only when* $x \approx 0.$

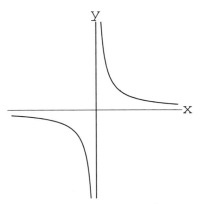

FIGURE 5.12: $y = 1/x$

To understand these rules, just think of p and q as "fixed," if large, and δ as being as small as you please (but not zero). It is not hard to give formal proofs from the definitions above, but this intuitive understanding is more important. The last rule can be "seen" on the graph of $y = 1/x$. Look at the graph and move down near the values $x \approx 0$.

EXAMPLE 5.13. $y = x^3 \Rightarrow dy = 3x^2 \, dx$, *for finite* x

The error term in the increment of $f(x) = x^3$, computed above is

$$\varepsilon = (\delta x[3x + \delta x])$$

If x is assumed finite, then $3x$ is also finite by the first rule above. Since $3x$ and δx are finite, so is the sum $3x + \delta x$ by that rule. The third rule, that says an infinitesimal times a finite number is infinitesimal, now gives $\delta x \times$ finite $= \delta x[3x + \delta x] =$ infinitesimal, $\varepsilon \approx 0$. This justifies the local linearity of x^3 at finite values of x, that is, we have used the approximation rules to show that

$$f(x + \delta x) = f(x) + f'[x] \, \delta x + \varepsilon \, \delta x$$

with $\varepsilon \approx 0$ whenever $\delta x \approx 0$ and x is finite, where $f(x) = x^3$ and $f'[x] = 3 \, x^2$.

EXERCISE 5.14. *Small Increment of* $f(x) = x^2$
Let $f(x) = x^2$, *compute a small increment of the function, and put it in the form*

$$f(x + \delta x) = (x + \delta x)^2 = ?$$
$$f(x + \delta x) = f(x) + f'[x] \, \delta x + \varepsilon \, \delta x$$

Why is $\varepsilon \approx 0$?

The Mathematical Background chapter on Computation of Limits contains many more drill problems using the rules of Theorem 5.11.

5.4. Rigorous Technical Summary

Before we go on with the most important part of the chapter, computing increments of various functions, we want to give the exact official definition of smoothness. Your work in the last chapter showed you that the intuitive meaning of the exact technical definition is that when a function satisfies the symbolic properties, then 'sufficiently powerful' magnifications of its graph 'appear' straight. The details of the next definition are explained in the Mathematical Background Chapter on Epsilon and Delta. In particular, it is shown there why the infinitesimal computations are equivalent to the "uniform" limits of the graphs of approximations to the derivative.

DEFINITION 5.15. *Smoothness (See Mathematical Background for Details)*
A real function $f(x)$ is smooth on the whole real line if there is another real function $f'[x]$ called the derivative of $f(x)$ such that whenever x is finite and an infinitesimal increment of the independent variable is given, $\delta x \approx 0$, then the increment of the dependent variable is approximately linear, that is,

$$f(x + \delta x) = f(x) + m \cdot \delta x + \varepsilon \cdot \delta x$$

where the error $\varepsilon \approx 0$ and $m = f'[x]$.

Equivalently, $f(x)$ is smooth on the whole real line if there is a function $f'[x]$ such that for every bound $b > 0$, the function limit

$$\lim_{\Delta x \to 0} \frac{f(x + \Delta x) - f(x)}{\Delta x} = f'[x] \qquad \textit{uniformly for all } |x| \le b$$

When the $f(x)$ is smooth we summarize the local linear approximation by

$$y = f(x) \qquad \Rightarrow \qquad dy = f'[x]\, dx$$

and say $dy = f'[x]\, dx$ is the differential of $f(x)$ or $\frac{dy}{dx} = f'[x]$ is the derivative of $f(x)$. When we write this we are claiming that the approximation given in the definition is valid.

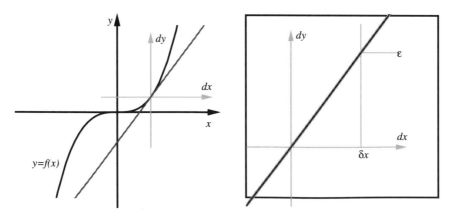

FIGURE 5.16: Infinitesimal Microscope

The geometric meaning of this definition is

THEOREM 5.17. *The Increment Principle*
If $f(x)$ is a smooth real function on the real line, then under an infinitely powerful microscope focused over a finite x, the graph $y = f(x)$ appears the same as the linear graph given in microscope (dx, dy)-coordinates by $dy = m \cdot dx$, where $m = f'[x]$ and the center of the microscope is focused at the (x, y)-point $(x, f(x))$.

Equivalently, for any size real screen pixel ,θ, and any bound $b > 0$ on x, there is a sufficiently small scale μ, so that if the magnification of the graph $y = f(x)$ is more than $1/\mu$ and focused at an $|x| \le b$, then the graph $y = f(x)$ is within one pixel of the linear graph $dy = f'[x] \, dx$.

5.5. Increment Computations

Given a function $f(x)$, we want the change $f(x + \delta x) - f(x)$ on a scale of δx in terms of a linear function of δx, $m \, \delta x$, where $m = f'[x]$ does not depend on δx. The first part of this computation just uses ordinary rules of algebra to rearrange formulas. The second part uses the approximation rules to show $\varepsilon \approx 0$. For example, if $f(x) = x^3$, rules of algebra (see the computation above) give

$$(x + \delta x)^3 = x^3 + 3x^2 \, \delta x + [\delta x \, (3x + \delta x)] \, \delta x$$

The second part uses the approximation rules in Theorem 5.11 to show that $[\delta x \, (3x + \delta x)] \approx 0$ (as shown above). Together we have the increment approximation of the definition of smoothness:

$$f(x + \delta x) = f(x) + f'[x] \, \delta x + \varepsilon \, \delta x$$

where $f'[x] = 3x^2$ and the error $\varepsilon = [\delta x \, (3x + \delta x)] \approx 0$.

EXAMPLE 5.18. *Exceptional Numbers and the Derivative of $y = \dfrac{1}{x}$*

We want to calculate the increment of another simple function to show that there will sometimes have to be exceptional cases. In these cases, we can not even get close to the bad values of x. This is an annoying, but necessary complication.
Small Increment of $f(x) = \frac{1}{x}$

$$f(x + \delta x) = \frac{1}{x + \delta x}$$

$$f(x + \delta x) - f(x) = \frac{1}{x + \delta x} - \frac{1}{x} = \frac{x - (x + \delta x)}{x(x + \delta x)}$$

$$= \frac{-1}{x(x + \delta x)} \cdot \delta x$$

It is "clear" that since $\delta x \approx 0$,

$$\frac{-1}{x(x + \delta x)} \approx \frac{-1}{x \cdot x} = \frac{-1}{x^2}$$

This is true unless $x \approx 0$. In particular, if $x = -\delta x$, then $\frac{-1}{x(x+\delta x)}$ is not even defined.

It is intuitively "clear" that $\frac{-1}{x(x+\delta x)} \approx \frac{-1}{x^2}$, but we can show this using the approximation rules. This problem is a little harder than the x^3 example because of the algebra needed. Make the error explicit by the formula

$$\frac{-1}{x(x+\delta x)} = \frac{-1}{x^2} + \varepsilon$$

in an unknown ε. Subtract and put the expression on a common denominator

$$\begin{aligned}
\varepsilon &= \frac{-1}{x(x+\delta x)} - \frac{-1}{x^2} \\
&= \frac{1}{x^2} + \frac{-1}{x(x+\delta x)} \\
&= \frac{x + \delta x}{x^2(x+\delta x)} + \frac{-x}{x^2(x+\delta x)} \\
&= \frac{\delta x}{x^2(x+\delta x)} \\
&= \delta x \cdot \frac{1}{x^2(x+\delta x)}
\end{aligned}$$

The approximation rules of Theorem 5.11 now show that $\varepsilon \approx 0$ as long as x is not infinitesimal.

We want to phrase the exceptions in terms of intervals.

DEFINITION 5.19. *Notation*

If a and b are numbers, we define open intervals as follows
a) $(a, b) = \{x : a < x < b\}$
b) $(-\infty, b) = \{x : x < b\}$
c) $(a, \infty) = \{x : a < x\}$

The symbol ∞ is neither a real nor hyperreal number, but a symbol for 'keeps on going toward infinity.'

We will say that the function $y = f(x) = 1/x$ is smooth on the intervals $(-\infty, 0)$ and $(0, \infty)$. The "approximation" to $f(x + dx) - f(x)$ given by $dy = m \cdot dx$ with $m = -1/x^2$ does not work if x is infinitesimal. It is clear that $x = 0$ is exceptional and the point of the computation above is that we must stay finitely away from the "bad" point.

The rigorous general way to exclude exceptional numbers is as follows.

DEFINITION 5.20. *Smooth Function on an Interval*
A real function $f(x)$ is smooth on the real interval (a, b) if there is another real function $f'[x]$, called the derivative of $f(x)$, such that whenever $a < x < b$ and x is finite and not infinitely near a or b, then an infinitesimal increment of the dependent variable is approximately linear, that is,

$$f(x + \delta x) = f(x) + f'[x]\,\delta x + \varepsilon\,\delta x$$

where the error $\varepsilon \approx 0$, whenever $\delta x \approx 0$.

Notice the extra complication when $f(x)$ is only smooth on an open interval $(a, b) = \{x : a < x < b\}$: We can only claim the microscopic increment approximation if x is NOT close to either a or b. (If a or b is $\pm\infty$, then x must be finite.) This happens for functions such as $f(x) = 1/x$ or $f(x) = \sqrt{x}$, which have no derivative at zero. We can not use the linear increment approximation for $x \approx 0$ in either of these cases. The increment principle with provisos says

THEOREM 5.21. *The Increment Principle for Intervals*
If $f(x)$ is a smooth real function on a real interval, then under an infinitely powerful microscope focused over a finite x inside the interval and not infinitely close to an endpoint, the graph $y = f(x)$ appears the same as the linear graph given in microscope (dx, dy)-coordinates by $dy = m\, dx$, where $m = f'[x]$.

EXAMPLE 5.22. *Exceptional Numbers and the Derivative of $y = \sqrt{x}$*

Here is another example of the algebraic part of computing increments. First we do algebra using the same rules as in high school, then we make approximations. Here is the first part.

Small Increment of $f(x) = \sqrt{x}$

$$f(x + \delta x) = \sqrt{x + \delta x}$$
$$f(x + \delta x) - f(x) = \sqrt{x + \delta x} - \sqrt{x}$$
$$= \frac{(\sqrt{x + \delta x} - \sqrt{x})(\sqrt{x + \delta x} + \sqrt{x})}{\sqrt{x + \delta x} + \sqrt{x}}$$
$$= \frac{1}{\sqrt{x + \delta x} + \sqrt{x}} \cdot \delta x$$

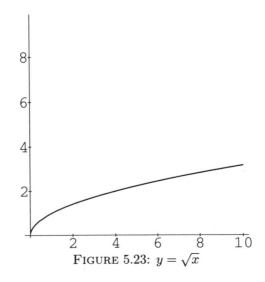

FIGURE 5.23: $y = \sqrt{x}$

We want to put these computations in the form

$$f(x + \delta x) = f(x) + f'[x]\, \delta x + \varepsilon\, \delta x$$

where $\varepsilon \approx 0$ and $m = f'[x]$ only depends on x, not δx. This requires use of the rules about the way infinitesimal and finite numbers interact (from Theorem 5.11) plus a little cleverness about square roots. The "obvious" approximation $1/[\sqrt{x + \delta x} + \sqrt{x}] \approx 1/[2\sqrt{x}]$ does not work if $x \approx 0$, because we could have $x + \delta x < 0$ negative, so that the square root $\sqrt{x + \delta x}$ is not defined. However, as long as $x > 0$ and x is not infinitesimal, this approximation does work and we ask you to show this by the rules of Theorem 5.11 in the second part of the next exercise.

EXERCISE 5.24.
Calculate the error terms ε in the following approximations and use rules to show that $\varepsilon \approx 0$. This proves the differentiation formulas indicated.

(1) *Prove that* $\quad y = \frac{1}{x} \quad \Rightarrow \quad dy = \frac{-1}{x^2}\, dx \quad$ *on* $(-\infty, 0)$ *and* $(0, \infty)$.
For x - not infinitesimal and any $\delta x \approx 0$

$$f(x) = \frac{1}{x}$$

$$f(x + \delta x) - f(x) = \frac{-1}{x(x + \delta x)} \cdot \delta x = \frac{-1}{x^2} \delta x + \varepsilon \cdot \delta x$$

Calculate ε and show that $\varepsilon \approx 0$ whenever x is not near zero and $\delta x \approx 0$. (Note that if x is infinitesimal, this fails when $x = -\delta x \approx 0$.)

(2) *Prove that* $\quad y = \sqrt{x} \quad \Rightarrow \quad dy = \frac{1}{2\sqrt{x}}\, dx \quad$, *on* $(0, \infty)$.
For x positive and not infinitesimal and any $\delta x \approx 0$

$$f(x) = \sqrt{x}$$

$$f(x + \delta x) - f(x) = \frac{1}{\sqrt{x + \delta x} + \sqrt{x}} \cdot \delta x = \frac{1}{2\sqrt{x}} \delta x + \varepsilon \cdot \delta x$$

with $\varepsilon \approx 0$, because

$$\frac{1}{\sqrt{x + \delta x} + \sqrt{x}} \approx \frac{1}{2\sqrt{x}}$$

(Note that if x is infinitesimal, even if it is positive, $\delta x = -2x$ makes $f(x + \delta x)$ undefined.) As in part (1), calculate ε and use the approximation rules to show that it is small. The algebra requires some tricks. Solve the equation

$$\frac{1}{\sqrt{x + \delta x} + \sqrt{x}} \cdot \delta x = \frac{1}{2\sqrt{x}} \delta x + \varepsilon \cdot \delta x$$

for ε, put the resulting expression on a common denominator and then use the expression you already know for

$$-(\sqrt{x + \delta x} - \sqrt{x})$$

to show that $\varepsilon \approx 0$.

The meaning of these computations is that each function $f(x)$ is locally linear and that $f'[x]$ has been found (at the specified x values),

$$f(x) = x^2 \Rightarrow f'[x] = 2x \qquad\qquad y = x^2 \Rightarrow dy = 2x \; dx$$

$$f(x) = x^3 \Rightarrow f'[x] = 3x^2 \quad \text{or} \quad y = x^3 \Rightarrow dy = 3x^2 \; dx$$

$$f(x) = \frac{1}{x} \Rightarrow f'[x] = \frac{-1}{x^2} \qquad\qquad y = \frac{1}{x} \Rightarrow dy = \frac{-1}{x^2} \; dx$$

$$f(x) = \sqrt{x} \Rightarrow f'[x] = \frac{1}{2\sqrt{x}} \qquad\qquad y = \sqrt{x} \Rightarrow dy = \frac{1}{2\sqrt{x}} \; dx$$

PROBLEM 5.25.
Use the δx method as above to find $f'[x]$ and write the whole increment approximation

$$f(x + \delta x) = f(x) + f'[x] \, \delta x + \varepsilon \, \delta x$$

showing that $\varepsilon \approx 0$, when

(1) $f(x) = x^4$
(2) $f(x) = x^n$
(3) $f(x) = \frac{1}{x^2}$
(4) $f(x) = \sqrt[3]{x}$

You can use Mathematica to check your symbolic computations. For example, open a new Mathematica window and type the program

$$f[x_{-}] := \frac{1}{x^2};$$

$f[x](Enter)$

$(Output)$

$f[x + dx](Enter)$

$(Output)$

$f[x + dx] - f[x](Enter)$

$(Output)$

$Together[f[x + dx] - f[x]](Enter)$

$(Output)$

$Together[f[x + dx] - f[x] + 2/x^3 dx](Enter)$

$(Output)$

5.6. Derivatives of Sine and Cosine

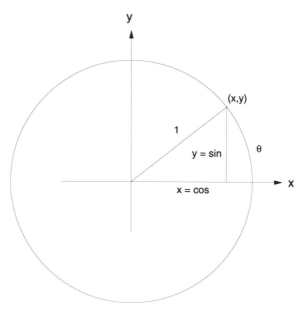

FIGURE 5.26: Sine and Cosine as a Point on the Unit Circle

The derivative of Sine in radians is Cosine and the derivative of Cosine in radians is -Sine. This important fact can be seen by magnifying the unit circle. We assume that you know the definition of radian measure of angles and the associated fact that $(\text{Cos}[\theta], \text{Sin}[\theta])$ is the (x, y)-point on the unit circle at the angle θ, measured counterclockwise from the x-axis. This important high school material is reviewed in the Mathematical Background Chapter on high school review.

We now wish to consider what happens as we move from a point $(\text{Cos}[\theta], \text{Sin}[\theta])$ to a nearby point $(\text{Cos}[\theta + \delta\theta], \text{Sin}[\theta + \delta\theta])$. We magnify the unit circle, noting that the more we magnify, the straighter the magnified portion of the circle appears. Here is the microscopic view of the circle that gives us the results.

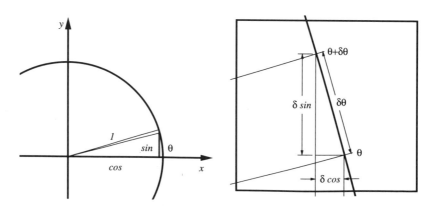

FIGURE 5.27: Derivatives of Sine and Cosine

The infinitesimal figure with $\delta\theta$ appears to be a triangle at magnification $1/\delta\theta$. The length of the long side of the apparent triangle is $\delta\theta$, because we use radian measure. (Degrees are not the distance along a unit circle.) The radii coming from the larger figure appear to meet it at right angles, so the apparent triangle is similar to the large triangle at the left with hypotenuse 1 and sides $\mathrm{Sin}[\theta]$, $\mathrm{Cos}[\theta]$. The sides of the apparent triangle are the differences in sine and cosine, with cosine decreasing, hence a negative sign.

Consider the apparent similarity, comparing the long sides of the two triangles,

$$\frac{\mathrm{Cos}[\theta]}{1} \approx \frac{\delta sin}{\delta\theta} = \frac{\mathrm{Sin}[\theta + \delta\theta] - \mathrm{Sin}[\theta]}{\delta\theta}$$

Since we only know the apparent triangle up to an infinitesimal, we only write approximate similarity. To be explicit, let the difference equal,

$$\varepsilon = \frac{\mathrm{Cos}[\theta]}{1} - \frac{\mathrm{Sin}[\theta + \delta\theta] - \mathrm{Sin}[\theta]}{\delta\theta}$$

Now do a little algebra to see,

$$\mathrm{Sin}[\theta + \delta\theta] = \mathrm{Sin}[\theta] + \mathrm{Cos}[\theta] \cdot \delta\theta + \varepsilon \cdot \delta\theta$$

with $\varepsilon \approx 0$ whenever $\delta\theta \approx 0$.

This proves half of the following: For θ in radians,

$$y = \mathrm{Sin}[\theta] \Rightarrow \frac{dy}{d\theta} = \mathrm{Cos}[\theta] \quad \text{or} \quad \mathrm{Sin}[\theta + \delta\theta] - \mathrm{Sin}[\theta] = \mathrm{Cos}[\theta]\delta\theta + \varepsilon \cdot \delta\theta$$

$$y = \mathrm{Cos}[\theta] \Rightarrow \frac{dy}{d\theta} = -\mathrm{Sin}[\theta] \quad\quad \mathrm{Cos}[\theta + \delta\theta] - \mathrm{Cos}[\theta] = -\mathrm{Sin}[\theta]\delta\theta + \varepsilon \cdot \delta\theta$$

with $\varepsilon \approx 0$ if $\delta\theta \approx 0$ (θ can take any value.)

EXERCISE 5.28. *Derivative of Cosine*
Use the infinitesimal microscope to prove that the differential of cosine is minus sine,

$$y = \mathrm{Cos}[\theta] \quad\quad \Rightarrow \quad\quad dy = -\mathrm{Sin}[\theta]\,d\theta$$

PROBLEM 5.29. *Derivative of Tangent*
Find the differential of the tangent function by examining an increment in the figures below. The segment on the line $x = 1$ between two rays from the circle is the increment of the tangent, because SOH-CAH-TOA with adjacent side of length 1 gives $\mathrm{Tan}[\theta]$ as the length of the segment on $x = 1$ between the x-axis and the ray.

(1) *The area of the triangle with $\delta y = \delta(\mathrm{Tan}[\theta])$ as its tiny vertical side is $\frac{1}{2}\delta\,\mathrm{Tan}[\theta]$, because the 'height' is 1 for the 'base' of $\delta\,\mathrm{Tan}[\theta]$.*

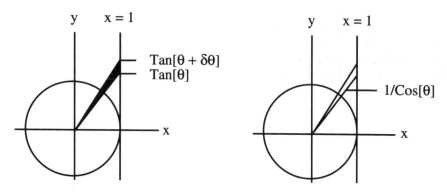

FIGURE 5.30: An Increment of Tangent and The Secant Function

(2) *The length of the ray at θ out to $x = 1$ is $1/\operatorname{Cos}[\theta]$. Why?*

(3) *The area of the shaded region bounded by a circular arc at the point on $x = 1$ at height $\operatorname{Tan}[\theta]$ is $\frac{\delta\theta}{2\operatorname{Cos}^2[\theta]} + \varepsilon \cdot \delta\theta$. Why? (You can use an approximate triangle or a circular sector formula. What is the length of the circular arc?) This shows*

$$\frac{1}{2}\frac{1}{\operatorname{Cos}^2[\theta]}\,\delta\theta \leq \frac{1}{2}\delta\operatorname{Tan}[\theta]$$

(4) *Similarly, show*

$$\frac{1}{\operatorname{Cos}^2[\theta]}\,\delta\theta \leq \delta\operatorname{Tan}[\theta] \leq \frac{1}{\operatorname{Cos}^2[\theta+\delta\theta]}\,\delta\theta$$

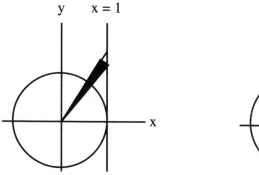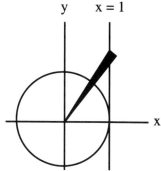

FIGURE 5.31: A Lower Estimate and an Upper Estimate

(5) *When is $\frac{1}{\operatorname{Cos}[\theta]} \approx \frac{1}{\operatorname{Cos}[\theta+\delta\theta]}$, for $\delta\theta \approx 0$?*

(6) *Prove that*

$$y = \operatorname{Tan}[\theta] \quad \Rightarrow \quad dy = \frac{1}{\operatorname{Cos}^2[\theta]}\,d\theta$$

provided $\operatorname{Cos}[\theta] \neq 0$.

EXAMPLE 5.32. *Increments of Sine*

The most important meaning of the increment formula for Sin and Cos is simply that a small piece of their graphs is given by the linear equation with slope $m = \text{Cos}[x]$ in the case of Sin. For example, suppose that we want to estimate the sine of 29 degrees. We know sine of 30 degrees and we can take the increment of -1 degree, using the 'microscope equation.' We must first convert to radian measure, because the increment formulas above are only valid in radian measure. We take $\theta = \frac{\pi}{6}$ and $\delta\theta = -\frac{\pi}{180}$,

$$f(x + \delta x) = f(x) + f'[x] \cdot \delta x + \varepsilon \cdot \delta x$$
$$\text{Sin}[\theta + \delta\theta] = \text{Sin}[\theta] + \text{Cos}[\theta] \cdot \delta\theta + \varepsilon \cdot \delta\theta$$
$$\text{Sin}[\theta + \delta\theta] = \frac{1}{2} + \frac{\sqrt{3}}{2} \cdot \frac{-\pi}{180} + \varepsilon \cdot \delta\theta$$
$$\text{Sin}[\theta + \delta\theta] \approx \frac{1}{2} + \frac{\sqrt{3}}{2} \cdot \frac{-\pi}{180}$$
$$\text{Sin}[\theta + \delta\theta] \approx 0.484885$$

Mathematica's approximation of sine of 29 degrees is 0.48481. That's not bad considering that one degree isn't really *infinitesimal*.

5.6.1. Differential Equations and Functional Equations. It is intuitively clear that magnified circles appear straighter and straighter, but complete justification of the local linearity of sine and cosine requires that we really show that the magnified increment of the circle is close to a triangle. For this we need to extend functions to the hyperreal numbers (or make some elaborate inequality estimates). We will not do this here except to mention two specific cases of the use of the Function Extension Axiom that are important in their own right. More details are contained in the Mathematical Background.

The formula

$$(\text{Sin}[\theta])^2 + (\text{Cos}[\theta])^2 = 1$$

simply says that sine and cosine lie on the unit circle. If $x = \text{Cos}[\theta]$ and $y = \text{Sin}[\theta]$, $x^2 + y^2 = 1$ is the equation of the unit circle. The Function Extension Axiom guarantees that this equation still holds for hyperreal angles like $\theta = \pi + \delta$ with $\delta \approx 0$.

The addition formulas for sine and cosine also hold for hyperreal numbers. Rather than using the increment approximation based on an infinitely magnified circle, we could use the exact addition formulas to obtain increments of trig functions. In the case of the sine,

$$\text{Sin}[\theta + \delta\theta] = \text{Sin}[\theta]\,\text{Cos}[\delta\theta] + \text{Cos}[\theta]\,\text{Sin}[\delta\theta]$$
$$\text{Sin}[\pi/6 + \delta\theta] = \frac{1}{2}\text{Cos}[\delta\theta] + \frac{\sqrt{3}}{2}\text{Sin}[\delta\theta]$$

These are exact formulas for the increments, but we need to obtain the differential approximations

$$\text{Sin}[\delta\theta] = \delta\theta + \varepsilon_1 \cdot \delta\theta$$
$$\text{Cos}[\delta\theta] = 1 + \varepsilon_2 \cdot \delta\theta$$

to complete the last step in obtaining the local linear approximation.

The point we wish to illustrate is this:

The differential

$$d(\text{Sin}[\theta]) = \text{Cos}[\theta]\ d\theta$$

is a sort of simplified infinitesimal version of the functional identity

$$\text{Sin}[\theta + \delta\theta] - \text{Sin}[\theta] = \text{Sin}[\theta]\,\text{Cos}[\delta\theta] - \text{Sin}[\theta] + \text{Cos}[\theta]\,\text{Sin}[\delta\theta]$$
$$= \text{Cos}[\theta]\delta\theta + \text{Cos}[\theta](\text{Sin}[\delta\theta] - \delta\theta) + \text{Sin}[\theta](\text{Cos}[\delta\theta] - 1)$$
$$= \text{Cos}[\theta]\delta\theta + \varepsilon \cdot \delta\theta$$

EXERCISE 5.33.

(1) *Write an exact formula for* $\text{Cos}[x + \delta x] = \text{Cos}[x]\,\text{Cos}[\delta x]\ldots$ *using the extended addition formula for cosine. Compare the exact formula with the increment approximation obtained from the microscopic view of the circle. What is the exact formula for* ε?

(2) *Approximate the value of cosine of 46 degrees using the linear increment approximation, discarding the* $\varepsilon\,\delta x$ *term. Give your answer in terms of exact constants such as* π *and* $\sqrt{2}$ *as well as numerically. Compare your approximation with Mathematica or your scientific calculator's approximation.*

The important functional identities of exponential functions are as follows

THEOREM 5.34. *Laws of Exponents*
For a positive base $a > 0$ *and any real (or hyperreal) numbers* p *and* q

$$a^{-p} = \frac{1}{a^p} \qquad a^p \cdot a^q = a^{p+q}$$
$$a^{1/p} = \sqrt[p]{a} \qquad (a^p)^q = a^{p \cdot q}$$

The Mathematical Background Exercise 3.2 shows you a simple numerical way to approximate the derivatives of exponentials.

EXERCISE 5.35.
Write an exact formula for the difference $a^{x+\delta x} - a^x$ *in terms of* a^x *times an expression,*
$a^{x+\delta x} - a^x = a^x \cdot [?].$

We would like to find an expression for the difference of exponentials in the previous exercise of the form

$$a^{x+\delta x} - a^x = a^x \cdot k_a \cdot \delta x + \varepsilon \cdot \delta x$$

where k_a is constant depending only on a. The 'natural' base $e \approx 2.71828 \cdots$ plays an important role in solving the problem. In fact, the number $e \approx 2.71828$ is the unique number that makes

$$\frac{e^{\delta x} - 1}{\delta x} \approx 1 = k_e$$

for $\delta x \approx 0$, that is, $k_a = 1$ when $a = e$. It turns out that the mysterious expression [?] in the previous exercise is of the form $[k_a + \varepsilon]\,\delta x$ for the constant $k_a = \text{Log}[a]$ and $\varepsilon \approx 0$, but this approximation is difficult to establish.

EXERCISE 5.36. *Logs*
Use properties of logs to show that

$$\frac{\text{Log}[x + \delta x] - \text{Log}[x]}{\delta x} = \frac{\text{Log}[1 + \frac{\delta x}{x}]}{\delta x}$$

Suppose you are given that

$$\text{Log}[1 + \eta] = \eta + \varepsilon \cdot \eta$$

with $\varepsilon \approx 0$ when $\eta \approx 0$. (In other words, that the derivative at 1 is 1. Make the substitutions Log[] for $f()$, $x = 1$, $\delta x = \eta$ and $f'[1] = 1$ in the increment equation $f(x + \delta x) = f(x) + f'[x] \cdot \delta x + \varepsilon \cdot \delta x$.) Use these two facts to prove that for all noninfinitesimal positive x,

$$\text{Log}[x + \delta x] = \text{Log}[x] + \frac{\delta x}{x} + \varepsilon \cdot \delta x$$

with $\varepsilon \approx 0$ when $\delta x \approx 0$.

EXERCISE 5.37. *What is wrong with the following computation?*

$$\frac{\text{Log}[x + \delta x] - \text{Log}[x]}{\delta x} = \frac{\text{Log}[x + \delta x - x]}{\delta x} = \frac{\text{Log}[\delta x]}{\delta x}$$

Once we know the derivative of the natural exponential and rules of differentiation, we can find the differentials of all exponentials. For this reason, the natural log and exponential play a major role in science and mathematics. Just as radian measure makes the calculus of trig functions 'natural,' the $e \approx 2.71828 \cdots$ base for logs and exponentials makes their calculus 'natural.' We postpone further discussion until Chapter 7, but give the results now. You may use these results as needed.

THEOREM 5.38. *Derivatives of Logs and Exponentials*
The derivative of the natural base exponential function is

$$y = e^x \qquad \Rightarrow \qquad \frac{dy}{dx} = y$$

or, written in terms of the independent variable,

$$\frac{de^x}{dx} = e^x$$

The derivative of the natural base logarithm is

$$\frac{d\,\text{Log}[x]}{dx} = \frac{1}{x}$$

Here is some practice using these derivatives:

EXERCISE 5.39. *Natural Increments*
1) Use the formula for the derivative of the natural exponential to write the increment approximation for $y = e^x$,

$$e^{x + \delta x} = e^x + [?] \cdot \delta x + \varepsilon \cdot \delta x$$

2) Use the formula for the derivative of the natural logarithm to estimate $\text{Log}[2.8] = \text{Log}[e + (2.8 - 2.71828 \cdots)]$.

5.7. Continuity and the Derivative

We saw in the Exercise 4.16 that a function can be continuous, but still not smooth or differentiable. An official definition of continuity is the following

DEFINITION 5.40. *Continuity of $f[x]$*
A real function $f[x]$ is continuous at the real point a if for every $x \approx a$, $f[x]$ is defined and

$$f[x] \approx f[a]$$

Intuitively, this just means that $f(x)$ is close to $f(a)$ when x is close to a.

EXERCISE 5.41.

(1) *Suppose that $f[x]$ is smooth on an interval around a, so that the 'microscope' increment equation is valid. Suppose that $x \approx a$ so that $x = a + \delta x$ for $\delta x \approx 0$. Show that $f[x] = f[a + \delta x] \approx f[a]$, in other words, show that smooth real functions are continuous at real points.*
(2) *Consider the real function $f[x] = 1/x$, which is undefined at $x = 0$. We could extend the definition by simply assigning $f[0] = 0$. Show that this function is not continuous at $x = 0$, but is continuous at every other real x.*
(3) *Give an intuitive graphical description of the definition of continuity in terms of infinitesimal microscopes and explain why it follows that smooth functions must be continuous.*
(4) *The function $f[x] = \sqrt{x}$ is defined for $x \geq 0$, there is nothing wrong with $f[0]$. However, our increment computation for $\sqrt{x}$ above was not valid at $x = 0$ because a microscopic view of the graph focused at $x = 0$ looks like a vertical ray (or half-line). Explain why this is so, but show that $f[x]$ is still continuous 'from the right,' that is, if $0 < x \approx 0$, then $\sqrt{x} \approx 0$, but $\frac{\sqrt{x}}{x}$ is infinitely large.*

THEOREM 5.42. *Continuity of $f(x)$ and $f'[x]$*
If $f(x)$ is smooth on the interval $h < x < k$ and $x_1 \approx x_2$ with neither infinitely near h or k, then $f(x_1) \approx f(x_2)$ and $f'(x_1) \approx f'(x_2)$.

PROOF FOR f(x):
Proof of continuity of f is easy algebraically, but is obvious geometrically: a graph that is indistinguishable from linear clearly only moves a small amount in a small x-step. Symbolically, take $x = x_1$ and $\delta x = x_2 - x_1$ and use the approximation $f(x_2) = f(x + \delta x) = f(x_1) + f'(x_1)\delta x + \varepsilon \delta x$ where $[f'(x_1) + \varepsilon]\delta x$ is finite x infsml = infsml, so $f(x_1) \approx f(x_2)$. That's the proof. Draw the picture.
PROOF FOR f'(x):
Proof of continuity of $f'[x]$ requires us to view the increment from both ends. First take $x = x_1$ and $\delta x = x_2 - x_1$ and use the approximation

$$f(x_2) = f(x + \delta x) = f(x_1) + f'(x_1)\delta x + \varepsilon_1 \delta x.$$

Next let $x = x_2$, $\delta x = x_1 - x_2$ and use the approximation

$$f(x_1) = f(x + \delta x) = f(x_2) + f'(x_2)\delta x + \varepsilon_2 \delta x.$$

The different x-increments are negatives, so we have

$$f(x_1) - f(x_2) = f'(x_2)(x_1 - x_2) + \varepsilon_2(x_1 - x_2)$$

and

$$f(x_2) - f(x_1) = f'(x_1)(x_2 - x_1) + \varepsilon_1(x_2 - x_1)$$

Adding, we obtain

$$0 = \{[f'(x_2) - f'(x_1)] + [\varepsilon_2 - \varepsilon_1]\}(x_1 - x_2)$$

Dividing by the non-zero $(x_1 - x_2)$ we see that

$$f'(x_2) = f'(x_1) + [\varepsilon_1 - \varepsilon_2], \quad \text{so} \quad f'(x_2) \approx f'(x_1)$$

NOTE:

The derivative defined in many calculus books is a weaker pointwise notion than the notion of smoothness we have defined. The weak derivative function need not be continuous. (The same approximation does not apply at both ends with the weak definition.) This is explained in the Mathematical Background Chapter on "Epsilon - Delta" Approximations.

5.8. Instantaneous Rates of Change

We don't have to use the definition of the derivative in many instances. (What a relief!) And we don't even have to compute derivatives by rules in many important cases. Rates of change are one important way to view derivatives. This section contains two examples.

5.8.1. The Dead Canary. The derivative $\frac{dT}{dt}$ of a function $T(t)$ is the instantaneous rate at which T changes as t changes. In the chapter on linearity we told you about Jonnie putting sister Susie's canary cage out into an Iowa winter night. The initial temperature was 75° F and ten minutes later the temperature was 60° F. The rate of cooling would be fast when the temperature is high and gradually become slower as the temperature gets closer to zero. When the temperature reaches zero, the cage wouldn't cool any more. The simplest relationship with this property is that the rate of cooling is proportional to the temperature or

$$\text{The rate of cooling} = kT$$

The poor canary has long since died, but that makes the mathematical model simpler anyway (since he no longer even generates a little heat.) We want to finish the story, mathematically speaking, before we bury the canary. The simplest way to measure the rate of cooling of the canary cage would be to measure the rate at which the temperature *decreases* as time increases. In other words,

$$\text{The rate of cooling} = -\frac{dT}{dt}$$

so the cooling law above becomes Newton's law of cooling

$$\frac{dT}{dt} = -kT$$

and we know that $T = 75$ when $t = 0$ and $T = 60$ when $t = 10$.

PROBLEM 5.43. *Use the idea of the Mathematica NoteBook**SecondSIR.ma** to compute the temperature as a function of time. Type a NoteBook like the one following this problem. Run it with various values of k until* $T = 60$ *when* $t = 10$.

Once you determine k in this way, change 'tfinal' to ten hours $= 600$ *(use* $dt = 1$ *with such a big* T*) and plot the whole canary's plight.*

```
k = 0.03;
tInitial = 0;
TiNITIAL = 75;
Tdata = { { tInitial,TiNITIAL} };
tfinal = 10;
dt = 0.25;
T = TiNITIAL
Do[
Trate = - k T;
T = T + Trate dt;
(*This is the computer's T(t + dt) = T(t) + T'(t) dt + ε dt, dropping ε.*)
AppendTo[ Tdata, { t,T} ];
, { t,0,tfinal,dt} ]
ListPlot[Tdata, PlotJoined -> True]
```

The reason that $-\frac{dT}{dt}$ is the instantaneous rate of decrease in temperature can be seen from the increment approximation. The reason the *Mathematica* NoteBook in the Exercise above approximates the temperature also follows from the increment approximation. We will postpone the second bit of theory until we get to 'Euler's method' in the study of differential equations later in the semester. Here is why $\frac{dT}{dt}$ is the rate of change of temperature - instantaneously. The change in temperature from time t to a small time later $t + \delta t$ is

$$T(t + \delta t) - T(t)$$

(in units of degrees) so the rate at which it changes during this time interval of duration δt is the ratio

$$\frac{T(t + \delta t) - T(t)}{\delta t} = \frac{\delta T}{\delta t} \approx \frac{dT}{dt}$$

(in units of degrees per minute). The increment approximation for $T(t)$ is

$$T(t + \delta t) = T(t) + \frac{dT}{dt} \cdot \delta t + \varepsilon \cdot \delta t$$

subtracting gives

$$T(t + \delta t) - T(t) = \frac{dT}{dt} \cdot \delta t + \varepsilon \cdot \delta t$$

dividing gives

$$\frac{T(t + \delta t) - T(t)}{\delta t} = \frac{dT}{dt} + \varepsilon$$

where $\varepsilon \approx 0$ when $\delta t \approx 0$. In an 'instant' δt becomes zero, the left hand side of the previous equation no longer makes sense mathematically, but 'tends' to the instantaneous rate of change, and ε becomes zero. This gives

$$\text{The rate of change of } T \text{ in an instant of } t = \frac{dT}{dt}$$

Exactly what an instant is may not be clear, but the discussion before the exercise makes it clear that it is useful.

PROBLEM 5.44. *Computer Solution of Differential Equations*
Explain the main idea in the computer's computation of the temperature using only the rate of change of temperature and the initial temperature. Write a few sentences covering the main ideas without giving a lot of technical details, but rather only giving a description and a few sample computations. In other words, convince us that you understand the computer program.

5.8.2. The Fallen Tourist. You have a chance to visit Pisa, Italy and are fascinated by the leaning tower. You decide to climb up and throw some things off to see if Galileo was right. Luckily, you have a camera and your glasses to help with the experiment.

You drop your camera and glasses simultaneously and are amazed that they smash simultaneously on the sidewalk below. Unconvinced that a heavy object doesn't fall faster, you race down and retrieve a large piece and a small piece of the stuff that's left.

Dropping the pieces again produces the same simultaneous smash and you also notice, of course, that the objects speed up as they fall. Not only that, but both the heavy object and the light object speed up at the same rates. You know that a feather would be affected by air friction, but you theorize that in vacuum, there must be just one 'law of gravity' to govern falling objects.

About then the Italian police arrive and take you to a nice quiet room to let you think about the grander meaning of your experiments while an interpreter arrives from the American consulate.

PROBLEM 5.45. *Galileo's First Conjecture*
Since all objects speed up as they fall, and at the same rate neglecting air friction, you theorize that

> *The speed of a falling object is proportional to the distance it has fallen.*

Let D be the distance the object has fallen from the top of the tower at time t (in the units of your choice.)
1) What is the speed in terms of the rate of change of D with respect to t?
2) Formulate your 'law of gravity' as a differential equation about D.
3) What is the initial value of D in your experiment?
4) Use Mathematica or mathematics to decide if your 'law of gravity' is correct.

5.9. Projects

5.9.1. Hubble's Law and the Increment Equation.

$$R(t + \delta t) = R(t) + R'(t)\,\delta t + \varepsilon\,\delta t$$

Evidence of an expanding universe is one of the most important astronomical observations of this century. Light received from a distant galaxy is "old" light, generated millions of years ago at a time t_e when it was emitted. When this old light is compared to light generated at the time received t_r it is found that the characteristic colors, or spectral lines, do not have the same wavelengths. All the current wavelengths are longer, $\lambda(t_e) < \lambda(t_r)$. This means that light is "redder" now; this is the famous red shift.

The Scientific Project on Hubble's Law shows you an explanation for the expanding universe that is based just on using the increment approximation.

5.9.2. Small Enough Real Numbers or "Epsilons and Deltas". The increment approximation used to estimate Sin['29 degrees'] was very close, but how do we know that the increment approximation gets close for real increments, not just infinitely close for infinitesimal increments? The Mathematical Background chapter on "epsilons and deltas" answers this question.

In general, if we can establish an infinitesimal error estimate with real functions, a 'small enough' real limit estimate follows from expressing our real error need in terms of inequalities. Usually, we do not need to write the inequalities out, but simply see that 'small enough' real numbers exist to satisfy our practical error tolerance. Both rigorous infinitesimals and 'epsilon - delta' limits are explained in the Mathematical Background chapter on 'Epsilon - Delta' approximations.

CHAPTER 6

Symbolic Differentiation

Calculus gives remarkably simple symbolic rules for finding formulas for derivatives if we are given formulas for the functions. This chapter presents the 'method of computing' or 'calculus' of derivatives. You must use techniques of high school algebra and trig, but you do not have to establish the increment approximation directly as we did in the last chapter. When we compute a derivative, we want to know that the increment approximation is valid. Graphically, this tells us that infinitesimal increments of the function are "approximately linear" in the sense that the explicit graph "looks" linear in an infinitesimal microscope. The rules of calculus are theorems which guarantee that this approximation is valid, provided the resulting formulas are defined on intervals.

The symbolic form of the increment approximation is given in Definition 5.15. It is the exact symbolic statement of what it means to appear linear in a microscope. The basic formula is

$$f(x + \delta x) = f(x) + f'[x]\, \delta x + \varepsilon\, \delta x$$

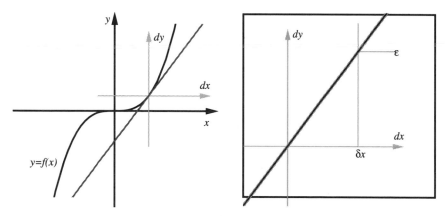

FIGURE 6.1: Infinitesimal Microscope

where the error $\varepsilon \approx 0$ whenever $\delta x \approx 0$. The linear approximation is called the differential, $dy = f'[x] \cdot dx$. The linear part of the formula above is $m \cdot \delta x$ with $m = f'[x]$. This is the differential $dy = f'[x]\, dx$ evaluated at $dx = \delta x$.

The rules of this chapter guarantee that the increment approximation holds when the resulting formulas are valid on intervals. It is very important that you learn the increment approximation equation, because this formula is what will let you understand the symbolic meaning of smoothness. Of course, your first task now is to learn the rules themselves. In the next chapter we will return to some basic applications of the increment approximation.

6.1. Rules for Special Functions

The algebra of exponents together with the derivatives we have computed suggest a simplified rule that includes the examples of Exercise 5.25 and before. To understand why this rule covers the cases of roots and reciprocals, you must understand the laws of exponents for power functions. These topics are reviewed in the Mathematical Background Chapter, "High School Review with *Mathematica*."

THEOREM 6.2. *The Power Rule*
For any constant p,

$$y = x^p \quad \Rightarrow \quad \frac{dy}{dx} = p\, x^{p-1}$$

In other words, functions that can be expressed as powers are locally linear with derivative as above – provided the formulas on both sides of the implication make sense on an interval.

EXAMPLE 6.3. $\frac{d(\sqrt{x})}{dx}$

We showed directly in the last chapter that if $y = \sqrt{x}$ then $dy = \frac{1}{2\sqrt{x}}\, dx$. This is one special case of the Power Rule with $p = 1/2$, because

$$y = \sqrt{x} = x^{\frac{1}{2}}, \quad \text{so} \quad \frac{dy}{dx} = \frac{1}{2}\, x^{\frac{1}{2}-1} = \frac{1}{2}\, x^{-\frac{1}{2}}$$
$$= \frac{1}{2}\, \frac{1}{x^{\frac{1}{2}}} = \frac{1}{2}\, \frac{1}{\sqrt{x}} = \frac{1}{2\sqrt{x}}.$$

Notice that our final formula is only valid on the open interval $(0, \infty) = \{x : 0 < x < \infty\}$. The open interval of validity is part of the power rule, but you compute first and then think. Don't forget the second step.

EXERCISE 6.4. *You can't divide by zero or take even roots of negative numbers (as real functions). Show that the Power Rule does not apply at $x = 0$ for $p = \frac{1}{3}$ or at $x = -2$ for $p = \frac{1}{4}$.*

Use Mathematica to graph the two functions $y = x^{1/3} = \sqrt[3]{x}$ and $y = x^{1/4} = \sqrt[4]{x}$ for $-3 \leq x \leq 3$, type and enter the calculation:

```
y := x^(1/3)
Plot[ y , {x,-3,3} ]
```

and explain your analytical result above in terms of the graph.

We will not prove the Power Rule Theorem now, but ask you to check it for the cases we already know in the next exercise.

EXERCISE 6.5.

(1) *Show that the power rule agrees with all the derivatives we computed directly before Exercise 5.25 as well as those that you computed in Exercise 5.25 by direct infinitesimal computations. They are the following:*

$$y = x \Rightarrow \frac{dy}{dx} = 1 = 1x^{1-1} = x^0 \qquad y = \frac{1}{x} = x^{-1} \Rightarrow \frac{dy}{dx} = \frac{-1}{x^2} = -1x^{-2}$$

$$y = x^2 \Rightarrow \frac{dy}{dx} = 2x^1 = 2x^{2-1} \qquad y = \frac{1}{x^2} = x^{-2} \Rightarrow \frac{dy}{dx} = \frac{-2}{x^3} = -2x^{-3}$$

$$y = x^3 \Rightarrow \frac{dy}{dx} = 3x^2 = 3x^{3-1} \qquad y = \sqrt{x} = x^{\frac{1}{2}} \Rightarrow \frac{dy}{dx} = \frac{1}{2\sqrt{x}} = \frac{1}{2}x^{-\frac{1}{2}}$$

$$y = x^n \Rightarrow \frac{dy}{dx} = nx^{n-1} \qquad y = \sqrt[3]{x} = x^{\frac{1}{3}} \Rightarrow \frac{dy}{dx} = \frac{1}{3\sqrt[3]{x^2}} = \frac{1}{3}x^{-\frac{2}{3}}$$

(2) *Differentiate the following by first converting to power form and then applying the Power Rule. Convert your derivatives back to radical notation.*

a) $y = \frac{1}{x^5}$ b) $y = \sqrt[5]{x}$ c) $y = \frac{1}{\sqrt[3]{x^2}}$

(3) *The derivative of a constant function is zero. Why? The power rule also includes a case of this in the form,*

$$y = x^0 \quad \Rightarrow \quad \frac{dy}{dx} = 0 = 0\,x^{0-1}$$

What is the value of x^0? Is 0^0 defined?

PROBLEM 6.6. *When the graph of $y = x^{2/3}$ is magnified at $x = y = 0$, what do we see? The power rule does not apply at $x = 0$, why? Still, we can either use infinitesimals directly, or look at microscopic views for smaller and smaller values of δx. The question is: In an infinitesimal microscope, do we see a "VEE" or a vertical straight line segment? (HINT: Run the animation in the Mathematica NoteBook **Zoom.ma**, then explain what you see analytically.)*

PROBLEM 6.7. *The formula for the volume of a sphere is $V(r) = \pi \frac{4}{3} r^3$ and for the surface area is $S(r) = \pi 4 r^2$. Compute the derivative $\frac{dV}{dr}$ and explain.*

We proved the sine and cosine using the infinitesimal microscopic view of the circle in the last chapter. The angles must be measured in radians in order to compare differences in sine and cosine with length along the unit circle. You already used this result to approximate Cos["46 degrees"] in the last chapter.

THEOREM 6.8. *The Sine and Cosine Rules*
For θ in radians,

$$y = \text{Sin}[\theta] \quad \Rightarrow \quad \frac{dy}{d\theta} = \text{Cos}[\theta]$$

$$y = \text{Cos}[\theta] \quad \Rightarrow \quad \frac{dy}{d\theta} = -\text{Sin}[\theta]$$

The sine and cosine rules are valid for all real θ. This means that the increment approximation holds on $(-\infty, \infty)$, as we showed in the last chapter.

We will postpone the proof of the following rules, but include them here because they are the only other special function rules you need to learn.

THEOREM 6.9. *The Log and Exponential Rules*

$$y = e^x \qquad \Rightarrow \qquad \frac{dy}{dx} = y = e^x$$

$$x = \text{Log}[y] \qquad \Rightarrow \qquad \frac{dx}{dy} = \frac{1}{y}$$

The exponential rule is valid for all real x, or in other words, the increment approximation holds on $(-\infty, \infty)$. The natural logarithm rule only makes sense if the log and the formula for the derivative are both defined. In other words, the increment approximation for $\text{Log}[y]$ is valid on $(0, \infty)$.

It is important to be able to apply the differentiation rules to functions defined in terms of letters other than x. At first, it is simplest to learn the manipulations with one letter, that's true, but it is also important to move beyond that.

EXERCISE 6.10. *Other Variables*

a) $y = x^2 \quad \Rightarrow \quad \frac{dy}{dx} = ?$ b) $u = \frac{1}{v^2} \quad \Rightarrow \quad \frac{du}{dv} = ?$

c) $y = \sqrt{x} \quad \Rightarrow \quad \frac{dy}{dx} = ?$ d) $u = \frac{1}{\sqrt{v}} \quad \Rightarrow \quad \frac{du}{dv} = ?$

e) $y = \text{Sin}[x] \quad \Rightarrow \quad \frac{dy}{dx} = ?$ f) $u = \text{Cos}[v] \quad \Rightarrow \quad \frac{du}{dv} = ?$

g) $y = \text{Log}[x] \quad \Rightarrow \quad \frac{dy}{dx} = ?$ h) $u = e^v \quad \Rightarrow \quad \frac{du}{dv} = ?$

The Power Rule, The Sine and Cosine Rules and The Log and Exponential Rules are the only particular function rules you need to learn. The other general rules of differentiation allow you to use these to build up a host of formulas that you can differentiate.

The general functional rules can be viewed several ways: simply as symbolic rules, as theorems we prove by showing that the increment equations for the components imply the increment equation for the combination, and as generalizations of everyday facts in various applications. You should try to understand them from all these points of view. The general function combination rules take up the next three sections.

6.2. The Superposition Rule

The physical Superposition Principle says that the response to a sum of stimuli is the sum of the responses to the separate stimuli. Another way to express the rule is:

Output[stimulus 1 + stimulus 2] = Output[stimulus 1] + Output[stimulus 2]

This is a simple property that is often violated in real life. For example, the combined effect of a cup of coffee and an aspirin is not the same as the two separate effects. Systems that satisfy the Superposition Principle are often called 'linear systems,' because you can apply the 'output' to a linear combination $af(x) + bg(x)$ or form the same linear combination of the separate outputs,

$$\text{Out}[af(x) + bg(x)] = a\text{Out}[f(x)] + b\text{Out}[g(x)]$$

Next semester we will see important applications where physical superposition does apply and 'linearity' of the derivative is at the heart of the matter. In the case of differentiation, "Output" means derivative and "stimulus" means input function.

THEOREM 6.11. *The Superposition Rule (or Linearity of Differentiation)*
If $f(x)$ and $g(x)$ are smooth real functions for $\alpha < x < \beta$ and a and b are real constants, then the function $h(x) = a\,f(x) + b\,g(x)$ is also smooth for $\alpha < x < \beta$ and

$$\frac{d\,[a\,f(x) + b\,g(x)]}{dx} = a\frac{df(x)}{dx} + b\frac{dg(x)}{dx}$$

In words the theorem says that the derivative of a linear combination of functions is the same linear combination of their derivatives.

EXAMPLE 6.12. $\frac{d(5\cdot\sqrt{x}-\pi\cdot x^2)}{dx}$

Let

$$f(x) = \sqrt{x}, \qquad a = 5, \qquad g(x) = x^2, \qquad b = -\pi$$

then

$$a\,f(x) + b\,g(x) = 5 \cdot \sqrt{x} - \pi \cdot x^2$$

and the derivatives of the pieces are $\frac{df}{dx} = \frac{1}{2\sqrt{x}}$ and $\frac{dg}{dx} = 2x$, so

$$\frac{d\,[a\,f(x) + b\,g(x)]}{dx} = a\frac{df(x)}{dx} + b\frac{dg(x)}{dx}$$
$$\frac{d(5 \cdot \sqrt{x} - \pi \cdot x^2)}{dx} = 5 \cdot \frac{1}{2\sqrt{x}} - \pi \cdot 2x$$

A shortcut way to write this computation is

$$\frac{d(5\sqrt{x} - \pi \cdot x^2)}{dx} = 5\frac{d(\sqrt{x})}{dx} - \pi\frac{d(x^2)}{dx} = 5\frac{d(x^{1/2})}{dx} - \pi\frac{d(x^2)}{dx}$$
$$= 5 \cdot \frac{1}{2} \cdot x^{\frac{1}{2}-1} - \pi \cdot 2 \cdot x^{2-1} = \frac{5}{2}x^{-\frac{1}{2}} - 2\pi x$$
$$= \frac{5}{2}\frac{1}{x^{1/2}} - 2\pi x$$
$$= \frac{5}{2\sqrt{x}} - 2\pi x$$

EXAMPLE 6.13. *The Constant Multiple Rule,* $\frac{d(a\,f(x))}{dx} = a\frac{df}{dx}(x)$

We may take $b = 0$ and $g(x) = 0$ in the Superposition Rule. If $y = \frac{e^\pi}{\sqrt[3]{x}}$, let $a = e^\pi$ and $f(x) = \frac{1}{x^{1/3}}$, $b = g(x) = 0$, so

$$\frac{d(\frac{e^\pi}{\sqrt[3]{x}})}{dx} = e^\pi \frac{d(\frac{1}{\sqrt[3]{x}})}{dx} = e^\pi \frac{d(\frac{1}{x^{1/3}})}{dx} = e^\pi \frac{d(x^{-1/3})}{dx}$$

$$= e^\pi \frac{-1}{3} x^{-\frac{1}{3}-1} = e^\pi \frac{-1}{3} x^{-\frac{1}{3}-\frac{3}{3}} = e^\pi \frac{-1}{3} x^{-\frac{4}{3}}$$

$$= -\frac{e^\pi}{3} \frac{1}{x^{\frac{4}{3}}} = -\frac{e^\pi}{3} \frac{1}{\sqrt[3]{x^4}}$$

Notice that e^π is a constant, $e^\pi = (2.71828\cdots)^{(3.14159\cdots)} \approx 23.1407$

EXERCISE 6.14. *Basic Superposition Drill*
Find $\frac{dy}{dx}$ for each of the following functions $y = y[x]$. *(The letters a, b, c and h denote constants or parameters, and e is the natural base for logs and exponentials.)*

a) $y = 7\sqrt[5]{x} - 3\frac{1}{x^4}$ b) $y = 7e^x - 3\operatorname{Log}[x]$

c) $y = e\sqrt[5]{x} - \pi\frac{1}{x^4}$ d) $y = f(x) = 7\operatorname{Cos}[x] - 3\operatorname{Sin}[x]$

e) $y = f(x) = a + bx + cx^2 + hx^3$ f) $y = 3\operatorname{Sin}[x] - \operatorname{Cos}[7e]$

PROOF OF SUPERPOSITION FOR DIFFERENTIATION:
The general proof of the Superposition Rule is little more than algebra. For finite x not infinitely near α or β, we have the Increment Formula for $f(x)$ and $g(x)$, so

$$h(x + \delta x) - h(x) = [a\,f(x + \delta x) + b\,g(x + \delta x)] - [a\,f(x) + b\,g(x)]$$
$$= a\{f(x + \delta x) - f(x)\} + b\{g(x + \delta x) - g(x)\}$$
$$= a\{f'(x)\delta x + \varepsilon_1 \delta x\} + b\{g'(x)\delta x + \varepsilon_2 \delta x\}$$
$$= [a\,f'(x) + b\,g'(x)]\,\delta x + [a\varepsilon_1 + b\varepsilon_2]\delta x$$

Since $\varepsilon_1 \approx 0$ and $\varepsilon_2 \approx 0$ (by the Increment Formula for f and g) we know by the rule 'infsml x finite = infsml' that $[a\varepsilon_1 + b\varepsilon_2] \approx 0$, thus $h'(x) = a\,f'(x) + b\,g'(x)$ and the theorem is proved.

The proof of 'linearity of differentiation' is easy provided you understand the function notation. See the next exercise if the function notation is mysterious to you, because you do need to understand the more abstract formulas. That takes effort, but the exercise should help. One payoff to understanding the function formulas is that you will be able to write better *Mathematica* programs, so it even has 'applications.'

EXERCISE 6.15. *Let $f(x) = \sqrt{x}$, let $g(x) = x^2$, let $a = 5$ and $b = \pi$. Write out each step of the proof of the Superposition Rule that appears above, except write the steps with these specific functions and constants.*

The next exercise shows two practical ways that superposition of derivatives arise. You need to link those 'obvious' applications with the symbolic expressions.

PROBLEM 6.16. *Express the conditions in the following scenarios in terms of three functions yielding positions as a function of time and the time derivatives of these functions. Write the general function rules that yield the answer to the questions and verify that the mathematical rules agree with the intuitively obvious answers.*

(1) *A man and a woman are travelling on a train which is travelling at the rate of 75 mph. Inside the train the woman is walking forward at the rate of 4 mph and the man is walking backward at the rate of 3 mph. How fast is the man travelling relative to the ground? How fast is the woman travelling relative to the ground? Let $T(t)$ equal the distance (in miles) that the train has traveled along the ground. Let $m(t)$ equal the distance the man has traveled forward on the train. Let $w(t)$ equal the distance the woman has traveled forward on the train. How much are $\frac{dT}{dt}$, $\frac{dm}{dt}$ and $\frac{dw}{dt}$ including sign? What does the function $f(t) = T(t) + m(t)$ represent? What is $\frac{df}{dt}$? How does this compare to $\frac{dT}{dt} + \frac{dm}{dt}$? What are the similar constructions for the woman?*

(2) *A U.S. tourist is driving in Canada at 90 kilometers per hour, but her odometer and speedometer read in the archaic English units of her home country. We want to see the functional relationship between English and metric measurements of speed and distance. Let the odometer reading be the numerical function $E(t) =$ distance traveled in miles, where $t = $ time in hours. The distance traveled in kilometers is a function $M(t)$. There are approximately 1.609 kilometers in a mile. Express E in terms of a constant and M. Express $\frac{dE}{dt}$ in terms of this same constant and $\frac{dM}{dt}$. What is her speed in miles per hour?*

It is important to be able to apply the differentiation rules to functions defined in terms of letters other than x.

EXERCISE 6.17. *Other Variables*

a) $v = u^2 + 2u + 2 \quad \Rightarrow \quad \frac{dv}{du} = ?$ 　　　b) $u = 4\sqrt{v} - \pi\frac{1}{v^2} \quad \Rightarrow \quad \frac{du}{dv} = ?$

c) $y = 3u\sqrt{u} + 5/u^2 \quad \Rightarrow \quad \frac{dy}{du} = ?$ 　　d) $u = v^{1/4} - \frac{1}{v^{1/3}} \quad \rightarrow \quad \frac{du}{dv} = ?$

e) $y = \cos[\theta] - \sin[\theta] \quad \Rightarrow \quad \frac{dy}{d\theta} = ?$ 　　f) $u = \log[v] - e^v \quad \Rightarrow \quad \frac{du}{dv} = ?$

6.3. Symbolic Differentiation with *Mathematica*

Mathematica can 'do' all the symbolic differentiation exercises in this chapter. At first this fact might discourage you, but it shouldn't. We want you to learn to use the rules of differentiation well enough to be confident that you understand them. *Mathematica* can't think or understand the meaning of the result of these computations and the input syntax needed to make it 'do' the exercises is troublesome itself. Once you understand the rules, *Mathematica* can become a mental 'lever,' because it can do complicated symbolic computations for you with great reliability. You are then left with the important and interesting job of formulating the problem, programming it into *Mathematica* and interpreting the result.

The *Mathematica* NoteBook **DiffRules.ma** contains exercises to show you how the rules of differentiation can be used to build a *Mathematica* program for differentiation. The *Mathematica* NoteBook **Dfdx.ma** shows you how to use *Mathematica's* built in differentiation.

EXERCISE 6.18. *Differentiation by Computer*
*Run the **DiffRules.ma** NoteBook and work through it line by line so that you can see how the addition of each of the Superposition Rule, Product Rule and Chain Rule makes symbolic differentiation more powerful. We can't differentiate $\sqrt{x}\,\text{Cos}[x]$ without the Product Rule and neither can the computer.*

You are also welcome to use *Mathematica* to check all your work from this chapter. Use the built in *Mathematica* D[.] command once you have learned all the differentiation rules. The **DiffRules.ma** NoteBook is only intended to show you what can not be done without some of the rules. Knowing what can't be done is part of understanding the strength of the general functional rules.

PROBLEM 6.19. *Using only superposition, the power rule and derivatives of sine, cosine, natural log and exponential, find the following or write brief explanations why they can't be done this way. The letters a, b, c and m denote parameters (or unknown constants).*
*Once you have the Product Rule, you will be able to do some additional parts of the problem and when you also have the Chain Rule, you will be able to do all the parts. The Mathematica NoteBook **DiffRules.ma** can be used to check your work. Only enter the specific function rules and the Superposition Rule. If Mathematica can not find the derivative with these rules, it will return your question as its output.*

a) $y = mx + b$, $y' = ?$ b) $y = uv + w$, $\frac{dy}{dv} = ?$

c) $y = uv + w$, $\frac{dy}{dw} = ?$ d) $f(x) = 1 + \frac{1}{x} + \frac{1}{x^2} + \frac{1}{x^3}$, $f'(x) = ?$

e) $f(x) = a + bx + cx^2 + mx^3$, $f'(x) = ?$ f) $u = 3\,\text{Sin}(\theta) - \text{Cos}(\theta)$, $\frac{du}{d\theta} = ?$

g) $y = \frac{1}{\sqrt[3]{x^5}}$, $\frac{dy}{dx} = ?$ h) $y = \sqrt{x} + \text{Sin}(x)$, $\frac{dy}{dx} = ?$

i) $\sqrt{x}\,\text{Sin}(x)$, $\frac{dy}{dx} = ?$ j) $y = \sqrt{\text{Sin}(x)}$, $\frac{dy}{dx} = ?$

k) $y = \text{Sin}(\sqrt{x})$, $\frac{dy}{dx} = ?$ l) $y = \sqrt{x}$, $f(x) = \frac{dy}{dx}$, $f'(x) = ?$

m) $y = e^x + \text{Log}[x]$, $\frac{dy}{dx} = ?$ n) $y = e^x\,\text{Sin}[x]$, $\frac{dy}{dx} = ?$

What is the slope of each of the above graphs when the independent variable equals -1? 0? (What are the tricks in this question?)

6.4. The Product Rule

Let m, n, b and c denote unknown constants or parameters. We cannot differentiate the product $y = (mx + b)(nx + c)$ directly using the rules we have so far. However, we could do some algebra,

$$(mx + b)(nx + c) = mnx^2 + mcx + bnx + bc = mnx^2 + (mc + nb)x + bc$$

so,

$$\frac{dy}{dx} = 2mnx + mc + nb$$

To verify that this agrees with The Product Rule given below, take $f(x) = mx + b$ and $g(x) = nx + c$. The derivatives are

$$\frac{df}{dx}(x) = \frac{d(mx + b)}{dx} = m \quad \& \quad \frac{dg}{dx}(x) = \frac{d(nx + c)}{dx} = n$$

The formula from the Product Rule below gives

$$\frac{dy}{dx} = \frac{df(x)}{dx} \cdot g(x) + f(x) \cdot \frac{dg(x)}{dx}$$
$$= m \cdot (nx + c) + (mx + b) \cdot n$$

This is the same answer because

$$m(nx + c) + (mx + b)n = mnx + mc + mnx + nb = 2mnx + mc + nb$$

Algebra can rearrange products of many power functions and linear combinations of power functions into forms we can differentiate, but often it is easier to use the Product Rule. A product like $\sqrt{x}\,\text{Cos}[x]$ really requires a new rule.

THEOREM 6.20. *The Product Rule*
If $f(x)$ and $g(x)$ are smooth for $\alpha < x < \beta$, then the product $P(x) = f(x) \cdot g(x)$ is also locally linear for $\alpha < x < \beta$ and

$$\frac{d[f(x)\,g(x)]}{dx} = \frac{df(x)}{dx} \cdot g(x) + f(x) \cdot \frac{dg(x)}{dx}$$

In words, the Product Rule says,'Differentiate the terms of a product one at a time, multiply by the other undisturbed term and add these together.'

EXAMPLE 6.21. $\frac{d\{\sqrt{x}\,\text{Cos}[x]\}}{dx}$

Let $f(x) = \sqrt{x}$ and $g(x) = \text{Cos}[x]$, then $\frac{df}{dx} = \frac{1}{2\sqrt{x}}$ and $\frac{dg}{dx} = -\text{Sin}[x]$, so

$$\frac{d[f(x)\,g(x)]}{dx} = \frac{df(x)}{dx} \cdot g(x) + f(x) \cdot \frac{dg(x)}{dx}$$
$$\frac{d\{\sqrt{x}\,\text{Cos}[x]\}}{dx} = \frac{1}{2\sqrt{x}} \cdot \text{Cos}[x] - \sqrt{x} \cdot \text{Sin}[x]$$

We can also combine the Superposition Rule and Product Rule.

EXAMPLE 6.22. $\frac{d\{(\frac{a}{\sqrt{x}} - b\,\text{Log}[x])(e^x - c)\}}{dx}$

Let $f(x) = \frac{a}{\sqrt{x}} - b\,\text{Log}[x]$ and $g(x) = e^x - c$, for constants a, b and c. Then superposition says

$$\frac{df}{dx} = a\frac{d(\frac{1}{\sqrt{x}})}{dx} - b\frac{d(\text{Log}[x])}{dx} = a\frac{d(x^{-1/2})}{dx} - b\frac{d\,\text{Log}[x]}{dx}$$
$$= -\frac{a}{2}x^{-3/2} - \frac{b}{x} = -\frac{1}{x}\left(\frac{a}{\sqrt{x}} + b\right)$$

and

$$\frac{dg}{dx} = \frac{d(e^x)}{dx} - \frac{dc}{dx} = e^x + 0 = e^x$$

so

$$\frac{d[f(x)\,g(x)]}{dx} = \frac{df(x)}{dx} \cdot g(x) + f(x) \cdot \frac{dg(x)}{dx}$$

$$= \left[-\frac{1}{x}\left(\frac{a}{\sqrt{x}} + b \right) \right] \cdot [e^x - c] + \left[\frac{a}{\sqrt{x}} - b\,\text{Log}[x] \right][e^x]$$

EXERCISE 6.23. *Drill on Products*
Break each of the following expressions into a product of two terms and apply the Product Rule to find $\frac{dy}{dx}$ for each of the following functions $y = y[x]$. (The letters a, b, c and h denote constants or parameters, and e is the natural base for logs and exponentials.)

a) $y = \frac{\text{Log}[x]}{\sqrt[5]{x}}$ b) $y = (e^x - 3)(\frac{1}{x^4})$

c) $y = x^2\,e^x$ d) $y = f(x) = 7\,e^x\,\text{Cos}[x]$

e) $y = f(x) = (a + b\,x)(c\,x^2 + h\,x^3)$ f) $y = 3\,\text{Sin}[x] \cdot \text{Cos}[7e]$

g) $y = \left(x^{1/2} + x^{1/3} + x^{1/4} \right)\left(x^{-1/2} + x^{-1/3} + x^{-1/4} \right)$

PROOF OF THE PRODUCT RULE:
The proof of the product rule is another straightforward computation with infinitesimal increments. All we do is add and subtract a term.

$$h(x + \delta x) - h(x) = f(x + \delta x)g(x + \delta x) - f(x)g(x)$$
$$= f(x + \delta x)g(x + \delta x) - f(x)g(x + \delta x) + f(x)g(x + \delta x) - f(x)g(x)$$
$$= [f(x + \delta x - f(x)]\,g(x + \delta x) + f(x)\,[g(x + \delta x) - g(x)]$$
$$= [f'(x)\delta x + \varepsilon_1 \delta x]\,g(x + \delta x) + f(x)\,[g'(x)\delta x + \varepsilon_2 \delta x]$$
$$= [f'(x)\delta x + \varepsilon_1 \delta x]\,[g(x) + \varepsilon_3] + f(x)\,[g'(x)\delta x + \varepsilon_2 \delta x]$$
$$= f'(x)g(x)\delta x + f(x)g'(x)\delta x + \delta x[\varepsilon_4]$$
$$= [f'(x)g(x) + f(x)g'(x)]\,\delta x + \delta x[\varepsilon_4]$$

EXERCISE 6.24. *Verify that $\varepsilon_3 \approx 0$ and $\varepsilon_4 \approx 0$ in the proof of the product rule above.*

EXERCISE 6.25. *Let $f(x) = x^2$ and $g(x) = x^3$ and write out the steps of the proof of the product rule given above for these specific functions.*

PROBLEM 6.26. *The 'rule' that says the derivative of a product is the product of the derivatives might make things simpler, but it is false. Show this by finding a counterexample from among functions you know how to differentiate directly from infinitesimal computations (such as simple powers of x.) For example, let $f[x] = x^2$ and $g[x] = x$, so $\frac{df}{dx} = 2x$ and $\frac{dg}{dx} = 1$. How much is $\frac{df}{dx} \times \frac{dg}{dx}$? How much is $\frac{d(f \cdot g)}{dx} = \frac{d(x^3)}{dx}$? How much is $\frac{df}{dx} \cdot g + f \cdot \frac{dg}{dx}$?*

Explain why the derivative of a product is not necessarily the product of the derivatives.

The function notation in the general proof of the Product Rule may seem obscure, but the idea is not. The next two problems show you why.

PROBLEM 6.27. *A Concrete Instance of the Product Rule*

(1) *A contractor's crew is making forms to lay a rectangular concrete floor. One member of the crew measures the length l (feet) and makes an error Δl (feet), another measures the width w (feet) and makes a separate error Δw (feet). The contracted area is $A = lw$, but the actual area is $A + \Delta A$ where the error in area is $\Delta A = $ _?* *(Write a formula in terms of l, w, Δl and Δw. See the next figure for help.)*

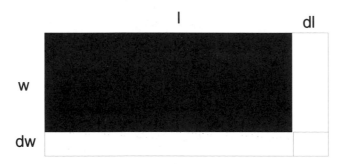

FIGURE *6.28: Three Error Rectangles*

(2) *Check your formula with a numerical example. Suppose that the floor has design dimensions of 20 by 30 feet, the error in length is 2 inches and the error in width is 1 inch; both too long. How much too large is the floor? Is the error larger or smaller than a desktop?*

(3) *With the same dimensions as in the previous part and same error in length, suppose the error in width is 2 inches too short. Use $\Delta w = -2/12$ in your formula and verify that it gives the correct result. Is the floor too large or too small? What is the error in area?*

(4) *Suppose that the length and width are changing with time (instead of a measurement error), so $\frac{\Delta l}{\Delta t}$ and $\frac{\Delta w}{\Delta t}$ are the rates of change of length and width during the time increment Δt. (Imagine the floor expanding with a change in temperature.) Show that*

$$\frac{\Delta A}{\Delta t} = \frac{\Delta w}{\Delta t} \cdot l + w \cdot \frac{\Delta l}{\Delta t} + \frac{\Delta w}{\Delta t} \cdot \frac{\Delta l}{\Delta t} \cdot \Delta t$$

What is the Product Rule for $A = l(t)\, w(t)$? How do these expressions differ when Δt is very small?

The Scientific Projects Chapter on The Role of Rules for Derivatives shows another way that the Product Rule arises in everyday discussions. The problem on the expanding economy shows what happens as both a population and economy expand.

It is important to be able to apply the differentiation rules to functions defined in terms of letters other than x.

EXERCISE 6.29. *Other Variables*

(1) $v = (u^3 + 3\,u + 3)(\text{Sin}[u] + \text{Cos}[u]) \quad \Rightarrow \quad \frac{dv}{du} = \,?$

(2) $u = (4\sqrt{v} - \pi\,\frac{1}{v^2})(v^{1/4} - \frac{1}{v^{1/3}}) \quad \Rightarrow \quad \frac{du}{dv} = \,?$

(3) $y = (\text{Cos}[v] - \text{Sin}[v])(\text{Log}[v] - e^v) \quad \Rightarrow \quad \frac{dy}{dv} = \,?$

EXERCISE 6.30. *Which parts of Exercise 6.19 can you do now using the Product Rule that you couldn't do without it? (Again, you can check your work with the **DiffRules.ma** NoteBook, by only entering the rules up to the Product Rule, but not entering the Chain Rule.)*

We can not differentiate $y = \text{Sin}[2\,x]$ with the rules we have developed so far. Even the simple expression $2\,x$ inside the sine makes this problem outside our present rules.

EXAMPLE 6.31. *An Impossible Problem Made Possible*

We can use the addition formula for sine to show that $\text{Sin}[2\,x] = 2\,\text{Sin}[x] \times \text{Cos}[x]$,

$$\text{Sin}[\alpha + \beta] = \text{Sin}[\alpha]\,\text{Cos}[\beta] + \text{Sin}[\beta]\,\text{Cos}[\alpha]$$
$$\text{Sin}[x + x] = \text{Sin}[x]\,\text{Cos}[x] + \text{Sin}[x]\,\text{Cos}[x]$$
$$\text{Sin}[2\,x] = 2\,\text{Sin}[x]\,\text{Cos}[x]$$

This form of the expression can be differentiated using the Product Rule as follows.

EXAMPLE 6.32. $\frac{d\{\text{Sin}[x]\,\text{Cos}[x]\}}{dx}$

Let $f(x) = \text{Sin}[x]$ and $g(x) = \text{Cos}[x]$, so $\frac{df}{dx} = \text{Cos}[x]$ and $\frac{dg}{dx} = -\,\text{Sin}[x]$, and

$$\frac{d[f(x)\,g(x)]}{dx} = \frac{df(x)}{dx} \cdot g(x) + f(x) \cdot \frac{dg(x)}{dx}$$
$$\frac{d\{\text{Sin}[x]\,\text{Cos}[x]\}}{dx} = \text{Cos}[x] \cdot \text{Cos}[x] - \text{Sin}[x] \cdot \text{Sin}[x]$$

EXERCISE 6.33. *Show that the derivative of $\text{Sin}[\theta] \times \text{Cos}[\theta]$ is $\text{Cos}[2\theta]$ using the addition formula for cosine. (You can look that formula up in the Mathematical Background Chapter, "High School Review with Mathematica."*

Combine this fact with the previous example to show that

$$\frac{d(\text{Sin}[2\,x])}{dx} = \frac{d(2\,\text{Sin}[x]\,\text{Cos}[x])}{dx} = 2\,\text{Cos}[2\,x]$$

(without using the Chain Rule!)

We will be able to verify this more simply once we have the Chain Rule. The point is that there are often several ways to apply the rules of algebra and trig in combination with the rules of differentiation.

6.5. The Expanding House

The Scientific Projects contain a Chapter on the "Expanding House." The wood in your house expands during the course of a normal day's warming. A 40 foot long house only expands about 0.03 inches on a normal Fall day, but there is a substantial increase in the volume of the house. Do you think it is about a thimble, a bucket or a bathtub full? The numerical calculation might surprise you, but the ideas of this calculation are similar to the way that the rules of calculus are built. The project starts with simple but surprising arithmetic and progresses to a symbolic formulation of volume expansion. A final section uses the Chain Rule from the next section to give a direct solution. We recommend that you at least skim through the Expanding House Project now.

6.6. The Chain Rule

The Chain Rule for derivatives shows how to differentiate functions that are 'hooked together in a chain.' Mathematically this occurs in expressions like $y = \text{Sin}[x^2 + 1]$. If we let $u = x^2 + 1$ and let $y = \text{Sin}[u]$, then the original formula is what we would get if we start with x, compute u, then use that answer for u to compute y.

$$u = x^2 + 1 \quad \rightarrow \quad y = \text{Sin}[u]$$

The functional notation for this 'chaining' if $y = f(u)$ and $u = g(x)$ is

$$y = f(g(x)) = \text{Sin}[x^2 + 1]$$

Of course, we could have broken the final formula down in other ways. For example,

$$v = x^2 \quad \rightarrow \quad y = \text{Sin}[v + 1]$$

or $y = f(g(x))$, where $f(v) = \text{Sin}[v + 1]$ and $v = g(x) = x^2$. The advantage of the first decomposition, however, is that we can differentiate each of the component pieces with rules already at our disposal,

$$\frac{du}{dx} = 2x \quad \& \quad \frac{dy}{du} = \text{Cos}[u]$$

Notice the importance of being able to differentiate with respect to a letter other than x. This is one reason that we emphasized "other letters' in the early sections of the chapter.

The Chain Rule given next tells us how to use the derivatives of the components to find the derivative of the whole composition. In this case,

$$\begin{aligned}
\frac{dy}{dx} &= \frac{dy}{du} \cdot \frac{du}{dx} \\
&= \text{Cos}[u] \cdot 2x \\
&= \text{Cos}[x^2 + 1] \cdot 2x \\
&= 2x \, \text{Cos}[x^2 + 1]
\end{aligned}$$

We removed the 'link' variable u in our final expression for $\frac{dy}{dx}$, because we only introduced it to help solve the problem. (In some applications the link variables actually have important separate meanings.)

THEOREM 6.34. *The Chain Rule*
If $y = f(u)$ is smooth on the range of $u = g(x)$ and g is smooth for $\alpha < x < \beta$, then the chained composition $y = h(x) = f(g(x))$ is smooth for $\alpha < x < \beta$ and $h'(x) = f'(g(x)) \cdot g'(x)$ or

$$\frac{dy}{dx} = \frac{dy}{du} \cdot \frac{du}{dx}$$

EXAMPLE 6.35. $\frac{d(\sqrt{x^2+1})}{dx}$

When you want to differentiate an expression like $v = \sqrt{x^2 + 1}$, you need to find a decomposition of the formula satisfying,

(1) Each piece of the decomposition can be differentiated by known rules.
(2) When chained back together, the pieces 'compose' the original formula.

In this case, we let $u = x^2 + 1$ and $v = \sqrt{u}$, because substituting this value for u into the u-expression for v makes $v = \sqrt{x^2 + 1}$. The Power Rule and Superposition Rule tell us

$$\frac{du}{dx} = 2x \quad \& \quad \frac{dv}{du} = \frac{1}{2\sqrt{u}}$$

so the Chain Rule above says

$$\frac{dv}{dx} = \frac{dv}{du} \cdot \frac{du}{dx} = \frac{1}{2\sqrt{u}} \cdot 2x$$
$$= \frac{x}{\sqrt{x^2 + 1}}$$

EXERCISE 6.36. *Drill on Chains*
Break each of the following expressions into a composition of two functions and apply the Chain Rule to find $\frac{dy}{dx}$. (The letters a, b, c and h denote constants or parameters, and e is the natural base for logs and exponentials.)

a) $y = f(x) = (a + bx)^{33}$ b) $y = e^{x^2}$

c) $y = 3(\text{Sin}[x])^3$ d) $y = 3\,\text{Sin}[x^3]$

e) $y = \text{Log}[x \sqrt[5]{x}]$ f) $y = f(x) = \text{Cos}[7\,e^x]$

PROOF OF THE CHAIN RULE:
    The general proof is only a little more complicated use of the increment approximation.

$$f(g(x + \delta x)) - f(g(x)) = f\left([g(x) + g'(x)\delta x + \varepsilon_1 \cdot \delta x]\right) - f(g(x))$$
$$= f\left(g(x) + L_1\delta x\right) - f(g(x))$$
$$= [f(g(x)) + f'(g(x))L_1\delta x + \varepsilon_2 \cdot L_1\delta x] - f(g(x))$$
$$= f'(g(x))g'(x)\delta x + \varepsilon_3\delta x$$

where $L_1 = [g'(x) + \varepsilon_1]$ is finite, so that $L_1 \delta x \approx 0$ and we may apply the increment approximation to f at $g(x)$ with change $L_1 \delta x$. Also, $\varepsilon_3 = [f'(g(x))\varepsilon_1] + \varepsilon_2 L_1 \approx 0$, since $f(u)$ is smooth on the range of $g(x)$.

EXERCISE 6.37. *The product* $\text{Sin}[\theta]\,\text{Cos}[\theta] = \frac{1}{2}\,\text{Sin}[2\theta]$. *Show this using the addition formula for the sine.* (*Hint:* $\text{Sin}[\theta + \theta] = ?$) *Use the Chain rule to show that the derivative of the product is* $\text{Cos}[2\theta]$ *by differentiating the right side of the equality. Check your work using the Product Rule on the left side of the equality.*

6.6.1. The Everyday Meaning of the Chain Rule.

Several applications of the Chain Rule are given in the next chapter of the core text. We already mentioned the Expanding House in the Scientific Projects.

EXERCISE 6.38. *The sine function in degrees can be thought of this way*

$$sIN[D] = \text{Sin}[\frac{\pi}{180} D]$$

where $\text{Sin}[u]$ *denotes the radian measure sine function. Use the chain rule to show that the derivative of the sine in degrees is* $\frac{\pi}{180}$ *times the cosine in degrees,*

$$y = sIN[D] \Rightarrow \frac{dy}{dD} = \frac{\pi}{180}cOS[D]$$

EXERCISE 6.39. *Which parts of Exercise 6.19 can you do now using the chain rule that you couldn't do without it?*

6.7. Derivatives of Other Exponentials by the Chain Rule

The first section in the Mathematical Background chapter on exponentials shows you a direct way to approximate the derivative of a general exponential function. The next topic shows you an exact symbolic method. This is important in the theory of calculus, because it reduces the calculus of other bases to the natural base.

Suppose we have $y = e^{ct}$ for some constant c. The chain rule says

$$y = e^u \qquad u = ct$$
$$\frac{dy}{du} = e^u \qquad \frac{du}{dt} = c$$

$$\frac{d(e^{ct})}{dt} = \frac{dy}{du} \times \frac{du}{dt} = c\,e^{ct}$$

If we want to differentiate $y = 2^t$ we can use the preceding chain rule computation and two other important facts about exponentials. We know the general exponential law

$$(a^c)^x = a^{cx}$$

so we try to find a constant c that satisfies

$$2 = e^c$$

Once we find this c we have $2^t = e^{ct}$ for all t, because of the law of exponents. We know how to differentiate $y = e^{ct}$, and with this value of c, we learn how to differentiate 2^t,

$$\frac{d(2^t)}{dt} = \frac{d(e^{ct})}{dt} = ce^{ct} = c2^t$$

Now we solve for c using natural log. Natural log and exponential are inverse functions. This simply means:

$$\text{Log}(e^t) = t \quad and \quad e^{\text{Log}(s)} = s, \quad s > 0$$

We apply this to our problem by taking logs of both sides of

$$2 = e^c$$
$$\text{Log}(2) = \text{Log}(e^c) = c$$

Thus $c = \text{Log}(2)$ and we see that

$$\frac{d(2^t)}{dt} = \text{Log}(2) \times 2^t$$

EXERCISE 6.40. *For $y(t)$ as follows, find $\frac{dy}{dt}(t)$*

a) $y = 3^t$

b) $y = 10^t$

c) $y = a^t$

d) $y = t^t$

e) $y = 2^{\cos(t)}$

f) $y = te^t$

g) $y = e^{\frac{1}{t}}$

h) $y = 3^{\sqrt{t}}$

i) $y = \sqrt{3^t - 2^t}$

EXERCISE 6.41. *If the number of algae cells is $N(t) = N_0 2^{t/6}$, how long does it take to triple the number of cells?*

EXERCISE 6.42. *What is wrong with the following nonsensical differentiation?*

$$\frac{dx^x}{dx} = xx^{x-1} = x^1 x^{x-1} = x^{1+x-1} = x^x$$

6.8. Derivative of The Natural Logarithm

The inverse of $x = \text{Log}[y]$ is $y = e^x$ and has derivative $\frac{dy}{dx} = e^x = y$, therefore $dy = e^x \, dx$ or

$$dy = y \, dx \quad \Rightarrow \quad \frac{dx}{dy} = \frac{1}{y}$$

This computation is explored in more detail in the Mathematical Project on Inverse Functions. The point is that the derivative of the inverse function is the reciprocal of the derivative of the function.

EXERCISE 6.43. *Differentiate $y = \text{Log}[x^3]$ using $u = x^3$ and $y = \text{Log}[u]$. Also use the log identity, $\text{Log}[x^p] = p \, \text{Log}[x]$ and differentiate $y = \text{Log}[x^3] = 3 \, \text{Log}[x]$ without the Chain Rule. Compare the two answers.*

EXERCISE 6.44. *Practice problems using the log and exp rules*
For $y(t)$ as follows, find $\frac{dy}{dt}(t)$

a) $y = (\text{Log}[t])^3$ b) $y = \text{Log}[\text{Cos}[x]]$ c) $y = t\,\text{Log}[t] - t$

d) $y = \text{Log}[\text{Log}[x]]$ e) $y = \text{Log}[t^{1/t}]$ f) $y = \text{Log}[t^2 + 2x]$

Assume first that x is independent of t? Second, if x is a function of t, but we forgot to give you a formula for $x = x(t)$, express your answers in terms of x and $\frac{dx}{dt}$. What is $\frac{dx}{dt}$ if x is independent of t?

6.9. Combined Symbolic Rules

EXERCISE 6.45. *Use all the rules of differentiation to find the following:*

a) $y = \frac{1}{2x-1}$, $\frac{dy}{dx} =?$ b) $y = \frac{1}{ax+b}$, $\frac{dy}{dx} =?$ $\left(\begin{array}{l} u = ax + b \\ y = \frac{1}{u} = u^{-1} \end{array} \right)$

c) $y = \text{Cos}(x^2)$, $\frac{dy}{dx} =?$ d) $y = \text{Cos}^2(x)$, $\frac{dy}{dx} =?$ *(unchain two ways)*

e) $y = \sqrt{1 - x^2}$, $\frac{dy}{dx} =?$ f) $y = \frac{1}{\sqrt{1-x^2}}$, $\frac{dy}{dx} =?$

g) $y = (12 - 3x^7)^8$, $\frac{dy}{dx} =?$ h) $y = \text{Sin}(x^2 + x^3)$, $\frac{dy}{dx} =?$

i) $y = [(ax + b)^{-1} + c]^{-1}$, $\frac{dy}{dx} =?$

j) $y = \text{Sin}(x)\,\text{Cos}(x)$, $\frac{dy}{dx} =?$ k) $y = (2x^3 + 4)(x^2 - \sqrt{x})$, $\frac{dy}{dx} =?$

l) $y = \text{Sin}(2x)$, $\frac{dy}{dx} =?$ m) $y = \frac{1}{\text{Cos}(3x)}$, $\frac{dy}{dx} =?$

n) $y = \text{Sin}(x) \cdot [\text{Cos}(x)]^{-1} = \frac{\text{Sin}(x)}{\text{Cos}(x)} - \text{Tan}(x)$, $\frac{dy}{dx} -?$

o) $y = \frac{2-3x}{1+2x} = (2 - 3x)[(1 + 2x)^{-1}]$, $\frac{dy}{dx} =?$

Check your work with *Mathematica* , for example,
y[x_] := Cos[x∧2];
y[x]
D[y[x] , x]

EXERCISE 6.46. *Differentiate $y = \sqrt{x^2 + 2x + 1}$ using the chain rule, power rule and linearity rule. The graph of this function is rather simple, as we saw in Exercise 4.16. Where does your symbolic answer not make sense? Can you sketch the graph? Why doesn't the symbolic answer work at the bad point?*

EXAMPLE 6.47. *Custom Rules*

You can make some additional rules of differentiation for general cases. Suppose you often need to differentiate a cube of a product of functions, $y = (f[x] \cdot g[x])^3$, for various smooth functions $f[x]$ and $g[x]$. We use the Chain Rule with unknown functions:

$$y = u^3 \qquad\qquad u = f[x] \cdot g[x]$$

$$\frac{dy}{du} = 3\,u^2 \qquad\qquad \frac{du}{dx} = \frac{df}{dx} \cdot g + f \cdot \frac{dg}{dx}$$

$$\frac{dy}{dx} = \frac{dy}{du} \cdot \frac{du}{dx} = 3\,u^2 \left(\frac{df}{dx} \cdot g + f \cdot \frac{dg}{dx} \right)$$

$$= 3(f[x] + g[x])^2 \left(\frac{df}{dx}[x] \cdot g[x] + f[x] \cdot \frac{dg}{dx}[x] \right)$$

EXERCISE 6.48. *Use the Superposition Rule and Product Rule repeatedly to show in general that if $f(x)$, $g(x)$ and $h(x)$ are smooth on an interval, then so are their sum and product and*

$$\frac{d(f(x) + g(x) + h(x))}{dx} = \frac{df}{dx}(x) + \frac{dg}{dx}(x) + \frac{dh}{dx}(x)$$

$$\frac{d[f(x)g(x)h(x)]}{dx} = \frac{df}{dx}(x) \cdot g(x) \cdot h(x) + f(x) \cdot \frac{dg}{dx}(x) \cdot h(x) + f(x) \cdot g(x) \cdot \frac{dh}{dx}(x)$$

(Hint: Let $G(x) = [g(x) \cdot h(x)]$ and apply the Product Rule to $f(x) \cdot G(x)$.)

6.9.1. The Quotient Rule.

EXERCISE 6.49. *The Quotient Rule*
Derive the quotient rule: If $q(x) = \frac{f(x)}{g(x)}$, where f and g are smooth and $g(x) \neq 0$ for $\alpha < x < \beta$, then $q(x)$ is also smooth for $\alpha < x < \beta$ and

$$\frac{d(\frac{f(x)}{g(x)})}{dx} = \frac{\frac{df(x)}{dx}g(x) - f(x)\frac{dg(x)}{dx}}{[g(x)]^2}$$

Use the Chain Rule and Product Rule on the formula

$$\frac{f(x)}{g(x)} = f(x) \times [g(x)]^{-1} = f(x) \times h(x)$$

You will have to put your answer on a common denominator to get the formula above.

6.9.2. The Relative Growth Rule.
We often make relative measurements stating the error (or accuracy) as a fraction of the amount (or stating a percentage). A similar notion is the relative rate of change given by

$$f^*(x) = \frac{1}{f(x)} \frac{df}{dx}(x)$$

EXERCISE 6.50. *Give a general symbolic rule for the relative rate of change of a product in terms of $f^*(x)$ and $g^*(x)$.*

$$(f(x)\,g(x))^* = ?$$

*(Hint: Substitute a product into the rule for * and re-write using ordinary rules.)*

6.10. Test Your Differentiation Skills

The Mathematical Background Chapter, "Differentiation Drill," contains more practice problems on differentiation and sample skill exams to test your proficiency. You can also use *Mathematica* to check your work on differentiation drill problems. We want you to understand the basic rules of differentiation, but not worry about too many technical details, knowing that you have the computer to back you up. Now it is important to get on with important uses of all the formulas.

CHAPTER 7

Basic Applications of Differentiation

This chapter uses the symbolic ideas of Chapter 6 and the approximation of Chapter 5 in some basic applications of differentiation. The main new topic is an application of the Chain Rule called "related rates problems" given in the second section. When functions are chained or composed, the rate of change of the first output variable changes the second output variable - their rates are related. This idea is generalized to implicitly linked variables later in that section.

Implicitly linked variables change with one another, but there are no explicit functions connecting them, only a formula involving both variables. Implicit differentiation is often an easier way to solve related rate, max - min or other problems later in the course. Essentially, this method is easier because implicit differentiation 'treats all variables equally.' If you skip this chapter now, you may want to return to implicit differentiation later when it arises in another application.

Later sections of the chapter review the combined knowledge of Chapters 5 and 6 in an effort to help you organize your knowledge. Most of this chapter is independent of the next few, so it could be skipped now in favor of other topics.

7.1. Differentiation with Parameters and Other Variables

In the Chain Rule, we asked you to use a different variable name u and find $\frac{dy}{du}$ with formulas you just learned for $\frac{dy}{dx}$. A few of the exercises were also written in terms of other variables. Usually students don't like this at first and that is an understandable reaction. You are just getting used to the $\frac{dy}{dx}$ versions of the rules. However, it is important to get used to working with parameters (letters you treat as constants) and other variable names. This section has examples to show you why.

It is customary in physics to let the greek letter omega, ω, denote frequency and T denote absolute temperature. Certainly the capital T suggests the word that the variable measures. Whether you like ω or not, it is almost impossible to read the physical literature without working with it. Here is an example to test your skills.

EXERCISE 7.1. *Planck's radiation law can be written*

$$I = \frac{a\omega^3}{e^{b\omega/T} - 1}$$

*for constants a and b. This expresses the intensity I of radiation at frequency ω for a body at absolute temperature T. Suppose T is also fixed. Express I as a chain or composition of functions (of the variable ω with parameters) and products of functions to which the rules of this chapter apply. For example, you can use the Exponential Rule, $\frac{d(e^u)}{du} = e^u$. What is the formula for $\frac{dI}{d\omega}$? Check your work with the **Dfdx.ma** NoteBook.*

We will return to Planck's law below when we study the interaction between calculus and graphs and between calculus and maximization.

There are times mathematically when you have already used the variable x for something, but need to vary something else. Here is an oversimplified example to illustrate the point. Suppose we are finding a root of the quadratic equation

$$ax^2 + bx + c = 0$$

where the coefficient b is a measured quantity and not known with perfect accuracy. We want to know how sensitive the largest root of the equation is to errors in measuring b. The largest root of the quadratic equation above can be written as a function of b, including the parameters a and c:

$$x = \frac{-b + \sqrt{b^2 - 4ac}}{2a}$$

EXERCISE 7.2. *Approximate Roots*
1) Compute the derivative $\frac{dx}{db}$ (considering a and c as parameters). When is this defined? In the cases when it is not defined, what is going on in the original root-finding problem? Consider some special cases to help such as $(a, b, c) = (1, -3, 2)$, $(a, b, c) = (1, -2, 1)$, $(a, b, c) = (0, -2, 1)$.
2) Consider the case $(a, b, c) = (1, -3, 2)$ and denote the error in measuring b by db. Suppose the magnitude of db could be as large as 0.01. Use the differential approximation to estimate the resulting error in the root.
The following are for practice differentiating with respect to different letters:
3) Compute the derivative $\frac{dx}{da}$ (considering b and c as parameters). When is this defined?
4) Compute the derivative $\frac{dx}{dc}$ (considering a and b as parameters). When is this defined?
5) Check your differentiation with Mathematica by typing and entering:

```
x := ( -b + Sqrt[ -b + 4 a c ])/(2 a);
x
D[ x , b ]
D[ x , a ]
D[ x , c ]
```

This problem is solved with implicit differentiation in the next section so that you can compare the explicit and implicit methods.

7.2. Linked Variables and Related Rates

Many applications of calculus involve different quantities that vary with time, but are 'linked' with one another; the time rate of change of oné variable determines the time rate of change of the other. We illustrate the idea with the following example.

EXAMPLE 7.3. *A lighthouse is one mile off a straight shore with its beacon revolving 3 times per minute. How fast does the beam of light sweep down the beach at the points that are 2 miles from the lighthouse?*

The 'chain' in this exercise is that the angle of the beacon depends on time and the distance to the point of contact down the beach depends on the angle. We will assume that the light revolves clockwise in the diagram below. We want to know the rate of change of distance with respect to time.

We introduce the variables shown in the diagram below:

t = the time from $t = 0$ when the beam is perpendicular to the shore (*minutes*)

θ = the angle the beam makes from the perpendicular line (*radians*)

D = the distance along the shore from the perpendicular point   (*miles*)

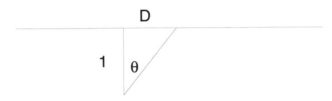

FIGURE 7.4: A Lighthouse Beam Sweeps The Shore

The link between the angle and the distance is

$$\text{Tan}[\theta] = D$$

because $D/1$ is the opposite over the adjacent sides. (SOH-CAH-TOA).

The link between the time and the angle can be expressed as an explicit function, but what is important is to know the rate of change,

$$\frac{d\theta}{dt} = 6\pi = \frac{3 \times 2\pi}{1} = \frac{3 \text{ revolutions in radians}}{\text{one minute}}$$

Since the derivative is constant, this is equivalent to $\theta = 6\pi t$, because $t = 0$ when $\theta = 0$.

The question in this problem is:

FIND:

$$\frac{dD}{dt} \quad \text{when } D = 2$$

SOLUTION: Distance as an explicit function of time is

$$D = \text{Tan}[\theta(t)] \qquad \text{where} \qquad \theta = \theta(t) = 6\pi t$$

The Chain Rule gives us

$$\frac{dD}{dt} = \frac{dD}{d\theta} \cdot \frac{d\theta}{dt}$$
$$= \frac{d(\text{Tan}[\theta])}{d\theta} \cdot \frac{d\theta}{dt}$$
$$= \frac{1}{(\text{Cos}[\theta])^2} \cdot \frac{d\theta}{dt} = \frac{6\pi}{(\text{Cos}[\theta])^2}$$

This formula tells us the speed of the light's motion in terms of θ. Notice that θ is not the independent variable t, but the link variable or output from the first function in the chain. We could use the first function $\theta = 6\pi t$ to express the speed in terms of t, but this is neither necessary nor desirable (unless we want to know when the light gets to the point 2 miles down the beach.) In fact, it is not even necessary to know the derivative $\frac{d\theta}{dt}$ at any other time; it need not be constant as long as we know that $\frac{dD}{dt} = \frac{1}{(\text{Cos}[\theta])^2} \cdot \frac{d\theta}{dt}$.

The specific position when $D = 2$ is shown in the next diagram.

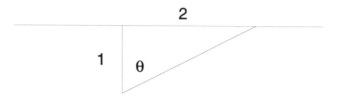

FIGURE 7.5: The Angle when $D = 2$

The Pythagorean Theorem applied to the figure above tells us that the hypotenuse of the right triangle shown is $\sqrt{2^2 + 1^2} = \sqrt{5}$. SOH-CAH-TOA tells us that

$$\text{Cos}[\theta] = \frac{2}{\sqrt{5}}$$

Finally, when $D = 2$

$$\frac{dD}{dt} = \frac{6\pi}{(\text{Cos}[\theta])^2} = \frac{6\pi \cdot 2^2}{(\sqrt{5})^2} = \frac{24\pi}{5} \approx 15.0796 \ (\text{mi/min}) \approx 900 \text{ mph}$$

COMMON ERROR: A common student error in related rate problems is to fix a quantity too soon. In the example above, we want the speed $\frac{dD}{dt}$ when $D = 2$, but if we fix this position at $D = 2$ miles down the shore before we differentiate, we get $\text{Tan}[\theta] = 2$, so θ is constant and there is nothing to differentiate. You must think about the drill problems in their general variable cases, differentiate and then fix the quantities at the specific values.

EXERCISE 7.6.
A child is blowing up a balloon by adding air at the rate of 2 cubic inches per second. Well before it bursts its radius is 6 inches. How fast is the surface stretching at this point? Assume that the balloon is a perfect sphere.

HINTS: Give variables for surface area, volume, time and whatever else you need, and express this scenario in terms of your variables and their derivatives. The 'chain' in this exercise is that volume depends on time (through the child's efforts), say $V = V(t)$ and surface area depends on volume, say $S = S[V]$. We want to know how area changes with time, $\frac{dS}{dt}$, where $S = S[V(t)]$. Express $\frac{dS}{dt}$ as a product of derivatives using the Chain Rule, and explain physically and geometrically what the terms mean. (Recall the earlier Exercise 3.3 to express surface area as a function of volume.)

This exercise is also briefly solved by the method of implicit differentiation later in this section so that you can compare the explicit and implicit methods. There are related rate drill problems at the end of the section that you can solve with either the implicit or explicit method.

7.2.1. Implicit Differentiation. Implicit equations have many powerful uses and we can differentiate them directly simply by treating all variables equally.

A unit circle in the plane is given by the set of (x, y)-points satisfying the implicit equation

$$x^2 + y^2 = 1$$

This equation is called "implicit" because, if we treat x as given, then y is only implicitly given by the equation. Two values of y satisfy the equation, but the equation does not give a direct way to compute them. The explicit equation

$$y = -\sqrt{1 - x^2}$$

gives a direct way to compute y on the lower half-circle, while the implicit equation does not.

The differential of $u^2 + b$, when u is the variable and b is a parameter, is $2\,u\ du$. Treating x as the variable and y^2 as a parameter in $x^2 + y^2$, we have the differential $2\,x\ dx$. Treating y as the variable and x^2 as a parameter in $x^2 + y^2$, we have the differential $2\,y\ dy$. Adding the two, we obtain $2\,x\ dx + 2\,y\ dy$. Since the differential of the constant 1 is zero, the total differential of the implicit equation becomes

$$x^2 + y^2 = 1 \qquad \Rightarrow \qquad 2\,x\ dx + 2\,y\ dy = 0$$

We may view the total differential as an implicit equation for the tangent line to the circle in the local variables (dx, dy) when (x, y) is a fixed point on the circle.

If we solve the equation $2\,x\ dx + 2\,y\ dy = 0$ for the slope of the tangent line,

$$\frac{dy}{dx} = -\frac{x}{y}$$

we obtain a valid formula for the slope of the tangent to a circle. However, this expression uses both variables, so that to use it, we need to know both x and y. For example, the point $(1/2, -\sqrt{3}/2)$ lies on the lower half of the circle. At this point the slope is

$$\frac{dy}{dx} = -\frac{x}{y} = \frac{1/2}{\sqrt{3}/2} = \frac{1}{\sqrt{3}}$$

FIGURE 7.7: $y = -\sqrt{1 - x^2}$ and $dy = dx/\sqrt{3} \Leftrightarrow y + \frac{\sqrt{3}}{2} = (x - \frac{1}{2})/\sqrt{3}$

Let us compare this computation of the slope of the circle at $(1/2, -\sqrt{3}/2)$ to the computation with the explicit equation.

$$y = -\sqrt{1 - x^2}$$

$$y = -\sqrt{u} = -u^{1/2} \qquad u = 1 - x^2$$

$$\frac{dy}{du} = -u^{-1/2} = -\frac{1}{2\sqrt{u}} \qquad \frac{du}{dx} = -2x$$

so the Chain Rule gives

$$\frac{dy}{dx} = \frac{dy}{du} \cdot \frac{du}{dx} = \frac{x}{2\sqrt{u}} = \frac{x}{2\sqrt{1 - x^2}}$$

and when $x = 1/2$

$$\frac{dy}{dx} = \frac{x}{2\sqrt{1 - x^2}} = \frac{1/2}{2\sqrt{1 - (1/2)^2}} = \frac{1}{\sqrt{3}}$$

The computation

$$x^2 + y^2 = 1 \qquad \Rightarrow \qquad 2x \, dx + 2y \, dy = 0$$

is certainly simpler.

The Mathematical Background Chapter on Differentiation Drill has a section on general implicit differentiation. The idea is to differentiate everything with respect to x and multiply by dx, then differentiate everything with respect to y and multiply by dy, and finally add all the differentials together. This description is a little vague, but the next exercise may be enough to give you the idea. Also see the implicit solutions of the related rate problems following the exercise. If you need more detail, see the Mathematical Background.

EXERCISE 7.8. *Implicit Drill*
Verify the following implicit differential calculations:

$$3x^2 + y^2/5 = 1 \qquad \Rightarrow \quad 6x \, dx + \frac{2}{5}y \, dy = 0$$

$$xy = 1 \qquad \Rightarrow \quad y \, dx + x \, dy = 0$$

$$x + xy = 2y \qquad \Rightarrow \quad dx + x \, dy + y \, dx = 2 \, dy$$

$$y + \sqrt{y} = x + x^2 \qquad \Rightarrow \quad dy + \frac{1}{2\sqrt{y}} \, dy = dx + 2x \, dx$$

$$e^x = \text{Log}[y] \qquad \Rightarrow \quad e^x \, dx = \frac{1}{y} \, dy$$

$$x = \text{Sin}[xy] \qquad \Rightarrow \quad dx = y \, \text{Cos}[xy] \, dx + x \, \text{Cos}[xy] \, dy$$

$$x = \text{Log}[\text{Cos}[3x + 5y]] \quad \Rightarrow \quad dx = -3 \, \text{Tan}[3x + 5y] \, dx - 5 \, \text{Tan}[3x + 5y] \, dy$$

EXAMPLE 7.9. *Implicit Solution of The Balloon Exercise* 7.6

In Exercise 7.6 you calculated the rate of change of surface area of a balloon that was being blown up so that its volume was increasing at the rate of 2 cubic inches per second. The point of mentioning the exercises again is that we know that the volume determines the radius and the radius determines the surface area, so the volume determines the surface area. We want to know how the rate of change of volume effects the rate of change of surface area. The implicit solution of that problem does not require that we find the explicit function $S = S[V]$. Here is the implicit solution of the problem:

$$V = \frac{4}{3}\pi\, r^3 \qquad\qquad S = 4\pi\, r^2$$
$$dV = 4\pi\, r^2\, dr \qquad\qquad dS = 8\pi\, r\, dr$$

We solve the volume differential for dr, $dr = dV/(4\pi r^2)$ and substitute into the surface differential, $dS = 8\pi\, r\, dV/(4\pi\, r^2)$, obtaining

$$dS = \frac{2}{r}\, dV$$
$$\frac{dS}{dt} = \frac{2}{r}\frac{dV}{dt}$$
$$\frac{dS}{dt} = \frac{2}{6}\cdot 2 = \frac{2}{3} \qquad (m^2/sec)$$

Compare this solution with your explicit solution from Exercise 7.6 that first solved

$$S[V] = (36\pi\, V^2)^{1/3}$$

and used the Chain Rule to compute

$$\frac{dS}{dt} = \frac{dS}{dV}\cdot\frac{dV}{dt} = 2\left(\frac{4\pi}{3\,V}\right)^{1/3}\frac{dV}{dt}$$

EXAMPLE 7.10. *Implicit Solution of Exercise* 7.2

After you solve Exercise 7.2, you should compare your solution to the following implicit method. We begin with the equation

$$a\,x^2 + b\,x + c = 0$$

We are told that a and c are known exactly, but that b is measured and may have some error when we determine the largest root x satisfying the equation above. In other words, the independent variable b implicitly determines the variable x. The variables in this problem are b and x and x is the dependent variable.

The total differential of the equation is

$$(2\,a\,x + b)\, dx + x\, db = 0$$

The first term is the familiar derivative with respect to x and the second is the derivative with respect to b, treating all other letters, including x, as parameters. The rate of change of x with respect to b is obtained by solving,

$$(2\,a\,x + b)\ dx + x\ db = 0$$
$$(2\,a\,x + b)\ dx = -x\ db$$
$$dx = -\frac{x}{2\,a\,x + b}\ db$$
$$\frac{dx}{db} = -\frac{x}{2\,a\,x + b}$$

When $(a, b, c) = (1, -3, 2)$, the largest root of the original equation is $x = 2$, so as b varies from -3,

$$\frac{dx}{db} = -\frac{x}{2\,a\,x + b} = -\frac{2}{4 - 3} = -2$$

For example, if $b + db = -3 + .001 = -2.99$, then $dx \approx -2\ db = -2 \times .001 = -.002$ and $x + dx = 1.998$. (The exact solution is $(2999 + \sqrt{994001})/2000 \approx 1.99799799397791$.)

When $(a, b, c) = (1, -2, 1)$, the only root of the original equation is $x = 1$. In this case, the implicit differential breaks down

$$(2\,a\,x + b)\ dx + x\ db = 0$$
$$(2 - 2)\ dx + db = 0$$
$$0\ dx + db = 0$$

This equation can not be solved for dx. Implicit differentiation of $\frac{dx}{db}$ fails in this case. We still can graph the line $0\ dx + db = 0$ in the (db, dx)-plane. The equation $db = 0$ is the vertical dx-axis and vertical lines do not have slopes.

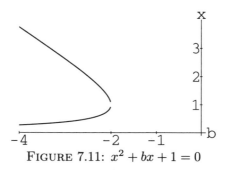

FIGURE 7.11: $x^2 + bx + 1 = 0$

Of course, this degeneracy is only a hint of the trouble in the original implicit equation. If we change b to -1.99, there are no real solutions to the original problem. The equation "branches" here between the positive discriminant and negative one, $x = -b/2 + \sqrt{b^2 - 4}/2$ and $x = -b/2 - \sqrt{b^2 - 4}/2$. The vertical tangent we just computed is the tangent touching these two branches shown in the figure above.

Differentiating the explicit function $x(b)$ is more complicated, as you saw in when you solved Exercise 7.2

$$x = \frac{-b \pm \sqrt{b^2 - 4ac}}{2a}$$

$$\frac{dx}{db} = \frac{1}{2a}\left[-1 \pm \frac{b}{\sqrt{b^2 - 4ac}}\right]$$

$$dx = \frac{1}{2a}\left[-1 \pm \frac{b}{\sqrt{b^2 - 4ac}}\right] db$$

You should show that the implicit and explicit formulas for dx are the same when $x = \frac{-b \pm \sqrt{b^2 - 4ac}}{2a}$.

EXAMPLE 7.12. *The Fast Ladder*

A ladder of length L rests against a vertical wall. The bottom of the ladder is pulled out horizontally at a constant rate r (m/sec). What is the vertical speed of the tip that rests against the wall?

We introduce the variables shown in the diagram below:

x = the horizontal distance from the corner to the bottom of the ladder (*meters*)

y = the vertical distance from the corner to the top of the ladder (*meters*)

t = the time from some starting time $t = 0$ (*seconds*)

The length L is a parameter, fixed, but not known to us. We wish to
Find: $y'[t] = \frac{dy}{dt}$ in terms of $x'[t] = \frac{dx}{dt}$ and the other variables.

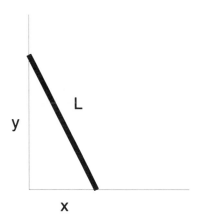

FIGURE 7.13: Ladder & Wall

The fact that the length of the ladder is fixed, together with the Pythagorean Theorem gives us the implicit linkage between x and y:

$$x^2 + y^2 = L^2 \qquad \text{a constant}$$

The fact that there is some implicit relationship is clear physically. If we place the base of the ladder at a given x, we know that there is a unique place, y, where it will rest against the wall. We could solve the equation above for y, since only the positive solution makes physical sense, but we will leave that approach to the problem to an exercise below.

The total differential of the implicit equation is

$$2\,x\,dx + 2\,y\,dy = 0$$

which we can solve for $dy = -\frac{x}{y}\,dx$.

In this problem, we also know that x and y are functions of time. This means that the differentials can be expanded

$$dx = x'[t]\,dt \qquad \& \qquad dy = y'[t]\,dt$$

Substitution of the time expressions of the differentials into the equation above yields the solution of our question,

$$dy = -\frac{x}{y}\,dx$$

$$y'[t]\,dt = -\frac{x[t]}{y[t]}\,x'[t]\,dt$$

$$y'[t] = -\frac{x[t]}{y[t]}\,x'[t]$$

The time rate of change of y equals the time rate of change of x times x over y and the motion is down, or $y'[t]$ is negative.

Suppose we have an enormous ladder of length 5 m. If the base rests 3 m. from the corner, then $y = 4$. If we are pulling the base out from the wall at the rate of 1/7 m/sec, then the top is moving down the wall at the rate

$$y'[t] = -\frac{x[t]}{y[t]}\,x'[t]$$

$$= -\frac{3}{4}\cdot\frac{1}{7}\quad (m/sec)$$

If the base of this same ladder is moved to 4 m. from the corner, so that $y = 3$, and if the base is still pulled out at the speed $x'[t] = 1/7$, then the speed down the wall is

$$y'[t] = -\frac{x[t]}{y[t]}\,x'[t]$$

$$= -\frac{4}{3}\cdot\frac{1}{7}\quad (m/sec)$$

A LIMITING CASE

As we pull the base of the ladder toward the point $x = L$ away from the wall, what happens to the speed with which the other tip moves down the wall? We can see from the diagram that when $x \to L$, $y \to 0$. The formula for the vertical speed tends to minus infinity.

$$y'[t] = -\frac{x[t] \to L}{y[t] \to 0}\quad r \to -\infty$$

Humm, this calculation is a little mysterious. Here is another way to look at it. Solve the equation $x^2 + y^2 = L^2$ for $y = \sqrt{L^2 - x^2}$ and substitute into the expression for vertical speed,

$$y'[t] = -\frac{x[t]}{y[t]} \; x'[t]$$

$$= -\frac{x[t]}{\sqrt{L^2 - (x[t])^2}} \; x'[t]$$

Clearly, as x tends to L, $\frac{x}{\sqrt{L^2 - x^2}}$ tends to infinity. This is correct mathematically, and we want you to verify it by another approach now.

EXERCISE 7.14. *The Explicit Fast Ladder*
Solve the equation $x^2 + y^2 = L^2$ for $y = \sqrt{L^2 - x^2}$. And use the fact that $x = x[t]$ is a function of time, together with the Chain Rule to show that

$$y'[t] = \frac{dy}{dt} = \frac{dy}{dx} \cdot \frac{dx}{dt} = \frac{-x[t]}{\sqrt{L^2 - (x[t])^2}} \cdot x'[t]$$

Verify that this model predicts that as x approaches L, the speed of the tip resting on the wall tends to infinity.

The result of the previous exercise is wrong for a real ladder. The tip of such a ladder can not go faster than the speed of light. There is a physical condition that this simple mathematical model does not take into account. The real falling ladder is explored in the Scientific Projects Chapter on Mechanics. It uses Galileo's law of gravity from Chapter 9.

EXERCISE 7.15. *Related Rates Drill*
1) Each edge of a cube is expanding at the rate of 3 inches per second. How fast is the volume expanding at the point where the volume equals 64 cubic inches?
2) Each edge of a cube is expanding at the rate of 3 inches per second. How fast is the surface area expanding at the point where the volume equals 64 cubic inches?
3) A 6 foot tall man walks away from an 8 foot tall lamp at the rate of 5 feet per second. How fast is his shadow growing at the point where he is 7 feet from the lamp?
4) A snowball melts at a rate proportional to its surface area, that is, it loses volume per unit time in this proportion. Say the constant of proportionality is k. What is $\frac{dr}{dt}$, the rate of change of radius with respect to time?

7.2.2. Derivatives of Inverse Functions. The method of implicit differentiation applies to inverse functions. This case is treated in detail in the Mathematical Background Chapter on Inverse Functions.

EXAMPLE 7.16. *Derivative of* $\mathrm{Log}[y]$

Consider an example for the inverse pair of equations

$$y = e^x \qquad \Leftrightarrow \qquad x = \mathrm{Log}[y]$$

The differential of the exponential equation is $dy = e^x \, dx$ but we may use the fact that $e^x = y$ to write

$$dy = y \, dx$$

Solving for the derivative of x with respect to y gives us the derivative of the logarithm,

$$\frac{d(\text{Log}[y])}{dy} = \frac{dx}{dy} = \frac{1}{y}$$

EXAMPLE 7.17. *Derivative of* ArcTan[y]

Computation of inverse function derivatives this way can present computational difficulties such as the following.

$$y = \text{Tan}[x] \qquad \Leftrightarrow \qquad x = \text{ArcTan}[y]$$

You computed the derivative of tangent directly in Chapter 5 and using rules in Chapter 6, obtaining

$$dy = \frac{1}{(\text{Cos}[x])^2} \, dx$$

so we have a formula for the derivative of the arctangent:

$$\frac{d(\text{ArcTan}[y])}{dy} = \frac{dx}{dy} = (\text{Cos}[x])^2$$

Unfortunately, this form of the equation is in terms of the dependent variable for arctangent, so some trig tricks are needed to put it in the form:

$$\frac{d(\text{ArcTan}[y])}{dy} = \frac{1}{1 + y^2}$$

See the Mathematical Background chapter on inverse functions.

Further examples are included in the Mathematical Background along with complete justification of this method of computing derivatives of inverse functions. The justification is in the form of a procedure you can use to compute the actual nonlinear inverse function. In other words, it is a 'practical' proof.

7.3. Review - Inside the Microscope

Calculus lets us "see" inside a powerful microscope without actually magnifying the nonlinear graph. We know that the curve looks like its tangent line at high magnification. You should try to separate the mechanics of how we "see" from the idea of what we see. The "rules" of differentiation are the way we "see," so the next exercise separates the step of using the rules from "looking."

EXERCISE 7.18. *A Partial View of the 'Bell Shaped' Curve*
The derivative of the function $f(x) = e^{-x^2}$ is $f'(x) = -2x \cdot e^{-x^2}$ (as you may verify using rules of differentiation.) This question asks: So what? (or what does this tell us mathematically?) You answer it as follows. Draw microscopic views of the graph $y = e^{-x^2}$ when the microscope is focused on the graph over the x-points, $x = 0, \pm 0.1, \pm 1, \pm 10$. Give the numerical values of the derivatives and sketch the slopes to scale on equal axes.

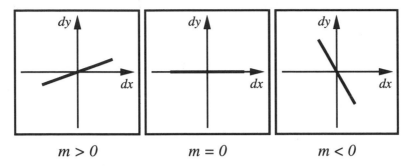

FIGURE 7.19: Possible Microscopic Views

The next exercise asks you to do all the steps involved in "looking" in an infinitesimal microscope. This is a question that requires you to summarize the steps in writing. This should help you combine the facts you have learned.

PROBLEM 7.20.
1) Sketch a pair of (x, y)-axes and plot the point $(1, -1)$. Let x run from 0 to 3, and y run from -2 to 1.
2) The point $(x, y) = (1, -1)$ lies on the explicit curve $y = x^2 \text{Cos}[\pi/x]$. Verify this.
3) Add a pair of (dx, dy)-axes at the the (x, y)-point $(1, -1)$. How are these axes related to the (x, y) axes?
4) Use rules of calculus to show that

$$y = x^2 \text{Cos}[\pi/x] \qquad \Rightarrow \qquad dy = (2x \text{ Cos}[\pi/x] + \pi \text{ Sin}[\pi/x]) \ dx$$

5) Substitute $x = 1$ into your differential to show that

$$dy = -2 \ dx$$

at the (x, y)-point $(1, -1)$ or the (dx, dy)-point $(0, 0)$.
6) Plot the line $dy = -2 \ dx$ on your (dx, dy)-axes.
7) What would you see if you looked at the graph of $y = x^2 \text{Cos}[\pi/x]$ under a very powerful microscope?
*8) Use the Mathematica NoteBook **Microscope1-D.ma** to plot the function, its differential and make an animation of a microscope zooming in on the graph at the (x, y)-point $(1, -1)$.*
*9) Explain how the Differential cell of the **Microscope1-D.ma** NoteBook is actually solving parts (2) - (7) of this exercise. How does calculus let us "see" a graph in an infinitesimal microscope?*

7.4. Review - Numerical Increments

When a function is smooth we summarize the local linear approximation by

$$y = f(x) \qquad \Rightarrow \qquad dy = f'[x] \ dx$$

The differential $dy = f'[x] \ dx$ is a linear function of the local variable dx, with dependent variable dy. This is the linear equation in microscope variables. The variable x in $f'[x]$ is considered fixed until we move the point where we focus our microscope. The quantity dy is an approximation to the change $f(x + dx) - f(x)$ in the actual function.

You should memorize Definition 5.15 and strive to understand its algebraic and geometric consequences. Functions given by formulas are important in science and mathematics, but they are not the only kind of functions. Calculus goes well beyond the scope of such formulas in the study of functions given by differential equations, infinite series and other 'approximation' procedures. To understand the development of the more advanced parts of the course, you will need to use the symbolic increment approximation. The following exercises let you try out the symbolic form of the microscope approximation on the simple kind of functions given by formulas.

The 'rules' of calculus are theorems that guarantee that the local linear approximation is valid. These rules are remarkably easy to use compared with the direct verification of the approximations as in Chapter 5. You simply compute and look at the answers. We used the symbolic in Chapter 5 to estimate Sin[46°], but there we had verified the approximation directly.

Contrast what we learn about a function from the approximation with the simplicity of the computation which guarantees that the approximation holds. If

$$y = x^{-3}, \quad \text{then} \quad dy = -\frac{3}{x^4} \, dx$$

according to the rules. Both of these formulas are valid when $x \neq 0$, so the increment approximation defining the derivative holds and the change in f is approximated by dy with $dx = \delta x$,

In general,

$$f(x + \delta x) = f(x) + f'[x] \cdot \delta x + \varepsilon \cdot \delta x$$

or

$$f(x + dx) - f(x) = f'[x] \cdot dx + \varepsilon \cdot dx$$
$$f(x + dx) - f(x) = dy + \varepsilon \cdot dx \approx dy$$

In this case,

$$\frac{1}{(x + dx)^3} - \frac{1}{x^3} = \frac{-3}{x^4} \cdot dx + \varepsilon \cdot dx \approx \frac{-3}{x^4} \cdot dx$$

$$\frac{1}{(3 + .01)^3} - \frac{1}{27} \approx -\frac{3}{81} \times 0.01$$

$$\frac{1}{(3.01)^3} - \frac{1}{27} \approx -\frac{.01}{27}$$

so

$$\frac{1}{(3.01)^3} \approx .99 \times \frac{1}{27} \approx 0.0366667$$

Mathematica gives $1/(3.01)^3 = 0.0366691$, so the increment approximation is quite close even though the x increment $\delta x = 0.01$ is not infinitesimal.

EXERCISE 7.21. *Differential Approximation*
Approximate $\sqrt[3]{1,000,000,000,000,002} - \sqrt[3]{1,000,000,000,000,000}$ *using the differential increment of the function* $f(x) = \sqrt[3]{x}$, $x = 1,000,000,000,000,000$ *and* $\delta x = 2$. *(Two is not infinitesimal, but it is small compared to* 10^{15}. *Computers have a very hard time with this kind of computation because they work in fixed length decimal approximations.)*

We want a simple way to estimate $f(x+\delta x) - f(x)$, since we know $f(x) = 100,000$ when $x = 10^{15}$.

The increment equation $f(x + \delta x) = f(x) + f'(x)\,\delta x + \varepsilon\,\delta x$ may be re-written with differentials, $f(x + dx) - f(x) = df(x) + \varepsilon\,dx$ where $df(x) = f'(x)\,dx$ In other words, $f(x + dx) - f(x) \approx df(x)$, with an error small even compared to dx when dx is small. To make differential approximations, do the following steps.

(1) Compute $f'(x)$ by rules. The rules guarantee the approximation when $dx \approx 0$.

(2) Substitute the fixed x ($x = 10^{15}$, in this case) to find the numerical value of the differential slope $m = f'(x)$.

(3) Compute $df = m\,dx$ when $dx =$ your perturbation ($dx = 2$ in this case). This is the approximate **change** in f.

(4) Compute $f(x + dx)$ using this approximation and the value of $f(x)$, $f(x + \delta x) \approx f(x) + df(x)$.

How many decimals of your approximation are accurate in this case? (A tough question, because the computer can't help directly. Try it with your calculator or *Mathematica*; you'll get the wrong answer unless you work with very very high precision arithmetic. The differential is much better.)

PROBLEM 7.22. *You are interested in the accuracy of your speedometer and perform the following experiment on a stretch of flat, straight, deserted Interstate highway. You drive at constant speed with your speedometer reading 60 mph, crossing between two consecutive mile markers in 57 seconds as measured by your quartz watch. For constant speed, we know the formula, "distance equals rate times time," $D = R \times T$, so $R = D/T$ when the units are correct.*

(1) *Express the rate of speed R in miles per hour as a function of the distance D in miles and the time t in **seconds**.*

(2) *Compute the differential $dR =? \times dt$ using the appropriate rules, assuming that $D = 1$ is measured exactly.*

(3) *Approximate the speed of your car in the above experiment using the differential to approximate the increment ΔR. (See Exercise 7.21.)*

(4) *Use Mathematica to compare the actual rate with the differential approximation as follows, using your formula for r:*

```
r[t_ ] := ???
Plot[{r[t], 60 - (t - 60)} , {t, 50, 70}]
Do[Print[{ N[ r[t] ] , 60 - (t - 60)},{t, 50 , 70 }]
```

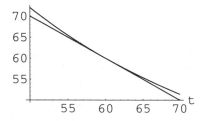

FIGURE 7.23: Rate and Approximation

Summarize the idea of this problem in a few sentences.

7.5. Differentials and The (x, y)-Equation of the Tangent Line

The equation of the tangent line to $y = f(x)$ in local coordinates is simply the differential $dy = f'(x) \, dx$, but there is possible confusion when we try to convert back to regular coordinates, because we are treating x as fixed in the local dx-dy-equation. Here is a way to find the equation of the tangent line to $y = x^2$ when $x = -1/3$. We know $dy = 2x \, dx$ and $x = -1/3$, so the slope is $2 \cdot (-1/3) = -2/3$. When $x = -1/3$, $y = x^2 = 1/9$, so the line goes through $(-1/3, 1/9)$ and has slope $-2/3$. Using the change form of a line (or the point-slope formula),

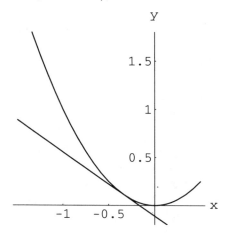

$$\frac{\Delta y}{\Delta x} = m$$

$$\frac{y - b}{x - a} = m$$

$$\frac{y - 1/9}{x + 1/3} = -2/3$$

$$y = -\frac{2}{3} x - \frac{1}{9}$$

FIGURE 7.24: $y = x^2$ & $y = -\frac{2}{3} x - \frac{1}{9}$

This can be plotted with *Mathematica* as follows:

```
y[x_ ] := x∧2;
l[x_ ] := -2 x/3 -1/9;
y[x]
l[x]
Plot[{ y[x],l[x]} , { x,-2,1} ]
```

To find the (x, y) equation of the tangent line, do the following steps.

(1) Compute $f'(x)$ by rules.
(2) Substitute the fixed x ($x = -1/3$, in this case) to find the numerical value of the slope $m = f'(x)$.
(3) Compute the specific y value, $y = f(x)$ at the point of tangency. This gives you a specific point $(x, y) = (a, b)$ that lies on both the curve and tangent line.
(4) Use the point-slope form of a line $\frac{y-b}{x-a} = m$ and simplify to the form $y = m x + i$.

EXERCISE 7.25. *Find the (x, y)-equation of the tangent to $y = x^3$ at $x = -2$.*
Find the (x, y)-equation of the tangent to $y = \text{Sin}[x]$ at $x = \pi/3$.
Find the (x, y)-equation of the tangent to $y = \text{Log}[x]$ at $x = 1$.
Plot these pairs of curves and tangent lines with Mathematica.

Which step in our procedure for finding the tangent line in (x, y) coordinates requires us to fix the specific slope of the derivative?

PROBLEM 7.26. *What is wrong with the following "general formula" for the tangent to $y = x^2$ at the point $x = a$? We have $dy = 2x\, dx$ and we know that $dx = x - a$, while $dy = y - a^2$. So (false conclusion) the equation of the tangent line is $\frac{\Delta y}{\Delta x} = \frac{y - a^2}{x - a} = 2x$ and we can simplify to the form $y = m\, x + b$.*

The (x, y) equation of the tangent is often not needed (if you plot in (dx, dy) coordinates), but the idea of working in the correct coordinates is important. Here is another kind of example of tangency and keeping track of all the variables. A circle of radius 1 centered on the y axis is moved down the axis until it touches the parabola $y = x^2$, as shown in the next figure. Since the circle just touches the parabola, both curves are tangent at the point of contact.

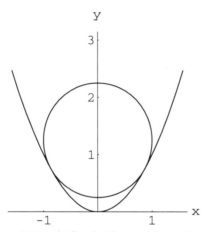

FIGURE 7.27: A Circle Tangent to a Parabola

PROBLEM 7.28. *Tangent Curves*
Write the equation of a circle of radius 1 centered on the y axis in terms of a parameter c for the unknown y coordinate of the center.

Calculate the differential of the equation of your circle, using the unknown parameter. Solve for $\frac{dy}{dx}$ and write the equation that says

"*the slope of the circle at (x, y) equals the slope of the parabola at (x, y)*"

Write the system of three equations that say "(x, y) is the point of tangency:"

$$y = x^2 \quad \text{"(x, y) lies on the parabola"}$$
$$? = ? \quad \text{"(x, y) lies on the circle through $(0, c)$"}$$
$$? = ? \quad \text{"the circle and parabola have the same slope at (x, y)"}$$

Finally, solve the three equations in three unknowns, using Mathematica if you like.

Here is a sample of the use of *Mathematica* to solve a set of equations:
equns = { y == x∧2 , x∧2 + (y - c)∧2 == 1 , x == 2 x (c - y) }
Solve[equns , { x , y , c }]

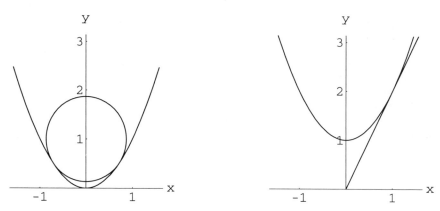

FIGURE 7.29: Tangent Curves

PROBLEM 7.30. *More Tangent Curves*
A circle with center $(0, 1)$ is expanded until it just touches the parabola $y = x^2$. What is it's radius? Where does it make contact?

A line passes through the origin and is tangent to $y = x^2 + 1$. What is the point of tangency?

CHAPTER 8

The Natural Logarithm and Exponential

This chapter contains some important ideas about logs and exponentials. It can be studied any time after Chapter 6. You might skip it now, but should return to it when your high school knowledge is not sufficient to solve a problem. For example, exponentials as solutions to differential equations are needed before you study continuous dynamical systems. Some of the 'orders of infinity' limits of Section 8.4 arise in graphing and max-min. The Mathematical Background Chapters on High School Review with *Mathematica* and on Exponentials contain further information and projects.

The "natural" base exponential function and its inverse

$$y = e^x \qquad \Leftrightarrow \qquad x = \ln(y)$$

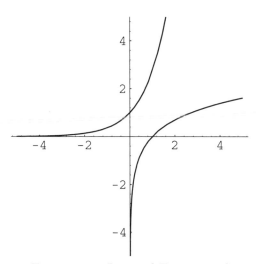

FIGURE 8.1: Log and Exponential

are two of the most important functions in mathematics. This is reflected in the fact that *Mathematica* has a built-in algorithms and separate names for them:

$$y = \text{Exp}[x] \qquad \Leftrightarrow \qquad x = \text{Log}[y]$$

We already know the differentiation rules for log and exponential, and the basic high school review material about logs and exponentials is contained in the Mathematical Background Chapter on High School Review with *Mathematica*. The main facts are:

$$\frac{de^t}{dt} = e^t \qquad\qquad\qquad \frac{d\,\mathrm{Log}(s)}{ds} = \frac{1}{s}$$

$$e^a \times e^b = e^{a+b} \qquad\qquad \& \qquad\qquad \mathrm{Log}(a \times b) = \mathrm{Log}(a) + \mathrm{Log}(b)$$

$$(e^c)^t = e^{c \cdot t} \qquad\qquad\qquad \mathrm{Log}(a^p) = p \times \mathrm{Log}(a)$$

$$\mathrm{Log}(e^t) = t \qquad\qquad\qquad e^{\mathrm{Log}(s)} = s, \ \ s > 0$$

Some of the graphical properties of these functions will be formulated as limits comparing them to power functions later in the chapter. The limits say:

- e^t goes to infinity faster than any power as $t \to \infty$
- $\mathrm{Log}(t)$ tends to infinity slower than any root as $t \to \infty$.
- e^{-t} is positive, but tends to zero faster than any reciprocal power as $t \to \infty$
- $\mathrm{Log}(s) \downarrow -\infty$ as $s \downarrow 0$

See the *Mathematica* NoteBooks **ExpGth.ma** and **LogGth.ma** for an intuitive understanding of these limits. Section 8.4 explains these "orders of infinity" more technically and shows how to build more limits from them.

This chapter answers some very basic questions, such as, Why is the peculiar number $e \approx 2.71828...$ considered "natural?" We have not given a proof of the formulas for the derivatives of logs or exponentials and this chapter will turn this fact on its head by defining the exponential to be the function whose derivative equals itself. No matter where we begin in terms of a basic definition, this is an essential fact. It is so essential that everything else follows from it. We call this the official theory.

The official theory is only really important when we get to a question we can not answer with facts from high school and simple differentiation formulas. What do we mean by e^π or even $3^{\sqrt{2}}$? Certainly, e^3 means "multiply e times itself 3 times," but you can't multiply e times itself π times - that makes no sense. You probably don't want to believe that, because you can use your calculator for an approximate answer.

When we use calculators to approximate e^π we raise the approximate base $e \approx 2.71828$ to the approximate power $\pi \approx 3.14159$. This implicitly assumes that the y^x-button on our calculator is continuous in both inputs. In other words, the small errors in both e and π only produce a small error in the approximate output to e^π. Does your calculator produce 6 significant digits of e^π when you put 6 digits of accuracy in for e and π? This is a tough question, because you have to decide what's exact. Similarly, an approach to exponentials based on

$$\lim_{x \to \pi,\ y \to e} y^x = e^\pi$$

as both $y \to e$ and $x \to \pi$ is a very difficult way build a basic theory. It is "natural" in some ways, but technically too hard. (You will use partial derivatives to prove it later.)

8.1. The Official Definition of the Natural Exponential

One of the most important ways that exponential functions arise in science and mathematics is as the solution to linear growth and decay laws. We have already seen these in Newton's Law of Cooling, Exercise 8.10 and the first (false) conjecture of Galileo on the law of gravity. These laws simply say, "The rate of change of a quantity is proportional to the amount present." This is the differential equation

$$\frac{dy}{dt} = k\,y$$

with a positive constant k if it represents growth and a negative constant if it represents decay.

SOMETHING NEW

The most noteworthy thing about the formulas in this chapter is this: The dependent variable y appears on both sides of the equation

$$\frac{dy}{dt} = k\,y$$

This is our first important differential equation, not just another differentiation formula like the ones in Chapter 6. (Those equations all have explicit functions of the independent variable t on the right hand side, for example, $y = 3\,t^5 \Rightarrow \frac{dy}{dt} = 15\,t^4$.)

The equation tells us how the quantity changes. If we also know an initial value of the quantity, we can solve for later values of $y(t)$. Where you start and how you change is the idea of a continuous dynamical system. It is intuitively clear that this determines where you go, though it may NOT be entirely clear HOW it determines it.

Mathematically, this continuous dynamical system is the 'operation' of going from the 'where you start and how you change' information to a function of t,

$$y[0] = Y_0$$
$$dy = k\,y\,dt$$
$$\longrightarrow \quad y[t], \quad t \in [0, \infty)$$

We will approximate this 'operation' by solving a discrete system that moves in small steps of size δt

$$y[0] = Y_0$$
$$y[t + \delta t] \approx y[t] + k\,y[t]\,\delta t$$
$$\longrightarrow \quad \{y[0], y[\delta t], y[2\delta t], y[3\delta t], \cdots\}$$

This approximation is called 'Euler's method' of approximating solutions of initial value problems. We have already seen this idea in several places, beginning with the **Second-SIR.ma** NoteBook in chapter 2. (Euler's method for more general dynamical systems is studied later in Chapter 21.)

You already have an idea of what $y = e^t$ means, so it may seem a little silly to introduce a 'definition' for it at this late stage of the game. We have pointed out some gaps in what you 'know' and want a definite place to fall back to when the problems get more difficult. (For example, later we will want to compute $e^{i\,t}$, where $i = \sqrt{-1}$.) Many important functions in higher mathematics are characterized by their differential equation, so this is the first time you will see something that is quite powerful.

DEFINITION 8.2. *The Official Natural Exponential Function*
The function

$$y = \text{Exp}[t]$$

is officially defined to be the unique solution of the initial value problem

$$y[0] = 1$$
$$dy = y \, dt$$

If we use the un-proved formula for the derivative, we can see that the natural exponential function $y(t) = e^t$ satisfies this differential equation and initial condition,

$$\frac{dy}{dt} = y \quad \text{with the initial condition} \quad y(0) = 1$$

since $\frac{dy}{dt} = \frac{de^t}{dt} = e^t = y$ and $e^0 = 1$.

The point of this section is NOT to use any un-proved facts about the natural exponential function, but rather to see that this definition gives us a way to work with the function which satisfies the conditions. We want you to experiment with the official definition as a method by which to compute e^t.

The general Euler's method is a simple idea once you know the increment approximation from the Definition 5.15 of the derivative. When our function is $f(x)$, we write this approximation

$$f(x + \delta x) = f(x) + f'[x] \, \delta x + \varepsilon \cdot \delta x$$

where the error $\varepsilon \approx 0$ is small when $\delta x \approx 0$ is small. Our function now is $y = y(t)$, so the approximation becomes

$$y(t + \delta t) = y(t) + y'[t] \, \delta t + \varepsilon \cdot \delta t$$

where $\varepsilon \approx 0$ is small when $\delta t \approx 0$ is small. We do not have a formula for $y(t)$, but we do have the value of $y(0) = 1$ and a formula for $y'[t] = y(t)$ given in terms of y. This gives us the new kind of equation:

APPROXIMATE SOLUTION OF $y(0) = 1$ & $\frac{dy}{dt} = y$:

$$y(t + \delta t) = y(t) + y(t) \, \delta t + \varepsilon \cdot \delta t$$
$$y(t + \delta t) \approx y(t) + y(t) \, \delta t = y(t)(1 + \delta t) \qquad \text{when } \delta t \approx 0$$

so

$$y(\delta t) \approx y(0) \cdot (1 + \delta t) = (1 + \delta t)$$
$$y(\delta t + \delta t) \approx y(\delta t) \cdot (1 + \delta t) = (1 + \delta t)^2$$
$$y(2\delta t + \delta t) \approx y(2\delta t) \cdot (1 + \delta t) = (1 + \delta t)^2 \cdot (1 + \delta t) = (1 + \delta t)^3$$

In general, we see that if $y(0) = 1$ and $\frac{dy}{dt} = y$, then

$$y(t) \approx (1 + \delta t)^{(t/\delta t)} \qquad \text{for } t = 0, \delta t, 2\delta t, 3\delta t, \cdots$$

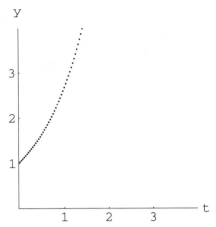

FIGURE 8.3: Euler's Approximation

A very basic fact of mathematics says

$$\lim_{\delta t \to 0} (1 + \delta t)^{1/\delta t} = e$$

This is a special case of the convergence of the solution of our discrete dynamical system to the solution of the continuous one.

EXERCISE 8.4. *Compare Mathematica's built in function* Exp[t] *to the Euler approximation of the official definition by typing and running the following Mathematica program for* $dt = 1/2, 1/4, 1/16, 1/256$ *How large is the difference between* Exp[1] *and* y[1] *when* $dt = 1/256$?

```
y[t_ ] := (1 + dt)∧(t/dt);
y[t]
dt = 1/2
Plot[{ y[t], Exp[t]},{ t, 0, 2} ]
Exp[ 1]
y[1]
```

Now we want you to use the idea that gave us the approximation $(1 + \delta t)^{t/\delta t} \approx e^t$ to approximate the solution of a more general exponential law.

EXERCISE 8.5. *Approximate Solution of* $\frac{dy}{dt} = k\, y$ *with* $y(0) = Y_0$:
Show that $y(t) = Y_0 \, (1 + k\, \delta t)^{t/\delta t}$ *(for* $t = 0, \delta t, 2\delta t, 3\delta t, \cdots$ *) is an approximate solution to the initial condition and differential equation*

$$y(0) = Y_0$$
$$\frac{dy}{dt} = k\, y$$

Test your approximate formula for the case

$$y(0) = 3$$
$$\frac{dy}{dt} = -2\, y$$

which has the exact solution $z(t) = 3\,Exp[-2\,t]$. Use Mathematica to make graphs and numerical comparisons as you did above for e^t.

Here is some help with the exercise. First and foremost, recall the microscope approximation of Definition 5.15 and apply it to the (unknown) function $y(t)$. Discarding the error term yields an approximation,

$$y(t + \delta t) = y(t) + ?? + \varepsilon \cdot \delta t \approx y(t) + ???$$

Next, use the fact that $y'[t] = k\,y(t)$ and substitute this into the microscope approximation,

$$y(t + \delta t) = y(t) \cdot (??)$$

We know $y(0) = Y_0$, the initial condition. To find $y(\delta t)$, use your approximation

$$y(\delta t) \approx Y_0 \cdot ??$$
$$y(2\delta t) \approx y(\delta t + \delta t) = y(\delta t) \cdot ??$$
$$y(3\delta t) \approx y(2\delta t + \delta t) = y(2\delta t) \cdot ??$$

Simplification yields the desired result.

We can summarize the section with the formula

$$y[t] = Y_0(1 + k\,\delta t)^{t/\delta t} \qquad \approx \qquad y[t] = Y_0\,e^{k\,t}$$

The formula on the left can be computed by hand for $t = 0, \delta t, 2\delta t, \cdots$ if necessary and it comes straight from the initial value problem.

8.2. Properties Follow from The Official Definition

Next semester we will prove a general existence and uniqueness result for continuous dynamical systems. The proof just amounts to showing that Euler's discrete dynamical system approximation converges as the step size Δt tends to zero. We will see then that the mathematical uniqueness of the solution to an initial value problem is what makes dynamical systems deterministic scientific models. Uniqueness is really what you are thinking of when you say, '**The** solution of this system models $\cdots$.' Uniqueness is also mathematically important. For now we will use the

THEOREM 8.6. *Unique Solution to a Linear Dynamical System*
For any real constants Y_0 and k, there is a unique real function $y[t]$ defined for all real t satisfying

$$y[0] = Y_0$$
$$\frac{dy}{dt} = k\,y$$

The theorem above assures us that a function $Exp[t]$ of our official definition exists. We know intuitively from our experience with NoteBooks like **SecondSIR.ma** that the computer approximations do converge. The next subsection shows how the important addition formula for exponentials follows from uniqueness of the solution to our official definition.

8.2.1. Proof that $e^c\,e^t = e^{(c+t)}$. Let $C = \text{Exp}[c]$ and consider the function $y[t] = C\,\text{Exp}[t]$. We know $y[0] = C$ and by the Superposition Rule for differentiation that $\frac{dy}{dt} = C\,\text{Exp}[t] = y$. This means that y is a solution to:

$$y[0] = C$$
$$\frac{dy}{dt} = y$$

Now consider the function $z[t] = \text{Exp}[c + t]$. Again, $z[0] = \text{Exp}[c] = C$ and the Chain Rule says $\frac{dz}{dt} = \text{Exp}[c+t] = z$, so z is also a solution to:

$$z[0] = C$$
$$\frac{dz}{dt} = z$$

This is the same dynamical system, just written with a different letter. Uniqueness means that both $y[t]$ and $z[t]$ are the same function

$$\text{Exp}[c]\,\text{Exp}[t] = \text{Exp}[c + t]$$

and proves the functional identity of the natural exponential function.

Our new definition causes us a technical problem. We do know how to compute rational powers of e. We want to show that the new definition extends what we already know.

PROBLEM 8.7. *Let e be the number $\text{Exp}[1]$, that is, the value of the solution of the dynamical system at time 1.*

(1) *Show that $e^2 = \text{Exp}[2]$.*
(2) *Show that $e^{1/3} = \text{Exp}[1/3]$ by letting $a = \text{Exp}[1/3]$ and computing $a \times a \times a$.*
(3) *Show that $e^q = \text{Exp}[q]$ for any rational $q = m/n$.*

8.3. e As a "Natural" Base for Exponentials and Logs

All exponential bases are not created equal. All exponential functions $y = b^t$ satisfy

$$y(0) = 1$$
$$\frac{dy}{dt} \propto y$$

but the base with constant of proportionality 1 is $b = e$. This makes e the "natural" base from the point of view of calculus.

EXERCISE 8.8.

(1) *Let $y = b^t$ for an unknown (but fixed) positive constant b. Use the chain rule (see Section 6.7) to show that $y(t)$ satisfies*

$$y(0) = 1$$
$$\frac{dy}{dt} = k\,y$$

What is the value of the constant k?

(2) *Show that $y = e^{kt}$ satisfies*

$$y(0) = 1$$
$$\frac{dy}{dt} = k\,y$$

If the constant k is the same as in the first part, how much is e^k in terms of b?
(3) *Solve the initial value problem*

$$y(0) = 5$$
$$\frac{dy}{dt} = y$$

(4) *Solve the initial value problem*

$$y(0) = 5$$
$$\frac{dy}{dt} = k\,y$$

where $k = \text{Log}[2]$. Show that your solution may also be written as $y = 5 \cdot 2^t$.

The moral of this exercise is this: We *could* write solutions of initial value problems

$$y(0) = Y_0$$
$$\frac{dy}{dt} = k\,y$$

as $y = Y_0 \cdot b^t$, where $b = e^k$ or $\text{Log}[b] = k$, but for the purposes of calculus, it is "more natural" to write them in the form $y = Y_0 e^{kt}$,

$$y[0] = Y_0$$
$$dy = k\,y\,dt$$

$$\rightarrow \qquad y[t] = Y_0 \cdot e^{kt}, \quad t \in [0, \infty)$$

We want you to put this to work in the next exercise.

8.3.1. The Canary's Postmortem.

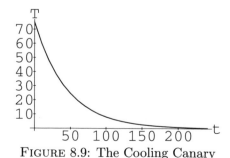

FIGURE 8.9: The Cooling Canary

EXERCISE 8.10. *The Canary's Postmortem*
Let $T = T_0 e^{-kt}$ for unknown positive constants T_0 and k. Show that

$$\frac{dT}{dt} = -kT$$

by using the chain rule, $T = T_0 e^u$ with $u = -kt$, so $\frac{dT}{du} = T_0 e^u$ and $\frac{du}{dt} = -k$. Express $\frac{dT}{dt}$ in terms of the dependent variable T.

The value of $e^0 = 1$, so $T = T_0$ when $t = 0$. Show that the function T solves the cooling problem of Exercise 5.43 for certain choices of the constants T_0 and k. How much is T_0 in that problem? How could you find the constant k so $T = 60$ when $t = 10$? (Hint: Solve $60 = 75\, e^{-10\,k}$ using log.)

Graph your specific function $T = T_0\, e^{-k\,t}$ using Mathematica and verify that the temperature at time 10 is 60.

8.3.2. Bugs Bunny's Law of Gravity. The story of the fallen tourist in Exercise 5.45 led to the differential equation

$$\frac{dD}{dt} = k\,D$$

for some positive constant k, where D is the distance an object has fallen and t is time. This differential equation says, 'The farther you fall, the faster you go.' in a specific way. We called this Galileo's first conjecture, because it gives rise to the Bugs Bunny Law of Gravity: If you don't look down, you don't fall. We want you to show why.

PROBLEM 8.11. *If You Don't Look Down, You Don't Fall*
How does Bugs' Law say, 'The farther you go, the faster you fall?' In a general sense this is true, but it can not be specifically as Bugs' claims. Prove that an object released from $D = 0$ at $t = 0$ does not fall under the Bugs Bunny Law of Gravity: $\frac{dD}{dt} = k\,D$, $k > 0$ constant.

8.4. Growth of Log and Exp Compared with Powers

Two important facts about the rate of growth of the natural log and exponential functions are:

THEOREM 8.12. *Orders of Infinity*
Let p be any positive real number. Then

$$\lim_{t \to \infty} \frac{e^t}{t^p} = \infty \qquad and \qquad \lim_{t \to \infty} \frac{t^p}{\text{Log}[t]} = \infty$$

or equivalently,

$$\lim_{t \to \infty} \frac{t^p}{e^t} = 0 \qquad and \qquad \lim_{t \to \infty} \frac{\text{Log}[t]}{t^p} = 0$$

The exponential beats any power and any power beats the logarithm to infinity.

PROOF BASED ON HIGH SCHOOL MATH:
First we will consider the ratio $\frac{t^p}{e^t}$ for integer values of t, $\frac{n^p}{e^n}$, for $n = 1, 2, \cdots$. It is sufficient to show that the pth root of the ratio tends to zero,

$$\frac{n}{e^{n/p}} \to 0$$

by continuity of positive powers at zero. Let $b = e^{1/p}$ and notice that $b > 1$ and so $b^{1/2} > 1$ as well. Write

$$b^{1/2} = 1 + a$$

for $a > 0$.

The binomial theorem says

$$b^{n/2} = (1 + a)^n = 1 + na + \cdots + a^n > na$$

since $a > 0$, so that $e^{n/p} = b^n > n^2 a^2$. Hence we have

$$\frac{n}{e^{n/p}} < \frac{n}{n^2 a^2} = \frac{1}{na^2} \to 0 \qquad \text{as} \quad n \to \infty$$

For continuous values of $t > 0$, there is always an integer satisfying $n - 1 < t \le n$. For these values we have

$$\frac{t^p}{e^t} < \frac{n^p}{e^{n-1}} = e \frac{n^p}{e^n} \to 0$$

EXERCISE 8.13. *Use logs to prove the other half of the Orders of Infinity theorem above. If we wish to show that* $\lim_{x \to \infty} \dfrac{x^p}{\text{Log}[x]} = \infty$, *then change variables by letting* $u = \text{Log}[x]$, *so* $x = e^u$. *We show show instead that*

$$\lim_{u \to \infty} \frac{(e^u)^p}{u} = \infty$$

by taking p^{th} *roots. Why does this establish the result?*

EXAMPLE 8.14. $2^\infty / \infty^3$

There are many other 'orders of infinity' or 'orders of infinitesimal' that we can deduce from the basic result given in the preceding theorem. For example, what is

$$\lim_{x \to \infty} \frac{2^x}{x^3} = ?$$

First, we can write $2 = e^k$, where $k = \text{Log}[2]$. (Simply take logs of both sides of $2 = e^k$.) So that our limit becomes,

$$\lim_{x \to \infty} \frac{e^{kx}}{x^3} = ?$$

for a positive k. ($\text{Log}[2] = k \approx 0.693147806$.)

We consider the kth root of our limit to obtain

$$\lim_{x \to \infty} \sqrt[k]{\frac{e^{kx}}{x^3}} = \lim_{x \to \infty} \left(\frac{e^{kx}}{x^3} \right)^{1/k}$$

$$= \lim_{x \to \infty} \frac{e^{kx/k}}{x^{3/k}}$$

$$= \lim_{x \to \infty} \frac{e^x}{x^p} = \infty$$

with $p = 3/k$, by the theorem. Since positive roots only tend to infinity when the quantity tends to infinity, we have

$$\lim_{x \to \infty} \frac{2^x}{x^3} = \infty$$

EXERCISE 8.15. *All Exponentials Beat All Powers*
Suppose $b > 1$ and $p > 0$. What are the values of the limits

$$\lim_{x \to \infty} \frac{b^x}{x^p} = ? \qquad and \qquad \lim_{x \to \infty} \frac{x^p}{b^x} = ?$$

EXAMPLE 8.16. $2^{-\infty}/\infty^3$ or $\lim_{x \to -\infty} \frac{2^x}{x^3} = 0$

Consider the limit

$$\lim_{x \to -\infty} \frac{2^x}{x^3} = ?$$

First, change to natural base.

$$\lim_{x \to -\infty} \frac{e^{k\,x}}{x^3} = ?$$

with $k = \text{Log}[2] > 0$. Next, replace x by $u = -x$,

$$\lim_{u \to +\infty} \frac{e^{-k\,u}}{(-u)^3} = -\lim_{u \to +\infty} \frac{e^{-k\,u}}{u^3}$$

Now notice that the limit is obvious by some arithmetic,

$$-\lim_{u \to +\infty} \frac{e^{-k\,u}}{u^3} = -\lim_{u \to +\infty} \frac{1}{u^3\,e^{k\,u}} = 0$$

because both terms in the denominator tend to infinity. (We can also see this limit by "plugging in" the limiting values.)

EXAMPLE 8.17. $-\infty^3 \cdot 2^{-\infty}$

The limit

$$\lim_{x \to -\infty} x^3\,2^x = ?$$

has one term growing and the other shrinking ($2^{\text{large negative number}}$). Replace $x = -u$ and $2 = e^k$, for $k = \text{Log}[2] > 0$, so we have

$$\lim_{u \to +\infty} (-u)^3\,2^{-u} = -\lim_{u \to +\infty} u^3\,2^{-u}$$

$$= -\lim_{u \to +\infty} \frac{u^3}{2^u}$$

$$= -\lim_{u \to +\infty} \frac{u^3}{(e^k)^u} = -\lim_{u \to +\infty} \frac{u^3}{e^{k\,u}}$$

$$= 0$$

since exponentials beat powers to infinity.

Similar tricks reduce various limits to the logarithmic comparison of the Orders of Infinity Theorem.

EXAMPLE 8.18. $\lim_{x \downarrow 0} x^x = 1$

We may write $x = e^{\text{Log}[x]}$, so $x^x = e^{x\,\text{Log}[x]}$ and

$$\lim_{x \downarrow 0} x^x = \lim_{x \downarrow 0} e^{x\,\text{Log}[x]} = e^{\lim_{x \downarrow 0} x\,\text{Log}[x]}$$

if the last limit exists. This is a fundamental limit which we compute next.

EXAMPLE 8.19. $\lim_{x \downarrow 0} x \, \text{Log}[x] = 0$

Replace $x = 1/u$ in the limit, so $u \to +\infty$ as $x \downarrow 0$, and the limit

$$\lim_{x \downarrow 0} x \, \text{Log}[x]$$

becomes

$$\lim_{u \to \infty} \frac{1}{u} \text{Log}[\frac{1}{u}] = \lim_{u \to \infty} \frac{1}{u} \text{Log}[u^{-1}]$$
$$= - \lim_{u \to \infty} \frac{\text{Log}[u]}{u}$$
$$= 0$$

since powers beat logs to infinity.

Now apply this result to the x^x limit,

$$\lim_{x \downarrow 0} x^x = e^{\lim_{x \downarrow 0} x \, \text{Log}[x]} = e^0 = 1$$

This limit is a reason to write $0^0 = 1$.

EXERCISE 8.20. *Drill Limits*
Compute the limits

a) $\lim_{x \to 0} x^p \, \text{Log}[x] = 0$
b) $\lim_{x \to \infty} x \, \text{Log}[x] = \infty$
c) $\lim_{x \to \infty} x^{1/x} = 1$
d) $\lim_{\delta x \to 0} (1 + k \, \delta x)^{1/\delta x} = e^k$
e) $\lim_{\delta x \downarrow 0} \delta x^{1/\text{Log}[\delta x]} = e$
f) $\lim_{\delta x \downarrow 0} \delta x^{(\delta x + k)/\text{Log}[\delta x]} = e^k$

8.5. *Mathematica* Limits

Mathematica can do many exact symbolic limits. For example, try the following:

y[x_] := x Log[x]
Limit[y[x] , x-> 0]

However, it doesn't know them all. Try

y[x_] := Log[x]/x
Limit[y[x] , x-> Infinity]
Limit[y[x] , x-> 0]

and variations.

8.6. Projects

8.6.1. Numerical Computation of $\frac{da^t}{dt}$. The Mathematical Background Chapter on Logs and Exponentials contains two small projects that will help you understand the natural base. The first project has you use *Mathematica* to directly compute the derivative of b^t. You will see that this leads to

$$\frac{db^t}{dt} \propto b^t$$

but it does not give an obvious value to the constant of proportionality. However, you can experiment with your computations and find the value of b that makes this constant one. This is $b = e \approx 2.71828 \cdots$.

8.6.2. The Canary Resurrected - Cooling Data. The second mathematical project in that chapter asks you to compare some actual cooling data with the prediction of Newton's law of cooling. You will use Logs to verify this exponential law, so we put it in the Mathematical Background. It is an interesting scientific project for you to measure this yourself and we believe that you can do better than the first student data we present if you wish to try. (We are not sure if the students warmed the cup before they started measurements. You will see how this shows up in their data.)

CHAPTER 9

Graphs and the Derivative

Mathematica draws beautiful graphs and doesn't force us to think very much, so why use calculus to draw graphs? There really are reasons. Calculus helps you set the scales which usually aren't so obvious in real applications. Calculus finds the qualitatively interesting range to plot and once this is found, *Mathematica* can make a quantitatively accurate picture. Calculus also finds formulas for geometric features of interest. Again our goal is for you to form a nonlinear combination

$$\text{Knowledge[calculus + computing]} >$$
$$\text{Knowledge[calculus]} + \text{Knowledge[computing]}$$

9.1. Planck's Radiation Law

Hot bodies emit radiation. No, this isn't the start of a racy novel; we're referring to a blacksmith shop where iron comes out of the fire white hot, cools to yellow, then red, and passes below the visible spectrum into the infrared. (Warm human bodies do emit infrared radiation which is detected by the 'motion' lights many of us have outside our houses.) Physicists wanted to be able to predict the 'color' of radiation from the temperature. Wien's empirical law says: 'When the absolute temperature of a radiating black body increases, the most powerful wave length decreases in such a way that

$$\lambda_{\max} T = W$$

is constant.' Why is this so? The measured constant $W \approx 0.28978$ (cm°K) is known as Wien's displacement constant.

Planck's Law is a formula for the intensity of each frequency radiating from a black body. The formula was the first triumph of quantum mechanics. The classical theory made a wrong prediction and the new theory gave the right one and shows why Wien's Law holds. We will show how Planck's Law gives a theoretical derivation of Wien's law and how calculus, *Mathematica* and graphing interact in seeing this. We think this should show you why you should learn the rest of the chapter.

Planck began his study with a graph of the measured intensities, an empirical curve. He then fit a formula to this curve in what his 1920 Noble Prize lecture described as "an interpolation formula which resulted from a lucky guess." The lucky guess led him to a derivation from physical laws. If you go to your local quantum mechanic's shop, just next door to the blacksmith, she will tell you that the intensity of radiation for the frequencies between ω and $\omega + d\omega$ of a black body at absolute temperature T is

$$I \, d\omega = \frac{\hbar \omega^3}{\pi^2 c^2 (e^{\hbar\omega/(kT)} - 1)} d\omega$$

where $\hbar = \frac{h}{2\pi}$ and $h \approx 6.6255 \times 10^{-27}$ (erg sec) is called Planck's constant, $c \approx 2.9979 \times 10^{10}$ (cm/sec) is the speed of light and $k \approx 1.3805 \times 10^{-16}$ (erg/deg) is called Boltzman's constant. Physicists treat this formula as a differential because you need to measure a small range of frequencies, not just the intensity at one frequency.

AN ERROR:

We want to be sloppy about the units in this example and just emphasize the difficulty we encounter because of the scales of the real physical constants. Normally, we will not accept this practice ourselves or as part of your homework, but we don't want to get side-tracked with the physics. The *Feynman Lectures on Physics* derive Planck's Law (in angular frequency form) with their usual mixture of brilliance, charm and mathematical giant steps (flim-flam?). We won't repeat that. If you wish to look that up, you can also straighten out the units and give careful definitions of the variables.

We want to express Planck's Law in terms of wavelength λ using the basic relation

$$\omega \lambda = 2\pi c$$

and its differential, so

$$\omega = \frac{2\pi c}{\lambda} \quad \& \quad d\omega = -\frac{2\pi c}{\lambda^2} d\lambda$$

giving

$$I \, d\lambda = 8\pi h c^2 \frac{1}{\lambda^5 (e^{hc/(k\lambda T)} - 1)} d\lambda$$

$$= a \frac{1}{\lambda^5} \frac{1}{e^{b/(\lambda T)} - 1} d\lambda$$

where $a = 8\pi h c^2$ and $b = hc/k$.

Let's take a hot object, say $T = 1000°$K (about $1300°$ F). The first lesson the computer learns from hand computations is what range to plot. Normally in a textbook problem to plot a function like $i = a/\lambda^5$, you might pick a range like $0 < \lambda < 10$. This is nonsense in our problem, because the interesting range of wavelengths is orders of magnitude smaller. Let's try it with *Mathematica* anyway:

$$a = 1.4966 \ \ 10^{-4}$$
$$b = 1.4388$$
$$T = 10^3$$
$$i := a/(l^5 (Exp[b/(l \ T)] - 1))$$
$$i$$
$$Plot[i, \{l, 0, 10\}]$$

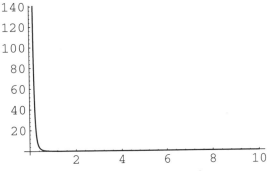

FIGURE 9.1: A useless plot

The wavelength of red cadmium light in air is used for calibrations and equals

$$\lambda = 6438\text{A} = 6438 \times 10^{-8}\text{cm} = 6.438 \times 10^{-5}\text{cm}$$

so let's try another range

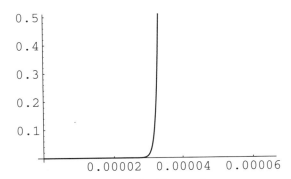

FIGURE 9.2: Also useless

This plot tells us something, at least. The interesting stuff must be between these ranges, so we try a range of $\{l, 0, .002\}$

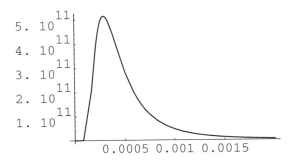

FIGURE 9.3: Intensity for $0 < \lambda < .002$

Now we see what we want - the maximum. It seems to be around 0.0003 as well as we can read the graph. We didn't really use calculus, but a little physics to look for a reasonable range. We were lucky to find the peak. What we wanted was the place where the curve peaks, so we could have used calculus to find it. And we can do that now. Let $B = \frac{b}{T}$, so $I = a\lambda^{-5}(e^{B/\lambda} - 1)^{-1}$ and use the Product Rule and Chain Rule:

$$\frac{dI}{d\lambda} = a\left(-5\lambda^{-6}(e^{B/\lambda} - 1)^{-1} + \lambda^{-5}(-1)(e^{B/\lambda} - 1)^{-2}(B(-\lambda^{-2}e^{B/\lambda})\right)$$

$$= \frac{a}{\lambda^6} \cdot \frac{5 - 5e^{b/\lambda T} + \frac{b}{\lambda T}e^{b/\lambda T}}{(e^{b/\lambda T} - 1)^2}$$

$$= \frac{1.4966 \ 10^{-4}}{\lambda^6} \cdot \frac{\left(5 - 5e^{1.4388 \ 10^{-3}/\lambda} + \frac{1.4388 \ 10^{-3}}{\lambda}e^{1.4388 \ 10^{-3}/\lambda}\right)}{(e^{1.4388 \ 10^{-3}/\lambda} - 1)^2}$$

It is horribly messy to do this computation with the explicit values of the constants. We want you to learn to use parameters, even in cases like this where they only make the calculations neater and thus easier to follow. We will look at the computation with parameters below for a more important reason. Actually, we asked you to differentiate Planck's formula in the last chapter and showed you the answer in the *Mathematica* NoteBook **Dfdx.ma**. All the computations in this section are on the *Mathematica* NoteBook **PlanckL.ma** and we encourage you to compare that with your Planck's Law exercise below.

The graph is horizontal when $\frac{dI}{d\lambda} = 0$ so the numerator must equal zero

$$5 - 5e^{1.4388 \ 10^{-3}/\lambda} + \frac{1.4388 \ 10^{-3}}{\lambda}e^{1.4388 \ 10^{-3}/\lambda} = 0$$

If we try to solve $\frac{dI}{d\lambda} = 0$ using *Mathematica* , we encounter nasty numerical problems because of the scale. A change of variables makes the problem appear simpler and scales it in a form that *Mathematica* can handle. Let $x = \frac{1.4388 \ 10^{-3}}{\lambda}$ and re-write the equation above to obtain the problem:

FindRoot[5 == (5 - x) Exp[x],{ x,?}]

We need a starting guess at a root (shown with '?' above). A Plot shows the following figure with a root near $x = 5$.

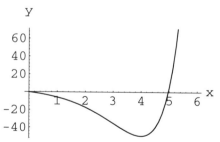

FIGURE 9.4: $y = 5 - (5 - x)Exp[x]$

Mathematica finds our root more accurately to be $x = \frac{1.4388 \ 10^{-3}}{\lambda} \approx 4.96511$, so $\lambda \approx \frac{1.4388}{4.96511} \times 10^{-3} = 2.89782 \times 10^{-4}$. The observed wavelength at $1000°$ K is 0.000289782 cm. Very nice start, but this is not the answer we are looking for. We have only shown what happens at $T = 1000$ and we want to know what happens as we vary T.

We could try another value, say $T = 900$ and work through the whole problem again, but that would just give us another specific number. We need a formula that tells us how the maximum point depends on the temperature T. We can find this by solving the problem as before, but keeping the letter T as an unknown constant or parameter rather than letting $T = 1000$.

The peak on each curve $I = I(\omega)$ where T is constant (but not fixed at some specific number) still occurs the point where the slope is zero or

$$\frac{dI}{d\lambda} = 0$$

We want to solve the equation with the parameter T in the expression. This isn't so hard given all our preliminary work. We know

$$\frac{dI}{d\lambda} = \frac{a}{\lambda^6} \cdot \frac{5 - 5e^{b/\lambda T} + \frac{b}{\lambda T}e^{b/\lambda T}}{(e^{b/\lambda T} - 1)^2}$$

and the derivative is zero if the numerator is zero, or

$$5 = (5 - \frac{b}{\lambda T})e^{\frac{b}{\lambda T}}$$

Substituting $x = \frac{b}{\lambda T}$, we are left with exactly the problem we solved numerically above,

$$5 = (5 - x)e^x$$

which holds when $x \approx 4.96511$.

What do we conclude? At the peak,

$$\frac{b}{\lambda T} \approx 4.96511$$

or

$$\lambda T \approx \frac{1.4388}{4.96511} = 0.289782$$

This is Wien's Law, $\lambda_{\max}T = W$ constant. We have shown that the empirical law follows theoretically from Planck's quantum mechanical law and derived $W = 0.289782$ vs. the measured $W = 0.28978$. The formulas are a little complicated and calculus plays a role in just finding plot ranges. It plays an indispensable role in finding the *formula* for the peak by differentiating and solving with parameters.

In summary, for fixed temperature, T, the maximum intensity occurs at the wavelength satisfying

$$\lambda_{\max}T = 0.289782$$

or $\lambda_{\max} = 0.289782/T$. The solution to the symbolic maximization problem gives us a function of the parameter T. This is a derivation of Wein's empirical law from Planck's theoretical law.

In the next exercise you can compare your work with frequency form of Planck's Law in **PlanckF.ma** to the *Mathematica* wavelength solution just described in the text.

EXERCISE 9.5. *Planck's Law written in terms of frequency does not lead to Wien's same constant. We want you to show this by finding the maximum of $I(w)$. Use the Mathematica*

NoteBook "PlanckF.ma" to help with the computations and sketch curves like the ones we have done for wavelength. The intensity formula with frequency is

$$I = \frac{\hbar\omega^3}{\pi^2 c^2 (e^{\hbar\omega/(kT)} - 1)}$$

$$= \frac{\alpha\omega^3}{e^{\beta\omega/T} - 1}$$

(1) *Sketch I vs. ω on an appropriate range. Here are two poor starting attempts:*

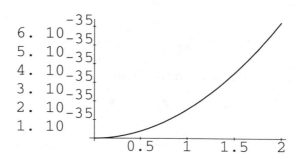

FIGURE *9.6: A Bad Range of Frequency*

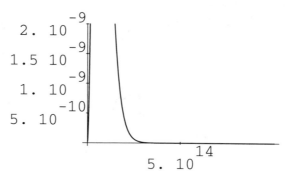

FIGURE *9.7: A Better Range*

(2) *Compute the derivative*

$$\frac{dI}{d\omega} = a\omega^2 \frac{(3 - \frac{b\omega}{T})e^{\beta\omega/T} - 3}{(e^{\beta\omega/T} - 1)^2}$$

and show that this leads to the properly scaled root finding problem

$$(3 - x)e^x = 3$$

(3) *Use Mathematica to solve this equation. (There is help in "PlanckF.ma")*
(4) *Use the root of the previous part to show that the maximum angular frequency occurs when $\frac{\beta\omega}{T} \approx 2.82144$ so $\lambda T \approx \frac{hc}{k2.82144} = 0.509951$, since $\lambda = \frac{hc}{\hbar\omega}$.*
(5) *(Optional) Ask your neighborhood quantum mechanic why maximum frequency gives a different Wien's law.*

9.1.1. Simple-minded Scales. Poor choice of scales can come up in innocent or simple-minded ways, not just as a result of large differences in the size of scientific constants as in Planck's Law. Here is an example. Which of the following is the graph of

$$y = x^5 + 4x^4 + x$$

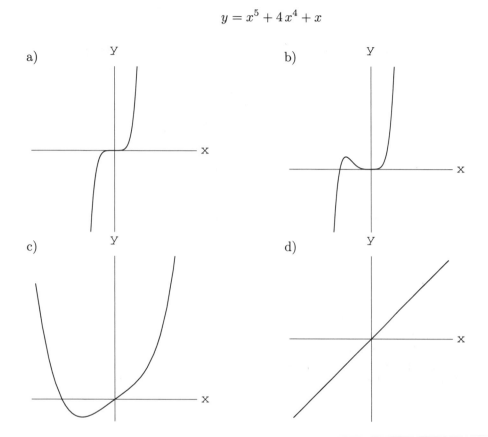

FIGURE 9.8: $y = x^5 + 4x^4 + x$ on Four Different Scales

The answer is: All of these are graphs of this same polynomial. Graph (a) is for $-100 < x < 100$. Graph (b) is for $-10 < x < 10$. Graph (c) is for $-1 < x < 1$. Graph (d) is for $-0.1 < x < 0.1$. They appear different because the wiggle in the medium scale graph is an insignificant part of the term x^5 when $x = 100$. The small unit scale misses some of the medium size 10 wiggle. Why is the tiny scale straight? (What is the view of the graph in a microscope?) These shape differences are obvious if you compute some sizes, but could surprise you if you are using *Mathematica* and blindly hoping for "good" graphs. Calculus can find the interesting shape information and the computer can then draw it accurately. The four graphs above are accurate, but they are drawn on different scales. Try these with *Mathematica* by typing

```
y[x_ ] := x∧5 + 4 x∧4 + x;
y[x]
s = 100
Plot[ y[x] , { x , -s , s} ];
```

EXERCISE 9.9. *Scale of the Plot*
Use Mathematica to make several plots on different scales. First, re-plot

$$y = x^5 + 4\,x^4 + x$$

at the 4 scales described above, but leave the default "Ticks" on, so that Mathematica puts the scales on the plots. This will show you more clearly what we described in the previous paragraph.

Second, plot the functions below on the different suggested scales:

$$y = x^x \qquad\qquad 0 < x < 2 \qquad and \qquad 0 < x < 5$$
$$y = 3x^4 - 4x^3 - 36x^2 \qquad -10 < x < 10 \qquad and \qquad -3 < x < 5$$

Explain why the pairs of graphs appear to be different, even though they are the same function.

9.2. Graphing and The First Derivative

Calculus lets us 'look' in an infinitesimal microscope at a graph before we have the whole graph. We must 'look' by computing derivatives. Of course, all we would see in an infinitesimal microscope is the graph of a straight line $dy = m\,dx$ in the microscope (dx, dy) coordinates (where $m = f'(x)$ with x fixed). This can only be one of the three qualitative shapes:

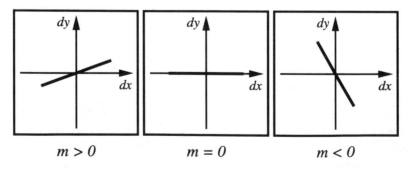

FIGURE 9.10: $dy = m\,dx$ for $m > 0$, $m = 0$, $m < 0$

If $f'(x)$ is undefined, something else may appear in the microscope, but this is an exceptional case. The exact shape of the slope can be measured on an auxiliary scale as follows, but usually this not necessary.

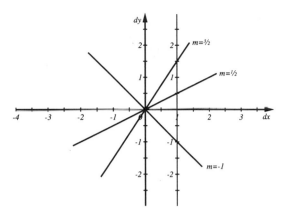

FIGURE 9.11: A Slope Scale

We begin with a simple case in the next exercise. You are given four choices for the graph and asked which one is best. Identify a shape feature related to the derivative (or what you would see in a microscope) and then check the formula to see if it matches. For example, graph (1) is decreasing for large magnitude negative numbers. $\frac{dy}{dx} = 15(x^2 - 1)x^2$ is positive if the magnitude of x is large, the squares remove the dependence on sign. What else can you eliminate?

EXERCISE 9.12. *Which of the following is the graph of* $y = 3x^5 - 5x^3$?

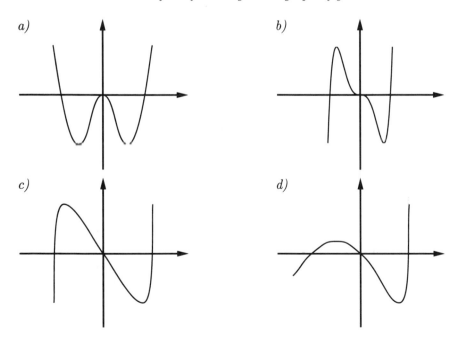

The previous exercise is a little artificial. It really means, 'which graph has all the shape information of the algebraic curve.' It also implicitly assumes that one of the figures is correct. The point is that calculus tells us the shape or all the 'ups and downs' of the curve.

The next exercise has a row of microscopic views filled out. Just looking across the y' row we see that the graph goes up - over - down - over -down - over - up. Plot the points given in the x and y row. Add little tangent segments given in the y' row at the (x, y) points. Then fill in a curve that goes through the points and is tangent to the segments.

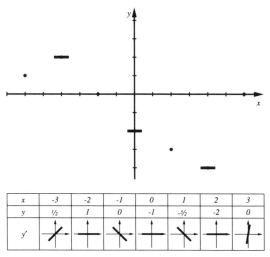

x	-3	-2	-1	0	1	2	3
y	½	1	0	-1	-¾	-2	0
y'							

FIGURE 9.13: Points and Microscopic Views

EXERCISE 9.14. *The table below the axes in the preceding figure contains x and y coordinates of the point as well as microscopic views of $y = f(x)$ at those dotted points. Sketch the graph.*

Reverse the procedure of the last exercise. Look at the next graph, fill out the x and y numbers and make microscopic views of the graph at these points.

EXERCISE 9.15. *Fill out the table below to correspond to the given graph.*

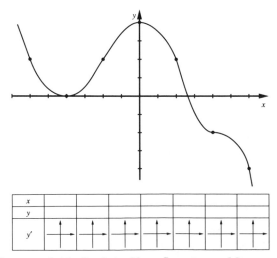

x							
y							
y'							

FIGURE *9.16: Look in Your Imaginary Microscope*

Our approach to graphing will be to fill out a table that looks like the blank graphing figure below. Here's the procedure:

9.2.1. Graphing $y = f(x)$ with the First Derivative.

(1) Compute $f'(x)$ and find **all** values of x where $f'(x) = 0$ (or $f'(x)$ does not exist). Record these in the microscope row as horizontal line segments (or $*'s$ if the derivative does not exist).
(2) Check the sign (+) or (-) of $f'(x)$ at values between each of the points from the first part. Record $f'(x) = (+) > 0$ as an upward sloping microscope line and $f'(x) = (-) < 0$ as a downward sloping microscope line.
(3) Compute a few key (x, y) pairs (using $f(x)$ to find values of y, not $f'(x)$) and record the numbers on the x and y rows. For example, you should at least compute the (x, y) pairs for the x values used in step 1.
(4) Plot the points and mark small tangent lines.
(5) Connect the points with a curve matching the tangents as you pass through the points and increasing or decreasing according to the table between the horizontal points.

Do not start by plotting points. Start with the 'shape' information of the microscope row so you first find out which points are interesting to plot.

If you are plotting with *Mathematica* , you can just work steps (1) and (2). This will tell you the range of x values to plot.

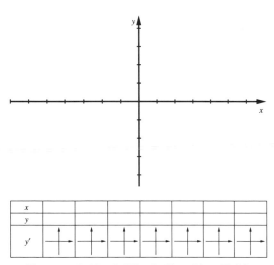

FIGURE 9.17: A Blank Graphing Table

EXAMPLE 9.18. *Graphing $y = f(x) = 3x^5 - 5x^3$ from Scratch*

First, compute $f'(x)$,

$$y' = 3 \cdot 5 \cdot x^4 - 5 \cdot 3 \cdot x^2 = 15x^4 - 15x^2$$
$$= 15x^2(x^2 - 1)$$

This derivative is defined for all real values of x.

Second, find all places, x, where $y' = 0$,

$$0 = 15\,x^2\,(x^2 - 1) \qquad \Leftrightarrow \qquad x = 0 \quad \text{or} \quad x = +1 \quad \text{or} \quad x = -1$$

The derivative is always defined and is zero only at these three points, nowhere else.

Third, we check the sign of y' for x-values between the three places where the slope is zero, $x = -2$, $x = -1/2$, $x = 1/2$ and $x = 2$. Notice that we are computing $f'(x)$, not $f(x)$.

$$x = -2, \qquad y' = 15 \cdot (-2)^2 \cdot [(-2)^2 - 1] = 15 \cdot 4 \cdot [4 - 1] = (+) \qquad \text{slope up}$$

$$x = -\frac{1}{2}, \qquad y' = 15 \cdot (-\frac{1}{2})^2 \cdot [(-\frac{1}{2})^2 - 1] = 15 \cdot \frac{1}{4} \cdot [\frac{1}{4} - 1] = (-) \qquad \text{slope down}$$

$$x = \frac{1}{2}, \qquad y' = 15 \cdot (\frac{1}{2})^2 \cdot [(\frac{1}{2})^2 - 1] = 15 \cdot \frac{1}{4} \cdot [\frac{1}{4} - 1] = (-) \qquad \text{slope down}$$

$$x = 2, \qquad y' = 15 \cdot (2)^2 \cdot [(2)^2 - 1] = 15 \cdot 4 \cdot [4 - 1] = (+) \qquad \text{slope up}$$

Fourth, we compute the (x, y) coordinates of several important points. The points with zero slope are important and we can get a reasonable idea of the shape of the curve with only these three values. Notice that now we are using the formula $y = f(x)$ and not $f'(x)$.

$$x = -1, \qquad y = 3 \cdot (-1)^5 - 5 \cdot (-1)^3 = -3 + 5 = 2 \qquad (-1, 2)$$

$$x = 0, \qquad y = 3 \cdot (0)^5 - 5 \cdot (0)^3 = 0 \qquad (0, 0)$$

$$x = 1, \qquad y = 3 \cdot (1)^5 - 5 \cdot (1)^3 = 3 - 5 = -2 \qquad (1, -2)$$

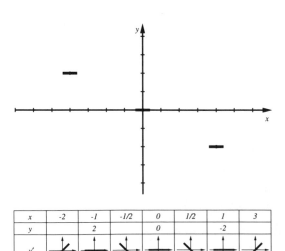

x	-2	-1	-1/2	0	1/2	1	3
y		2		0		-2	
y'							

FIGURE 9.19: Slope Information for $y = 3\,x^5 - 5\,x^3$

Finally, here is *Mathematica's* version of the graph.

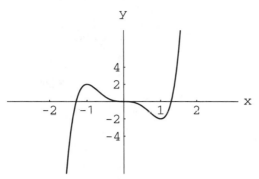

FIGURE 9.20: $y = 3\,x^5 - 5\,x^3$

EXAMPLE 9.21. *Graphing $y = x\,e^{-x}$*

First, use the Product Rule and Chain Rule to compute the derivative

$$y = f(x)\,g(x) \qquad\qquad f(x) = x \qquad\qquad \frac{df}{dx} = 1$$

$$g(x) = e^{-x} \qquad g = e^u \qquad\qquad u = -x$$

$$\frac{dg}{du} = e^u \qquad\qquad \frac{du}{dx} = -1$$

$$\frac{dg}{dx} = \frac{dg}{du}\frac{du}{dx}$$

$$= (e^u)(-1) = -e^{-x}$$

So the final derivative is

$$\frac{dy}{dx} = \frac{df}{dx}\,g + f\,\frac{dg}{dx} = 1\,e^{-x} - x\,e^{-x} = (1 - x)\,e^{-x}$$

This derivative is defined for all real x.

We know that $e^u \neq 0$ for any u, so $\frac{dy}{dx} = 0$ only if $1 - x = 0$ or $x = 1$.

Second, we check signs at x values between $-\infty$ and 1 and between 1 and $+\infty$, $x = -1$ satisfies $-\infty < -1 < 1$ and $x = 3$ satisfies $1 < 3 < +\infty$.

$$y'(-1) = (1 + 1)\,e^{+1} = 2 \cdot e \approx 5.4$$

$$y'(+3) = (1 - 3)\,e^{-3} = -2\,e^{-3} \approx -0.10$$

The values of the y-coordinate at these points are

$$y(-1) = (-1)\,e^{+1} = -e \approx -2.7$$

$$y(+1) = 1\,e^{-1} = 1/e \approx 0.368$$

$$y(+3) = 3\,e^{-3} = 3/e^3 \approx 0.15$$

We also know that

$$\lim_{x \to -\infty} x\,e^{-x} = -\infty \quad\&\quad \lim_{x \to \infty} \frac{x}{e^x} = 0$$

and that $y = x\,e^{-x} > 0$ for $x > 0$. (See Theorem 8.12.)

All this graphing information is recorded in the following table

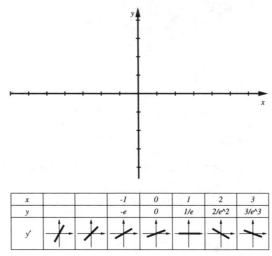

x			-1	0	1	2	3
y			$-e$	0	$1/e$	$2/e\textasciicircum2$	$3/e\textasciicircum3$
y'							

FIGURE 9.22: Slope Information for $y = x\,e^{-x}$

EXERCISE 9.23. *Graphing with Slopes Drill*

Use the above first derivative procedure to sketch the graphs of

a) $y = f(x) = 2x^2 - x^4$ b) $y = f(x) = x^3 - 3x$

c) $y = f(x) = 6x^5 - 10x^3$ d) $y = (x+2)^{\frac{2}{3}}$

e) $y = \mathrm{Sin}[x] + \mathrm{Cos}[x]$ f) $y = \mathrm{Sin}[x] \times \mathrm{Cos}[x]$

g) $y = x\,e^x$ h) $y = e^{-x^2}$

i) $y = x\,\mathrm{Log}[x]$ j) $y = x - \mathrm{Log}[x]$

You may check your graphs with *Mathematica* after you sketch by hand. Later you will need the skills you develop in this exercise together with *Mathematica* in order to understand complicated graphs like Planck's law above.

9.3. The Theorems of Bolzano and Darboux

How do we know that it is sufficient to just check one point between the zeros of $f'(x)$ in the graphing procedure of the last section? This is because derivatives have the property that they cannot change sign without being zero, provided that they are defined on an interval. If $f'(x)$ is not zero in an interval $a < x < b$, then $f'(x)$ cannot change sign. This is taken up in the Mathematical Background Chapter on Bolzano's Theorem, Darboux' Theorem and the Mean Value Theorem.

9.4. Graphing and the Second Derivative

Which is a better graph of graph of $y = 3x^5 - 5x^3$?

a)   b)

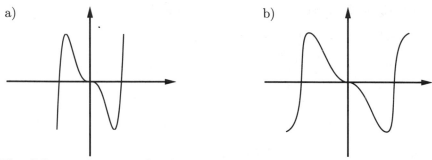

The difference between the choices is the bending, not the slope. This information can be obtained from asking whether the first derivative is increasing or decreasing. If the slope gets steeper and steeper, the curve bends up. The derivative $f'(x)$ is just a function and its derivative must be positive if it is increasing, $f''(x) = (+) > 0$.

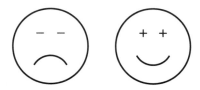

FIGURE 9.24: Negative on Frown - Positive on Smile

Negative second derivatives make the derivative decrease. If the slope decreases, the curve bends down.

EXAMPLE 9.25. *The Bends of $y = 3x^5 - 5x^3$*

We computed the slope information of $y = 3x^5 - 5x^3$ in Example 9.18 above. The derivatives are

$$y = 3x^5 - 5x^3$$
$$y' = 3 \cdot 5x^4 - 5 \cdot 3x^2 = 15(x^4 - x^2)$$
$$y'' = 15(4x^3 - 2x) = 30x(2x^2 - 1)$$

We already found the slope table and (x, y)-points at the places where the slope is zero. Now we want to find out where the curve bends up (or looks like part of a smile) and where it bends down.

$$y'' = 0 \quad \Leftrightarrow \quad 0 = 30x(2x^2 - 1) \quad \Leftrightarrow \quad x = 0 \quad \text{or} \quad 2x^2 - 1 = 0$$
$$\Leftrightarrow \quad x = 0 \quad \text{or} \quad x = \frac{1}{\sqrt{2}} \approx 0.707 \quad \text{or} \quad x = -\frac{1}{\sqrt{2}} \approx -0.707$$

The second derivative is always defined and only is zero at these three points.

Now we check the signs of the second derivative at values between the zeros.

$$x = -1, \qquad y'' = 30 \cdot (-1) \cdot [2(-1)^2 - 1] = -30 \cdot [4 - 1] = (-) \qquad \text{frown}$$

$$x = -\frac{1}{2}, \qquad y'' = 30 \cdot (-\frac{1}{2}) \cdot [2(-\frac{1}{2})^2 - 1] = -15 \cdot [\frac{1}{2} - 1] = (+) \qquad \text{smile}$$

$$x = \frac{1}{2}, \qquad y'' = 30 \cdot (\frac{1}{2}) \cdot [2(\frac{1}{2})^2 - 1] = 15 \cdot [\frac{1}{2} - 1] = (-) \qquad \text{frown}$$

$$x = 1, \qquad y'' = 30 \cdot (1) \cdot [2(1)^2 - 1] = 30 \cdot [4 - 1] = (+) \qquad \text{smile}$$

So we see that graph (a) above is a better representation of the curve $y = 3x^5 - 5x^3$, because the extra bends on graph (b) have the bending sequence smile - frown - smile - frown - smile - frown. We could eliminate graph (b) because the second derivative at a large positive number would need to be negative in order to have the right-most downward (frown) bend shown on graph (b). Similarly, the left-most upward bend would require that the second derivative is positive. Graph (a) above has the (-) - (+) - (-) - (+) or frown - smile - frown - smile sequence of signs to its second derivative.

Identify bends in the figures of the next exercise and then compute y'' from the given formula to see which graph is right.

EXERCISE 9.26. *Which graph is nearest* $y = 4x^2 - \frac{x^5}{5}$?

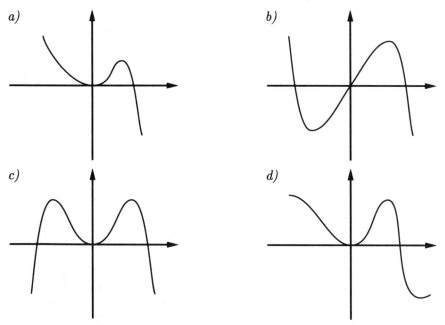

a)

b)

c)

d)

9.4.1. Plotting with the First and Second Derivatives.

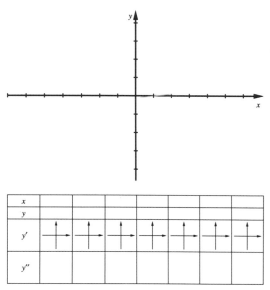

FIGURE 9.27: Blank Table for Plotting with x, y, y' and y''

(1) Compute $f'(x)$ and find **all** values of x where $f'(x) = 0$ (or $f'(x)$ does not exist). Record these in the microscope row as horizontal line segments (or $*'s$).

(2) Check the sign (+) or (-) of $f'(x)$ at values between each of the points from the first part. Record $f'(x) = (+) > 0$ as an upward sloping microscope line and $f'(x) = (-) < 0$ as a downward sloping microscope line.

(3) Compute $f''(x)$ and find **all** values of x where $f''(x) = 0$ (or $f''(x)$ does not exist). Record these in the y'' row of the table as $0's$ (or $*'s$).

(4) Check the sign (+) or (-) of $f''(x)$ at values between each of the points from the first part. Record $f''(x) = (+) > 0$ as a smile and $f'(x) = (-) < 0$ as a frown.

(5) Compute a few key (x, y) pairs (using $f(x)$ to find values of y, not $f'(x)$ or $f''(x)$) and record the numbers on the x and y rows. For example, you should at least compute the (x, y) pairs for the x values used in steps 1 and 3.

(6) Plot the points and mark small tangent lines.

(7) Connect the points with a curve matching the tangents as you pass through the points and increasing or decreasing and bending according to the table.

Do not start by plotting points. Start with the 'shape' information of the microscope row so you first find out which points are interesting to plot.

EXAMPLE 9.28. $y = x^3 + x$

The graph of $y = x^3 + x$ illustrates the additional information of the second derivative. We have $\frac{dy}{dx} = 3x^2 + 1$ which is always positive. The first derivative slope information just says the graph is increasing. However, $\frac{d^2y}{dx^2} = 6x$ which is zero at $x = 0$ and only there. When $x < 0$, $\frac{d^2y}{dx^2} < 0$, so the graph bends downward, but slopes upward. The left half of a frown slopes up, but bends down. When $x > 0$, $\frac{d^2y}{dx^2} > 0$, so the graph bends upward and slopes upward. The right half of the smile slopes up, and bends up. Fill out the slope and bend shape tables for this graph which is given next.

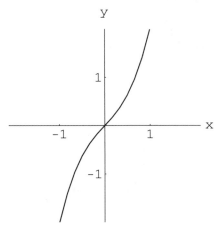

FIGURE 9.29: $y = x^3 + x$

EXAMPLE 9.30. *Graphing $y = x\,e^{-x}$ - con't*

We now add the second derivative information to the sketch of the graph of Example 9.21. We know from that example that $\frac{dy}{dx} = (1 - x)\,e^{-x}$, so the second derivative is

$$\frac{d^2 y}{dx^2} = (x - 2)\,e^{-x}$$

which is defined everywhere.

The second derivative is zero only if $x = 2$, since $e^u > 0$, for all u.

We check the values $x = 0$ with $-\infty < 0 < 2$ and $x = 3$ with $2 < 3 < +\infty$,

$$y''(0) = -2\,e^0 = -2 \qquad \text{frown}$$
$$y''(3) = (3 - 2)\,e^{-3} = 1/e^3 > 0 \qquad \text{smile}$$

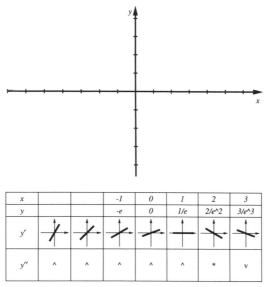

x			-1	0	1	2	3
y			-e	0	1/e	2/e^2	3/e^3
y'							
y''	^	^	^	^	^	*	v

FIGURE 9.31: Slope and Bend Information for $y = x\, e^{-x}$

The *Mathematica* graph is as follows

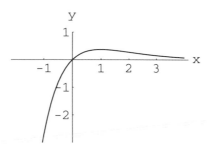

FIGURE 9.32: *Mathematica* graph of $y = x\, e^{-x}$

EXAMPLE 9.33. *Graph* $y = e^{-x^4}$

Begin with the Chain Rule,

$$y = e^u \qquad\qquad\qquad u = -x^4$$

$$\frac{dy}{du} = e^u \qquad\qquad\qquad \frac{du}{dx} = -4x^3$$

$$\text{so} \qquad \frac{dy}{dx} = \frac{dy}{du}\cdot\frac{du}{dx} = -4x^3\, e^{-x^4}$$

We have $\frac{dy}{dx} = 0$ only if $x = 0$ and the derivative is defined everywhere. Check $\frac{dy}{dx}$ at $x = \pm 1$ to see that the slope table is up - over - down.

Now use the Product Rule on $\frac{dy}{dx} = f(x)\,g(x)$,

$$f = -4x^3 \qquad\qquad\qquad g = e^{-x^4}$$

$$\frac{df}{dx} = -12x^2 \qquad\qquad\qquad \frac{dg}{dx} = -4x^3 e^{-x^4}$$

so

$$\frac{d^2y}{dx^2} = \frac{df}{dx}\,g + f\,\frac{dg}{dx} = -12x^2\,e^{-x^4} + 16x^6\,e^{-x^4}$$

$$\frac{d^2y}{dx^2} = \left(16x^6 - 12x^2\right)e^{-x^4}$$

The second derivative is defined for all x and only equals zero when $16x^6 - 12x^2 = 0$. This happens at $x = 0$ and $16x^4 = 12$ or $x = \pm\sqrt[4]{\frac{3}{4}} \approx \pm 0.9306$. Checking values we see that the bending table is smile - frown - frown - smile.

The limiting values as $x \to \pm\infty$ are simple since $-x^4 < 0$ tends to $-\infty$, $y \to 0$.

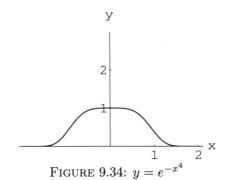

FIGURE 9.34: $y = e^{-x^4}$

EXAMPLE 9.35. *Graph $y = x\operatorname{Log}[x]$*

SOLUTION: The product rule makes the derivative

$$\frac{dy}{dx} = \operatorname{Log}[x] + x\frac{1}{x} = 1 + \operatorname{Log}[x]$$

Setting this equal to zero, we find that

$$\frac{dy}{dx} = 0 \quad\Leftrightarrow\quad x = e^{-1} = \frac{1}{e}$$

and the slope table on $(0,\infty)$ is down-over-up. This makes the minimum occur at $x = 1/e$. We also know that $x\operatorname{Log}[x]$ is negative for $x < 1$.

The rest of the graphing information is easy to get,

$$\frac{d^2y}{dx^2} = \frac{1}{x} > 0$$

for all $x > 0$, so the curve always bends upward.

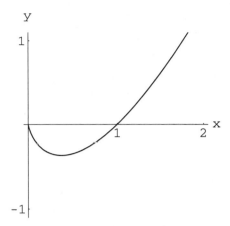

FIGURE 9.36: $y = x \operatorname{Log}[x]$

When $x = H$ an infinitely large number, $H \operatorname{Log}[H]$ is also infinitely large, so

$$\lim_{x \to \infty} x \operatorname{Log}[x] = \infty$$

The question is: What value does $x \operatorname{Log}[x]$ increase to as x decreases to zero?

$$\lim_{x \downarrow 0} x \operatorname{Log}[x] = ?$$

To compute the limit at zero, we make a change of variables. Let $z = 1/x$ and re-write our problem

$$\lim_{x \downarrow 0} x \operatorname{Log}[x] = \lim_{z \to \infty} \frac{\operatorname{Log}[1/z]}{z} = \lim_{z \to \infty} \frac{-\operatorname{Log}[z]}{z} = 0$$

(See Example 8.18.)

EXERCISE 9.37. *Use first and second derivative procedure to sketch graphs of*

a) $y = f(x) = 2x^2 - x^4$ b) $y = f(x) = x^3 - 3x$

c) $y = f(x) = 6x^5 - 10x^3$ d) $y = (x + 2)^{\frac{2}{3}}$

e) $y = \operatorname{Sin}[x] + \operatorname{Cos}[x]$ f) $y = \operatorname{Sin}[x] \times \operatorname{Cos}[x]$

g) $y = x \, e^x$ h) $y = e^{-x^2}$

i) $y = \dfrac{\operatorname{Log}[x]}{x}$ j) $y = x - \operatorname{Log}[x]$

k) $y = \dfrac{1}{1 + x^2}$ l) $y = \dfrac{1}{1 - x^2}$

Check your graphs with Mathematica , but remember that these are simple practice problems for you to work by hand. Later messy real-world problems(like Planck's law) will require this calculus effort before you start Mathematica .

Graphical analysis is very useful and often only a rough idea is enough. Here is an example of using one graph to help find another.

PROBLEM 9.38. *Graph $y = (x^3 - x^2 + 1)^2$. How does the graph of the simpler equation $z = x^3 - x^2 + 1$ help in finding **all** the places where $y'(x) = 0$?*

9.5. Another Kind of Graphing from the Slope

A differential equation like

$$\frac{dy}{dt} = y(3 - y)$$

can be thought of as a description of the slope of the curve $y = y(t)$, given that you know y. In this case, if we start at $y = 1$ when $t = 0$, the initial slope is $\frac{dy}{dt} = y(3-y) = 1(3-1) = 2$. We can begin to sketch the curve by putting our pencil at $(0, 1)$ and moving up along a line of slope 2. After we have sketched a small distance, both t and y will be larger and the slope will change accordingly.

We can move a specific small amount to y_1 and re-compute the slope from the differential equation, $y_1(3 - y_1)$, as we have the computer do in the **SecondSIR.ma** NoteBook and the computations on the dead canary, Exercise 5.43. The specific amounts are not essential to sketch the curve. Whenever y is between 1 and 3, the slope, $\frac{dy}{dt} = y(3-y)$ is positive. This means that as t increases, y increases. However, as we approach $y = 3$ from below, the term $3 - y$ tends to zero and the slope $\frac{dy}{dt} = y(3-y)$ also tends to zero. This means that the rate of increase slows down and y approaches 3. If we ever get to $y = 3$, the slope becomes zero and the curve ceases to increase. This rough reasoning gives a sketch of the solution of the differential equation when the solution starts at $y(0) = 1$. It does not require a formula for the solution.

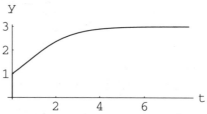

FIGURE 9.39: Solution to $y(0) = 1$ & $\frac{dy}{dt} = y(3-y)$

This simple reasoning is a graphical version of the idea of the **SecondSIR.ma** NoteBook and while it is not very accurate, it does give the behavior of solutions. We will return to this kind of curve sketching when we study differential equations later in the course.

One of the most fundamental "decay laws" in mathematics says that the rate of decrease of a quantity is proportional to the amount that is left. Radioactive substances have this property. The next exercise asks you to sketch the amount of such a substance as a function of time.

PROBLEM 9.40. *Exponential Decay*
Given that a quantity q satisfies $q(0) = 4$ and $\frac{dq}{dt} = -\frac{1}{2} q$, sketch the graph of q vs t. What happens as the amount of the quantity gets close to $q = 0$?

Can you find an analytical solution to the conditions $q(0) = 4$ and $\frac{dq}{dt} = -\frac{1}{2} q$? (This is not needed for the sketch.)

9.6. Projects

9.6.1. The Inverse Function Rule. The Mathematical Background has a Chapter on finding the derivative an inverse function such as ArcTan[y] and computing values of the inverse itself. This project relies on basic graphical understanding and the microscope idea.

9.6.2. Algebraic Formulations of Increasing and Bending. The Mathematical Background has a Chapter on "Taylor's Formula," which can be used to give algebraic proofs of the meaning of the slope and bending icons used in graphing.

9.6.3. Bolzano, Darboux, and The Mean Value Theorem. The Mathematical Background on these three theorems provides complete justification for our simple method of finding the slope and bend tables.

CHAPTER 10

Velocity, Acceleration and Calculus

This brief chapter explores the connection between first and second derivatives and basic mechanics. This is a non-graphical way to use and view these mathematical ideas, but they can also be related to graphs. The mechanical ideas are familiar in everyday experience, but now we want you to connect them with calculus. We have discussed several cases of this idea already. For example, recall the following (re-stated) Exercise 3.30 from chapter 3.

EXERCISE 10.1. *Make a qualitative rough sketch of a graph of the distance traveled s as a function of time t on the following hypothetical trip. You travel a total of 100 miles in 2 hours. Most of the trip is on rural Interstate highway at the 65 mph speed limit. You start from your house at rest and gradually increase your speed to 25 mph, then slow down and stop at a stop sign. You speed up again to 25, travel a while and enter the Interstate. At the end of the trip you exit and slow to 25, stop at a stop sign and proceed to your final destination.*

10.1. Velocity and the First Derivative

Physicists make an important distinction between speed and velocity. A speeding train whose speed is 75 mph is one thing, and a speeding train whose velocity is 75 mph on a vector aimed directly at you is the other. Velocity is speed plus direction, while speed is only the instantaneous time rate of change of distance traveled. When an object moves along a line, there are only two directions, so velocity can simply be represented by speed with a sign, + or −.

EXERCISE 10.2. *An object moves along a straight line such as a straight level railroad track. Suppose the time is denoted t, with t = 0 when the train leaves the station. Let s represent the distance the train has traveled. The variable s is a function of t, s = s(t). We need to set units and a direction. Why? Explain in your own words why the derivative $\frac{ds}{dt}$ represents the instantaneous velocity of the object (assuming that the technical mathematical conditions are met so that the correct approximations work for the function s(t).) What does a negative value of $\frac{ds}{dt}$ mean? Could this happen? How does the train get back?*

10.2. Acceleration and the Second Derivative

Acceleration is the physical term for 'speeding up your speed... ' Look up your old solution to Exercise 3.30 or Exercise 10.1 and add a graphing table like the ones from the last chapter with slope and bending. Fill in the parts of the table corresponding to $\frac{ds}{dt}$ and $\frac{d^2s}{dt^2}$ using the microscope and smile and frown icons, but add $+$ and $-$ signs, because we need to understand the mechanical interpretation of these derivatives, not just the graphical interpretation.

Remember that $\frac{d^2s}{dt^2}$ is the derivative of the function $\frac{ds}{dt}$, so for example, when it is positive, the function increases, and when it is negative, the function decreases.

EXERCISE 10.3. *This question seeks the everyday interpretation of the positive and negative signs of $\frac{d^2s}{dt^2}$ on the hypothetical trip from Exercise 10.1. Use your solution graph of time, t, versus distance, s, to analyse the following questions.*

Where is your speed increasing? Decreasing? Zero? The speed is shown on your graph as its slope, why? If speed (slope) is increasing,. what geometric shape must that portion of the graph have?

Is $\frac{ds}{dt}$ ever negative in your example? Could it be negative on someone's solution? Why does this mean that you are headed back home?

Why must $\frac{d^2s}{dt^2}$ be negative somewhere on everyone's solution?

Summarize both the mechanics and geometrical meaning of the sign of the second derivative $\frac{d^2s}{dt^2}$ in a few words.

10.2.1. The Fallen Tourist Revisited. Recall the tourist of Exercise 5.45. He threw his camera and glasses off the leaning tower of Pisa in order to confirm Galileo's law of gravity. The Italian police videotaped his crime and recorded the following information:

$t =$time (seconds)	$s =$distance fallen (meters)
0	0
1	4.90
2	19.6
3	44.1

We want to compute the average speed of the falling object during each second, from 0 to 1, from 1 to 2, and from 2 to 3? For example, at $t = 1$, the distance fallen is $s = 4.8$ and at $t = 2$, the distance is $s = 18.5$, so the change in distance is $18.5 - 4.8 = 13.7$ while the change in time is 1. Therefore, the average speed from 1 to 2 is 13.7 m/sec,

$$\text{Average speed} = \frac{\text{change in distance}}{\text{change in time}}$$

Time interval	Average speed $= \dfrac{\Delta s}{\Delta t}$
$[0,1]$	$v_1 = \dfrac{4.90 - 0}{1 - 0} = 4.90$
$[1,2]$	$v_2 = \dfrac{19.6 - 4.90}{2 - 1} = 14.7$
$[2,3]$	$v_3 = \dfrac{44.1 - 19.6}{3 - 2} = 24.5$

EXAMPLE 10.4. *The Speed Speeds Up*

These average speeds increase with increasing time. How much does the speed speed up during these intervals? (This is not very clear language, is it? How should we say, 'the speed speeds up'?)

Interval to interval	Rate of change in speed
$[0,1]$ to $[1,2]$	$a_1 = \dfrac{14.7 - 4.90}{?} = \dfrac{9.8}{?}$
$[1,2]$ to $[2,3]$	$a_2 = \dfrac{24.5 - 14.7}{?} = \dfrac{9.8}{?}$

The second speed speeds up 9.8 m/sec during the time difference between the measurement of the first and second average speeds, but how should we measure that time difference since the speeds are averages and not at a specific time?

The tourist's camera falls 'continuously.' The data above only represents a few specific points on the graph, which would look like the following graph of time vs. height or time vs. $s =$ distance fallen.

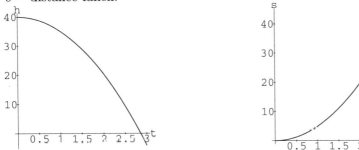

FIGURE 10.5: Continuous Fall of the Camera

The computation of the previous exercise finds $s(1) - s(0)$, $s(2) - s(1)$ and $s(3) - s(2)$. Which continuous velocities do these best approximate? The answer is $v(\frac{1}{2})$, $v(\frac{3}{2})$ and $v(\frac{5}{2})$ - the times at the midpoint of the time intervals. Sketch the tangent at time 1.5 on the graph of s vs. t and compare that to the segment connecting the points on the curve at time 1 and time 2. In general, the symmetric difference

$$\frac{f[x + \frac{\delta x}{2}] - f[x - \frac{\delta x}{2}]}{\delta x} \approx f'(x)$$

gives the best numerical approximation to the derivative of

$$y = f[x]$$

when we only have data for f. The difference quotient is best as an approximation at the midpoint. The Mathematical Background Chapter on Taylor's Formula shows algebraically and graphically what is happening. Graphically, if the curve bends up, a secant to the right is too steep, while a secant to the left is not steep enough. The average of one slope below and one above is a better approximation to the slope of the tangent. The average slope given by the symmetric secant, even though that secant does not pass through $(x, f[x])$. The general figure looks as follows:

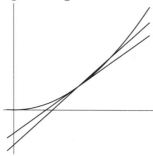 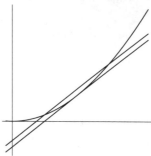

FIGURE 10.6: $[f(x + \delta x) - f(x)]/\delta x$ versus $[f(x + \delta x) - f(x - \delta x)]/\delta x$

The best times to associate to our average speeds in comparison to the continuous real fall are the midpoint times:

Time	Speed $= \dfrac{\Delta s}{\Delta t}$
0.50	$v_1 = v[0.50] = 4.90$
1.50	$v_2 = v[1.50] = 14.7$
2.50	$v_3 = v[2.50] = 24.5$

This interpretation gives us a clear time difference to use in computing the rates of increase in the acceleration:

Time	Rate of change in speed
Ave[0.50&1.50] = 1	$a_1 = \dfrac{14.7 - 4.90}{1.50 - 0.50} = 9.8$
Ave[1.50&2.50] = 2	$a_2 = \dfrac{24.5 - 14.7}{2.50 - 1.50} = 9.8$

We can summarize the whole calculation by writing the difference quotients in a table opposite the various midpoint times as follows:

Time t (sec)	Position s (m)	Velocity v (m/sec)	Acceleration a ([m/sec]/sec)
0	0		
0.50		4.90	
1	4.90		9.8
1.50		14.7	
2	19.6		9.8
2.50		24.5	
3	44.1		

There is more accurate data for the fall of the camera. In half second time steps, we have:

Time t (sec)	Position s (m)	Velocity v (m/sec)	Acceleration a ([m/sec]/sec)
0	0		
		?	
0.500	1.223		?
		?	
1.00	4.901		?
		?	
1.50	11.03		

$$\vdots$$

EXERCISE 10.7. *Compute the average speeds corresponding to the positions in the more accurate table above and write them next to the correct midpoint times, so that they correspond to approximate continuous velocities at those times.*

Then use your approximate velocities to compute rates of change of velocities, or accelerations at the proper midpoint times.

EXERCISE 10.8. *What are the best times, t_1 and t_2, corresponding to the accelerations of the camera in the previous exercise, $a(t_1)$, $a(t_2)$? Why are there 4 distances, 3 velocities and only 2 accelerations?*

10.3. Galileo's Law of Gravity

Data for a lead cannon ball dropped off a tall cliff is contained the *Mathematica* notebook **Gravity.ma**. The NoteBook contains time-distance pairs for $t = 0$, $t = 0.5$, $t = 1.0$, $\cdots$, $t = 9.5$, $t = 10$. A graph of the data is included here.

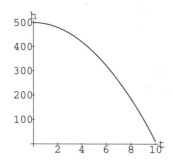 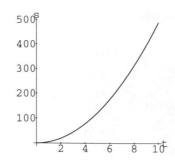

FIGURE 10.9: Free Fall without Air Friction

Galileo's famous observation turns out to be even simpler than the first conjecture of a linear speed law (which you will reject in an exercise below). He found that as long as air friction can be neglected, the rate of increase of speed is constant. Most striking, the constant is universal - it does not depend on the weight of the object.

EXERCISE 10.10. *Galileo's Law and $\frac{d^2 s}{dt^2}$*
Write Galileo's law, 'The rate of increase in the speed of a falling body is constant.' in terms of the derivatives of the distance function $s(t)$. What derivative gives the speed? What derivative gives the rate at which the speed increases? How do you express Galileo's law using the word "acceleration?"

We want you to verify Galileo's observation for the lead ball data in the **Gravity.ma** NoteBook.

EXERCISE 10.11. *Use Mathematica to make lists and graphs of the speeds from 0 to $\frac{1}{2}$ second, from $\frac{1}{2}$ to 1, etc. from the **Gravity.ma** NoteBook. Are the speeds constant? Should they be?*
Also use Mathematica to compute the rates of change in speed from 0 to $\frac{1}{4}$ seconds, from $\frac{1}{4}$ to $\frac{3}{4}$, etc. Are these constant? What does Galileo's law say about them?

In the Exercise 5.45 you formulated a model for the distance an object has fallen. You observed that the farther an object falls, the faster it goes, so tried the simplest such relationship, namely, the speed is proportional to the distance fallen. This is a reasonable first guess, but it is not correct. We want you to see why.

EXERCISE 10.12. *Try to find a constant to make the conjecture of Exercise 5.45 match the data in the NoteBook **Gravity.ma**, that is, make the differential equation*

$$\frac{ds}{dt} = k\,s$$

predict the position of the falling object. This will not work, but trying will show why.
There are several ways to approach this problem. You could work first from the data. Compute the speeds between 0 and $\frac{1}{2}$ seconds, between $\frac{1}{2}$ and 1 second, etc., then divide these numbers by s and see if the list is approximately the same constant. The differential equation $\frac{ds}{dt} = k\,s$ says it should be, because

$$v = \frac{ds}{dt} = k\,s \qquad \Leftrightarrow \qquad \frac{v}{s} = k \quad \text{is constant.}$$

*There is some help in the **Gravity.ma** NoteBook getting Mathematica to compute differences of the list. Remember that the time differences are $\frac{1}{2}$ seconds each. You need to add a computation to divide the speeds by s,*

$$\frac{s(t+\frac{1}{2}) - s(t)}{\frac{1}{2}} \approx \frac{ds}{dt} \quad at \ t + \frac{1}{4}$$

each divided by s(t). Notice that there is some computation error caused by our approximation to $\frac{ds}{dt}$ actually being best at $t + \frac{1}{4}$, but only having data for s(t). Be careful manipulating the lists with Mathematica, because the velocity list has one more term than the acceleration list.

Another approach to rejecting Galileo's first conjecture is to start with the differential equation. We can solve $\frac{ds}{dt} = k\,s$ with the initial $s(0) = 0$ by methods of Chapter 8, obtaining $s[t] = S_0\,e^{k\,t}$. What is the constant S_0 if $s[0] = 0$. How do you compute s(0.01) from this? See the Bugs Bunny's Law of Gravity and Exercise 8.10.

The zero point causes a difficulty as the preceding part of this problem shows. Let's ignore that for the moment. If the data actually is a solution to the differential equation, $s = S_0\,e^{k\,t}$, then

$$\mathrm{Log}[s] = \mathrm{Log}[S_0\,e^{k\,t}] = \mathrm{Log}[S_0] + \mathrm{Log}[e^{k\,t}] = \sigma_0 + k\,t$$

so the logarithms of the positions (after zero) should be linear. Compute the logs of the list of (non-zero) positions with Mathematica and plot them. Are they linear?

10.4. Projects

The Scientific Projects contain a Chapter on Mechanics that covers topics that go beyond this basic chapter.

10.4.1. The Falling Ladder. Example 7.12 from Chapter 6 introduces a simple mathematical model for a ladder sliding down a wall. The rate at which the tip resting against the wall slides tends to infinity as the tip approaches the floor. Could a real ladder's tip break the sound barrier? The speed of light? Of course not. That model neglects the physical mechanism that makes the ladder fall - Galileo's law of gravity. The project on the ladder asks you to correct the physics of the falling ladder model.

10.4.2. Linear Air Resistance. A feather does not fall off a tall cliff as fast as a bowling ball. The acceleration due to gravity is the same, but air resistance plays a significant role in counteracting gravity for a large light object. A basic project in the Mechanics Chapter explores the path of a wooden ball thrown off the same cliff as the lead ball we just studied in this chapter.

10.4.3. Bungee Diving and Nonlinear Air Resistance. Human bodies falling long distances are subject to air resistance, in fact, sky divers do not keep accelerating, but reach a 'terminal velocity.' Bungee divers leap off tall places with a big elastic band hooked to their legs. Gravity and air resistance act on the diver in his initial 'flight,' but once he reaches the length of the cord, it pulls up by an amount depending on how far it is stretched. The Bungee Diving Project has you combine all these forces to find out if a diver hits the bottom of a canyon or not.

10.4.4. Symmetric Differences. The Mathematical Background Chapter on Taylor's Formula contains a section that shows you why the best time for the velocity approximated by $[s(t_2) - s(t_1)]/[t_2 - t_1]$ is at the midpoint, $v([t_2 + t_1]/2)$. This is an important general numerical result that you should use any time that you need to estimate a derivative from data.

CHAPTER 11

Maxima and Minima in One Variable

Finding a maximum or minimum is clearly important in everyday experience. A manufacturer wants to maximize her profits, a contractor wants to minimize his costs subject to doing a good job, a physicist wants to find the wavelength that produces the maximum intensity of radiation. Even a manufacturer with a monopoly cannot maximize her profit by charging a very high price, because at some point consumers will stop buying. She seeks an optimum balance between supply and demand. In everyday experience these optima are sought in intuitive ways, probably incorrectly in many cases. The problems are not usually simple and often they are not even clearly formulated. Calculus can help. It can solve closed-form problems and offer guidance when the mathematical models are incomplete. Much of the success of science and engineering is based on finding symbolic optima for accurate models, but no one pretends to know closed form models for the national health profile and similar interesting, but complicated facts of everyday life.

This chapter starts with some basic mathematical theory and then looks at some 'simple' applications. They are simple in terms of life's big questions, but still offer some challenge mathematically. They may not be major 'real world' revelations, but are a start and have some practical significance. We are not retreating from our basic philosophy in this course which is to show you what good calculus is. We think you know that optimization is important, so we will begin with the easy part: the mathematical theory. (Ugh, you say, but it really is the easy part.) Read the theory quickly, try the applications and come back to the theory as needed.

You should think of the theory of max-min as a lazy approach to graphing. If you had a complete graph, you could look and see where the maximum and minimum occurred. Even then, as in Planck's derivation of Wien's Law in Chapter 9, you might want to compute the symbolic formula for the maximum or minimum. Many scientific results amount to the symbolic way a max or min depends on some parameter (such as the temperature in Wein's Law.) We will see other examples at the end of this chapter.

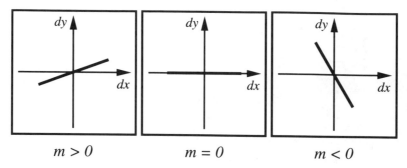

$$m > 0 \qquad m = 0 \qquad m < 0$$

FIGURE 11.1: Up, Over, Down

Usually you don't need the whole graph (which is frequently hard to compute) to find a max or min. For example, if you knew that a function $f(x)$ had a microscope table starting at $x = 1$ with an upward slope that continued until $x = 2$, and then had a downward slope until $x = 3$, you would know that the maximum of $f(x)$ for $1 \le x \le 3$ occurs at $x = 2$. Use the microscope table to sketch such a curve.

Where does the minimum of this function occur? You need more information, but only a little. It could be at either $x = 1$ or $x = 3$, but it could not be at any other x. Why? This is obvious. Think about it and sketch two graphs with the microscope table up-over-down, one with a min at $x = 1$, and the other with a min at $x = 3$. Endpoints or the lack of endpoints play an important role in max - min theory. You will learn to look for them in applications, because they simplify such problems.

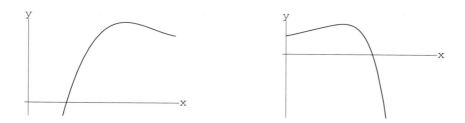

FIGURE 11.2: Minima at Different Endpoints

11.1. Critical Points

Mathematicians are fussy about the exact meaning of technical terms. When we say the function $f(x)$ attains its maximum for all real x at $x = 0$, we mean that $f(0) \ge f(x)$ for all x. That is reasonable enough. The graph below of $f(x) = \frac{1}{x^2+1}$ has its max at $x = 0$. Everyday talk might say that the minimum of this function is $y = 0$, but this is **not** the custom in mathematics. It is true that $f(x)$ tends to zero as x tends to either $+\infty$ or $-\infty$, and it is true that $f(x) > 0$ for all x, but since $f(x)$ never actually takes the value zero, we say $f(x)$ does not have a minimum. Perhaps it would be better to say it does not attain its minimum.

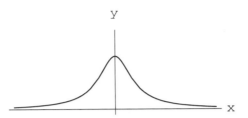

FIGURE 11.3: Max, but no min attained

Limiting values are not the difficulty entirely. The graph of $g(x) = \frac{x}{x^2+1}$ shown next has both a max and a min and also has limiting values. Precisely speaking, there are points x_{min} and x_{MAX} such that for all other x, $y_{min} = g(x_{min}) < g(x) < g(x_{MAX}) = y_{MAX}$. Find these points (x_{min}, y_{min}) and (x_{MAX}, y_{MAX}) on the graph.

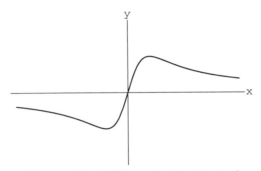

FIGURE 11.4: Max and min attained

Having pre-assigned endpoints does simplify max-min theory, because it eliminates limiting behavior. We discuss that in the next section. For now we want to explore the 'critical point condition.' The question for this section is, "What slope could the graph have at a max or min?" The answer is clear graphically, except for a proviso. What is the answer? This is really obvious in one case, and the next exercise shows you the proviso.

EXERCISE 11.5. *Find the maximum and minimum values of the function $f(x) = 3 - 2x$ for $-1 \leq x \leq 2$. What is the slope of the graph $y = f(x)$ at these points? Sketch the graph and mark the max and min.*

The next result rules out many possible places that might be maxima or minima. That's right, rules out points. It does not tell you where they are, but where they are not. It is usually stated in the following logically correct but confusing way, but it means that we only need to examine points x_0 such that $f'(x_0) = 0$ for the possibility that they are a max or min. These points are 'critical' in our investigation of extrema, but they may or may not be extrema.

THEOREM 11.6. *Interior Critical Points*
Suppose $f(x)$ is a smooth function on some interval. If $f(x)$ has a maximum or minimum at a point x_0 inside the interval, then $f'(x_0) = 0$.

PROOF: We give a geometric proof. Suppose $f'(x_0) \neq 0$. We know that we will see a straight line of slope $f'(x_0) \neq 0$ if we look at the graph under a powerful microscope. Now, **if** we can move even a small amount to either side of x_0, **then** we can make $f(x)$ both larger and smaller than $f(x_0)$ by moving both ways. If x_0 is inside an interval over which we maximize or minimize, then x_0 cannot be the max or min.

(This proof is correct, but we can elaborate on it some more. If $f'(x_0) \neq 0$, then $f(x)$ is either increasing or decreasing in an interval containing x_0. This is intuitively clear if you think of looking in a microscope and seeing a linear graph of slope $f'(x_0)$, but we prove it algebraically in a Theorem in the Mathematical Background chapter on Taylor's Formula.)

This completes the proof by use of the contrapositive. Yuck! The contrapositive? Why don't you find $f'(x)$, set it equal to zero and solve, then the solutions will be the max and min. Why not? Because that is wrong. Notice that the proof of the theorem is little more than a negative statement of where the extrema cannot be. Implications can be confusing and we want to make it clear that the theorem does **not** say that 'if $f'(x_0) = 0$, then x_0 is a max or a min.' Logically, A implies B does **not** mean that B implies A.

EXERCISE 11.7. *Find the max and min of $f(x) = x^3$ for $-1 \leq x \leq 2$. Graph the function. Is $f'(x) = 0$ at the max or at the min? Where is $f'(x) = 0$?*

Correct reasoning is important in everyday life as well as mathematics. The logic of implication can be confusing and that is what is behind the erroneous everyday conclusion that A implies B means that B also implies A. Let's see why this is so. 'If it is raining, then there are clouds.' is a correct statement.

$$A \Rightarrow B$$

where A stands for 'it is raining' and B stands for 'there are clouds.' Certainly

$$B \Rightarrow A \text{ is false}$$

because $B \Rightarrow A$ says 'if there are clouds, then it is raining.'

EXERCISE 11.8. *In The Interior Critical Point Theorem, A stands for the statement '$f(x)$ has an extremum at x_0 inside the interval' and B stands for '$f'(x_0) = 0$.' The logical statement is of the form $A \Rightarrow B$. Give an example function to show that $B \Rightarrow A$ is false in this case.*

A way to explain implication is

$$A \Rightarrow B \quad \equiv \quad (\text{not} A) \text{ or } B$$

This is a funny way to think of it, but it is equivalent. When A is true, the implication means that B must also be true. When A is false, B can be either true or false. The other statement has the same properties, so it is equivalent to the implication. If A is true, not$A$ is false, so the other part of the "or" must be true, B. If A is false, not$A$ is true, so B may be either true or false.

Now we can do some logical calculus to see the contrapositive

$$A \Rightarrow B \quad \equiv \quad (\text{not} B) \Rightarrow (\text{not} A)$$

because $(\text{not} A) \text{ or} B \equiv (\text{not not} B) \text{ or} (\text{not} A)$.

EXERCISE 11.9. *Let B be the statement "$f'(x_0) = 0$". Let A be the statement "$f(x_0)$ is an interior extremum for the interval I". The interior critical point theorem above is the statement $A \Rightarrow B$. State the contrapositive theorem $(\text{not} B) \Rightarrow (\text{not} A)$ in English.*

11.2. Max - min with Endpoints

Various intervals with and without endpoints arise in max-min problems. It is convenient to have some notation for the various cases. The basic notation is that round brackets (or) cut off 'just before' the endpoint, while square brackets [or] include the endpoint. Here we will use the love knot symbol ∞ to mean intervals 'keep going.' The infinity symbol ∞ cannot stand for an infinitely large hyperreal number, because it **violates** rules of algebra (contrary to the Algebra Axiom for hyperreals). For example, $\infty + \infty = \infty$ and $\infty \times \infty = \infty$ which is false for both real and hyperreal numbers. It is not a number, but it is a convenient symbol.

DEFINITION 11.10. *Notation: If a and b are numbers,*

a) $[a,b] = \{x : a \leq x \leq b\}$ b) $(a,b) = \{x : a < x < b\}$

c) $[a,b) = \{x : a \leq x < b\}$ d) $(a,b] = \{x : a < x \leq b\}$

e) $(-\infty, b] = \{x : x \leq b\}$ f) $(-\infty, b) = \{x : x < b\}$

g) $[a, \infty) = \{x : a \leq x\}$ h) $(a, \infty) = \{x : a < x\}$

This is simple, try it out:

EXERCISE 11.11. *Write a description of the following intervals and sketch them as segments on the real line:*

a) $(0, \infty)$ b) $(0, 1)$ c) $(-1, 3]$ d) $[-2, 5]$

The most important intervals are the ones of finite length that include their endpoints, $[a, b]$, for numbers a and b. These intervals are sometimes described as 'closed and bounded,' because they have the endpoints and have bounded length. A shorter name is 'compact' intervals.

THEOREM 11.12. *The Extreme Value Theorem*
If $f(x)$ is a continuous real function on the real compact interval $[a, b]$, then f attains its maximum and minimum, that is, there are real numbers x_m and x_M such that $a \leq x_m \leq b$, $a \leq x_M \leq b$, and for all x with $a \leq x \leq b$

$$f(x_m) \leq f(x) \leq f(x_M)$$

PROOF: We will show how to locate the maximum, you can find the minimum. The proof is somewhat like the proof of Bolzano's Intermediate Value Theorem **??**, we partition the interval into steps of size Δx,

$$a < a + \Delta x < a + 2\Delta x < \cdots < b$$

and define a function

$$M(\Delta x) = \text{the x of the form } x_1 = a + k\Delta x$$

so that

$$f(M(\Delta x) = f(x_1) = max[f(x) : x = a + h\Delta x, \ h = 0, 1, \cdots, n]$$

This function is the discrete maximum from among a finite number of possibilities, so that $M(\Delta x)$ has two properties: (1) $M(\Delta x)$ is one of the partition points and (2) all other partition points $x = a + h\Delta x$ satisfy $f(x) \leq f(M(\Delta x))$.

Next, we partition the interval into infinitesimal steps,

$$a < a + \delta x < a + 2\delta x < \cdots < b$$

and consider the natural extension of the discrete maximizing function $M(\delta x)$. By the Function Extension axiom we know that (1) $x_1 = M(\delta x)$ is one of the points in the infinitesimal partition and (2) $f(x) \leq f(x_1)$ for all other partition points x.

Since the hyperreal interval $[a, b]$ only contains finite numbers, there is a real number $x_M \approx x_1$ (standard part) and every other real number x_2 in $[a, b]$ is within δx of some partition point, $x_2 \approx x$.

Continuity of f means that $f(x) \approx f(x_2)$ and $f(x_M) \approx f(x_1)$. The numbers x_2 and x_M are real, so $f(x_2)$ and $f(x_M)$ are also real and we have

$$f(x_2) \approx f(x) \leq f(x_1) \approx f(x_M)$$

Thus, for any real x_2, $f(x_2) \leq f(x_M)$, which says f attains its maximum at x_M. This completes the proof.

You can make the computer mimic the proof for small partitions or large Δx. Of course, you cannot compute with infinitesimal step size, but you might hope to take Δx smaller and smaller and see what $M(\Delta x)$ converges toward. There really are serious problems with using this as a general purpose algorithm. It is very inefficient. It can oscillate wildly as you make Δx smaller if there are several places near the maximum. In other words, it might not converge. It is useless for symbolic computations. Still you might want to try the proof directly on some simple functions.

EXERCISE 11.13. *Write a Mathematica program to compute $M(\Delta x)$. Here is a start:*

```
f[x_] := x - x^3
a = -1;
b = 3;
dx = .1;
partition = Table[x,{x,a,b,dx}]
values = Map[ f[#]&, partition]
Max[values]
Position[values, Max[ values ]]
...
```

11.2.1. Summary of the Theory on Compact Intervals. I am not a candidate for governor. That means that you do not have to consider me in making your vote or in your betting pool. We have two theorems about differentiable functions $f(x)$. One says that interior points of $[a, b]$ cannot be extrema unless $f'(x) = 0$ and the other says that there must be a max and a min. So what's left?

CANDIDATES:

If $f(x)$ is a differentiable function on the real compact interval $[a, b]$, then there is a max and a min amongst the points:
1) Endpoints, $x = a$ and $x = b$
2) Critical points $x = c$ such that $f'(c) = 0$

To find a max and a min, isolate the candidates, examine their record and vote. That is,

ISOLATE THE CANDIDATES:

If you have a differentiable function $f(x)$ to extremize over a compact interval $[a, b]$:

1) Compute $f'(x)$. (Be sure it is defined on all of $[a, b]$.)

2) Find the critical points, that is, all solutions of $f'(c) = 0$ with $a < c < b$.

3) The candidates for extrema are the endpoints and critical points, so compute $f(x)$ for each of these points. It is usually convenient to make a table of these values.

4) Select the largest and smallest values from the values of the function at the candidate points.

It is often helpful to compute the microscopic slope table after you have found all the critical points in step (2). This is not essential when you have endpoints, but shows clearly where the graph is increasing and decreasing. The Extreme Value Theorem guarantees that there will be both a max and a min, while the Interior Critical Point Theorem says that we only need to check critical points and endpoints.

EXAMPLE 11.14. *Find the maximum and minimum of*

$$f(x) = x^3 - 6x^2 + 9x + 1$$

on the interval $[0, 5]$.

SOLUTION: First we isolate the possible candidates. The endpoints are

$$x = 0 \quad \text{and} \quad x = 5$$

The interior critical points are found by first computing $f'(x)$ and then finding all solutions of the equation $f'(x) = 0$.

$$\frac{df}{dx} = f'(x) = 3x^2 - 12x + 9$$
$$= 3(x - 1)(x - 3)$$

The derivative is always defined, so $f(x)$ is continuous and differentiable on $[0, 5]$.

The solutions of $f'(x) = 0$ are

$$3(x - 1)(x - 3) = 0 \quad \Leftrightarrow \quad x = 1 \quad \text{or} \quad x = 3$$

This isolates the candidates, so we compute their values:

Candidate	Value
$x =$	$f(x) =$
0	1
1	5
3	1
5	21

We see from the table that the maximum of f is 21 and that occurs at $x = 5$, $f(5) = 21$. The minimum is 1 and it occurs at two places, $f(0) = f(3) = 1$.

It is instructive to go on and compute the microscopic slope table. We see from the formula $f'(x) = 3x^2 - 12x + 9 = 3(x - 1)(x - 3)$, that $f'(x) > 0$ on $(-\infty, 1)$, $f'(x) < 0$ on $(1, 3)$ and $f'(x) > 0$ on $(3, 5)$

x	0	1		3		5
y	1	5		1		21
y'	╱	─	╲	─	╱	╱

FIGURE 11.15: Slope Table for $f = x^3 - 6x^2 + 9x + 1$

The slope table shows that the left endpoint is a possible min, the critical point $x = 1$ is a possible max, the critical point $x = 3$ is another possible min, and the endpoint $x = 5$ is a possible max. The table of values is needed to decide which possibilities occur. A *Mathematica* graph follows.

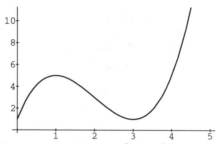

FIGURE 11.16: $y = x^3 - 6x^2 + 9x + 1$

EXERCISE 11.17. *Find the maximum and minimum of each of the following functions over the specified interval.*

(1) $f(x) = x^3 - 3x^2 + 3x + 2, \quad x \in [-2, 2]$

(2) $f(x) = x^3 - 6x + 4, \quad x \in [-3, 3]$

(3) $f(x) = x^3 + 7x + 4, \quad x \in [-10, 10]$

(4) $f(x) = x^3 + 7x + 4, \quad x \in [0, 10]$

(5) $f(x) = \frac{x}{x^2+1}, \quad x \in [-5, 5]$

It might be helpful to see how endpoint conditions arise in a simple application.

EXAMPLE 11.18. *We have a 12 inch square piece of thin material and want to make an open box by cutting small squares from the corners of our material and folding the sides up. The question is, Which cut produces the box of maximum volume?*

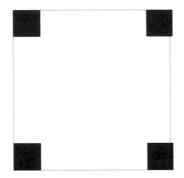

FIGURE 11.19: Max of Volume

First we assign variables

$x =$ the length of the cut on each side of the little squares

$V =$ the volume of the folded box

The length of the base after two cuts along each edge of size x is $12 - 2x$. The depth of the box after folding is x, so the volume is:

$$V = x(12 - 2x)(12 - 2x) = 4x^3 - 48x^2 + 144x$$

Endpoints arise mathematically from the geometry. A cut must have positive length, $0 \le x$, and even $x = 0$ means we have nothing to fold up. Two equal length cuts from 12 inches can each have a maximum length of 6 inches, and again $x = 6$ means we have no base left on our box.

MAXIMIZE

$$V = 4x^3 - 48x^2 + 144x \qquad x \in [0, 6]$$

The two endpoints $x = 0$ and $x = 6$ make $V(x) = 0$. Mathematically, these are minimum points, so of no interest in the application. All we are left with for maximum candidates are interior critical points.

$$\frac{dV}{dx} = 12x^2 - 96x + 144 = 12(x - 2)(x - 6)$$

and

$$\frac{dV}{dx} = 0 \qquad \Leftrightarrow \qquad x = 2 \qquad \text{or} \qquad x = 6$$

There is only one interior critical point, $x = 2$, where $V = 128$. This must be the maximum, because we have already checked all other candidates.

FIGURE 11.20: Folded Gutter

EXERCISE 11.21. *A homeowner has a long strip of 10 inch wide metal and wants to make a rain gutter by folding the sides up to form a rectangular cross section with an open top. Where should she fold in order to get the maximum cross sectional area?*

EXERCISE 11.22.

1) A farmer has 100 feet of fence and wants to make a rectangular holding pen. What dimensions should he make it?

2) Suppose the farmer makes the rectangular holding pen using an existing fence for one side. What dimensions should he use then to maximize the area within his 100 feet of new fence?

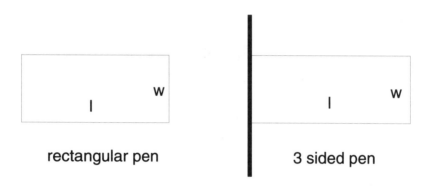

FIGURE 11.23: Holding Pens

EXERCISE 11.24. *A math professor moves to the country and wants to fence some property so that he will enclose a rectangle of exactly 625 square feet of land. He likes fencing, so find the maximum amount of fence he can use in doing this. (What's wrong with this question intuitively and mathematically? Real farmers would have more sense, of course.)*

What are the hypotheses of our max-min procedure?

EXERCISE 11.25. *Find the max and min of*

$$f(x) = \frac{\sqrt{x^2 - \frac{x^4}{2}}}{x} \qquad for \qquad x \in [-1, 1]$$

Be careful and use Mathematica to Plot[.] the graph if you have trouble analyzing the function.

Finding the interior critical points requires that we find all solutions of

$$f'(x) = 0$$

Our textbook exercises are mostly contrived to make the technical part of this problem fairly easy. Factor, use the quadratic formula, something basic usually works. In 'real world' problems this step can be quite difficult, but remember that *Mathematica* has FindRoot[.] and Solve[.] to help you.

11.3. Max - min without Endpoints

Without the hypotheses of the candidates procedure - that we are working with a differentiable function $f(x)$ on a compact intervàl $[a, b]$, there may not be either a max or a min. If the function is not differentiable, it might not be continuous and then there might not be a max or min. If the problem does not have the variable restricted to a compact interval, there may be problems as you approach an endpoint that isn't there.

The moral of this section is that you need to investigate more graphical properties when you have less theory.

EXAMPLE 11.26. *Find the max and min of*

$$f(x) = x + \frac{625}{x}$$

for all positive values, $x \in (0, \infty)$.

SOLUTION We know that we cannot have an interior max or min unless $f'(x) = 0$ or $f'(x)$ does not exist. In this problem

$$f'(x) = 1 - \frac{625}{x^2}$$

which is defined on the open interval $(0, \infty)$. The positive critical value is

$$f'(x) = 0 \qquad \Leftrightarrow \qquad x = 25$$

When $0 < x < 25$, $f'(x) < 0$ and when $x > 25$, $f'(x) > 0$, so the microscopic slope table is down-over-up:

x	0		25			+ infinity
y						
y'	*	⟍	—	⟋	⟋	*

FIGURE 11.27: Slope Table for $y = x + \frac{625}{x}$

The slope table proves that $f(25) = 50$ is a minimum. What about a maximum? There isn't any and now we have to reason directly. We need to know

$$\lim_{x \downarrow 0} f(x) \qquad \& \qquad \lim_{x \to \infty} f(x)$$

because the function increases as we move from 25 in either of these directions. How can we compute these limits? The idea of $\lim_{x \to 0} f(x)$ is to see what happens as x gets smaller and smaller. We cannot just plug in $x = 0$, because the function is undefined. We can take a positive infinitesimal $0 < \delta \approx 0$ and plug that in:

$$f(\delta) = \delta + \frac{625}{\delta} \approx \frac{625}{\delta} = 625\frac{1}{\delta}$$

The reciprocal of a positive infinitesimal is infinitely large and positive, so $f(\delta)$ is positive and infinitely large. In other words, $f(x)$ can be made as large as we please by taking x closer and closer to 0. We write this result

$$\lim_{x \downarrow 0} f(x) = \infty$$

What happens as x gets larger and larger? We write this $x \to \infty$, but recall that ∞ is not an ordinary real or hyperreal number. It is only a symbol to indicate 'larger and larger.' To compute $\lim_{x \to \infty} f(x)$, take an infinitely large positive number such as $H = \frac{1}{\delta}$ and compute

$$f(H) = H + \frac{625}{H} \approx H$$

which is infinitely large again, so

$$\lim_{x \to \infty} f(x) = \infty$$

This function has no maximum, you can make it as large as you please by either taking x near enough to zero or taking x very large. A *Mathematica* graph of $y = f(x)$ follows.

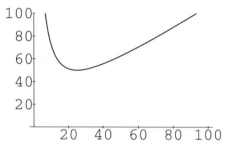

FIGURE 11.28: Min only Attained

Our next example comes up as part of an economic model later in the chapter.

EXAMPLE 11.29. *Find the max and min of*

$$T(p) = (p - 0.5)\left(\frac{1000}{1 + p^2}\right)$$

for $0.5 \le p$ or $p \in [0.5, \infty)$

SOLUTION

The first step is to find the derivative of $T(p) = 500(2p - 1)(1 + p^2)^{-1}$ using the Product Rule and Chain Rule (or quotient rule, if you prefer):

$$T'(p) = 500\frac{d(2p - 1)}{dp}(1 + p^2)^{-1} + 500(2p - 1)\frac{d\left([1 + p^2]^{-1}\right)}{dp}$$

$$= 1000(1 + p^2)^{-1} + 500(2p - 1)(-1)(1 + p^2)^{-2}(2p)$$

$$= 1000\frac{1 + p^2}{(1 + p^2)^2} - 500\frac{(2p - 1)(2p)}{(1 + p^2)^2}$$

$$= (-p^2 + p + 1)\frac{1000}{(1 + p^2)^2}$$

The critical points are where $-p^2 + p + 1 = 0$, so by the quadratic formula,

$$p = \frac{-1 \pm \sqrt{1 + 4}}{-2} = \frac{1}{2} \pm \frac{\sqrt{5}}{2}$$

or $p \approx 1.618$ and $p \approx -0.618$. We are only interested in values $p \in [0.5, \infty)$, so our only critical point is $p = \frac{1}{2} + \frac{\sqrt{5}}{2} \approx 1.618$.

The sign of $T'(p)$ for $0.5 \leq p < 1.618$ is $T'(p) > 0$. The sign for $p > 1.62$ is $T'(p) < 0$, so the microscopic slope table is up-over-down.

p	0.5		1.62			+ infinity
T						
T'	/	/	—	\	\	⋆

FIGURE 11.30: Slope Table for $T = (p - 0.5)(\frac{1000}{1+p^2})$

This proves that the maximum occurs at the critical value, but what about the minimum? At the endpoint $p = 0.5$, $T(0.5) = 0$. At all other values of $p \in [0.5, \infty)$, $T(p) > 0$, so zero is the minimum. What happens as p tends to infinity?

$$\lim_{p \to \infty} T(p) = ?$$

Take $p = P$ infinitely large and estimate

$$T(P) = (2P - 1)\frac{500}{1 + P^2}$$
$$= 500\frac{\left(\frac{2}{P} - \frac{1}{P^2}\right)}{\frac{1}{P^2} + 1}$$

The quantities $\frac{2}{P}$ and $\frac{1}{P^2}$ are infinitesimal when P is infinitely large, so

$$T(P) = 500\frac{\varepsilon}{\delta + 1}$$

for two infinitesimals ε and δ. The product $500\varepsilon \approx 0$ and $\frac{1}{\delta+1} \approx 1$, so

$$T(P) \approx 0 \qquad \& \lim_{p \to \infty} T(p) = 0$$

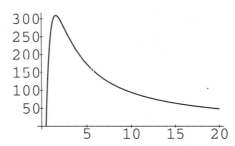

FIGURE 11.31: Min and Max

EXERCISE 11.32. *Find the maximum and minimum if they exist of the following functions for all real x where they are defined. Are the extrema attained or only limiting values?*

a) $f(x) = 3x^4 - 4x^3$

b) $f(x) = x^3 + 3x^2 + 4x + 5$

c) $f(x) = \dfrac{x}{x^2 + 4}$

d) $f(x) = \dfrac{x^2}{x^2 + 4}$

e) $f(x) = \dfrac{x^2}{x^3 + 4}$ f) $f(x) = \dfrac{x^3}{x^4 + 27}$

EXERCISE 11.33. *Find the maximum and minimum of the following functions for all real x where they are defined. Are the extrema attained or only limiting values? Use Mathematica if you have technical difficulties. You should be able to find the shapes, but perhaps not the exact limiting values.*

a) $f(x) = \dfrac{\mathrm{Sin}[x]}{x}$ b) $f(x) = \dfrac{\mathrm{Cos}[x]}{x}$

c) $f(x) = \dfrac{\mathrm{Log}[x]}{x}$ d) $f(x) = x\,\mathrm{Log}[x]$

e) $f(x) = xe^{x}$ f) $f(x) = xe^{-x^2}$

EXERCISE 11.34. *Complete the analysis of the max and min of*

$$f(x) = \dfrac{\sqrt{x^2 - x^4/2}}{x}$$

Where is it defined, what are its max and min, if any, are they attained or only approached?

11.4. Supply and Demand in Economics

In this section we will study simplified monopoly economies. 'Monopoly' means our manufacturer can choose whatever price she wants without fear of competition. Our marketplace will be populated by a large number of individual consumers with no organized cooperation. Their aggregate demand for the product will be determined by the price. If the price is low then the consumers will buy a great deal of the product. But, as the price rises, the demand will fall. Thus, the demand function will be some sort of decreasing function of price.

Our manufacturer wants to set the price that will maximize her total profit. She cannot simply make the price as high as she wishes, because demand will drop. A high price on a few units may well produce less profit than a smaller price on many. On the other hand, a very low price may sell many units, but produce no profit because of the costs of manufacturing.

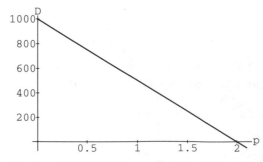

FIGURE 11.35: A Linear Demand Function

Our main simplification is in giving an explicit formula for the number of units of the product that consumers want to purchase or 'demand.' As an example we will begin by working with the linear demand function, $D(p) = 1000 - 500p$.

For this demand function, the maximum possible demand is 1000 units when the product is given away. The maximum price is 2 dollars after which no one wants the product. Thus, we are interested in a compact range of prices $0 \leq p \leq 2$.

Now, suppose that the manufacturing cost (i.e., raw materials and labor) for each unit of the product is 0.20. The unit profit at price p will be $(p - 0.20)$ and the total profit $T(p)$ at the price p will be the number of units sold at the price p multiplied by the profit per unit. In general, for any demand function $D(p)$,

$$T(p) = D(p)(p - \text{cost per item})$$

In this example

$$T(p) = (1000 - 500p)(p - 0.20)$$

This further restricts the range of prices to $0.20 \leq p \leq 2$, because the manufacturer will not sell at a loss. The next figure is a graph of this simple quadratic function.

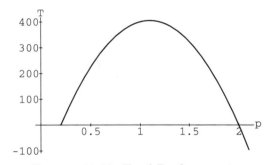

FIGURE 11.36: Total Profit vs. price

Our manufacturer wants to find the price p that maximizes her profit.

$$\text{Max}[T(p) : 0.20 \leq p \leq 2]$$

Notice that she loses money to sell for less than 20 cents and that no one will buy for 2 dollars or more.

She wants to find the maximum of the function $T(p) = (1000 - 500p)(p - 0.20)$ on the interval $[0.20, 2]$. We already know that the endpoints represent points of zero profit or the mathematically necessary, but economically uninteresting minima. The next step is to differentiate the function $T(p)$. (We use the Product Rule rather than expand the expression.)

$$T'(p) = (1000 - 500p)(1) + (p - 0.20)(-500)$$
$$= 1000 - 500p - 500p + 100$$
$$= 1100 - 1000p$$

Next, we solve the equation

$$T'(p) = 0$$
$$1100 - 1000p = 0$$
$$1000p = 1100$$
$$p = 1100/1000$$
$$p = 1.10$$

Thus, $T'(p)$ is zero at $p = 1.10$ and only there. Notice, in addition, that $T'(p)$ is positive if $p < 1.10$ and $T'(p)$ is negative if $p > 1.10$. This means that microscopic slope table is up-over-down, $T(p)$ is increasing from $p = 0.20$ to $p = 1.10$ and decreasing from $p = 1.10$ to $p = 2.00$. Of course, we spoiled this analysis by showing the graph above first. You can look at it to see the slope information. We could have done the slope analysis without the graph - and might need to with a complicated demand function.

We have already seen that $T(0.20) = 0$ (or $T(0) < 0$) and that $T(2.00) = 0$. There is only one critical point inside $(0.20, 2)$ so the Candidates procedure says the max must be at $p = 1.10$, even without the slope analysis or the graph.

The profit maximizing price is 1.10 and at this price the manufacturer will make a total profit of $T(1.10) = 405$.

EXERCISE 11.37. *Suppose that the unit manufacturing cost for the product we have been discussing rises to 0.30 per item. What price should the manufacturer charge to maximize her profit now? How much of the rise in the manufacturing cost is passed on to the consumer?*

EXERCISE 11.38. *Suppose the demand function for another product is $D(p) = 2000 - 500p$. What price should the manufacturer charge to maximize his profit in the following cases?*

  (1) *The cost of producing each item is 0.30.*
  (2) *The cost of producing each item is 0.50.*

How much of the rise in the manufacturing cost is passed on to the consumer?

EXERCISE 11.39. *Suppose the demand function for a third product is $D(p) = 1500 - 100p$. What price should the manufacturer charge to maximize profit in the following cases?*

  (1) *The cost of producing each item is 0.40.*
  (2) *The cost of producing each item is 0.50.*

How much of the rise in the manufacturing cost is passed on to the consumer?

A PARAMETRIC MAXIMUM

You may have noticed an interesting phenomenon in the three exercises above. In each case exactly half of the rise in the unit manufacturing cost was passed on to the consumer. Is it true in general that a manufacturer always passes half of a cost increase to her customers? Let's begin by analysing a general linear demand function. That is, use the demand function

$$D(p) = a - bp$$

where a and b are arbitrary positive constants. The graph of such a function looks like the next figure.

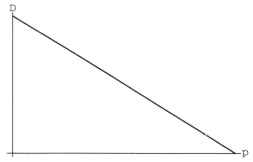

FIGURE 11.40: A General Linear Demand Function

where $D = a$ if $p = 0$ and $D = 0$ at $p = a/b$.

Let the constant c represent the unit manufacturing cost for the product. Thus, the profit per unit would be $(p - c)$ and the total profit would be

$$T(p) = (p - c)D(p) = (p - c)(a - bp)$$

where a, b and c are all positive constants.

Solving the optimization problem with the parameters a and b will tell us more than we would have known by solving only numerical examples. The numerical examples suggested that half the cost increases were passed on in the linear demand case, but the parameter solution will show it. Optimization with parameters often yields important insights beyond those that you can see from special numerical cases. The parameters may be confusing, so that working a numerical case helps to get started, but frequently they are indispensable for gaining scientific insight.

We want to to find the profit maximizing price as a function of the parameters. The Candidates procedure says to check the endpoints and interior critical points. We are to maximize $T(p)$ over the compact interval $[c, a/b]$, but we know that $T = 0$ at the endpoints $p = c$ and $p = a/b$. Hence we calculate

$$\begin{aligned} T'(p) &= (p - c)(-b) + (a - bp)(1) \\ &= -bp + bc + a - bp \\ &= -2bp + bc + a \end{aligned}$$

To find the maximum we set $T'(p) = 0$ and solve

$$\begin{aligned} -2bp + bc + a &= 0 \\ 2bp &= bc + a \\ p &= \frac{bc + a}{2b} \\ &= \frac{a}{2b} + \frac{c}{2} \end{aligned}$$

This price maximizes profit, because some point must by the Extreme Value Theorem and there is only one candidate left. (You could make a slope table, up-over-down, turning at this price.)

This answers our question for any linear demand function. If our first cost is c_1, our first maximizing price is

$$p_1 = \frac{a}{2b} + \frac{c_1}{2}$$

If our second cost is c_2, our second price is

$$p_2 = \frac{a}{2b} + \frac{c_2}{2}$$

and

$$p_2 - p_1 = \frac{a}{2b} + \frac{c_2}{2} - \frac{a}{2b} + \frac{c_1}{2} = \frac{c_2 - c_1}{2}$$

What about nonlinear demand functions?

EXERCISE 11.41. *Suppose that the demand for a particular commodity is given by the function*

$$D(p) = \frac{1000}{1 + p^2}$$

(See Example 11.29 for the max-min analysis of this exercise.) Find the profit maximizing price in the following cases:

(1) *The unit cost of production is $c = 0.50$*
(2) *The unit cost of production is $c = 1.00$*
(3) *The unit cost of production is $c = 1.50$*

EXERCISE 11.42. *In view of the cases of the previous exercise what can you say about our earlier question on how much of a cost of a materials increase is passed to consumers in the nonlinear case?*

In the next chapter we will investigate a dynamic model of price adjustment in an economy where producers do not have a monopoly. Instead, they are willing to produce goods in quantities depending on the price. That introduces the supply side.

11.5. Geometric Max-min Problems

To solve applied max-min exercises:

(1) Read the question and decide on a list of variables - main variables and auxiliary ones - that help in translation. List the variables with units.
(2) Re-read the question and translate each phrase into a statement about your variables.
(3) State the question in terms of your variables. In the case of a max-min problem, try to give a compact interval over which you seek the max or min. Look hard for endpoints, because they make the problem much easier. Unfortunately, it is not always possible to find a compact interval. In that case you have less theory to help you solve your problem and it may not even have a solution.
(4) Apply the max-min theory that you can, computing limiting values and graphs as needed. When you have a compact interval you only need to isolate the candidates as above. In the non-compact case you will have to examine slopes and perhaps even limits.
(5) Interpret your solution.

To solve real world max-min problems, you often need to formulate a clear statement of the question before you can begin to translate the information and state the problem mathematically.

We solved a compact interval max problem, the maximum volume of a folded box, in the section on the Extreme Value Theorem.

The Math Professor who tries farming in Exercise 11.24 is an example of a max-min problem over a non-compact interval. The problem is a little silly to help get your attention and to help you reason intuitively about what is wrong with the optimization question. If the length of the field is l (feet) and the width is w, then the area of 625 makes $l = 625/w$ and the length of the fence $p = 2l + 2w = 2(w + 625/w)$. In other words, we are asking to MAXIMIZE $f(w) = w + 625/w$ for all positive values of w. Of course, this function is not continuous at $w = 0$ and in fact, $f(w)$ tends to infinity as w tends to zero. We solved this in Example 11.26.

There is no maximum, the silly Math Professor can use as much fence as he wants and still only enclose 625 square feet. (One student suggested the solution was to have the math professor move back to town and rent an apartment.) The point is that lack of an endpoint, whether it is at infinity or zero, causes extra mathematical problems. The Extreme Value Theorem does not apply and we need to use additional information from microscopic shape tables and the resulting graphs.

EXERCISE 11.43. *What are the dimensions of the Math Professor's 625 square foot tract when he uses one mile of fence?*

11.5.1. Implicit Differentiation (revisited). Many max-min applications are easier to solve using the method called "implicit differentiation." This was first mentioned in Section 7.2.1. We want to use implicit differentiation to analyse contour graphs of functions of two variables later, anyway, so we introduce the method now in max min problems.

When we found the microscope equation $f(x + \delta x) = f(x) + f'(x)\delta x + \varepsilon \cdot \delta x$ we were comparing the graph of $y = f(x)$ to the local linear equation $dy = f'(x)\,dx$ where x was considered fixed. The linear function (of dx) $L(dx) = f'(x)\,dx$ is called the differential of f (with respect to x), often written $dy = f'(x)\,dx$. Differentials are convenient in many computations. To compute differentials, simply use rules for derivatives and write dx at the end if x is the original variable of differentiation. If u is the variable of differentiation, write du after the $f'(u)$ that you calculate with rules, $df = f'(u)\,du$. The fact that differentials treat variables on an equal footing is what makes them useful in circumstances where derivatives are a little more awkward. We begin with a very simple example of implicit differentiation to remind you of the material from Section 7.2.1.

EXAMPLE 11.44. *Implicit Slope of a Circle*

The circle of radius r (centered at the origin) is the set of (x, y) points satisfying

$$x^2 + y^2 = r^2$$

Here is the implicit method of finding the slope of the tangent: The differential of x^2 is $2xdx$, the differential of y^2 is $2ydy$, but since we are thinking of r as a constant, its differential is zero and

$$2xdx + 2ydy = 0$$

This may be solved for $\frac{dy}{dx}$ as follows.

$$2x\,dx + 2y\,dy = 0$$
$$x\,dx + y\,dy = 0$$
$$y\,dy = -x\,dx$$
$$dy = -\frac{x}{y}dx$$

$$\frac{dy}{dx} = -\frac{x}{y}$$

Because we have both x and y in the formula for the slope, we need to use the equation together with this formula in practice. For example, the point $(3,4)$ lies on the circle of radius 5, so the slope of the circle at that point is

$$\frac{dy}{dx} = -\frac{3}{4}$$

If we were only told: Find the slope at $x = 3$ on the circle of radius 5 centered at the origin, we would first have to compute the corresponding y value. The original implicit equation $x^2 + y^2 = 5^2$ is not set up to do this directly, but of course, can be made explicit (two ways)

$$x^2 + y^2 = r^2$$
$$y^2 = r^2 - x^2$$
$$y = \pm\sqrt{r^2 - x^2}$$

Let's take the positive solution and differentiate the explicit function for comparison. We have $y = \sqrt{u} = u^{\frac{1}{2}}$ with $u = r^2 - x^2$ and r constant. Thus,

$$y = \sqrt{r^2 - x^2} \qquad\qquad y = u^{1/2} \qquad\qquad u = r^2 - x^2$$
$$\frac{dy}{du} = \frac{1}{2}u^{-1/2} \qquad\qquad \frac{du}{dx} = -2/,x$$

$$\frac{dy}{dx} = \frac{dy}{du} \cdot \frac{du}{dx}$$
$$= \frac{1}{2\sqrt{u}}(-2\,x)$$
$$= \frac{-x}{2\sqrt{r^2 - x^2}} = \frac{-x}{y}$$

So it's a case of pay me now or pay me later, but differentiation is easier when we pay the algebraic penalty later.

EXERCISE 11.45. *Implicit Differentiation Drill*
Find the total differential and solve for $\frac{dy}{dx}$

a) $x^2 - y^2 = 3$
b) $y + \sqrt{y} = \frac{1}{x}$
c) $x\,y = 4$

d) $y = \text{Cos}[x\,y]$
e) $\text{Sin}[x]\,\text{Cos}[y] = \frac{1}{2}$
f) $y = \text{Sin}[x + y]$

g) $y = e^{xy}$ h) $e^x e^y = 1$ i) $e^x = x + y^2$

j) $x = \text{Log}[xy]$ k) $y = \text{Log}[x^2 y]$ l) $x = \text{Log}[x + y]$

11.5.2. The Distance Formula.

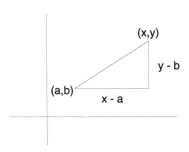

FIGURE 11.46: Coordinate Distance

Recall that the formula for the distance between the points with coordinates (x, y) and (a, b) is

$$\text{dist} = \sqrt{(x - a)^2 + (y - b)^2}$$

This is simply a coordinate expression of the Pythagorean theorem that the length of the hypotenuse is the square root of the sum of the squares of the lengths of the two legs of a right triangle. (English is an awkward way to express this.)

Solving the systems of equations that arise in max min problems can become quite technical, but don't forget the computer.

EXAMPLE 11.47. *Help from Mathematica*

Find the points on the ellipse

$$(\frac{x}{2})^2 + (\frac{y}{3})^2 = 1$$

that are nearest and farthest from the point $(1, 1)$.

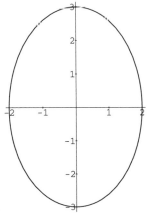

FIGURE 11.48: $(\frac{x}{2})^2 + (\frac{y}{3})^2 = 1$

SOLUTION

The distance is given by

$$\text{dist} = \sqrt{(x-1)^2 + (y-1)^2}$$

but it is easier to extremize the square of the distance,

$$D = (x-1)^2 + (y-1)^2$$

The differential of the squared distance is

$$dD = 2(x-1)dx + 2(y-1)dy$$

and the differential of the constraint equation $(\frac{x}{2})^2 + (\frac{y}{3})^2 = 1$ is the equation

$$2\left(\frac{x}{2}\right)^{2-1}(\frac{1}{2})dx + 2\left(\frac{y}{3}\right)^{2-1}(\frac{1}{3})dy = 0$$

or

$$\frac{1}{2}xdx + \frac{2}{9}ydy = 0$$

so

$$9xdx + 4ydy = 0$$

and

$$dy = -\frac{9x}{4y}dx$$

We substitute this expression for dy into the differential dD obtaining

$$dD = 2(x-1)dx + 2(y-1)\left(-\frac{9x}{4y}\right)dx$$

so

$$\frac{dD}{dx} = \frac{4y(x-1) - 9x(y-1)}{2y}$$

and we have $\frac{dD}{dx} = 0$ when the numerator is zero. Since the numerator also involves y because we used implicit differentiation, we must also satisfy the original ellipse equation $(\frac{x}{2})^2 + (\frac{y}{3})^2 = 1$. In *Mathematica* syntax,

$$\text{Solve}[\{4y(x-1) - 9x(y-1) == 0, (x/2) \wedge 2 + (y/3) \wedge 2 == 1\}]$$

Mathematica finds the simultaneous solutions to be

$$(x, y) \approx (1.82, 1.25) \qquad \text{and} \qquad (-.494, -2.91)$$

[and two complex roots].

EXERCISE 11.49. *Find the points on the unit circle $x^2 + y^2 = 1$ nearest and farthest from* $(1, 1)$. *Use implicit differentiation and check your answer with common sense geometric reasoning.*

Sometimes we can find extrema without calculus and it is a good idea for you to compare a simple example with several methods.

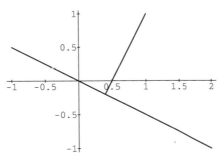

FIGURE 11.50: The Distance from a Point to a Line

PROBLEM 11.51. *A Geometric Minimum*

(1) *We have a line L and a point p not on the line. Draw an example. The point q on the line L that is closest to p is the one such that the segment pq is perpendicular to L. Add the segment to your figure. Any other point r on L must be farther from p than q. Use basic geometry to show why. Use the Pythagorean Theorem to show that any other point r on L is farther from p than q is.*

(2) *Suppose the line is given by the equation $y = -\frac{1}{2}x$ and the point has coordinates $(1, 1)$. Find the coordinates of the point q. Hints: The equation of the line through p and q has slope 2, the negative reciprocal of the slope of L and passes through $(1, 1)$. Find the equation. Make the algebraic statement that $q = (x, y)$ lies on both lines and solve.*

(3) *Use calculus to minimize the square of the distance from $(1, 1)$ to a point $r = (x, y)$ on the line $y = -\frac{1}{2}x$.*

EXERCISE 11.52. *Electronic Minima*
Find the point closest to $(x, y) = (1, 1)$ on each branch of the hyperbola

$$\left(\frac{x}{2}\right)^2 - \left(\frac{y}{3}\right)^2 = 1$$

Mathematica says:

In:
Solve[$\{4y(x - 1) + 9x(y - 1) == 0, (x/2)^2 - (y/3)^2 == 1\}$]
Out:
???
In:
N[%]
Out:
$(x, y) = (2.07215, 0.813035)$ & $(x, y) = (-2.03981, 0.601566)$
(Well, that's what it means!)
and 4 complex roots.

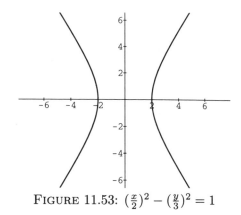

FIGURE 11.53: $(\frac{x}{2})^2 - (\frac{y}{3})^2 = 1$

Many things vary as only a function of the distance to an object. The intensity of radiation is proportional to one over the square of the distance from the source. Hence, it falls off fast as you move away from the source. For example, the apparent brightness of a planet being observed by a space probe is given by

$$I = k\frac{1}{D}$$

where $D = \left(\sqrt{(x-a)^2 + (y-b)^2}\right)^2 = (x-a)^2 + (y-b)^2$.

EXERCISE 11.54. *Find the time t and the place (x, y) when a planet at $(a, b) = (0, 1)$ appears brightest for the following rocket trajectories. (First express the simplest equivalent mathematical question you can. The parabolic trajectories correspond to different primary missions.)*

(1) $x = t - 1, \quad y = 0$
(2) $x = t - 1, \quad y = 2x^2 = 2(t-1)^2$
(3) $x = t - 1, \quad y = \frac{1}{3}x^2 = \frac{1}{3}(t-1)^2$

Sometimes it is not so easy to see where the endpoint conditions enter a problem. Here is another constrained maximization problem to illustrate this.

EXAMPLE 11.55. *Hidden Endpoints A farmer wants to build a silo with cylindrical sides and a hemispherical top. He has a fixed budget of 10,000 dollars. The sides cost 3 dollars per square foot, but the top costs 9 dollars per square foot because of the labor and materials used in making it spherical. What proportions should he make the silo?*

SOLUTION

The radius r and the height h (measured in feet) are shown on the figure:

FIGURE 11.56: Silo

The area of the cylindrical side is $A_{side} = 2\pi r h$, so the cost of the side is $C_{side} = 6\pi r h$. The area of the hemispherical top is $A_{top} = 2\pi r^2$ and it's cost is $C_{top} = 18\pi r^2$. The sum is the total cost, which we set equal to 10,000 dollars $6\pi r h + 18\pi r^2 = 10000$ and divide both sides by 6π to obtain

$$rh + 3r^2 = \frac{10000}{6\pi}$$

whose differential yields the equation

$$hdr + rdh + 6rdr = (6r + h)dr + rdh = 0$$

and

$$rdh = -(6r + h)dr$$

The volume of the cylindrical part of the silo is $V_{cyl} = \pi r^2 h$ and the volume of the hemispherical top is $V_{top} = \frac{2}{3}\pi r^3$, so the total volume is

$$V_{total} = \pi r^2 h + \frac{2}{3}\pi r^3$$

with differential

$$dV_{total} = \pi \left(2rhdr + r^2 dh + \frac{2}{3}3r^2 dr \right)$$
$$= \pi \left(\left(2rh + 2r^2 \right) dr + rrdh \right)$$

substituting $rdh = -(6r + h)dr$ we obtain

$$dV_{total} = \pi \left(rh - 4r^2 \right) dr$$

or

$$\frac{dV_{total}}{dr} = \pi r \left(h - 4r \right)$$

The critical values, $\frac{dV_{total}}{dr} = 0$ occur at $r = 0$ and $h = 4r$. The answer to the question is that the height should be four times the radius.

This method obscures the role of endpoints somewhat, because it does not make the dependence between h and r explicit. We can reason that there are "endpoints" for r if we choose it as the independent variable, however. First, if $r = 0$, the silo has no volume. The limit of V_{total} as r tends to zero should be zero, but this is not completely obvious because h must tend to infinity in order to hold the cost at 10,000 dollars. However, the microscope comes to our rescue again. The derivative $\frac{dV_{total}}{dr} = \pi r\,(h - 4r)$ is positive when r is small and positive and h is large and keeping the cost fixed. This means V_{total} increases for small values of r and cannot tend to a larger value than the one we found. (Can you write V_{total} explicitly in terms of r and find the limit of V_{total} as t tends to zero?)

At the other extreme, if $h = 0$, so the silo is a hemisphere, you can still spend all the money. Without solving, we can see that there is a maximal value of the radius.

EXERCISE 11.57. *A farmer wants to enclose a rectangular field whose area is 100 square feet with the least possible amount of fencing. What should the dimensions of the field be?*

EXERCISE 11.58. *A manufacturer wants to design a cylindrical can that will hold one liter of liquid (1 liter is 1000.0028 cubic centimeters or 0.02838 bushels). The total area of the outside of the can will be the sum of two disks at the top and bottom plus the cylindrical side. Express this in terms of radius and height. What should the dimensions of the can be in order to minimize the total area of the can and, hence, the cost of materials?*

11.6. Max-min with Parameters

Often scientific max-min problems ask how the maximum depends on some parameter. Wein's Law is derived from Planck's Law of radiation this way at the beginning of Chapter 8. Sharing half the cost of a price increase for linear demand was proved with parameters in economics section of this chapter. Here are some simple exercises of this type. Note that they are the general cases of the previous exercises. We will see other important examples of the use of parameters in the following sections.

EXERCISE 11.59. *A manufacturer wants to design a cylindrical can that will hold V cubic centimeters. The total area of the outside of the can will be $A = 2\pi r^2 + 2\pi rh$ where r is the radius of the cylinder and h is its height. What should the dimensions of the can be in order to minimize the total area of the can and, hence, the cost of materials?*

EXERCISE 11.60. *A farmer wants to enclose a rectangular field whose area is A square feet with the least possible amount of fencing.*

(1) *What should the dimensions of the field be?*
(2) *What is the minimal perimeter in terms of the parameter A?*
(3) *If p denotes the perimeter of a rectangular field of any dimensions whose area is A, prove the inequality*

$$4\sqrt{A} \leq p \qquad \text{with equality only for a square}$$

(4) *What is the largest area A of a field whose perimeter is fixed at a constant p?*

The arithmetic mean is the usual average

$$\frac{a + b}{2}$$

The geometric mean

$$(a \times b)^{\frac{1}{2}}$$

is a sort of 'multiplicative average' of positive numbers. Also notice that if $\alpha = \text{Log}[a]$ and $\beta = \text{Log}[b]$, then $\frac{\alpha+\beta}{2} = \frac{1}{2}(\text{Log}[a] + \text{Log}[b]) = \text{Log}[(a \times b)^{\frac{1}{2}}]$.

PROBLEM 11.61. *The Arithmetic and Geometric Means*

(1) *Use high school algebra to prove*

$$\sqrt{ab} \leq \frac{a+b}{2}$$

by squaring both sides and doing some algebra to put the inequality in the form

$$0 \leq a^2 - 2ab + b^2 = (a-b)^2$$

(2) *For which positive values of a and b is*

$$\sqrt{ab} = \frac{a+b}{2}$$

(3) *The inequality $\sqrt{ab} \leq \frac{a+b}{2}$ is equivalent to*

$$a \leq \frac{1}{b}\left(\frac{a+b}{2}\right)^2$$

Treat a as a constant and use calculus to minimize the function

$$f(b) = \frac{1}{b}\left(\frac{a+b}{2}\right)^2$$

and show that the minimum is a when $b = a$.

(4) *Use calculus to prove*

$$(abc)^{\frac{1}{3}} \leq \frac{a+b+c}{3}$$

for positive numbers, with equality only if $a = b = c$. First, this inequality is equivalent to

$$ab \leq \frac{1}{c}\left(\frac{a+b+c}{3}\right)^3$$

so treat a and b as parameters and minimize the function

$$f(c) = \frac{1}{c}\left(\frac{a+b+c}{3}\right)^3$$

showing that the minimum is greater than ab, unless $a = b = c$ when you get equality.

11.6.1. Resonance of a Linear Oscillator. Resonance of a vibrating system is a maximal response to forcing. Many old cars hum loudly at a speed like 46 mph, but are quieter at both slower and faster speeds. The vibration at 46 mph is maximal, at least for an interval of frequencies. (A local maximum.) You can hear the resonance.

Cars with worn out shock absorbers also oscillate. Next semester we will show that an idealized car's front end consisting of a spring of constant s, a shock absorber with damping constant c and mass m when forced with

$$\mathrm{Sin}[\omega t]$$

the pure sinusoid of frequency ω, has a response of amplitude

$$A(\omega) = \sqrt{\frac{1}{(s - m\omega^2)^2 + (c\omega)^2}}$$

EXERCISE 11.62. *Resonant Frequency of a Linear Oscillator*

Find the maximum of $A(\omega)$ for positive forcing frequencies ω. To simplify the problem, minimize the square of the denominator

$$B(\omega) = (s - m\omega^2)^2 + (c\omega)^2$$

by showing that it has critical values at zero and the roots of a simple quadratic equation.

The idealized front end will oscillate on its own at a certain 'natural frequency' when $4ms > c^2$. You do not get an interior maximum unless you have the condition $2ms > c^2$. Why is that?

The natural frequency is what you observe when a car with worn shocks goes over a bump and then back onto smooth (un-forced) pavement. The car keeps bouncing up and down for a long time. This natural frequency is

$$\omega_N = \sqrt{\frac{s}{m} - \frac{c^2}{4m^2}}$$

How does this compare to the maximal amplitude frequency (where B is minimized, $\omega_M = \sqrt{\frac{s}{m} - \frac{c^2}{2m^2}}$)? In particular, what happens in the particular cases

m	c	s
5	4	$\dfrac{5}{4}$
5	4	5

How do the frequencies compare when c is very small, that is, in the limit as your shocks become completely useless? What happens to the amplitude $A(\omega)$ in this limit?

11.7. Max-min in S-I-R Epidemics

The epidemic model of chapter 2 is given by the system of differential equations

(S-I-R DE's)
$$
\begin{aligned}
\frac{ds}{dt} &= -a\,s\,i \\
\frac{di}{dt} &= a\,s\,i - b\,i \\
r &= 1 - s - i
\end{aligned}
$$

and the values of the fractions of susceptibles, infectives and removed at the start of the epidemic. For example, we studied the case $s(0) = \frac{2}{3}$ and $i(0) = \frac{1}{30}$. Typically, the initial $i(0)$ is small and $s(0)$ is large, but not 1 unless no one has ever had the disease before.

We never found explicit formulas for the functions $s(t)$ and $i(t)$ and we want to make the point now that you can find max-min information anyway.

EXERCISE 11.63. *Prove that the function $s(t)$ is a decreasing function on $[0, \infty)$. What calculus criterion is needed?*

There are two 'peaks' of interest in an epidemic. (1) When is the disease spreading fastest? and (2) When are the most people sick?

PROBLEM 11.64. *The Peak of an S-I-R Epidemic*

(1) *When is the epidemic expanding in the sense that the number of sick people is increasing? In particular, if you are thinking in terms of absences from class, when does the epidemic 'peak?' Write your condition using a derivative and then express your answer in terms of s and the contact number $c = \frac{a}{b}$. Show that the variable increases before your condition and decreases afterward. Consider various cases of c and initial conditions in s and i including 'extreme' cases such as c very big and very small or $s(0)$ very big or very small. (Note: $c = 15$ for measles and $c = 4.6$ for polio. Nearly everyone is susceptible if $s(0) = 0.9$.)*

(2) *When is the disease spreading fastest in terms of the growth of new cases? Remember that $a\,s\,i$ is the rate of change of new cases. If we want to maximize this expression, we need to find zeros of its derivative*

$$
\begin{aligned}
\frac{d(s \cdot i)}{dt} &= \frac{ds}{dt} \cdot i + s \cdot \frac{di}{dt} \\
&= (-a\,s\,i)\,i + s\,(a\,s\,i - b\,i)
\end{aligned}
$$

Show that this is zero when $s - i = 1/c$.

(3) *When is the disease spreading fastest in terms of the growth of infectives? In other words, maximize $\frac{di}{dt} = a\,s\,i - b\,i$. Show that the critical point condition is:*

$$0 = \frac{d(a\,s\,i - b\,i)}{dt}$$

$$= a\,\frac{ds}{dt}\,i + a\,i\,\frac{di}{dt} - b\,\frac{di}{dt}$$

$$= -a^2\,s\,i^2 + a\,s^2\,i - a\,b\,s\,i - a\,b\,s\,i + b^2\,i$$

$$= a^2\,i\left(s^2 - s\,i - \frac{2}{c}\,s + \frac{1}{c^2}\right)$$

This equation is hard to analyse, but we can use the invariant from Chapter 2,

$$s + i - \frac{1}{c}\,\text{Log}[s] = a\ constant = k$$

Discuss various cases of the value of c and the initial values of s and i.

In Mathematica we can type:

```
dmax := s∧2 - s i - 2 s/c + 1/c∧2
invar := s + i + Log[s]/c;
c = 4.6; (*Polio*)
s = 0.75; i = 0.0; k = invar;
Print["Contact number c =",c," Invariant k = ",k]
Clear[s,i];
plot1 = ContourPlot[ dmax , { s, 0, 1},{ i, 0, 1},
            Contours -> { 0.0}, ContourShading -> False ];
plot2 = ContourPlot[ invar , { s, 0.01, 1},{ i, 0, 1},
            Contours -> { k}, ContourShading -> False ];
Show[ plot1, plot2 ];
```

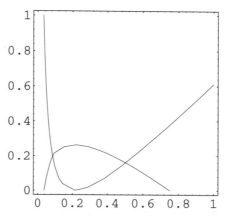

FIGURE 11.65: Intersection of the Critical Equation and Invariant

```
FindRoot[{ invar == k, dmax == 0 } ,{ s,0.5} ,{ i,0.15} ]
FindRoot[{ invar == k, dmax == 0 } ,{ s,0.1} ,{ i,0.2} ]
```

11.8. Projects

11.8.1. Optics and Least Time. Many interesting basic results in optics can be proved using minimization. Fermat's Principle says that light travels along the path that requires the least time. From this and max-min theory we can show that light reflects off mirrors at equal angles and is refracted according to Snell's Law which can be expressed:

$$\frac{\text{Sin}[\alpha]}{\text{Sin}[\beta]} = \frac{u}{v}$$

Using the computer, we can add many beautiful pictures showing light focusing at a point when reflecting off a parabolic mirror or creating a spherical aberration, or light caustic, when reflecting off a spherical mirror.

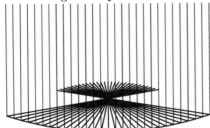 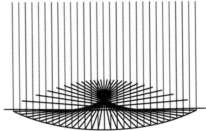

FIGURE 11.66: Parabolic and Spherical Mirrors

11.8.2. Monopoly Pricing. Optimal pricing is studied in a model of a monopoly which can charge different kinds of customers different prices. This is based on cable TV prices for separate homes and apartments.

CHAPTER 12

Discrete Dynamical Systems

Dynamical systems are mathematical models of 'how things move.' The motion of the bodies in the solar system comes to mind as a physical dynamical system. The moons and planets move in fairly complicated ways around the sun and one another, each exerting forces on the others. Mathematical dynamical systems are broadly applicable to other kinds of 'dynamics.' We will study discrete and continuous dynamical systems in this course. This introduction tries to answer the questions: What is a practical mathematical meaning of 'dynamical system?' What does 'discrete' or 'continuous' refer to?

In the first section we will study two economic models of price adjustment where the price is determined in discrete steps, $p[1]$, $p[2]$, $p[3]$, $\cdots$. The price at day $t+1$ is computed from the price at day t by an equation for the change in price. Time t moves in discrete steps, $t = 0, 1, 2, \cdots$. The change in the price is $p[t+1] - p[t]$ and what drives this change in our first model is the 'excess demand' or the amount by which demand exceeds supply.

$$p[t+1] - p[t] = k\{D(p[t]) - S(p[t])\}$$
$$= \alpha p[t] + \beta \quad \text{in the linear case}$$

Once we know an initial price $p[0]$, we can rewrite the change equation in the form of a recursive definition for the whole sequence with the unknown on one side and known values on the other,

$$p[t+1] = p[t] + \alpha p[t] + \beta$$
$$\text{so if}$$
$$p[0] = \text{given}$$
$$p[1] = p[0] + \alpha p[0] + \beta$$
$$p[2] = p[1] + \alpha p[1] + \beta$$
$$p[3] = p[2] + \alpha p[2] + \beta$$
$$p[4] = p[3] + \alpha p[3] + \beta$$

$$\vdots$$

201

The phrase 'recursive definition' means that in order to compute $p[4]$, you first must compute $p[1]$, $p[2]$ and $p[3]$. An equation like the ones above of the form

$$p[t+1] - p[t] = f(p[t])$$

is called an autonomous difference equation of first order, but discrete dynamical systems are more than just the difference equation. 'Autonomous' refers to the fact that the function $f(p)$ only depends on the price at time t and not the time itself. 'First order' refers to the fact that one time determines the next. (Whales take 9 years to produce babies, so the model of a whale population given in the Scientific Projects has order 9.)

The 'dynamics' in our economic models comes down to the question: Do prices tend to a limiting value or does the economy oscillate or 'blow up?' This really entails two mathematical questions. First, how the behavior of the sequence of prices depends on the initial price. Second, how the behavior - limiting or oscillating - depends on the parameters in the difference equation.

A practical working definition of 'discrete dynamical systems' then is the study of the behavior of solutions to a difference equation as a function of the initial condition $p[0]$ and the equation itself. A single system (of first order) is the function that has as input the initial condition $p[0]$ and as output the whole solution sequence $\{p[0], p[1], p[2], \cdots\}$

$$p[0] = P_0$$
$$p[t+1] = p[t] + f(p[t])$$
$$\rightarrow \qquad \{p[0], p[1], p[2], \cdots\}$$

Naturally, we will want to seek ways to summarize this infinite output such as by saying prices tend to an equilibrium value.

A continuous (first order, autonomous) dynamical system is the function that assigns a solution function $y[t]$ to an initial value by means of an equation that says how the quantity changes in continuous time. The rate of change of y is the derivative, so the analog of the difference equation is a formula for the rate of change

$$\frac{dy}{dt} = f(y[t])$$

or for the differential (hence the name 'differential equation')

$$dy = f(y[t]) \, dt$$

The whole continuous dynamical system becomes the function that assigns the solution function $y[t]$ for $t \in [0, \infty)$ to the initial condition through the differential equation

$$y[0] = Y_0$$
$$dy = f(y[t]) \, dt$$
$$\rightarrow \qquad y[t], \quad t \in [0, \infty), \qquad \text{with } y[0] = Y_0$$

Since the change in a differentiable function in a small time step δt is

$$y[t + \delta t] - y[t] = dy + \varepsilon \delta t$$
$$y[t + \delta t] - y[t] = f(y[t]) \, \delta t + \varepsilon \delta t$$

with $\varepsilon \approx 0$, or

$$y[t + \delta t] = y[t] + dy + \varepsilon \delta t$$
$$y[t + \delta t] = y[t] + f(y[t])\ \delta t + \varepsilon \delta t$$
$$y[t + \delta t] \approx y[t] + f(y[t])\ \delta t$$

a continuous dynamical system can be approximated on the computer by a discrete dynamical system that moves in steps of size δt

$$y[0] = Y_0$$
$$y[t + \delta t] \approx y[t] + f(y[t])\ \delta t$$
$$\longrightarrow \qquad \{y[0], y[\delta t], y[2\delta t], y[3\delta t], \cdots\}$$

This is called 'Euler's method' of approximating solutions of differential equations.

We have seen this connection before in our study of S-I-R epidemics in Chapter 2 and in the theory of natural logs and exponentials in Chapter 8. Now we will study the discrete case more systematically and later we will study the continuous case.

12.1. Two Models for Price Adjustment by Supply and Demand

In the last chapter we studied the maximum profit of a monopoly producer, but now we want to open up the production side of our model. The quantity of a product that producers are willing to make and the quantity that consumers are willing to buy both depend on the price of the product. The intuition is that producers will make more when the price of their product is high, but the consumers will buy less at a higher price. The quantity consumers demand $q = D(p)$ is a decreasing function of price and the quantity producers are willing to supply $q = S(p)$ is an increasing function of price. We will begin with linear supply and demand,

$$q = S(p) = a_S p - b_S$$
$$q = D(p) = b_D - a_D p$$

The next plot shows supply and demand when

$$a_S = 1,000$$
$$b_S = 400$$
$$a_D = 500$$
$$b_D = 1,000$$

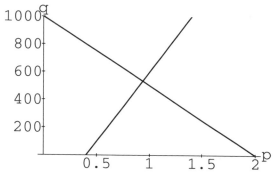

FIGURE 12.1: Linear Supply and Demand

In this example, the supply and demand curves cross when $p = 0.93$. This is the solution of the equation that says "supply equals demand."

$$1000p - 400 = 1000 - 500p$$
$$1500p = 1400$$
$$p = \frac{14}{15}$$
$$p \approx 0.93$$

12.1.1. Price Adjustment. What happens if supply is not equal to demand? There is more than one approach to this question, so we need more economic assumptions to formulate an answer. Suppose that we are considering a fast, but not continuous adjustment. One of the Project proposals asks you to explore a model of a baking economy. Each morning bakers make 'Byties' and sell them at the prevailing price. If the demand is very high, consumers buy them all up early in the day and suppliers respond the next day with more byties - and a higher price. (Since we assume that the aggregate of producers supplies strictly in accordance with the price.)

On the other hand, if today's Byties are selling at too high a price for consumer demand, some will go un-bought, spoil and be thrown out. Producers respond tomorrow by baking fewer Byties - and selling at a lower price. If things work nicely, prices tend to a place where producers and consumers agree. On the other hand, it is conceivable that daily adjustments are extreme. One day Byties are left over, so supplies and prices drop a lot and the next day everything is sold out by 8:00 am. Producers raise production and prices too much and consumers respond by not buying very many. Next, prices drop too far again...

How can we capture the economic dynamics of this situation and develop mathematical criteria for 'stability' of our economy? We can make the simplest possible assumption about how prices change, namely that the change in price is proportional to the amount by which demand exceeds supply,

$$p[t + 1] - p[t] = k(D(p) - S(p))$$

Notice that when demand exceeds supply $D[p] - S[p]$ is positive and the new price is higher than the old one (assuming k is positive). On the other hand, if supply exceeds demand, $D[p] - S[p]$ is negative and the new price is lower than the old one. In our specific linear supply and demand model, this gives us the dynamic model

$$p[0] = P_0$$
$$p[t+1] = p[t] + k((b_D + b_S) - (a_D + a_S)p[t])$$
$$\rightarrow \quad \{p[0], p[1], p[2], \cdots\}$$

or simply

$$p[0] = P_0$$
$$p[t+1] = p[t] + \alpha p[t] + \beta$$
$$\rightarrow \quad \{p[0], p[1], p[2], \cdots\}$$

with $\alpha = -k(a_D + a_S)$ and $\beta = k(b_D + b_S)$ Using the values above for a_D, etc. and $k = 0.0001$, beginning at a price of $p[0] = 0.20$ we obtain a solution whose graph is

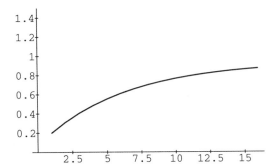

FIGURE 12.2: Prices Tend to Equilibrium from Below

With an initial price of $p = 1.20$ the solution looks like

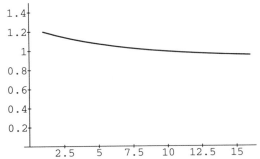

FIGURE 12.3: Prices Tend to Equilibrium from Above

The dynamics in this case is that the economy adjusts production so that the price tends to equilibrium.

Suppose we want to adjust prices faster by making $k = 0.00135$? The initial price $p = 0.20$ produces the following solution

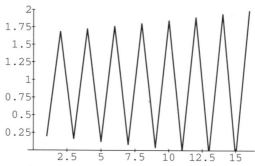

FIGURE 12.4: Bytie Prices Go Ape

   Stability of our price adjustment model apparently depends on the price adjustment constant k or the mathematical parameters α and β. We want you to explore this in an exercise below.

   12.1.2. Supply Adjustment. Another possible adjustment mechanism is suggested by the following different kind of economy. Farmers decide in the spring how much corn they will plant. Of course, each farmer does this separately, but the aggregate response is roughly an increasing supply for larger prices. In the fall after harvest (and all the uncertainties of farming which our model neglects), farmers sell in a market that adjusts prices very rapidly. For our purposes, suppose that they sell at the equilibrium price where supply equals demand. They are stuck with the price, but in spring, they can adjust supply by planting more or less. Again, for simplicity we suppose that supply and demand are linear

$$q = S(p) = a_S \cdot p - b_S$$
$$q = D(p) = b_D - a_D \cdot p$$

A simple equilibrium is the price p such that $S(p) = D(p)$

$$a_S \cdot p - b_S = b_D - a_D p$$
$$(a_S + a_D)p = b_D + b_S$$
$$p = \frac{b_D + b_S}{a_D + a_S}$$

but the dynamics of this model comes from the different times when decisions are made. Producers set supply by last year's prices

$$s[t + 1] = S(p[t])$$

whereas consumers decide on purchases by this year's prices

$$d[t + 1] = D(p[t + 1])$$

The equilibrium criterion becomes

$$s[t+1] = d[t+1]$$
$$a_S p[t] - b_S = b_D - a_D p[t+1]$$
$$p[t+1] = -\frac{a_S}{a_D} p[t] + \frac{b_S + b_D}{a_D}$$
$$p[t+1] = p[t] - (1 + \frac{a_S}{a_D}) p[t] + \frac{b_S + b_D}{a_D}$$

Notice that our basic difference equation is of the same mathematical form as in the first model, but that the economics is different, so your experiments will lead to different conjectures economically. In both models, the next price is an affine linear function of the current price $p[t+1] = p[t] + \alpha p[t] + \beta$ where $\alpha = -k(a_D + a_S)$ in the first economy and $\alpha = -(1 + \frac{a_S}{a_D})$ in the second. Some general mathematical results about the stability of affine linear dynamical systems

$$p[0] = P_0$$
$$p[t+1] = p[t] + \alpha p[t] + \beta$$
$$\rightarrow \quad \{p[0], p[1], p[2], \cdots\}$$

would apply to both of our economies (and several other applications.) For example, the general equation is stable when $\alpha = -\frac{3}{2}$ as well as other cases. This case means that the (mythical) Bytie economy is stable when $-\frac{3}{2} = -k(a_D + a_S) = -k(1500)$ or $k = 0.001$ and the (simplified) Corn economy is stable in the case $\frac{3}{2} = 1 + \frac{a_S}{a_D}$ or $a_S = \frac{a_D}{2}$. We would really like some general rules to tell us when we have stability in terms of the parameters of our various models. A mathematical simplification results if we first understand just the role of two parameters α and β.

EXERCISE 12.5. *Stability of* $p[t+1] = p[t] + \alpha p[t] + \beta$
Experiment with the FirstDynSys.ma NoteBook to formulate a conjecture about the role of α and β in determining both the stability and limiting value of the general affine linear dynamical system given above. Consider the cases:

(1) $\alpha = -0.2$, $\beta = 100$, $P_0 = 100, 500, 1000$
(2) $\alpha = +0.2$, $\beta = 100$, $P_0 = 100, 500, 1000$
(3) $\alpha = -0.2$, $\beta = 50$, $P_0 = 100, 500, 1000$
(4) $\alpha = +0.2$, $\beta = 50$, $P_0 = 100, 500, 1000$
(5) $\alpha = -0.1$, $\beta = 100$, $P_0 = 100, 500, 1000$
(6) $\alpha = +0.1$, $\beta = 100$, $P_0 = 100, 500, 1000$
(7) $\alpha = -1.1$, $\beta = 110$, $P_0 = 100, 500, 1000$
(8) $\alpha = -2.0$, $\beta = 200$, $P_0 = 100, 500, 1000$
(9) $\alpha = -2.2$, $\beta = 220$, $P_0 = 100, 500, 1000$

Apply the results of your conjectures to make predictions about the Bytie and Corn economy models above, that is, re-formulate your conjectures in terms of the parameters k, a_D, a_S, b_D, b_S, in both cases.

12.2. Function Iteration, Equilibria and Cobwebs/indexequilibrium, discrete

There is a helpful graphical device to use in computing iterates of a discrete dynamical system. The resulting figures look something like cobwebs, although they have nothing to do with real spiders.

We will program *Mathematica* to compute the sequences of a discrete dynamical system by iteration of a function. The general dynamical system

$$p[0] = P_0$$
$$p[t + 1] = p[t] + f(p[t])$$

recursively generates the sequence $p[0]$, $p[1] = p[0] + f(p[0])$, $p[2] = p[1] + f(p[1])$, $p[3] = p[2] + f(p[2])$, $\cdots$, but written in terms of $p[0]$ is $p[1] = p[0] + f(p[0])$, $p[2] = \{p[0] + f(p[0])\} + f(p[0] + f(p[0]))$, $p[3] = [\{p[0] + f(p[0])\} + f(p[0] + f(p[0]))] + f(\{p[0] + f(p[0])\} + f(p[0] + f(p[0])))$, $\cdots$. This looks pretty messy, but is just what we get by plugging the previous value into the equation for change. This has a simpler expression. Let

$$g(p) = p + f(p)$$

Then the recursion is

$$p[t + 1] = g(p[t])$$

and the sequence of values is

$$p[0] = P_0$$
$$p[1] = g(p[0])$$
$$p[2] = g(g(p[0]))$$
$$p[3] = g(g(g(p[0])))$$
$$\vdots$$

Mathematica has a command 'NestList[.]' to perform this repeated function iteration and we will use it for both discrete and continuous dynamical system computations.

We have not used the basic formulation

$$p[t + 1] = g(p[t])$$

for our dynamical systems, because we want to emphasize the meaning of the equation as a prescription for change in the form

$$p[t + 1] = p[t] + f(p[t])$$

The connection is simply $g(p) = p + f(p)$, but we want to formulate our results in terms of the function $f(p)$ in order to make the clearest connection between discrete and continuous systems.

12.2.1. Equilibria. The first example of the difference between using $f(p)$ vs. $g(p)$ is in saying what an equilibrium value is dynamically. If P_e is an equilibrium value, then when you start with $p[0] = P_e$ or get to P_e somehow, the sequence doesn't change any more.

$$p[1] = p[0] + f(p[0])$$
$$= P_e + f(P_e)$$
$$= P_e + 0$$

in other words, the change $f(P_e) = 0$.

DEFINITION 12.6. *Equilibrium Point of A Discrete Dynamical System*
The number P_e is an equilibrium point for

$$p[t+1] = p[t] + f(p[t])$$

if $f(P_e) = 0$. In this case the constant sequence $\{P_e, P_e, P_e, \cdots\}$ is a solution of the difference equation.

EXERCISE 12.7. *Could the constant sequence $\{P_0, P_0, P_0, \cdots\}$ be a solution to $p[t+1] = p[t] + f(p[t])$ if $f(P_0) \neq 0$?*

It might not always be most convenient to write a difference equation in this change form, but we can always do it by adding and subtracting $p[t]$. For example,

$$p[t+1] = 2.5p[t](1 - \frac{p[t]}{100})$$
$$= p[t] + (2.5p[t](1 - \frac{p[t]}{100}) - p[t])$$
$$= p[t] + (1.5p[t] - \frac{25p^2[t]}{1000})$$
$$= p[t] + 1.5p[t](1 - \frac{25p[t]}{1500})$$
$$= p[t] + 1.5p[t](1 - \frac{p[t]}{60})$$

We prefer this form because the change is zero at an equilibrium. In this case this means

$$1.5P_e(1 - \frac{P_e}{60}) = 0$$

so either $P_e = 0$ or $P_e = 60$.

EXERCISE 12.8.

(1) *Find the equilibria of the system*

$$p[t+1] = p[t] + 2p[t](1 - \frac{p[t]}{50})$$

(2) *Write the difference equation*

$$p[t+1] = 3.5p[t](1 - \frac{p[t]}{1000})$$

in the change form

$$p[t+1] = p[t] + f(p[t])$$

What is the function $f(p) = ?$ What are the equilibrium points of this system?
(3) *Find the equilibrium point of the general affine linear discrete dynamical system*

$$p[t+1] = p[t] + \alpha p[t] + \beta$$

How many equilibria can such a system have? What happens if $\alpha = 0$? How does this result compare with the conjecture you made in Exercise 12.5?

12.2.2. Iteration. Here is a way to 'draw' the function iteration

$$g(g(g(\cdots g(p[0]) \cdots)))$$

Graph $q = g(p)$ and $q = p$ on the same axes. Begin at the point $q = P_0$ on the line $q = p$. Move vertically to the graph $q = g(p)$ over $p = P_0$. The value $q = g(P_0) = p[1]$ by the form of the dynamical system $p[t+1] = g(p[t])$.

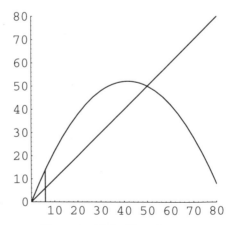

FIGURE 12.9: First Iterate

Now move horizontally from the point $q = p[1]$ on the graph of $q = g(p)$ over to the line $q = p$. This point is still $p[1]$, but next we view it as input to

$$p[2] = g(p[1])$$

and move vertically to the graph $q = g(p)$ along $p = p[1]$. This gives us $q = g(p[1]) = p[2]$.

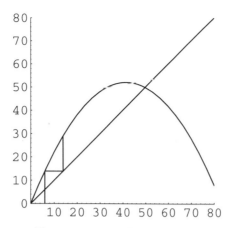

FIGURE 12.10: Two Iterates

Continue the process alternately moving vertically from $q = p$ to $q = g(p)$ and horizontally back to $q = p$. This is simply a graphical representation of

$$g(g(g(\cdots g(P_0)\cdots)))$$

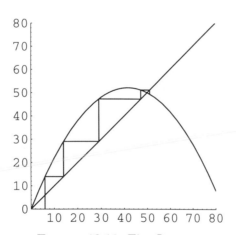

FIGURE 12.11: Five Iterates

A stable equilibrium point looks like a spider working inward toward a place where the line $q = p$ crosses $q = g(p)$ and an unstable system looks like a spider working outward.

EXERCISE 12.12. *Sketch the first three iterates of the 'cobweb' for the dynamical systems*

(1) *First initial condition:*

$$p[0] = 10$$

$$p[t + 1] = p[t] + 2p[t](1 - \frac{p[t]}{50})$$

(2) *Second initial condition:*

$$p[0] = 70$$

$$p[t+1] = p[t] + 2p[t](1 - \frac{p[t]}{50})$$

The *Mathematica* NoteBook FirstDynSys.ma draws cobwebs as well as conventional graphs of solutions to dynamical systems. Check your solutions to the previous exercise with *Mathematica*.

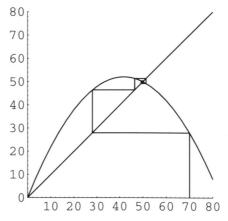

FIGURE 12.13: Six Iterates

EXERCISE 12.14. *The line $q = p$ crosses $q = g(p)$ when $P_e = g(P_e)$. Show that this is the same condition as $f(P_e) = 0$.*

12.3. The Linear System

The simplest (non-trivial) dynamical system we can study is

$$p[t+1] - p[t] = \alpha p[t]$$

or

$$p[t+1] = p[t] + \alpha p[t]$$

This has already come up in another form in Exercise 3.18 where we studied algae doubling its population in a pond every 3 hours. If t represents the number of three hour periods, then $r = 1$ and

$$p[t+1] = p[t] + p[t]$$

It is not hard to find a formula for the solution of

$$p[0] = P_0$$
$$p[t+1] = p[t] + \alpha p[t]$$

Plug in and do some algebra:

$$p[0] = P_0$$
$$p[1] = p[0] + \alpha p[0] = (1 + \alpha)P_0$$
$$p[2] = p[1] + \alpha p[1] = (1 + \alpha)p[1] = (1 + \alpha)[(1 + \alpha)P_0] = (1 + \alpha)^2 P_0$$
$$p[3] = p[2] + \alpha p[2] = (1 + \alpha)p[2] = (1 + \alpha)[(1 + \alpha)^2 P_0] = (1 + \alpha)^3 P_0$$

$$\vdots$$

$$p[t] = P_0(1 + \alpha)^t \ ; \quad t = 0, \ 1, \ 2, \ 3, \ \cdots$$

EXERCISE 12.15. *Show that the only equilibrium of the system $p[t + 1] = p[t] + \alpha p[t]$ is $P_e = 0$, draw the cobwebs and write the closed formula solutions in the cases*
a) $\alpha = -\frac{1}{2}$ b) $\alpha = +\frac{1}{2}$ c) $\alpha = -1$ d) $\alpha = +1$

The stability of the system

$$p[0] = P_0$$
$$p[t + 1] = p[t] + \alpha p[t])$$
$$\to p[t] = P_0(1 + \alpha)^t$$

is simply a question of

$$\text{What is } \lim_{t \to \infty} P_0(1 + \alpha)^t?$$

and this has a simple answer in five cases.

EXERCISE 12.16. *Show the following:*

$$\lim_{t \to \infty} P_0(1 + \alpha)^t = 0 \quad \text{if } |1 + \alpha| < 1 \quad \text{or } -2 < \alpha < 0$$
$$\lim_{t \to \infty} P_0(1 + \alpha)^t = \infty \quad \text{if } 1 + \alpha > 1 \quad \text{or } 0 < \alpha$$
$$p[t] \quad \text{oscillates and grows in magnitude if } 1 + \alpha < -1 \quad \text{or } \alpha < -2$$
$$p[t] = P_0 \quad \text{for all } t \text{ if } \alpha = 0$$
$$\{p[t]\} = \{P_0, -P_0, P_0, -P_0, \cdots\} \quad \text{if } \alpha = -2$$

(Hint: In the positive cases, you can take $\text{Log}[P_0(1 + \alpha)^t] = t \, \text{Log}[(1 + \alpha)] + \text{Log}[P_0]$.)

This proves a basic stability result:

THEOREM 12.17. *The Linear Stability Theorem*
If $\alpha \neq 0$, the linear dynamical system

$$p[0] = P_0$$
$$p[t + 1] = p[t] + \alpha p[t]$$

has the unique equilibrium point $P_e = 0$. If $-2 < \alpha < 0$, then 0 is a globally stable attractor, that is, every solution satisfies

$$\lim_{t \to \infty} p[t] = 0$$

EXERCISE 12.18. *State and prove two kinds of Linear Instability Theorems, one in the case $\alpha > 0$, and the other in the case $\alpha < -2$.*

12.3.1. The Affine Linear Equation. Now we want to extend our stability result to the dynamical systems of the form

$$p[0] = P_0$$
$$p[t + 1] = p[t] + \alpha p[t] + \beta$$

EXERCISE 12.19. *Show that the above system has* $P_e = -\frac{\beta}{\alpha}$ *as its unique equilibrium point if* $\alpha \neq 0$.

We will make a change of variables to convert our affine system into a truly linear one. This is a very useful trick in many mathematical problems. Let

$$q[t] = p[t] - P_e$$

the local variable measured from equilibrium. We know that $\alpha P_e + \beta = 0$, because P_e is an equilibrium, so

$$p[t + 1] = p[t] + \alpha p[t] + \beta$$
$$p[t + 1] - P_e = p[t] - P_e + \alpha p[t] + \beta$$
$$p[t + 1] - P_e = p[t] - P_e + \alpha p[t] + \beta + \alpha P_e - \alpha P_e$$
$$p[t + 1] - P_e = p[t] - P_e + \alpha p[t] - \alpha P_e + (\beta + \alpha P_e)$$
$$p[t + 1] - P_e = p[t] - P_e + \alpha (p[t] - P_e) + 0$$
$$q[t + 1] = q[t] + \alpha q[t]$$

The dynamical system for $q[t]$ is completely equivalent to the one for $p[t]$, except that we must add or subtract P_e.

EXERCISE 12.20. *Give a closed formula for the solution of*

$$p[0] = P_0$$
$$p[t + 1] = p[t] + \alpha p[t] + \beta$$

Write your answer in terms of t, α, β *and* P_0 *only. Note that the solution to*

$$q[0] = Q_0$$
$$q[t + 1] = q[t] + \alpha q[t]$$

is $q[t] = Q_0 (1 + \alpha)^t$.

The change of variables also proves

THEOREM 12.21. *The Affine Stability Theorem*
If $\alpha \neq 0$, *the affine dynamical system*

$$p[0] = P_0$$
$$p[t + 1] = p[t] + \alpha p[t] + \beta$$

has the unique equilibrium point $P_e = -\frac{\beta}{\alpha}$. *If* $-2 < \alpha < 0$, *then* P_e *is a globally stable attractor, that is, every solution satisfies*

$$\lim_{t \to \infty} p[t] = P_e$$

PROOF:

Apply the linear stability to $q[t]$ and un-change the variables.

EXERCISE 12.22. *Give the equilibria and their stability for the dynamical systems in Exercise 12.5*

12.4. Nonlinear Models

The population of species with distinct generations can be modeled by discrete dynamical systems. For example, $p[t]$ might measure the peak population of May flies in year t. May flies hatch, live for a short period to lay eggs and die. The first model we consider neglects environmental limitations such as space or food and only concentrates on basic fertility. The model is only realistic for the early growth of 'small' populations. Review the continuous model of algae growth in Exercise 3.18 and the *Mathematica* NoteBook ExpGth.ma for comparison.

Our model is simply

12.4.1. The Basic Fertility Model.

$$p[0] = P_0$$
$$p[t+1] = p[t] + \alpha p[t] = (1 + \alpha)p[t]$$

where α represents the per capita fertility (or 'percent' fertility as a fraction). If $\alpha = 0$, each bug just manages to replace him or herself. This is an average rate, so there may be many more females than males. If $\alpha = 1$, each bug replaces itself and also produces a new bug. The next year there are twice as many bugs. The second year there are 4 times as many, since the doubled population doubles, and so on. The solution to this model is

$$p[t] = P_0(1 + \alpha)^t \quad \text{for } t = 1, \ 2, \ 3, \ \cdots$$

This kind of prolific growth cannot continue very long. In the *Mathematica* NoteBook ExpGth.ma you saw that algae doubling every three hours would exceed the mass of all of Lake Michigan in a short period. Whenever $\alpha > 0$ this model eventually grows very fast.

Notice that we can express this in terms of instability of the equilibrium point $P_e = 0$. When $\alpha > 0$, we know that

$$\lim_{t \to \infty} p[t] = \lim_{t \to \infty} P_0(1 + \alpha)^t = \infty$$

If $\alpha < 0$ there is an annual deficit in fertility. For example, if $\alpha = -.05$, then the next year the population is only 95% of this year's, the second only $.95 \times .95 = .9025$ or 90.25 % of the starting population. As time progresses, the population goes to extinction. Mathematically,

$$\lim_{t \to \infty} P_0(1 + \alpha)^t = 0 \quad \text{when } -2 < \alpha < 0$$

EXERCISE 12.23. *Do the mathematically stable cases where $-2 < \alpha < -1$ make sense biologically in the above model? Why?*

A population could survive with $\alpha = -.05$ with migration. Imagine bugs on a desert island where survival is difficult, but where new bugs come in on birds that visit. This could be modeled by

12.4.2. The Migration Model.

$$p[0] = P_0$$
$$p[t+1] = p[t] + \alpha p[t] + \beta = (1 + \alpha)p[t] + \beta$$

EXERCISE 12.24.
1) Suppose $\alpha = -.05$ in the migration model above. If 100 new bugs migrate in each year, what is the equilibrium population?
2) Is the equilibrium value the dynamic limit of the system if $P_0 \neq P_e$?
3) What is the answer to this question in terms of more general parameter values, α and β?
4) What is the closed formula solution to the above system?

Bugs survive in the migration model without filling the entire universe, but the biological assumptions are quite restricted. Desert islands visited by bug carrying birds is a narrow scope. We want to put an increasing fertility at small populations together with a limited ability of the environment to support large populations. The next model has these features. The parameter C is a positive constant and we are biologically interested in the case when $\alpha > 0$.

12.4.3. The Logistic Growth Model.

$$p[0] = P_0$$
$$p[t+1] = p[t] + \alpha p[t](1 - \frac{p[t]}{C})$$

When $p[t]$ is small, the term $(1 - \frac{p[t]}{C})$ is near 1, so the model behaves like our Basic Fertility Model, but as $p[t]$ increases, the term $(1 - \frac{p[t]}{C})$ tends to zero and stops growth. In fact, if the population ever exceeds this value $p[t] > C$ the term is negative $(1 - \frac{p[t]}{C}) < 0$, so $p[t]$ 'grows' negatively, in other words, $p[t]$ declines. Biologically speaking, we want the population to persist, that is $\lim_{t \to \infty} p[t] \neq 0$ and we also seek stability, the population in harmony with the environment rather than undergoing wild oscillations in population.

EXERCISE 12.25. *Nonlinear Stability Experiments*
1) Find the two equilibrium values of the Logistic Growth Model in terms of the parameters of the model.
2) Formulate conjectures about the stability of these two equilibria by doing experiments with the FirstDynSys.ma NoteBook in various cases, for example:

$\alpha = 1.5$	$C = 50$	$P_0 = 10, \ 60, \ 70$
$\alpha = 2.5$	$C = 50$	$P_0 = 10, \ 60, \ 70$

12.5. Local Stability - Calculus and Nonlinearity

You probably found it rather difficult to formulate a general conjecture about even the Logistic Growth model's stability. That model is about as simple as a nonlinear model can be. We need some additional tools.

The main idea of differential calculus is simply that smooth nonlinear functions 'look' linear under an infinitesimal microscope. What would we 'see' if we focused our microscope at an equilibrium of a nonlinear cobweb diagram? Simple, we'd see a linear cobweb diagram.

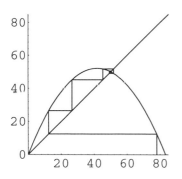
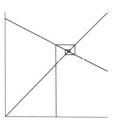

FIGURE 12.26: A Microscopic View of a Cobweb

We would expect the nonlinear cobweb to behave just like the linear magnification as long as we stay close to the equilibrium. We know that the stability of the linear diagram is completely covered by the Linear Stability Theorem above. How can we support the intuitive idea of microscopic stability with symbolic computations? What is the intuitive idea in graphical terms?

EXERCISE 12.27. *Re-phrase the Linear Stability Theorem in terms of the cobweb diagram. 'If the line $q = (1 + \alpha)\, p$ crosses $q = p$ at slope ?? or angle ??, then the linear system is stable...' Once you have done this, formulate a conjecture about the microscopic stability of a smooth nonlinear system.*

Our formulation of the stability theorem is the following symbolic result.

THEOREM 12.28. *Local Nonlinear Stability*
Suppose that $f(p)$ is a differentiable function and P_e is an equilibrium value of the dynamical system

$$p[0] = P_0$$
$$p[t + 1] = p[t] + f(p[t])$$

that is, $f(P_e) = 0$. If $-2 < f'(P_e) < 0$, then when P_0 is sufficiently close to P_e,

$$\lim_{t \to \infty} p[t] = P_e$$

that is, P_e is a stable attractor for near enough initial conditions P_0.

PROOF:
We use two ideas. The microscope or increment approximation

$$f(p + \delta p) = f(p) + f'(p)\delta p + \varepsilon \delta p$$

and the local variable trick that we used in the Affine Stability Theorem. The increment equation will be used with $p = P_e$ and $\delta p = p[t] - P_e$, so it becomes

$$f(p[t]) = f(P_e + \{p[t] - P_e\}) = f(P_e) + f'(P_e)\{p[t] - P_e\} + \varepsilon\{p[t] - P_e\}$$
$$f(p[t]) = 0 + f'(P_e)\{p[t] - P_e\} + \varepsilon\{p[t] - P_e\}$$
$$f(p[t]) = \{f'(P_e) + \varepsilon\}\{p[t] - P_e\}$$

First we look in the microscope, and then we show that we can look outside. If $p[t] \approx P_e$,

$$p[t+1] = p[t] + f(p[t])$$
$$= p[t] + \{f'(P_e) + \varepsilon\}\{p[t] - P_e\}$$

Now we localize our variable by subtracting P_e on both sides

$$p[t+1] - P_e = p[t] - P_e + \{f'(P_e) + \varepsilon\}\{p[t] - P_e\}$$
$$\{p[t+1] - P_e\} = \{p[t] - P_e\} + \{f'(P_e) + \varepsilon\}\{p[t] - P_e\}$$
$$\{p[t+1] - P_e\} = \{1 + f'(P_e) + \varepsilon\}\{p[t] - P_e\}$$

Any real number like m $= \text{Ave}[\, 1, |1 + f'(P_e)|] = \frac{1 + |1 + f'(P_e)|}{2}$, satisfies

$$|1 + f'(P_e) + \varepsilon| < m < 1 \quad \text{for all } \varepsilon \approx 0$$

because we have assumed that the real number $f'(P_e)$ lies between -2 and 0.

If we take absolute values of the estimate for the change in our dynamical system and further estimate the hyperreal term $|1 + f'(P_e) + \varepsilon|$ by m, then we see that for any $p[t] \approx P_e$,

$$|p[t+1] - P_e| < m|p[t] - P_e|$$

Since this is a real inequality with no hyperreals, the Function Extension Axiom guarantees that for $p[t]$ a sufficiently close real distance from P_e, we still have

$$|p[t+1] - P_e| < m|p[t] - P_e|$$

It follows by successively plugging in and estimating, that

$$|p[t] - P_e| < m^t|P_0 - P_e|$$

and clearly

$$0 \le \lim_{t \to \infty} |p[t] - P_e| \le \lim_{t \to \infty} m^t|P_0 - P_e| = 0$$

so $p[t] \to P_e$.

EXAMPLE 12.29. *Local Stability for Logistic Systems We will apply the theorem to the systems*

$$p[0] = P_0$$
$$p[t+1] = p[t] + \alpha p[t](1 - \frac{p[t]}{C})$$

where $f(p) = \alpha p(1 - \frac{p}{C})$.

SOLUTION

The equilibria are $P_e = 0$ and $P_e = C$. The derivative

$$f'(p) = \alpha(1 - \frac{p}{C}) + \alpha p(-\frac{1}{C})$$
$$= \alpha(1 - 2\frac{p}{C})$$

At the equilibrium $P_e = 0$, $f'(0) = \alpha$, so 0 is a locally stable equilibrium when $-2 < \alpha < 0$.

At the equilibrium $P_e = C$, $f'(C) = -\alpha$, so C is locally stable when $0 < \alpha < 2$.

EXERCISE 12.30. *Classify the equilibria of the examples in Exercise 12.25.*

Now put your knowledge together on some new examples.

EXERCISE 12.31. *For the following difference equations, sketch the cobweb figure - $q = p$ and $q = g(p)$ on the same graph. You will need to find all equilibria. Where do these appear on your figures? Apply the Local Stability Theorem to all of the equilibria. Which ones are locally stable?*

(1) $p[t+1] = \frac{1}{3}(p^2[t] - p[t] + 3)$
(2) $p[t+1] = \frac{1}{3}(p^2[t] + 5p[t] - 3)$
(3) $p[t+1] = p[t] + \frac{1}{3}(p^3[t] - p[t])$
(4) $p[t+1] = p[t] + \frac{1}{3}(p[t] - p^3[t])$
(5) $p[t+1] = 2\,\mathrm{Sin}[p[t]]$

12.5.1. Chaos. We cannot formulate a simple global repeller theorem for nonlinear systems, even for logistic equations. Microscopically, things may be forced away, but once out of the region where our local analysis applies, could come back in, then go out, then back in and so on. The system

$$p[0] = P_0$$
$$p[t+1] = p[t] + 2.58p[t] - 3.58p^2[t]$$
$$= 3.58p[t](1 - p[t])$$

exhibits chaotic behavior, 'Chaos' in the mathematical sense.

EXERCISE 12.32. *Show that the system above fails the local stability criterion for both of its equilibria. Experiment with its behavior by using the Mathematica NoteBook FirstDynSys.ma*

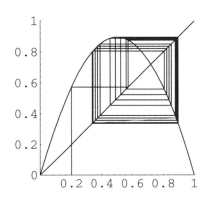
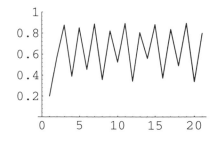

FIGURE 12.33: Chaos as a Cobweb and an Explicit Graph

EXERCISE 12.34. *House Elder Bugs*
In dry conditions, the population of mythical House Elder Bugs in southwestern Iowa City

on the 15th of September t years from now can be accurately estimated by taking 1,000,000 times q[t], where

$$q[0] = \text{the population last September 15th in millions}$$

$$q[t+1] = q^2[t]\frac{3 - q[t]}{2}$$

Show that there are 3 equilibria corresponding to $Q_e = 0$, $Q_e = 1$ and $Q_e = 2$. Which of these equilibria are locally stable?

12.6. Projects

12.6.1. A Model Economy. A price adjustment model economy is explored in the Scientific Projects.

12.6.2. Sustained Harvest of Sei Whales. One of the Scientific Projects deals with a real scientific model of Sei Whales. The question there is, 'How many whales per year can we harvest without sending the population to extinction?' Unfortunately, for the whales, the answer depends on their not being 'overexploited' initially. Mathematically, there is a positive local attracting equilibrium, but that is not the whole dynamic story. The *Mathematica* NoteBook Whales.ma offers you some help with the simulation, because the dynamical system is order 9 corresponding to the number of years it takes to produce a baby whale. The mathematical model was derived by J.R. Beddington in a article in the Report of the International Whaling Commission in 1978.

12.6.3. Computation of Inverse Functions. The inverse of a function $y = f(x)$ is the function $g(y)$ that 'un-does' what $f(x)$ does. That is, given a value of $y = y_1$, the value of $x = x_1$ that makes $f(x_1)$ equal the given y_1 is $x_1 = g(y_1)$. For example, ArcTan[y] gives the angle whose tangent equals y. A method for you to use yourself to compute the inverse of complicated functions like $y = x^x$ is given in the Mathematica Background. The method is simply a discrete dynamical system that converges to an equilibrium where $x_e^{x_e} = y_1$, for a given value of y_1.

Part 2

Integration in One Variable

CHAPTER 13

Basic Integration

This chapter contains the fundamental theory of integration. We begin with some problems to motivate the main idea: approximation by a sum of slices. The chapter confronts this squarely, and the next chapter concentrates on the basic rules of calculus that you use after you have found the integrand.

Both the geometric and physical integral approximations work in the following way. First find a formula for the quantity (volume, area, length, distance, etc.) in the case of either a constant or linear function. Next, approximate the nonlinear quantity by a sum of 'slices' using the constant or linear formula for each slice. The primary difficulty is usually in expressing the variable sizes of the approximating pieces. In the geometric problems, this step is analytical geometry - finding the formula that goes with the picture you want. The Fundamental Theorem of Integral Calculus will give us a simple way to exactly compute the limit of the sum approximations.

In order for sum approximations to tend to an integral we need to write them in the form

$$f(a)\,\Delta x + f(a + \Delta x)\,\Delta x + f(a + 2\Delta x)\,\Delta x + \cdots + f(b - \Delta x)\,\Delta x$$

where Δx is the 'thickness of the slice' and $f(x)$ is the variable 'amount of the slice.' This symbolic expression is an important part of the way the formulas are expressed in integration; without the symbolic expression the more or less obvious approximations could not be computed exactly in a common way. *Mathematica* has a sum command that computes this expression

$$\text{Sum}[f\,(x)\,\Delta x, \{x, a, b - \Delta x, \Delta x\}]$$

Mathematica summation is studied in detail in section 13.4 below. The integral is the limit of this sum as Δx tends to zero,

$$\int_a^b f(x)\,dx = \lim_{\Delta x \to 0} \text{Sum}[f\,(x)\,\Delta x, \{x, a, b - \Delta x, \Delta x\}]$$

The slicing approximation is the forest that you must strive to see through a tangle of technical trees. The first problem is in finding the symbolic sum.

13.1. Geometric Approximations by Sums of Slices

The volume of a right circular cone with height h and base of radius r is

$$V = \frac{\pi}{3}r^2 h$$

We want you to see how to derive this formula by approximating a cone by a sum of cylindrical 'disks.' Once you understand the step from a constant radius to a linearly varying radius, you will be able to set up integral formulas for very general solids of revolution. This is an important generalization of this classical formula for a cone.

The volume of a right cylinder is

$$V = A h$$

the area of the base times the height. If the base is a circle of radius r, this becomes

$$V = \pi r^2 h$$

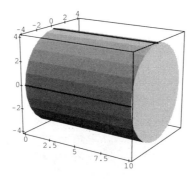

FIGURE 13.1: $V = \pi r^2 h$

This is the formula for the case of a constant radius function. The radius $R(x) = r$ is the same all along the cylinder.

13.1.1. The Volume of a Cone. Now we think of a cone as a figure with circular cross sections that vary linearly, starting with zero radius and increasing to radius r when the distance from the tip is h. Let the variable x run down the axis of the cone with $x = 0$ at the tip. The expression for the radius of the cross section can be expressed as a function of x.

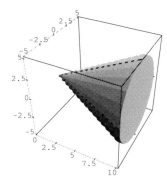

FIGURE 13.2: A Cone Along the x-Axis

EXERCISE 13.3. *Give a formula $R(x) =$? for the radius of the cross section at x for a cone with tip at $x = 0$ and having base radius r at height $x = h$.*

EXAMPLE 13.4. *The Volume of a Particular Cone*

Our approximation to the volume can be obtained by slicing the cone in steps of Δx, using the formula for the volume of a cylinder for each slice, $V_{slice} = \pi R^2(x)\Delta x$. For example, suppose we have a cone of height 10 and radius 5 at the base (in the same length units.)In this case the formula for the radius x units below the apex is

$$R(x) = \frac{1}{2}\,x$$

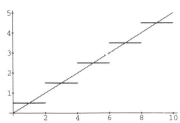

FIGURE 13.5: Generating Lines for The Cone Approximation

If we let $\Delta x = 2$, or slice the cone 5 times, we obtain 5 disks with radii at $x = 1$, $x = 3$, $\cdots$, $x = 9$. The disks are generated by the horizontal segments shown in the figure above, generate the rough approximating pile of disks shown below, and have volumes given as follows:

(1) x from 0 to 2, $R(1) = \frac{1}{2}$, thickness of cylindrical disk $= \Delta x = 2$,
$V_{slice1} = \pi \left(\frac{1}{2}\right)^2 2$

(2) x from 2 to 2, $R(3) = \frac{3}{2}$, thickness of cylindrical disk $= \Delta x = 2$,
$V_{slice2} = \pi \left(\frac{3}{2}\right)^2 2$

(3) x from 4 to 6, $R(5) = \frac{5}{2}$, thickness of cylindrical disk $= \Delta x = 2$,
$V_{slice3} = \pi \left(\frac{5}{2}\right)^2 2$

(4) x from 6 to 8, $R(7) = \frac{7}{2}$, thickness of cylindrical disk $= \Delta x = 2$,
$V_{slice4} = \pi \left(\frac{7}{2}\right)^2 2$

(5) x from 8 to 10, $R(9) = \frac{9}{2}$, thickness of cylindrical disk $= \Delta x = 2$,
$$V_{slice5} = \pi \left(\frac{9}{2}\right)^2 2$$

The total volume approximation is

$$\frac{\pi}{2}(1^2 + 3^2 + \cdots + 9^2) = \frac{\pi\, 165}{2} \approx 259.$$

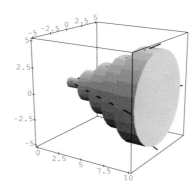

FIGURE 13.6: A Cone Approximated by 5 'Disks'

EXERCISE 13.7. *Show that the volume approximation for the cone of height* 10 *and base radius* 5 *when cut into* 10 *slices with thickness* 1 *and radii at the midpoints,* $\frac{1}{2}$, $\frac{3}{2}$, $\cdots$, $\frac{19}{2}$ *is*

$$\frac{\pi}{4^2}(1^2 + 3^2 + 5^2 + \cdots + 19^2) = \frac{\pi\, 1330}{4^2} \approx 261.$$

Begin by writing the volume approximation as a sum of the volumes of approximating disks as in the above list for 5 *slices.*

We need to find the formula that expresses the sum of the volumes of the slices in terms of a variable for the thickness of the slice, Δx. The previous exercise suggests an interesting special purpose algebraic approach, but that is only likely to work for this specific cone. We can express the approximation in terms of the formula for the radius $R(x)$ of the cross section at x and a thickness Δx as

$$\pi[(R[\Delta x/2])^2\Delta x + (R[3\Delta x/2])^2\Delta x + (R[5\Delta x/2])^2\Delta x + \cdots + (R[10 - \Delta x/2])^2\Delta x]$$
$$= \pi \operatorname{Sum}\left[(R[x])^2\Delta x, \{x, \Delta x/2, 10 - \Delta x/2, \Delta x\}\right]$$

EXERCISE 13.8. *Verify that your* 10 *slice approximation sum is a special case of this formula*

$$\pi \operatorname{Sum}\left[(R[x])^2\Delta x, \{x, \Delta x/2, 10 - \Delta x/2, \Delta x\}\right]$$

in the case where $\Delta x = 1$*, by writing out the steps of the sum command.*

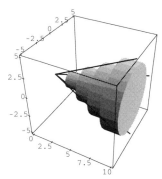

FIGURE 13.9: A Cone Approximated by 5 'Disks'-Radii at Left

If we measured our radii at the left end of the slice points instead of the midpoint, the approximation would be less accurate, because the slices fit completely inside the cone, but the formula would be simpler,

$$\pi[(R[0])^2\Delta x + (R[1\Delta x])^2\Delta x + (R[2\Delta x])^2\Delta x + \cdots + (R[10 - \Delta x])^2\Delta x]$$
$$= \pi\text{Sum}\left[(R[x])^2\Delta x, \{x, 0, 10 - \Delta x, \Delta x\}\right]$$

This has the form we described in the introduction above

$$f(a)\,\Delta x + f(a + \Delta x)\,\Delta x + f(a + 2\Delta x)\,\Delta x + \cdots + f(b - \Delta x)\,\Delta x$$

where $f(x) = \pi R^2(x)$, $a = 0$ and $b = 10$.

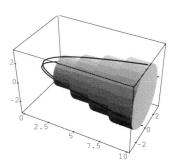

FIGURE 13.10: A Parabolic Rocket Nose Cone

13.1.2. A Parabolic Nose Cone. The simple linear formula for $R(x)$ in the case of the cone is easily generalized to more complicated shapes. A parabolic rocket nose cone may be described by the volume swept out as we rotate the curve $y = \sqrt{x}$ for $0 \le x \le 10$ about the x-axis.

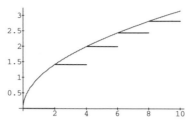

FIGURE 13.11: Slice Generators for Parabolic Nose Cone

The formula for the radius of a cross section is $R(x) = \sqrt{x}$ and the volume of a slice becomes $\pi R^2(x)h = \pi(\sqrt{x})^2 \Delta x = \pi\, x\, \Delta x$ making the approximating sum with the radius at the left side of the slices

$$\pi\left(0\ \Delta x + \Delta x\ \Delta x + 2\Delta x\ \Delta x + \cdots + (10 - \Delta x)\ \Delta x\right)$$
$$= \pi \mathrm{Sum}[x\ \Delta x, \{x, 0, 10 - \Delta x, \Delta x\}]$$

13.1.3. A Spear Tip. A slender spear tip may be described by the volume swept out by the region between $y = \frac{x^2}{50}$ and the x-axis as the region is revolved about the x-axis. In this case the radius of a cross section at x is $R(x) = \frac{x^2}{50}$, the volume of one slice is $\pi R^2(x)h = \pi\left(\frac{x^2}{50}\right)^2 \Delta x$ and the left radius sum by approximating disks is

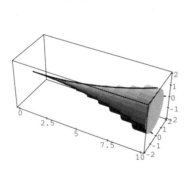 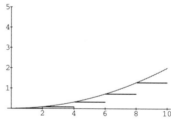

FIGURE 13.12: A Cusped Spear Point and Slice Generators

$$\pi\left(0\ \Delta x + \frac{\Delta x^4}{2500}\ \Delta x + \frac{(2\Delta x)^4}{2500}\ \Delta x + \cdots + \frac{(10 - \Delta x)^4}{2500}\ \Delta x\right)$$
$$= \frac{\pi}{2500}\mathrm{Sum}[x^4\ \Delta x, \{x, 0, 10 - \Delta x, \Delta x\}]$$

13.1.4. The Area Between Two Curves. We need two formulas to build an approximation to the complicated areas between curves. First, the distance between points with real coordinates y_1 and y_2. This is $|y_2 - y_1|$, the signed quantity $y_2 - y_1$ gives the directed distance from y_1 to y_2 where a minus sign means moving opposite the direction from zero to one. (The absolute value will cause us technical difficulties in calculus.) Second, the area of a rectangle is the height times the width.

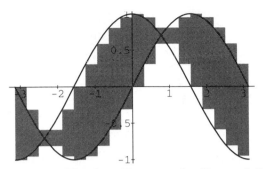

FIGURE 13.13: The Area Between the Sine and Cosine

EXERCISE 13.14. *Slice the region between the sine and cosine curves from $-\pi$ to π into strips of width Δx as shown in the above figure. Show that the height of the slice with left point at x is given by $|\mathrm{Sin}[x] - \mathrm{Cos}[x]| = |\mathrm{Cos}[x] - \mathrm{Sin}[x]|$ and that the area of the slice is $|\mathrm{Cos}[x] - \mathrm{Sin}[x]| \, \Delta x$. Show that the sum of all the approximating slices is*

$$
\begin{aligned}
(|\mathrm{Cos}[-\pi] - \mathrm{Sin}[-\pi]| \, \Delta x + \\
|\mathrm{Cos}[-\pi + \Delta x] - \mathrm{Sin}[-\pi + \Delta x]| \, \Delta x + \\
|\mathrm{Cos}[-\pi + 2\Delta x] - \mathrm{Sin}[-\pi + 2\Delta x]| \, \Delta x + \\
\cdots + |\mathrm{Cos}[\pi - \Delta x] - \mathrm{Sin}[\pi - \Delta x]| \, \Delta x) = \\
\mathrm{Sum}[|\mathrm{Cos}[x] - \mathrm{Sin}[x]| \, \Delta x, \{x, -\pi, \pi - \Delta x, \Delta x\}]
\end{aligned}
$$

13.1.5. The Length of a Curve. The way you 'slice' sometimes matters. The approximation to the length of a curve by sloping line segments that connect $(x, f(x))$ and $(x + \Delta x, f(x + \Delta x))$ produces a good approximation, but 'slices' that just run horizontally out from the slice points do not approximate the length. The horizontal slices are a good approximation to the area, but not the length.

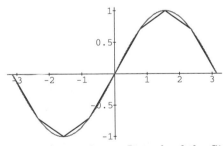

FIGURE 13.15: Approximate Length of the Sine Curve

EXERCISE 13.16. *Use the coordinate geometry version of the Pythagorean Theorem to show that the sum of the lengths of the segments in Figure 13.1.5 is*

$$
\mathrm{Sum}[\sqrt{(x + \Delta x - x)^2 + (\mathrm{Sin}[x + \Delta x] - \mathrm{Sin}[x])^2}, \{x, -\pi, \pi - \Delta x, \Delta x\}]
$$

or more generally if the curve is $y = f(x)$,

$$Sum[\sqrt{(x + \Delta x - x)^2 + (f[x + \Delta x] - f[x])^2}, \{x, a, b - \Delta x, \Delta x\}]$$

When we replace Δx *by a positive infinitesimal increment* δx, *use the differential approximation of Definition 5.15 to show that this sum becomes*

$$Sum[\sqrt{(\delta x)^2 + (f'[x] + \varepsilon)^2 (\delta x)^2}, \{x, a, b - \delta x, \delta x\}]$$
$$= Sum[\sqrt{1 + (f'[x] + \varepsilon)^2}\, \delta x, \{x, a, b - \delta x, \delta x\}]$$

with $\varepsilon \approx 0$.

Finally, show that

$$\sqrt{1 + (f'[x] + \varepsilon)^2} = \sqrt{1 + (f'[x])^2} + \iota$$

with $\iota \approx 0$. *HINT:*

$$\sqrt{1 + (f'[x] + \varepsilon)^2} - \sqrt{1 + (f'[x])^2} = \frac{?}{\sqrt{1 + (f'[x] + \varepsilon)^2} + \sqrt{1 + (f'[x])^2}}$$

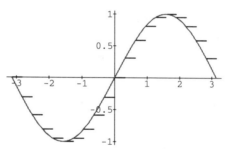

FIGURE 13.17: An Incorrect 'Approximation' to the Length

EXERCISE 13.18. *Find a sum expression for the attempt at approximating the length of the sine curve by horizontal slices. Use Mathematica to evaluate this sum, then explain why it is a bad approximation.*

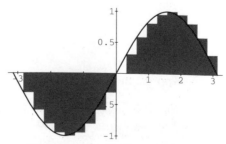

FIGURE 13.19: Area - A Correct Approximation

13.2. Extension of the Distance Formula, $D = R \cdot T$

If we drive 50 mph for an hour and a half, we compute the distance traveled by $D = RT$, 'distance equals the rate times the time,' $D = 50 \cdot \frac{3}{2} = 75$ miles. If we vary our speed, we can not use this formula. Suppose get on the highway and accelerate. We speed up cautiously and drive 25 mph for 1 minute, 26 mph for 1 minute, 27 mph for 1 minute, $\cdots$, 49 mph for 1 minute and 50 mph for the remainder of the hour and a half. How far do we go?

The distance traveled at 25 mph must be computed in the correct units,

$$D_1 = 25 \cdot \frac{1}{60} \approx 0.4167 \qquad (\text{mph X hours} = \text{miles})$$

The distance traveled at 26 mph is,

$$D_2 = 26 \cdot \frac{1}{60} \approx 0.4333 \qquad (\text{miles})$$

The distance traveled at 27 mph is,

$$D_3 = 27 \cdot \frac{1}{60} \approx 0.45 \qquad (\text{miles})$$

Each minute's distance can be computed for 25 minutes giving a sum of

$$\frac{25}{60} + \frac{26}{60} + \cdots + \frac{49}{60} \approx 15.42$$

The last part of the trip is $\frac{50 \cdot (90-25)}{60} \approx 54.17$ for a total of 69.59 miles.

There are several questions:

(1) What have we done symbolically?
(2) How can we interpret the computation geometrically?
(3) Why should we interpret the distance computation geometrically?
(4) How can we extend this to continuously varying speed?

Symbolically, we have speed as a function of time in minutes, $S(1) = 25$, $S(2) = 26$, $S(3) = 27, \cdots$ and

$$\text{Distance traveled during acceleration} = \text{Sum}[S(m) \cdot \frac{1}{60}, \{m, 1, 25, 1\}]$$

while we have

$$\text{Distance traveled at 50 mph} = 50 \cdot \frac{65}{60}$$

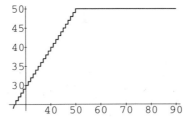

FIGURE 13.20: Speed vs. Time in Minutes

We may also think of this as a sum with a rate function defined at each minute, $S(m) = 50$ for all the minutes from 26 to 90. However, it is better to also use the correct units $t =$ elapsed time in hours and give the speed by the function

$$R(t) = 25 \qquad \text{for } 0 \le t < \frac{1}{60},$$

$$R(t) = 26 \qquad \text{for } \frac{1}{60} \le t < \frac{2}{60},$$

$$R(t) = 27 \qquad \text{for } \frac{2}{60} \le t < \frac{3}{60},$$

$$R(t) = 28 \qquad \text{for } \frac{3}{60} \le t < \frac{4}{60},$$

$$\vdots$$

$$R(t) = 50 \qquad \text{for } \frac{3}{60} \le t < \frac{90}{60},$$

In this case we can combine the two pieces into one sum

$$\text{Distance traveled} = \text{Sum}[R(t) \cdot \frac{1}{60}, \{t, 0, \frac{3}{2} - \frac{1}{60}, \frac{1}{60}\}]$$
$$= \text{Sum}[R(t) \cdot \Delta t, \{t, a, b - \Delta t, \Delta t\}]$$

with $\Delta t = \frac{1}{60}$, $a = 0$ and $b = \frac{3}{2}$.

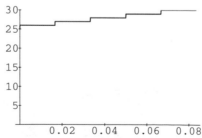

FIGURE 13.21: Speed vs. Time in Hours for 5 Min

The products $25 \cdot \frac{1}{60}$, $26 \cdot \frac{1}{60}$, $27 \cdot \frac{1}{60}$, etc. can be associated with rectangles of height 25, 26, 27, etc. and width $\frac{1}{60}$ hours or one minute. Distance is not an area, but in this case the distance traveled at 25 mph is a product and that product is also an area of the rectangle under the segment on the graph of the speed vs. time. This means that we can represent the distance traveled as the total area between the speed curve and the x-axis - provided our units are hours on the x-axis and miles per hour on the y-axis.

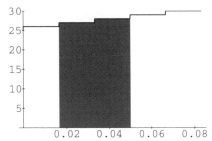

FIGURE 13.22: Distance represented as Area

The area representation helps us see what happens if we accelerate continuously, instead of going exactly 25 mph for exactly 1 minute, 26 mph for another minute, etc. Suppose we accelerate linearly from 25 mph to 50 mph in 25 minutes, so that our speed is given by

$$R(t) = 25 + 60\,t \qquad \text{for } 0 \le t < \frac{25}{60},$$

and

$$R(t) = 50 \qquad \text{for } \frac{3}{60} \le t < \frac{90}{60},$$

We could calculate the distance traveled in each second using the speed at the beginning of the second,

$$25 \cdot \frac{1}{3600} = \frac{1}{144} \approx 0.006944$$

$$\left(25 + 60\,\frac{2}{3600}\right) \cdot \frac{1}{3600} = \frac{1501}{216000} \approx 0.006949$$

$$\left(25 + 60\,\frac{3}{3600}\right) \cdot \frac{1}{3600} = \frac{751}{108000} \approx 0.006954$$

$$\left(25 + 60\,\frac{4}{3600}\right) \cdot \frac{1}{3600} = \frac{167}{24000} \approx 0.006958$$

This computation is no fun by hand, but *Mathematica* computes

$$\text{Sum}[R(t) \cdot \Delta t, \{t, 0, 1.5 - \Delta t, \Delta t\}] = \frac{20099}{288} \approx 69.788$$

when $\Delta t = 1/3600$.

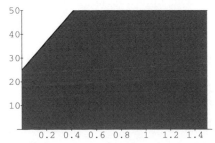

FIGURE 13.23: Linear Acceleration and Distance

The *Mathematica* computation is a waste in this case, however, because we can associate the distance traveled with the area under the speed curve and use the formula for a trapezoid for $0 \le t < 25/60$ and for a rectangle for $25/60 \le t \le 90/60$,

$$\text{Area} = \frac{1}{2}(25 + 50) \cdot \frac{25}{60} + 50 \cdot \frac{65}{60}$$
$$= \frac{1675}{24}$$
$$\approx 69.7917$$

If speed varies by a more complicated rule than linearly in pieces, we can still associate the area under the speed curve with the distance traveled, but now we have no formula to find the area. We will have to find that area and distance by definite integration. For a short interval of time, where speed does not change very much, the distance is approximately the product $R(t) \cdot \Delta t$.

$$\text{Distance traveled from time } t \text{ to time } t + \Delta t \approx \approx R(t) \cdot \Delta t$$

The approximation is actually close, even compared to Δt and this makes total distance an integral. We will see exactly what the approximation is and verify that it holds as long as $R(t)$ is a continuous function. This should be plausible, because we could also approximate by

$$\text{Distance traveled from time } t \text{ to time } t + \Delta t \approx \approx R(t') \cdot \Delta t$$

for any t' between t and $t + \Delta t$. Continuity of $R(t)$ means that $R(t) \approx R(t')$.

13.3. The Definition of the Definite Integral

We want to record the definition of the integral that was motivated by ideas like the problem of computing distance from variable speeds, and by the geometric area, length, and volume 'slicing' problems.

DEFINITION 13.24. *Let $f(x)$ be a continuous function defined on the interval $[a, b]$. The definite integral of $f(x)$ over $[a, b]$, is given by the following limit:*

$$\int_a^b f(x)\,dx = \lim_{\Delta x \to 0} \text{Sum}[f(x)\,\Delta x, \{x, a, b - \Delta x, \Delta x\}]$$
$$\approx \text{Sum}[f(x)\,\delta x, \{x, a, b - \delta x, \delta x\}], \quad \delta x \approx 0$$

Four steps in an example of this limit are shown next:

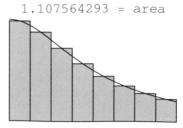

1.10878711 = area   1.107564293 = area

1.107252816 = area   1.107174755 = area

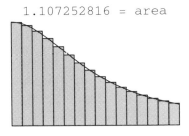   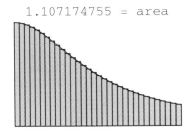

FIGURE 13.25: Approximations to the Area ArcTan[2] ≈ 1.10715

We need to use the algebraic properties of the summation function to deduce results about the integral. In either the limiting case $\Delta x \to 0$ or when we extend the Sum function to infinitesimal increments $\delta x \approx 0$, algebra of general sums extends to integrals. The most basic question is, 'What is the role of continuity of $f(x)$ in the definition of $\int_a^b f(x)\,dx$?' The answer is, 'Continuity guarantees us that the limit exists, or equivalently, that we get nearly the same value for each infinitesimal δx.' Extension of the integral to certain discontinuous functions is possible, but involves extra mathematical complications.

One of the most important algebraic properties of summation, the Telescoping Sum Theorem, plus the differential approximation (or microscope equation) for a smooth function will give us half of the Fundamental Theorem of Integral Calculus. That theorem tells us how to find exact (symbolic) integrals without summing or taking a limit. We build lots of 'techniques of integration' around that theorem because it gives us these exact answers. Summation is still important, however, because the sums of little strips idea is where nearly all applications of integration come from.

EXERCISE 13.26. *Run the Mathematica Notebooks **NumIntAprx.ma** and **GraphIntAprx.ma**. Type answers to the questions in the notebooks and save them to your account.*

13.4. *Mathematica* Summation

Suppose that $F(x)$ is a function of x. To form the sum

$$F(a) + F(a + \Delta x) + F(a + 2\Delta x) + \cdots + F(b)$$

we may use the single *Mathematica* command

$$\mathrm{Sum}[F(x), \{x, a, b, \Delta x\}]$$

For example, if $F(x) = 1$, $a = 0$, $b = 1$ and $\Delta x = 1$

$$\mathrm{Sum}[1, \{x, a, b, \Delta x\}] = F(0) + F(1) = 1 + 1 = 2$$

If we change to $\Delta x = \frac{1}{2}$, then

$$\mathrm{Sum}[1, \{x, a, b, \Delta x\}] = F(0) + F\left(\frac{1}{2}\right) + F(1) = 3$$

The way $\mathrm{Sum}[F(x), \{x, a, b, \Delta x\}]$ works on *Mathematica* is this: We start with $x = a$ and start with $\mathrm{Sum} = F(x) = F(a)$. Next, we add Δx to x and ask whether $x > b$. If not, we add $F(x) = F(a + \Delta x)$ to Sum, increment x again, check $x > b$ and add $F(x)$ to Sum and continue until $x > b$.

Suppose we have $F(x) = x$, $a = 0$, $b = 1$ and $\Delta x = \frac{1}{2}$, then

$$\text{Sum}[x, \{x, a, b, \Delta x\}] = F(0) + F\left(\frac{1}{2}\right) + F(1) = 0 + \frac{1}{2} + 1 = \frac{3}{2}$$

Keeping the other values, but changing to $\Delta x = \frac{1}{4}$ gives

$$\text{Sum}[x, \{x, a, b, \Delta x\}] = F(0) + F\left(\frac{1}{4}\right) + F\left(\frac{2}{4}\right) + F\left(\frac{3}{4}\right) + F(1)$$

$$= 0 + \frac{1}{4} + \frac{2}{4} + \frac{3}{4} + 1 = \frac{10}{4}$$

Values of Δx that don't exactly divide $b - a$ are also allowed, so when $\Delta x = .15$,

$$\text{Sum}[x, \{x, a, b, \Delta x\}] = 0 + .15 + .30 + .45 + .60 + .75 + .90 = 3.15$$

In integration we will take Δx smaller and smaller and still reach a limiting value in our sums by having a factor Δx in the summand. The next exercise shows you how this works.

EXERCISE 13.27. *Write out all the terms of the sums $Sum[F(x), \{x, a, b, \Delta x\}]$ defined by the following functions and increments, always using $a = 0$ and $b = 1$*

(1) $F(x) = x$, $\Delta x = \frac{1}{5}$
(2) $F(x) = x\Delta x$, $\Delta x = \frac{1}{2}$
(3) $F(x) = x\Delta x$, $\Delta x = \frac{1}{3}$
(4) $F(x) = x\Delta x$, $\Delta x = \frac{1}{4}$
(5) $F(x) = x\Delta x$, $\Delta x = \frac{1}{5}$

Compare the summed values of $Sum[x, \{x, a, b, \Delta x\}]$ and $Sum[x\Delta x, \{x, a, b, \Delta x\}]$ for $\Delta x = \frac{1}{2}$, $\Delta x = \frac{1}{3}$, $\Delta x = \frac{1}{4}$, $\Delta x = \frac{1}{5}$. (Use Mathematica if you wish.)

EXERCISE 13.28. *Run the Mathematica Notebook **Sums.ma**. Summarize your observations briefly in a text cell.*

13.5. The Algebra of Summation

Our first result is an algebraic method of computing exact sums.

THEOREM 13.29. *The Telescoping Sum Theorem*
If $F(x)$ is defined for $a \le x \le b$, then the sum of differences below equals the last value minus the first,

$$Sum[(F(x + \Delta x) - F(x)), \{x, a, b - \Delta x, \Delta x\}] = F(b') - F(a)$$

where b' is the last value of x of the form $a + n\Delta x$ which is less than or equal to b. If Δx divides $b - a$ exactly, $b' = b$.

PROOF:

$$Sum[(F(x + \Delta x) - F(x)), \{x, a, b, \Delta x\}] =$$
$$[F(a + \Delta x) - F(a)] + [F(a + 2\Delta x) - F(a + \Delta x)]$$
$$+ [F(a + 3\Delta x) - F(a + 2\Delta x)] + [F(a + 4\Delta x) - F(a + 3\Delta x)]$$
$$+ \cdots + [F(b') - F(b' - \Delta x)]$$

The part $F(a + \Delta x)$ from the first term cancels the part $-F(a + \Delta x)$ from the second term. The part $F(a + 2\Delta x)$ from the second term cancels the part $-F(a + 2\Delta x)$ from the third term, and so on. Each term has a positive and part that cancels the corresponding negative part from the next term. The negative part from the first term and the positive part from the last term are never canceled, so the sum 'telescopes' to $F(b') - F(a)$.

Notice that we sum to $b - \Delta x$ (in steps of Δx) so that the last term is $F(b)$ when Δx divides $b - a$ (rather than $[F(b + \Delta x) - F(b)]$)

EXAMPLE 13.30. *Finding the Difference*

The difficulty in using the Telescoping Sum Theorem is in finding an expression of the form $F(x + \Delta x) - F(x)$ for the summand. Consider the sum

$$\text{Sum}[\frac{\Delta x}{x(x + \Delta x)}, \{x, a, b - \Delta x, \Delta x\}]$$

Assuming that neither $x = 0$, nor $x + \Delta x = 0$, we can write

$$\frac{\Delta x}{x(x + \Delta x)} = \frac{1}{x} - \frac{1}{x + \Delta x}$$

because, putting the right hand side on a common denominator gives,

$$\frac{1}{x} - \frac{1}{x + \Delta x} = \frac{x + \Delta x}{x(x + \Delta x)} - \frac{x}{x(x + \Delta x)}$$
$$= \frac{x + \Delta x - x}{x(x + \Delta x)} = \frac{\Delta x}{x(x + \Delta x)}$$

This makes computation of the sum easy,

$$\text{Sum}[\frac{\Delta x}{x(x + \Delta x)}, \{x, a, b - \Delta x, \Delta x\}] = \text{Sum}[\frac{1}{x} - \frac{1}{x + \Delta x}, \{x, a, b - \Delta x, \Delta x\}]$$
$$= \text{Sum}[-\left(\frac{1}{x + \Delta x} - \frac{1}{x}\right), \{x, a, b - \Delta x, \Delta x\}]$$
$$= F(b') - F(a) = -\left(\frac{1}{b'} - \frac{1}{a}\right)$$
$$= \frac{1}{a} - \frac{1}{b'}$$

EXERCISE 13.31. *What's the difference?*
Use the Telescoping Sum Theorem to give exact answers to the following sums in the case where Δx divides $b - a$.

(1) $Sum[\Delta x, \{x, a, b - \Delta x, \Delta x\}]$
(2) $Sum[2x\Delta x + \Delta x^2, \{x, a, b - \Delta x, \Delta x\}]$
(3) $Sum[3x^2 \Delta x + 3x\Delta x^2 + \Delta x^3, \{x, a, b - \Delta x, \Delta x\}]$

The question really is, 'What $F(x + \Delta x) - F(x)$ equals the expressions that you are asked to sum?'

THEOREM 13.32. *Superposition of Summation*
Let α and β be constants, $F(x)$ and $G(x)$ be functions defined on $[a, b]$, then

$$Sum[\alpha F(x) + \beta G(x), \{x, a, b, \Delta x\}] =$$

$$\alpha Sum[F(x), \{x, a, b, \Delta x\}] + \beta Sum[G(x), \{x, a, b, \Delta x\}]$$

PROOF:

$$[\alpha F(a) + \beta G(a)] + [\alpha F(a + \Delta x) + \beta G(a + \Delta x)]$$
$$+ [\alpha F(a + 2\Delta x) + \beta G(a + 2\Delta x)] + \cdots + [\alpha F(b') + \beta G(b')]$$
$$=$$
$$\alpha[F(a) + F(a + \Delta x) + F(a + 2\Delta x) + \cdots + F(b')]$$
$$+ \beta[G(a) + G(a + \Delta x) + G(a + 2\Delta x) + \cdots + G(b')]$$

EXAMPLE 13.33. $Sum[x\Delta x, \{x, a, b - \Delta x, \Delta x\}] = \frac{b^2 - a^2}{2} - \Delta x \frac{b-a}{2}$

Given that

$$Sum[\Delta x, \{x, a, b - \Delta x, \Delta x\}] = Sum[([x + \Delta x] - x), \{x, a, b - \Delta x, \Delta x\}] = b - a$$
$$Sum[(x + \Delta x)^2 - x^2, \{x, a, b - \Delta x, \Delta x\}] = b^2 - a^2$$
and $(x + \Delta x)^2 - x^2 = 2x\Delta x + \Delta x^2$

we can use superposition to find

$$Sum[x\Delta x, \{x, a, b - \Delta x, \Delta x\}] = ?$$

We have

$$b^2 - a^2 = Sum[(x + \Delta x)^2 - x^2, \{x, a, b - \Delta x, \Delta x\}]$$
$$= Sum[2x\,\Delta x + \Delta x^2, \{x, a, b - \Delta x, \Delta x\}]$$
$$= 2\,Sum[x\,\Delta x, \{\cdots\}] + \Delta x\,Sum[\Delta x, \{\cdots\}]$$
$$= 2\,Sum[x\,\Delta x, \{\cdots\}] + \Delta x\,[b - a]$$

so

$$2\,Sum[x\,\Delta x, \{\cdots\}] = [b^2 - a^2] - \Delta x\,[b - a]$$
$$Sum[x\,\Delta x, \{x, a, b - \Delta x, \Delta x\}] = \frac{1}{2}\left([b^2 - a^2] - \Delta x\,[b - a]\right)$$

EXERCISE 13.34.
(1) Use telescoping sums, superposition and known sums to find

$$Sum[x^2\Delta x, \{x, a, b - \Delta x, \Delta x\}] = ?$$

Notice that

$$Sum[\Delta x, \{x, a, b - \Delta x, \Delta x\}] = [b - a]$$
$$Sum[x\Delta x, \{x, a, b - \Delta x, \Delta x\}] = \frac{1}{2}\left([b^2 - a^2] - \Delta x\,[b - a]\right)$$
$$and\ (x + \Delta x)^3 - x^3 = 3\,x^2\,\Delta x + 3\,x\,\Delta x^2 + \Delta x^3$$

(2) Use telescoping sums, superposition and known sums to find

$$Sum[x^3 \Delta x, \{x, a, b - \Delta x, \Delta x\}] = ?$$

THEOREM 13.35. *The Triangle Inequality*
For $F(x)$ defined on $[a, b]$,

$$|Sum[F(x), \{a, b - \Delta x, \Delta x\}]| \leq Sum[|F(x)|, \{a, b - \Delta x, \Delta x\}]$$

PROOF Terms of opposite sign inside the sum $|F(a) + F(a + \Delta x) + \cdots + F(b)|$ could cancel and make that sum smaller than the sum with all positive terms, $|F(a)| + |F(a + \Delta x)| + \cdots + |F(b)|$

EXERCISE 13.36. *Give an example of a sum with the cancellation described in the proof of the triangle inequality.*

THEOREM 13.37. *Additivity of Summation*
Suppose $F(x)$ is defined on $[a, c]$, Δx divides $b - a$ and $a < b < c$, then

$$Sum[F(x), \{a, b - \Delta x, \Delta x\}] + Sum[F(x), \{b, c, \Delta x\}] = Sum[F(x), \{a, c, \Delta x\}]$$

PROOF

$$[F(a) + F(a + \Delta x) + \cdots + F(b - \Delta x)] + [F(b) + F(b + \Delta x) + \cdots + F(c')]$$
$$= [F(a) + F(a + \Delta x) + \cdots + F(b - \Delta x) + F(b) + F(b + \Delta x) + \cdots + F(c')]$$

EXERCISE 13.38. *Write out the three sums in the proof of Additivity for $a = 0$, $b = 1$, $c = 2$, $\Delta x = \frac{1}{2}$ and $F(x) = x\Delta x$.*

THEOREM 13.39. *Monotony of Summation*
If $F(x) \leq G(x)$, then

$$Sum[F(x), \{a, b - \Delta x, \Delta x\}] \leq Sum[G(x), \{a, b - \Delta x, \Delta x\}]$$

PROOF: Obvious, because each term in the left sum is smaller than the corresponding term in the right sum.

EXERCISE 13.40. *Write out the two sums in the proof of Monotony for $a = 0$, $b = 1$, $\Delta x = \frac{1}{2}$, $F(x) = x^2 \Delta x$ and $G(x) = x\Delta x$.*

THEOREM 13.41. *Orientation of Summation*
Suppose $f(x)$ is defined on $[a, b]$ with $a < b$ and Δx divides $b - a$, then

$$Sum[f(x) \Delta x, \{a, b, \Delta x\}] = -Sum[f(x) \Delta x, \{b, a, -\Delta x\}]$$

PROOF: To go from b to a we must take negative steps.

EXERCISE 13.42. *Write out the two sums in the proof of Orientation for $a = 0$, $b = 1$, $\Delta x = \frac{1}{2}$ and $F(x) = x\Delta x$.*

13.6. The Algebra of Infinite Summation

For a fixed choice of the real function $f(x)$, the expression

$$\Sigma\left(a,b,\Delta x\right) = \mathrm{Sum}[f\left(x\right)\Delta x, \{a, b - \Delta x, \Delta x\}]$$

is a real function of three variables, a, b and Δx. Therefore, we may apply the Function Extension Axiom to Σ and use input values $\delta x \approx 0$. Since functional identities are preserved for the natural extensions of real functions we have the following properties of infinite sums: when $\delta x \approx 0$ divides $b - a$,

$$\mathrm{Sum}[\left(F\left(x + \delta x\right) - F\left(x\right)\right), \{x, a, b - \delta x, \delta x\}] = F\left(b\right) - F\left(a\right)$$

$$\begin{aligned}\mathrm{Sum}[\alpha F\left(x\right) &+ \beta G\left(x\right), \{x, a, b, \delta x\}] \\ &= \alpha \mathrm{Sum}[F\left(x\right), \{x, a, b, \delta x\}] + \beta \mathrm{Sum}[G\left(x\right), \{x, a, b, \delta x\}]\end{aligned}$$

$$\left|\mathrm{Sum}[F\left(x\right), \{a, b - \delta x, \delta x\}]\right| \leq \mathrm{Sum}[\left|F\left(x\right)\right|, \{a, b - \delta x, \delta x\}]$$

$$\mathrm{Sum}[F\left(x\right), \{a, b - \delta x, \delta x\}] + \mathrm{Sum}[F\left(x\right), \{b, c, \delta x\}] = \mathrm{Sum}[F\left(x\right), \{a, c, \delta x\}]$$

$$\mathrm{Sum}[f\left(x\right)\delta x, \{a, b, \delta x\}] = -\mathrm{Sum}[f\left(x\right)\delta x, \{b, a, -\delta x\}]$$

Since we know that

$$\int_a^b f\left(x\right)\,dx \approx \mathrm{Sum}[f\left(x\right)\delta x, \{x, a, b - \delta x, \delta x\}]$$

these infinite sum properties show that integrals have the following properties:

THEOREM 13.43. *Superposition*

$$\int_a^b \left(\alpha f\left(x\right) + \beta g\left(x\right)\right)\,dx = \alpha \int_a^b f\left(x\right)\,dx + \beta \int_a^b g\left(x\right)\,dx$$

THEOREM 13.44. *Triangle Inequality*

$$\left|\int_a^b f\left(x\right)\,dx\right| \leq \int_a^b \left|f\left(x\right)\right|\,dx, \left(a < b\right)$$

THEOREM 13.45. *Monotony*

$$\int_a^b f\left(x\right)\,dx \leq \int_a^b g\left(x\right)\,dx, \quad if f\left(x\right) \leq g\left(x\right)$$

THEOREM 13.46. *Additivity*

$$\int_a^b f\left(x\right)\,dx + \int_b^c f\left(x\right)\,dx = \int_a^c f\left(x\right)\,dx$$

THEOREM 13.47. *Orientation*

$$\int_b^a f(x)\, dx = -\int_a^b f(x)\, dx$$

EXAMPLE 13.48. *Integral Computation the Hard Way*

We have seen that

$$\text{Sum}[x\Delta x, \{x, a, b - \Delta x\}] = \frac{1}{2}[(b^2 - a^2) - \Delta x(b - a)]$$

so by natural extension of the sum function to $\delta x \approx 0$,

$$\int_a^b x\, dx \approx \text{Sum}[x\delta x, \{x, a, b - \delta x\}]$$

$$= \frac{1}{2}[(b^2 - a^2) - \delta x(b - a)]$$

$$\approx \frac{1}{2}(b^2 - a^2)$$

and the two real quantities must be equal,

$$\int_a^b x\, dx = \frac{1}{2}(b^2 - a^2)$$

EXERCISE 13.49.

1) Compute $\int_a^b x^2\, dx$ by extending the exact formula you computed earlier for

$$\text{Sum}[x^2\Delta x, \{x, a, b - \Delta x\}] \qquad \text{to } \delta x \approx 0$$

2) Compute $\int_a^b x^3\, dx$ by extending the exact formula for

$$\text{Sum}[x^3\Delta x, \{x, a, b - \Delta x\}] \qquad \text{to } \delta x \approx 0$$

THEOREM 13.50. *Estimation of Sums*

$$\text{Sum}[|F(x, \delta x)|\ \delta x, \{a, b - \delta x, \delta x\}] \le |F(x_{Max}, \delta x)| \times (b - a)$$

PROOF: This is the last technical result we need from algebra. We use the extension of the finite maximum function. Define a function like *Mathematica's* sum

$$\text{Max}[F(x, \Delta x), \{a, b - \Delta x, \Delta x\}] =$$
$$\text{maximum}[F(a, \Delta x), F(a + \Delta x, \Delta x), F(a + 2\Delta x, \Delta x), \ldots, F(b', \Delta x)]$$

We know that
$$\text{Max}[F(x, \Delta x), \{a, b - \Delta x, \Delta x\}] = F(x_M, \Delta x)$$

for some real x_M of the form $x_M = a + n\Delta x$, $a \le x_M \le b - \Delta x$. By natural extension of the Max function to infinitesimal δx,

$$\text{Max}[F(x, \delta x), \{a, b - \delta x, \delta x\}] = F(x_M, \delta x)$$

for some hyperreal x_M of the form $x_M = a + n\delta x$, $a \le x_M \le b$.

The formula we need is the formal statement that we can estimate by making all the terms of a sum larger,

$$\text{Sum}[|F(x, \delta x)| \delta x, \{a, b - \delta x, \delta x\}]$$
$$\leq \text{Max}[|F(x, \delta x)|, \{a, b - \delta x, \delta x\}] \times \text{Sum}[\delta x, \{a, b - \delta x, \delta x\}]$$
$$= |F(x_M, \delta x)| \times \text{Sum}[\delta x, \{a, b - \delta x, \delta x\}] = |F(x_M, \delta x)| \times (b - a)$$

EXERCISE 13.51. *Estimation*
We showed above that

$$\text{Sum}[\frac{\Delta x}{x(x + \Delta x)}, \{x, a, b - \Delta x, \Delta x\}] = \frac{1}{a} - \frac{1}{b'}$$

provided that neither $x = 0$ nor $x + \Delta x = 0$ for any term of the sum.
  Compute the integral

$$\int_a^b \frac{1}{x^2}\, dx = \frac{1}{a} - \frac{1}{b}$$

by estimating the difference between the unknown infinite sum

$$\text{Sum}[\frac{\delta x}{x^2}, \{x, a, b - \delta x, \delta x\}]$$

and the extension of the sum above,

$$\text{Sum}[\frac{\delta x}{x(x + \delta x)}, \{x, a, b - \delta x, \delta x\}] = \frac{1}{a} - \frac{1}{b'}$$

HINT: Calculate the difference $\left(\frac{1}{x^2} - \frac{1}{x(x+\delta x)} \right) \delta x$ to estimate

$$\text{Sum}[\frac{\delta x}{x^2}, \{x, a, b - \delta x, \delta x\}] - \text{Sum}[\frac{\delta x}{x(x + \delta x)}, \{x, a, b - \delta x, \delta x\}]$$

Is your computation valid if $a < 0 < b$?

13.7. The Fundamental Theorem of Integral Calculus, Part 1

This section gives the theory, the next chapter gives techniques for using the theory.
  The reader should recall the differential approximation for a smooth function 5.15 at this time, because we are about to use it to prove

THEOREM 13.52. *First Half of the Fundamental Theorem*
Given an integrand $f(x)\, dx$ suppose that we can find a function $F(x)$ such that $dF(x) = f(x)\, dx$ (or $\frac{dF}{dx} = f(x)$) for all $a \leq x \leq b$, then

$$\int_a^b f(x)\, dx = F(b) - F(a)$$

We also sometimes write

$$\int_a^b f(x)\, dx = \int_a^b dF(x) = F(x)\,|_a^b = F(b) - F(a)$$

The notation $F(x)\,|_a^b$ simply means $F(b) - F(a)$.

PROOF: The differential approximation 5.15 (microscope equation) for $F(x)$ is

$$F(x + \delta x) - F(x) = f(x)\,\delta x + \varepsilon \cdot \delta x$$

for all (hyperreal) x satisfying $a \le x \le b$, where $\varepsilon \approx 0$ when $\delta x \approx 0$. Recall that this forces $f(x)$ to be a continuous function (for the ordinary derivative defined in Definition 5.15. See the Mathematical Background Chapter on "Epsilons and Deltas" for further details.)

The telescoping sum and superposition properties say

$$\begin{aligned}
F(b) - F(a) &= \text{Sum}[F(x + \delta x) - F(x), \{a, b - \delta x, \delta x\}] \\
&= \text{Sum}[f(x)\,\delta x + \varepsilon \cdot \delta x, \{a, b - \delta x, \delta x\}] \\
&= \text{Sum}[f(x)\,\delta x, \{a, b - \delta x, \delta x\}] + \text{Sum}[\varepsilon \cdot \delta x, \{a, b - \delta x, \delta x\}] \\
&\approx \int_a^b f(x)\,dx + \text{Sum}[\varepsilon \cdot \delta x, \{a, b - \delta x, \delta x\}]
\end{aligned}$$

We conclude the proof by showing that

$$\text{Sum}[\varepsilon \cdot \delta x, \{a, b - \delta x, \delta x\}] \approx 0$$

The triangle inequality says

$$|\text{Sum}[\varepsilon \cdot \delta x, \{a, b - \delta x, \delta x\}]| \le \text{Sum}[|\varepsilon|\delta x, \{a, b - \delta x, \delta x\}]$$

The term ε in the sum actually depend on x and δx, but we know from the differential approximation that $\varepsilon(x, \delta x) \approx 0$ for all x in $[a, b]$ as long as $\delta x \approx 0$. We finish by the simple observation

$$\begin{aligned}
&\text{Sum}[|\varepsilon(x, \delta x)|\delta x, \{a, b - \delta x, \delta x\}] \\
&\le \text{Max}[|\varepsilon(x, \delta x)|, \{a, b - \delta x, \delta x\}] \times \text{Sum}[\delta x, \{a, b - \delta x, \delta x\}] \\
&= |\varepsilon(x_0, \delta x)| \times (b - a) \\
&\approx 0
\end{aligned}$$

This shows that the two real quantities satisfy

$$F(b) - F(a) \approx \int_a^b f(x)\,dx$$

which forces them to be equal and proves the theorem.

Strange though it sounds, we have shown two things with the proof of the first half of the Fundamental Theorem: (1) the integral exists! and (2) its value is $F(b) - F(a)$. (Instructor Note: The usual continuity hypothesis is hidden in our ability to find a function $F(x)$ satisfying the (uniform) microscope equation 5.15 with $f(x) = F'(x)$ for all x in $[a, b]$.)

EXAMPLE 13.53. *Computation of $\int_a^b x^2\,dx$ the Easy Way*

Compare the following use of the Fundamental Theorem with your direct computation from the last section. We have $f(x) = x^2$, so if we take $F(x) = x^3/3$, then $dF(x) = x^2\,dx$ and

$$\int_a^b x^2\,dx = \frac{1}{3}x^3\Big|_a^b = \frac{1}{3}\left(b^3 - a^3\right)$$

EXAMPLE 13.54. *Integral of Cosine*

$$\int_a^b \mathrm{Cos}\,(\theta)\,d\theta = \mathrm{Sin}\,(\theta)\,\big|_a^b = \mathrm{Sin}\,(b) - \mathrm{Sin}\,(a)$$

It would be quite hard to compute this integral without the help of the Fundamental Theorem.

If $f(x)$ is not continuous everywhere on $[a,b]$, the Fundamental theorem does not apply and we shall see that using $F(b) - F(a)$ anyway will lead to errors, even if $\frac{dF}{dx} = f(x)$ at all but one point of $[a,b]$.

FIGURE 13.55: Area Under the Curve $y = 1/x^2$

EXERCISE 13.56. *Is the following computation correct?*

$$\int_{-1}^{2} \frac{1}{x^2}\,dx = \int_{-1}^{2} x^{-2}\,dx$$

$$= -x^{-1}\big|_{-1}^{2} = \frac{1}{x}\big|_2^{-1}$$

$$= -1 - \frac{1}{2} = -\frac{3}{2}$$

Note that if $F(x) = -x^{-1}$, then $dF(x) = x^{-2}\,dx$, but how could the "area" under a positive curve be a negative number?

Draw a picture and compute the areas $\int_{.0001}^{1} \frac{1}{x^2}\,dx$, $\int_{-1}^{-.0001} \frac{1}{x^2}\,dx$ and $\int_{+1}^{2} \frac{1}{x^2}\,dx$. What do you think the area $\int_{-.0001}^{.0001} \frac{1}{x^2}\,dx$ should be?

Incorrect programs can produce incorrect results.

EXERCISE 13.57. *Garbage In, Garbage Out*
What is wrong with the following "Mathematica computation" of $\int_0^2 \frac{1}{\mathrm{Cos}^2[x]}\,dx$?

In[1]
```
f:= 1/(Cos[x])∧2;
Integrate[f, {x, 0, 2}]     < Enter >
```
Out[1]
$$\frac{\mathrm{Sin}[2]}{\mathrm{Cos}[2]}$$
In[2]
```
N[%]
```
Out[2]
-2.18504

Can the integral of a positive function be negative? Is the function $F(x) = \text{Sin}[x]/\text{Cos}[x]$ an antiderivative for the function $f(x) = 1/(\text{Cos}[x])^2$ over the whole interval $0 \leq x \leq 2$? That is, does $dF(x) = f(x)\,dx$ for all these values of x? Use Mathematica to plot the integrand function over the interval $0 \leq x \leq 2$.

13.8. The Fundamental Theorem of Integral Calculus, Part 2

When we cannot find an antiderivative $F(x)$ for a given $f(x)$, we sometimes still want to work directly with the definition of the integral. The **NumIntAprx.ma** *Mathematica* Notebook even shows us more efficient ways to estimate the limit of sums directly. Continuity of $f(x)$ is needed to show that the limit actually 'converges.' With discontinuous integrands, it is possible to make the sums oscillate as $\Delta x \to 0$. In these cases *Mathematica's* NIntegrate[] command will likely give the wrong answer.

We need naked existence of the limit for the second half of the Fundamental Theorem. The first half of the Fundamental Theorem is used often, whereas the second half is seldom used. This is why we have postponed a general proof of existence (without antiderivatives).

THEOREM 13.58. *Existence of the Definite Integral*
Let $f(x)$ be a continuous function of $[a, b]$, then there is a real number I such that

$$Sum[f(x)\,\Delta x, \{x, a, b - \Delta x, \Delta x\}] \to I$$

as $\Delta x \to 0$, or equivalently,

$$Sum[f(x)\,\delta x, \{x, a, b - \delta x, \delta x\}] \approx I$$

for any $\delta x \approx 0$.

PROOF: First, by the Extreme Value Theorem for Continuous Functions, $f(x)$ has a min, m, and a Max, M, on the interval $[a, b]$. Monotony of summation tells us

$$m \times (b - a) \leq \text{Sum}[f(x)\,\delta x, \{x, a, b - \delta x, \delta x\}] \leq M \times (b - a)$$

So that $\text{Sum}[f(x)\,\delta x, \{x, a, b - \delta x, \delta x\}]$ is a finite number and thus near some real value $I(\delta x) \approx \text{Sum}[f(x)\,\delta x, \{x, a, b - \delta x, \delta x\}]$. What we need to show is that if we choose a different infinitesimal, say δu, then $I(\delta x) = I(\delta u)$ or

$$\text{Sum}[f(x)\,\delta x, \{x, a, b - \delta x, \delta x\}] \approx \text{Sum}[f(u)\,\delta u, \{u, a, b - \delta u, \delta u\}]$$

Draw a picture of two different rectangular approximations to get this idea more clearly. If we superimpose the different infinitesimal partitions and consider two overlapping rectangles, the areas only differ by an infinitesimal on a scale of the increments, because $f(x) \approx f(u)$, by continuity, for $x \approx u$.

The proof that we do get the same real number is taken up in detail in the Mathematical Projects.

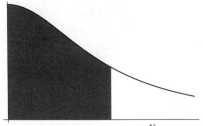

FIGURE 13.59: $A(X) = \int_a^X f(x)\, dx$

THEOREM 13.60. *Second Half of the Fundamental Theorem*
Suppose that $f(x)$ is a continuous function on an interval containing a and we define a new function by 'accumulation,'

$$A(X) = \int_a^X f(x)\, dx$$

Then $A(X)$ is smooth and $\frac{dA}{dX}(X) = f(X)$, in other words,

$$\frac{d}{dx}\int_a^X f(x)\, dx = f(X)$$

PROOF: We show that $A(X)$ satisfies the differential approximation $A(X + \delta X) = A(X) + f(X)\,\delta X + \varepsilon \cdot \delta X$

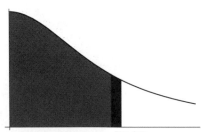

FIGURE 13.61: $\int_a^{X+\delta X} f(x)\, dx$

By definition of A and the additivity property of integrals, we have

$$A(X + \delta X) = \int_a^{X+\delta X} f(x)\, dx = \int_a^X f(x)\, dx + \int_X^{X+\delta X} f(x)\, dx$$

$$= A(X) + \int_X^{X+\delta X} f(x)\, dx$$

So we need to show that

$$\int_X^{X+\delta X} f(x)\, dx = f(X)\,\delta X + \varepsilon \cdot \delta X$$

with $\varepsilon \approx 0$, when $\delta X \approx 0$.

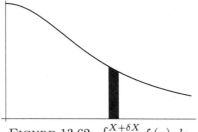

FIGURE 13.62: $\int_X^{X+\delta X} f(x)\, dx$

This is where we use the continuity hypothesis about $f(x)$. The Extreme Value Theorem for Continuous Functions 11.12 says that $f(x)$ has a max and a min on the interval $[X, X + \delta X]$, say $m = f(X_m) \leq f(x) \leq M = f(X_M)$ for all $X \leq x \leq X + \delta X$. Monotony of the integral gives us the estimates

$$m\delta X = \int_X^{X+\delta X} m\, dx \leq \int_X^{X+\delta X} f(x)\, dx \leq \int_X^{X+\delta X} M\, dx = M \cdot \delta X$$

Since both X_m and X_M lie in the interval $[X, X + \delta X]$ we know that $X_m \approx X_M$, when $\delta X \approx 0$. Continuity of $f(x)$ means that $f(X_m) \approx f(X_M) \approx f(X)$ in this case, so

$$\int_X^{X+\delta X} f(x)\, dx = f(X)\delta X + \varepsilon \cdot \delta X$$

using upper and lower estimates of the integral based on the max and min of the function over the small subinterval.

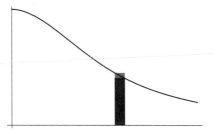

FIGURE 13.63: Upper and Lower Estimates

This proves the theorem, because we have verified the microscope equation from Definition 5.15

$$A(X + \delta X) = A(X) + A'[X]\delta X + \varepsilon \cdot \delta X$$

with $A'[X] = f(X)$.

EXERCISE 13.64. *Use the Second Half of The Fundamental Theorem to explain how YOU could compute* Log(x) *and* ArcTan(x) *directly using a numerical integration program. We know that*

$$\frac{d\,\mathrm{Log}(x)}{dx} = \frac{1}{x} \quad and \quad \frac{d}{dX} \int_1^X \frac{1}{x}\, dx = \frac{1}{X}$$

and

$$\frac{d\,\mathrm{ArcTan}(x)}{dx} = \frac{1}{1+x^2} \quad and \quad \frac{d}{dX} \int_0^X \frac{1}{1+x^2}\, dx = \frac{1}{1+X^2}$$

*Use the Numerical Integration Notebook **NumIntAprx.ma** to compute the various numerical integral approximations for these functions and compare your results with Mathematica's built-in algorithms for these functions. For example, how do the following compare:*

$$\text{Log}\,[7.38905] \approx NIntegrate[1/x, \{x, 1, 7.38905\}]$$

and

$$4 \times \text{ArcTan}\,[1] \approx 4\,NIntegrate[1/(1 + x \wedge 2), \{x, 0, 1\}]$$

Explain in terms of the exact symbolic computations why we have chosen these particular numbers (7.38905 in log and the 4 in arctangent).

Sometimes it is not possible to find an antiderivative in terms of elementary functions, but even in these cases, the integral defines a function and the numerical approximations give a direct way to compute the functions.

EXERCISE 13.65. *Non-Elementary Integrals*
Use Mathematica's numerical integral "NIntegrate[]" to approximate the following integrals:

a) $\int_0^1 e^{-x^2}\,dx = ?$ b) $\int_0^1 \frac{\text{Sin}[x]}{x}\,dx = ?$ c) $\int_0^1 \text{Sin}[x^2]\,dx =$

*Also use Mathematica's symbolic integration "Integrate[]" to find expressions for these integrals. These expressions are not combinations of elementary functions. See the symbolic integration NoteBook **SymbolicIntegr.ma** for details. The point of the exercise is that numerical integration can still be used to approximate the integrals even though they do not have elementary antiderivatives.*

CHAPTER 14

Symbolic Integration

This chapter shows you the main symbolic tricks of the 'integration' trade. These basic methods are important, but there are many more tricks that are useful in special situations. We will encounter a few of the special tricks later when they arise in important contexts. For example, the cable of a suspension bridge can be described by a differential equation which can be antidifferentiated with the "hyperbolic cosine." If you decide to work on that project, you will want to learn that trick. "Partial fractions" is another integration trick that arises in the logistic growth model, the S-I-S disease model, and the linear air resistance model in a basic form. We will take that method up when we need it. Remember that *Mathematica* has very sophisticated symbolic integration routines for you to use on very complicated problems. The basic techniques will help you set problems up and use *Mathematica*.

The Fundamental Theorem of Integral Calculus gives us an indirect way to exactly compute the limit of approximations by sums of the form

$$f(a)\,\Delta x + f(a+\Delta x)\,\Delta x + f(a+2\Delta x)\,\Delta x + \cdots + f(b-\Delta x)\,\Delta x$$
$$= \text{Sum}[f(x)\,\Delta x, \{x, a, b-\Delta x, \Delta x\}]$$

We have

$$\int_a^b f(x)\,dx = \lim_{\delta x \to 0} \text{Sum}[f(x)\,\Delta x, \{x, a, b-\Delta x, \Delta x\}]$$

but the limit can be computed without forming the sum. The Fundamental Theorem says that if we can find $F(x)$ so that $dF(x) = f(x)\,dx$, for all $a \le x \le b$, then

$$\int_a^b f(x)\,dx = \lim_{\Delta x \to 0} \text{Sum}[f(x)\,\Delta x, \{x, a, b-\Delta x, \Delta x\}] = F(b) - F(a)$$

We can skip from the left side of the above equations to the right side without going through the limit in the middle. This indirect computation of the integral works any time we can find a trick to to figure out an antiderivative. There are many such tricks or "techniques" and *Mathematica* knows them all. Your main task in this chapter is to understand the fundamental techniques and their limitations, rather than to develop skill at very elaborate integral computations. (The Mathematical Background has a Chapter on Integration Drill if you need more practice.)

The rules of integration are harder than the rules of differentiation because they amount to trying to use the rules of differentiation in reverse. You must learn all the basic techniques to understand this, but we won't wallow very deep into elaborate examples and esoteric techniques. That is left for *Mathematica*.

14.1. Indefinite Integrals

The first half of the Fundamental Theorem means that we can often find an integral in two steps. (1) Find an antiderivative. (2) Compute the difference in values of the antiderivative. It makes no difference which antiderivative we use, as we shall see (and was implicit in the proof.)

One antiderivative of $3x^2$ is x^3, since $\frac{dx^3}{dx} = 3x^2$, but another antiderivative of $3x^2$ is $x^3 + 273$, since the derivative of the constant 273 (or any other constant) is zero. The integral $\int_a^b 3x^2 dx$ may be computed with either antiderivative.

$$\int_a^b 3x^2 dx = x^3|_a^b = b^3 - a^3$$

$$= [x^3 + 273]|_a^b = [b^3 + 273] - [a^3 + 273] = b^3 - a^3$$

Our next result is the converse of the result that says the derivative of a constant is zero. It says that if the derivative is zero, the function must be constant. This is geometrically obvious - draw the graph of a function with a zero derivative!

THEOREM 14.1. *The Zero Derivative Theorem*
Suppose $F(x)$ and $G(x)$ are both antiderivatives for the same function $f(x)$ on the interval $[a, b]$, that is, $\frac{dF}{dx}(x) = \frac{dG}{dx}(x) = f(x)$ for all x in $[a, b]$. Then $F(x)$ and $G(x)$ differ by a constant for all x in $[a, b]$.

PROOF: The function $H(x) = F(x) - G(x)$ has zero derivative on $[a, b]$. We have

$$H(X) - H(a) = \int_a^X 0 \, dx$$
$$= 0$$

by the first half of the Fundamental Theorem and direct computation of the integral of zero.

If X is any real value in $[a, b]$, $H(X) - H(a) = 0$. This means $H(X) = H(a)$, a constant. In turn, this tells us that $F(X) - G(X) = F(a) - G(a)$, a constant for all X in $[a, b]$.

DEFINITION 14.2. *Notation for the Indefinite Integral*
The indefinite integral of a function $f(x)$ denoted $\int f(x) \, dx$ is equal to the collection of all functions $F(x)$ with differential $dF(x) = f(x) \, dx$ or derivative $\frac{dF}{dx}(x) = f(x)$.

We write '+c' after an answer to indicate all possible antiderivatives. For example,

$$\int \text{Cos}[\theta] \, d\theta = \text{Sin}[\theta] + c$$

14.2. Specific Integral Formulas

$$\int x^p \; dx = \frac{1}{p+1} x^{p+1} + c, \quad p \neq -1$$

$$\int \frac{1}{x} \; dx = \text{Log}[x] + c, \quad x > 0$$

$$\int e^x \; dx = e^x + c$$

$$\int \text{Sin}[x] \; dx = -\text{Cos}[x] + c$$

$$\int \text{Cos}[x] \; dx = \text{Sin}[x] + c$$

EXAMPLE 14.3. *Guess and Correct*

Suppose we want to find

$$\int 3 \, x^5 \; dx = F(x)?$$

We know that if we differentiate a power function, we reduce the exponent by one, so we guess and check our answer,

$$F_1(x) = x^6 \qquad\qquad F_1'(x) = 6 \, x^5$$

The constant is wrong, but would be correct if we chose

$$F(x) = \frac{1}{2} \, x^6 \qquad\qquad F'(x) = \frac{1}{2} \cdot 6 \, x^5 = 3 \, x^5$$

Here is another example of guessing and correcting the guess by adjusting a constant. Find

$$\int 7 \, \text{Cos}[3 \, x] \; dx = G(x)$$

Begin with the first guess and check

$$G_1(x) = \text{Sin}[3 \, x] \qquad\qquad G_1'(x) = 3 \, \text{Cos}[3x]$$

Adjusting our guess gives

$$G(x) = \frac{7}{3} \, \text{Sin}[3 \, x] \qquad\qquad G'(x) = \frac{7}{3} \, 3 \, \text{Cos}[3x] = 7 \, \text{Cos}[3 \, x]$$

It is best to check your work in any case, so you only need to remember the 5 specific basic formulas above and use them to adjust your guesses. We will learn general rules based on each of the rules for differentiation, but used in reverse. First, here is some basic drill.

EXERCISE 14.4. *Basic Drill on Guessing and Correcting*

a) $\int 7\sqrt{x} \; dx = \;?$

b) $\int 5 \, x^3 \; dx = \;?$

c) $\int \frac{3}{x^2} \; dx = \;?$

d) $\int x^{\frac{3}{2}} \; dx = \;?$

e) $\int (5 - x)^2 \; dx = \; ?$ *f)* $\int \text{Sin}[3\,x] \; dx = \; ?$

g) $\int e^{2\,x} \; dx = \; ?$ *h)* $\int \frac{-7}{x} \; dx = \; ?$

You can't do anything you like with indefinite integrals and expect to get the intended function. In particular, the "dx" in the integral tells you the variable of differentiation for the intended answer.

EXERCISE 14.5. *Explain what is wrong with the following nonsense:*

$$\int x^2 \; dx = \int x \cdot x \; dx = x \int x \; dx$$

$$= x \left[\frac{1}{2} x^2 + c \right] = \frac{1}{2} x^3 + c\,x$$

and

$$\int_0^1 x^2 \; dx = \int_0^1 x \cdot x \; dx = x \int_0^1 x \; dx$$

$$= x \left[\frac{1}{2} x^2 \big|_0^1 \right] = \frac{1}{2} x$$

14.3. Superposition of Antiderivatives

EXERCISE 14.6. *Superposition of Derivatives in Reverse*
Prove the superposition rule

$$\int a\,f(x) + b\,g(x) \; dx = a \int f(x) \; dx + b \int g(x) \; dx$$

by letting $F(x) = \int f(x) \; dx$, $G(x) = \int g(x) \; dx$ *and writing out what the claim for indefinite integrals means in terms of these functions. Do the arbitrary constants of integration matter?*

Now use your rule to break linear combinations of integrands into simpler pieces.

EXERCISE 14.7. *Superposition of Antiderivatives Drill*

a) $\int 5\,x^3 - 2 \; dx = 5 \int x^3 \; dx - 2 \int 1 \; dx = \; ?$

b) $\int \frac{3}{x^2} - \sqrt{x} \; dx = 3 \int \frac{1}{x^2} \; dx - \int \sqrt{x} \; dx \; ?$

c) $\int 5\,\text{Sin}[x] - e^{2\,x} \; dx = \; ?$

d) $\int \text{Cos}[5\,x] - \frac{5}{x} \; dx = \; ?$

Remember that the computer can be used to check your work on basic skills.

EXERCISE 14.8. *Run the Mathematica NoteBook* **SymbolicIntegr.ma** *and use Integrate[.] to check your work from the previous exercise.*

14.4. Change of Variables or 'Substitution'

One way to find an indefinite integral is to change the problem into a simpler one. Of course, you want to change it to an equivalent problem. Change of variables can be legitimately done as follows. First, let $u =$ part of the integrand. Next, calculate $du = \cdots$. If the remaining part of the integrand is du, make the substitution and otherwise, try a different substitution.

The point is that we must look for both an expression and its differential. Here is a very simple example.

EXAMPLE 14.9. *A Change of Variable and Differential*

Find

$$\int 2x \sqrt{1+x^2} \, dx = \int \sqrt{[1+x^2]} \{2x \, dx\}$$

Begin with

$$u = [1+x^2] \qquad\qquad du = \{2x \, dx\}$$

We replace the expression for u and du obtaining the simpler problem: Find

$$\int \sqrt{[u]} \{du\} = \int u^{\frac{1}{2}} \, du$$

$$= \frac{1}{1+\frac{1}{2}} \, u^{1+\frac{1}{2}} + c$$

$$= \frac{2}{3} \, u^{\frac{3}{2}} + c$$

The expression $\frac{2}{3} u^{\frac{3}{2}} + c$ is not an acceptable answer to the question, "What functions of x have derivative $2x \sqrt{1+x^2}$?" However, if we remember that $u = 1+x^2$, we can express the answer as

$$\frac{2}{3} \, u^{\frac{3}{2}} + c = \frac{2}{3} \, [1+x^2]^{\frac{3}{2}} + c$$

Checking the answer will show why this method works. We use the chain rule:

$$y = \frac{2}{3} \, u^{\frac{3}{2}} \qquad\qquad u = 1+x^2$$

so

$$\frac{dy}{du} = \frac{2}{3} \frac{3}{2} \, u^{\frac{3}{2}-1} \qquad\qquad \frac{du}{dx} = 2x$$

and

$$\frac{dy}{dx} = \frac{dy}{du} \frac{du}{dx} = u^{\frac{1}{2}} \, 2x = 2x\sqrt{1+x^2}$$

or

$$dy = 2x \sqrt{1+x^2} \, dx$$

EXAMPLE 14.10. *Another Change*

You MUST substitute for the differential du associated with your change of variables $u = \cdots$. Sometimes this is a little complicated. For example, suppose we try to compute

$$\int 2x\,\sqrt{1+x^2}\,dx$$

with the change of variable and differential

$$v = \sqrt{1+x^2} \qquad\qquad dv = \frac{x}{\sqrt{1+x^2}}\,dx$$

Our integral becomes

$$\int 2v[\text{ rest of above}]$$

with the rest of the above equal to $x\,dx$. We can find $dv = \frac{x}{\sqrt{1+x^2}}dx$ by multiplying numerator and denominator by $\sqrt{1+x^2}$, so our integral becomes

$$\int 2\sqrt{1+x^2}\,x\,\frac{\sqrt{1+x^2}}{\sqrt{1+x^2}}\,dx = \int 2\left(\sqrt{1+x^2}\right)^2[\frac{x}{\sqrt{1+x^2}}\,dx]$$

$$= 2\int v^2 dv = \frac{2}{3}v^3 + c$$

$$= \frac{2}{3}\left[1+x^2\right]^{\frac{3}{2}}$$

This is the same answer as before, but the substitution was more difficult on the dv piece.

EXAMPLE 14.11. *A Failed Attempt*

Sometimes an attempt to simplify an integrand by change of variables will lead you to either a more complicated integral or to a situation where you can not make the substitution for the differential. In these cases, scratch off your work and try another change.

Suppose we try a grand simplification of

$$\int 2x\,\sqrt{1+x^2}\,dx$$

taking

$$w = x\,\sqrt{1+x^2} \qquad\qquad dw = \left(\sqrt{1+x^2} + \frac{x^2}{\sqrt{1+x^2}}\right)dx$$

We might substitute w for the whole integrand, but there is nothing left to substitute for dw and we can not complete the substitution. We simply have to try a different method.

EXERCISE 14.12. *Change of Variable and Differential Drill*

a) $\int \frac{1}{(3x-2)^2}\,dx = $?

b) $\int \frac{2t}{\sqrt{1-t^2}}\,dt = $?

c) $\int x(3+7x^2)^3\,dx = $?

d) $\int \frac{3y}{(2+2y^2)^2}\,dy = $?

e) $\int \text{Cos}^3[x]\,\text{Sin}[x]\,dx = $?

f) $\int \frac{\text{Cos}[\log[x]]}{x}\,dx = $?

g) $\int e^{\text{Cos}[\theta]}\,\text{Sin}[\theta]\,d\theta = $?

h) $\int \text{Sin}[ax+b]\,dx = $?

14.4.1. Change of Limits of Integration - Pay Me Now or Pay Me Later. When we want an antiderivative such as $\int 2x\sqrt{1+x^2}\,dx$, we have no choice but to re-substitute the expressions for new variables back into our answer. In the first example of the subsection above,

$$\int 2x\sqrt{1+x^2}\,dx = \frac{2}{3}u^{\frac{3}{2}} + c = \frac{2}{3}[1+x^2]^{\frac{3}{2}}$$

However, when we want to compute a definite integral like

$$\int_5^7 2x\sqrt{1+x^2}\,dx$$

we can change the limits of integration along with the change of variables. For example,

$$u = 1+x^2 \qquad\qquad\qquad du = 2x\,dx$$
$$u(5) = 26 \qquad\qquad\qquad u(7) = 50$$

so the new problem is to find

$$\int_{26}^{50} u^{\frac{1}{2}}\,du$$

which equals

$$\frac{2}{3}u^{\frac{3}{2}}\Big|_{26}^{50} = \frac{2}{3}[50^{\frac{3}{2}} - 26^{\frac{3}{2}}] \approx 235.702 - 88.383 - 147.319$$

We recommend that you do your definite integral changes of variable this way. You CAN compute the antiderivative in terms of x and then use the original limits of integration, but there is a danger that you will loose track of which variable you started with. If you are careful, both methods give the same answer, for example,

$$\int_5^7 2x\sqrt{1+x^2}\,dx = \frac{2}{3}[1+x^2]^{\frac{3}{2}}\Big|_5^7 = \frac{2}{3}\left([1+7^2]^{\frac{3}{2}} - [1+5^2]^{\frac{3}{2}}\right) \approx 147.319$$

EXERCISE 14.13. *Change of Variables with Limits Drill*

a) $\int_0^1 \frac{x}{1+x^2}\,dx = ?$

b) $\int_0^{\pi/6} \mathrm{Sin}[3\theta]\,d\theta = ?$

c) $\int_2^3 \frac{2x}{(x^2-3)^2}\,dx = ?$

d) $\int_0^1 ax+b\,dx = ?$

e) $\int_0^1 \frac{3x^2-1}{1+\sqrt{x-x^3}}\,dx = 0$ *because of the new limits.*

f) $\int_0^1 x\,e^{-x^2}\,dx = ?$

g) $\int_0^{\pi/2} \mathrm{Cos}[\theta]\,\mathrm{Sin}[\theta]\,d\theta = ?$

EXAMPLE 14.14. *A Less Obvious Substitution with $u = \sqrt{x}$*

We may be slipping into the symbol swamp, but a little wallowing can be fun. Here is a change of variable and differential with a twist:

$$\int \frac{\sqrt{x}}{1+x}\, dx$$

$$u = \sqrt{x} \quad \Leftrightarrow \quad u^2 = x \quad (x > 0)$$

$$du = \frac{1}{2\sqrt{x}}\, dx \quad \Leftrightarrow \quad 2\,u\, du = dx$$

$$\int \frac{u}{1+u^2}\, 2\,u\, du = 2 \int \frac{u^2}{1+u^2}\, du = 2 \left[\int \left(1 - \frac{1}{1+u^2} \right) du \right]$$

$$= 2 \int du - 2 \int \frac{1}{1+u^2}\, du = 2\,u - 2\,\text{ArcTan}[u] + c$$

$$\int \frac{\sqrt{x}}{1+x}\, dx = 2\sqrt{x} - 2\,\text{ArcTan}[\sqrt{x}] + c$$

Now you try it.

EXERCISE 14.15. *Use the change of variable* $u = \sqrt{x}$ *and associated change of differential to convert* $\int \sqrt{x}\, \text{Cos}[\sqrt{x}]\, dx$ *into a multiple of* $\int u^2\, \text{Cos}[u]\, du$.

If you have checked several indefinite integration problems that you computed with a change of variable and differential, you probably realize that the Chain Rule lies behind the method. Perhaps you can formalize your idea.

PROBLEM 14.16. *The Chain Rule in Reverse*
Use the chain rule for differentiation to prove the indefinite integral form of 'Integration by Substitution.' Once you have this indefinite rule, use the Fundamental Theorem to prove the definite rule.

14.4.2. Integration with Parameters. Change of variables can make integrals with parameters, which are important in many scientific and mathematical problems, into specific integrals. For example, suppose that ω and a are constant, then

$$\int \text{Sin}[\omega\, t]\, dt$$

$$u = \omega\, t$$

$$du = \omega\, dt \quad \Leftrightarrow \quad \frac{1}{\omega}\, du = dt$$

$$\int \text{Sin}[\omega\, t]\, dt = \int \text{Sin}[u]\, \frac{1}{\omega}\, du = \frac{1}{\omega} \int \text{Sin}[u]\, du$$

and

$$\int \frac{1}{a^2 + x^2}\, dt = \int \frac{1}{a^2(1 + (x/a)^2)}\, dx = \frac{1}{a^2} \int \frac{1}{1 + (x/a)^2}\, dx$$

$$u = \frac{x}{a}$$

$$du = \frac{1}{a}\, dx \quad \Leftrightarrow \quad a\, du = dx$$

$$\int \frac{1}{a^2 + x^2}\, dt = \frac{1}{a^2} \int \frac{1}{1 + u^2}\, a\, du = \frac{1}{a} \int \frac{1}{1 + u^2}\, du$$

This type of change of variable and differential may be the most important kind for you to think about, because the computer can give you specific integrals, but you may want to see how an integral depends on a parameter.

PROBLEM 14.17. *Pull the Parameter Out*
Show that

$$\int_0^3 \sqrt{9 - x^2}\, dx = 3 \int_0^3 \sqrt{1 - (x/3)^2}\, dx = 9 \int_0^1 \sqrt{1 - u^2}\, du$$

Show that the area of a circle of radius r is r^2 times the area of the unit circle. First, change variables to obtain the integrals below and then read on to interpret your computation.

$$\int_0^r \sqrt{r^2 - x^2}\, dx = r^2 \int_0^1 \sqrt{1 - u^2}\, du$$

14.5. Trig Substitutions (Optional)

Study this section if you have a personal need to compute integrals with one of the following expressions (and don't have your computer):

$$\sqrt{a^2 - x^2}, \quad a^2 + x^2 \quad \text{or} \quad \sqrt{a^2 + x^2}$$

You should read over the section lightly even if you do not wish to develop this skill, because the positive sign needed to go between

$$(\mathrm{Cos}[\theta])^2 = 1 - (\mathrm{Sin}[\theta])^2 \qquad \leftrightarrow \qquad \mathrm{Cos}[\theta] = \sqrt{1 - (\mathrm{Sin}[\theta])^2}$$

can cause errors in the use of a symbolic integration package. In other words, *Mathematica* may use the symbolic square root where you intend for it to use the negative.

EXAMPLE 14.18. *A Trigonometric Change of Variables*

The following integral comes from computing the area of a circle using the integral. A sine substitution makes it one we can antidifferentiate with two tricks from trig.

$$\int_0^1 \sqrt{1 - u^2}\, du = \int_0^{\pi/2} \sqrt{1 - \mathrm{Sin}^2(\theta)}\, \mathrm{Cos}(\theta)\, d\theta$$

because we take the substitution

$$u = \mathrm{Sin}(\theta) \qquad du = \mathrm{Cos}(\theta)\, d\theta$$

$$u = 0 \Leftrightarrow \theta = 0 \qquad u = 1 \Leftrightarrow \theta = \pi/2$$

We know from high school trig that $1 - \mathrm{Sin}^2(\theta) = \mathrm{Cos}^2(\theta)$, so when cosine is positive, $\sqrt{1 - \mathrm{Sin}^2(\theta)} = \mathrm{Cos}(\theta)$ and

$$\int_0^{\pi/2} \sqrt{1 - \mathrm{Sin}^2(\theta)} \, \mathrm{Cos}(\theta) \, d\theta = \int_0^{\pi/2} \mathrm{Cos}^2(\theta) \, d\theta$$

We also know from high school trig (or looking at the graph and thinking a little) that $\mathrm{Cos}^2(\theta) = \frac{1}{2}[1 + \mathrm{Cos}(2\theta)]$, so

$$\int_0^{\pi/2} \mathrm{Cos}^2(\theta) \, d\theta = \frac{1}{2}[\int_0^{\pi/2} d\theta + \int_0^{\pi/2} \mathrm{Cos}(2\theta) \, d\theta]$$

$$= \frac{\pi}{4} + \frac{1}{4} \int_0^{\pi} \mathrm{Cos}(\phi) \, d\phi$$

because we use another change of variables

$$\phi = 2\theta \qquad d\phi = 2d\theta$$

$$\phi = 0 \Leftrightarrow \theta = 0 \qquad \phi = \pi \Leftrightarrow \theta = \pi/2$$

Finally,

$$\frac{1}{4} \int_0^{\pi} \mathrm{Cos}(\phi) \, d\phi = \frac{1}{4}[\mathrm{Sin}(\phi) \, |_0^{\pi} \,] = 0$$

as we could have easily seen by sketching a graph of $\mathrm{Cos}(\phi)$ from 0 to π, with equal areas above and below the ϕ-axis.

Putting all these computations together, we have

$$\int_0^1 \sqrt{1 - u^2} \, du = \frac{\pi}{4}$$

EXERCISE 14.19. *Geometric Proof that $\int_0^1 \sqrt{1 - u^2} \, du = \frac{\pi}{4}$*
Sketch the graph $y = \sqrt{1 - x^2}$ for $0 \leq x \leq 1$. What geometrical shape is this? What is the area of one fourth of a circle of unit radius?

One explanation of why the change of variables in the previous example works is that sine and cosine yield parametric equations for the unit circle. More technically speaking, the (Pythagorean Theorem) identity

$$(\mathrm{Sin}[\theta])^2 + (\mathrm{Cos}[\theta])^2 = 1 \qquad \Leftrightarrow \qquad (\mathrm{Cos}[\theta])^2 = 1 - (\mathrm{Sin}[\theta])^2$$

becomes the identity $(\mathrm{Cos}[\theta])^2 = 1 - u^2$ when we let $u = \mathrm{Sin}[\theta]$. There is an important algebraic detail when we write

$$\mathrm{Cos}[\theta] = \sqrt{1 - (\mathrm{Sin}[\theta])^2}$$

This is false when cosine is negative. For the definite integral above, we wanted $0 \leq u \leq 1$ and chose $0 \leq \theta \leq \pi/2$ to put $\mathrm{Sin}[\theta]$ in this range. It is also true that $\mathrm{Cos}[\theta] \geq 0$ for this range of θ, so that

$$\mathrm{Cos}[\theta] = \sqrt{1 - u^2}$$

If the cosine were not positive in the range of interest, the integration would not be valid. More information on this difficulty is contained in the Mathematical Background Chapter on Differentiation Drill.

An expression like $\sqrt{a^2 - x^2}$ can first be reduced to a multiple of $\sqrt{1-u^2}$ by taking $u = x/a$ and writing $\sqrt{a^2 - x^2} = a\sqrt{1-(x/a)^2}$. This means that the expression $\sqrt{a^2-x^2}$ can be converted to $\mathrm{Cos}[\theta]$ with the substitutions $u = (x/a)^2$ and $u = \mathrm{Sin}[\theta]$ (provided cosine is positive on the interval.)

Notice that the substitution $u = \mathrm{Cos}[\theta]$ converts $\sqrt{1-u^2}$ into $\sqrt{1-(\mathrm{Cos}[\theta])^2} = \mathrm{Sin}[\theta]$, provided sine is positive. Also note that sine is positive over a different range of angles than cosine.

Another trig identity says

$$(\mathrm{Tan}[\theta])^2 + 1 = (\mathrm{Sec}[\theta])^2 \qquad \Leftrightarrow \qquad \left(\frac{\mathrm{Sin}[\theta]}{\mathrm{Cos}[\theta]}\right)^2 + \left(\frac{\mathrm{Cos}[\theta]}{\mathrm{Cos}[\theta]}\right)^2 = \left(\frac{1}{\mathrm{Cos}[\theta]}\right)^2$$

$$\Leftrightarrow \qquad (\mathrm{Sin}[\theta])^2 + (\mathrm{Cos}[\theta])^2 = 1$$

If we make a change of variable $u = \mathrm{Tan}[\theta]$, then $u^2 + 1$ becomes $(\mathrm{Sec}[\theta])^2$ and $\sqrt{1+u^2} = \mathrm{Sec}[\theta]$ provided secant is positive.

EXERCISE 14.20. *Compute the integral $\int_0^9 \sqrt{9 - x^2}\, dx$. (Check your symbolic computation geometrically.)*

Use the change of variable $x = \mathrm{Sin}[\theta]$ with an appropriate differential to show that

$$\int_0^v \frac{1}{\sqrt{1-x^2}}\, dx = \int d\theta = \theta... = \mathrm{ArcSin}[v]$$

How large can we take v?

Use the change of variable $x = \mathrm{Tan}[\theta]$ with an appropriate differential to show that

$$\int_0^v \frac{1}{1+x^2}\, dx = \int d\theta = \theta... = \mathrm{ArcTan}[v]$$

How large can we take v?

EXAMPLE 14.21. *The Sine Substitution Without Endpoints*

The "cost" of not changing limits of integration in Example 14.18 is the following. The same tricks as above for indefinite integrals yield

$$\int \sqrt{1-u^2}\, du = \frac{1}{2}\theta + \frac{1}{4}\mathrm{Sin}[\phi]$$

where $u = \mathrm{Sin}[\theta]$ and $\phi = 2\theta$. If we really want the antiderivative of $\sqrt{1-u^2}$, then we must express all this in terms of u.

The first term is easy, we just use the inverse trig function

$$u = \mathrm{Sin}[\theta] \qquad \Leftrightarrow \qquad \theta = \mathrm{ArcSin}[u]$$

The term $\mathrm{Sin}[\phi] = \mathrm{Sin}[2\theta]$ must first be written in terms of functions of θ, (recall the addition formula for sine)

$$\mathrm{Sin}[2\theta] = 2\,\mathrm{Sin}[\theta]\,\mathrm{Cos}[\theta]$$

Now we use a triangle that contains the $u = \text{Sin}[\theta]$ idea. From SOH-CAH-TOA, if we take a right triangle with hypotenuse 1 and opposite side u, then $u = \text{Sin}[\theta]$. We also know that $\text{Cos}[\theta]$ is the adjacent side of this triangle. Using the Pythagorean theorem, we have $\text{Cos}[\theta] = \sqrt{1 - u^2}$

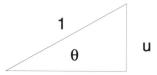

FIGURE 14.22: $\text{Cos}[\theta] = \text{adj/hyp} = \sqrt{1 - u^2}$

Combining these facts we have,

$$\text{Sin}[2\,\theta] = 2\,\text{Sin}\,\theta]\,\text{Cos}[\theta] = 2\,u\,\sqrt{1 - u^2}$$

and

$$\int \sqrt{1 - u^2}\,du = \frac{1}{2}\,\theta + \frac{1}{4}\,\text{Sin}[\phi]$$

$$= \frac{1}{2}\,\text{ArcSin}[u] + \frac{u\,\sqrt{1 - u^2}}{2}$$

EXERCISE 14.23. *Working Back from Trig Substitutions*
Suppose we make the change of variable $u = \text{Sin}[\theta]$ (as in the integration above). Express $\text{Tan}[\theta]$ in terms of u by using the figure above and TOA.

A triangle is shown below for a change of variable $v = \text{Cos}[\theta]$. Express $\text{Sin}[\theta]$ and $\text{Tan}[\theta]$ in terms of v by using the figure and the Pythagorean Theorem.

A triangle is shown below for a change of variable $w = \text{Tan}[\theta]$. Express $\text{Sin}[\theta]$ and $\text{Cos}[\theta]$ in terms of w by using the figure and the Pythagorean Theorem.

FIGURE 14.24: Triangles for $v = \text{Cos}[\theta] = \text{CAH}$ and $w = \text{Tan}[\theta] = \text{TOA}$

We are beginning to wallow in trig a little. The point of the previous example could simply be: Change the limits of integration when you change variable and differential. However, it is possible to change back to u and the previous exercise gives you a start at the trig skills needed to do this. Additional details are contained in the Mathematical Background Chapter on Integration Drill. (There are some important details that we have omitted about the various "branches" of the inverse trig functions that are taken up in the Background. It is possible for *Mathematica* to choose the wrong branch of a trig substitution.)

PROBLEM 14.25. *A Constant*
Use the change of variable $x = \text{Cos}[\theta]$ with an appropriate differential to show that

$$\int \frac{1}{\sqrt{1-x^2}} \, dx = -\int d\theta = -\theta... = -\text{ArcCos}[x] + c$$

Also use the change of variable $x = \text{Sin}[\theta]$ with an appropriate differential to show that

$$\int \frac{1}{\sqrt{1-x^2}} \, dx = \int d\theta = \theta... = \text{ArcSin}[x] + c$$

Is $\text{ArcCos}[x] = -\text{ArcSin}[x]$? Ask Mathematica to Plot $\text{ArcSin}[x]$ and $-\text{ArcCos}[x]$. Why do they look alike? How do they differ? Do the graphs of $-\text{Cos}[\theta]$ and $\text{Sin}[\theta]$ look alike? How do they differ?

14.6. Integration by Parts

Integration by Parts is very important theoretically, but it begins as another disgusting-looking trick. It is more than that, but first you should learn the trick.

The formulas for the 'technique' are:

$$\int_a^b u\,(x)\ dv\,(x) = u\,(x)\,v\,(x)\,|_a^b - \int_a^b v\,(x)\ du\,(x)$$

or suppressing the x dependence,

$$\int_{x=a}^b u\ dv = uv|_{x=a}^b - \int_{x=a}^b v\ du$$

The idea is to break up an integrand into a function $u(x)$ and a differential $dv(x)$ where you can find the differential $du(x)$ (usually easy), the antiderivative $v(x)$ (sometimes harder), and finally, where $\int v(x)\ du(x)$ is an easier problem. Unfortunately, often the only way to find out if the new problem is easier is to go through all the substitution steps for the terms.

We encourage you to block off the four terms in this formula. First break up your integrand into u and dv,

$$u = \qquad\qquad\qquad dv =$$

then compute the differential of u and the antiderivative of dv

$$du = \qquad\qquad\qquad v =$$

Here is an example of the use of Integration by Parts.

EXAMPLE 14.26. *Integration by Parts for $\int_a^b x \log(x)\,dx = ?$*

Use the "parts"

$$u = \text{Log}\,(x) \qquad\qquad dv = x\,dx$$
$$\text{so} \qquad\qquad\qquad \text{and}$$
$$du = \frac{1}{x}\,dx \qquad\qquad v = \frac{1}{2}x^2$$

making the integrals

$$\int_a^b u\,dv = uv\big|_a^b - \int_a^b v\,du$$

$$\int_a^b x\,\text{Log}\,(x)\,dx = \frac{1}{2}x^2\ \text{Log}\,(x)\,\big|_a^b - \int_a^b \frac{1}{2}x^2\frac{dx}{x}$$

$$\int_a^b x\,\text{Log}\,(x)\,dx = \frac{1}{2}x^2\ \text{Log}\,(x)\,\big|_a^b - \int_a^b \frac{1}{2}x\,dx$$

$$\int_a^b x\,\text{Log}\,(x)\,dx = \frac{1}{2}x^2\ \text{Log}\,(x)\,\big|_a^b - \frac{x^2}{4}\big|_a^b$$

$$\int_a^b x\,\text{Log}\,(x)\,dx = \frac{1}{2}[b^2\,\text{Log}\,(b) - a^2\,\text{Log}\,(a)] - \frac{1}{4}[b^2 - a^2]$$

provided that both a and b are positive. (Otherwise Log is undefined.)

CHECK: Notice that the indefinite integral of the calculation above is

$$\int x\,\text{Log}\,(x)\,dx = \frac{1}{2}\,x^2\ \text{Log}[x] - \frac{1}{4}\,x^2 + c$$

We check the correctness of this antiderivative by differentiating the right side of the equation. First, we use the Product Rule

$$\frac{d(f(x)\cdot g(x))}{dx} = \frac{df}{dx}\cdot g + f\cdot\frac{dg}{dx}$$

$$f(x) = x^2 \qquad\qquad g(x) = \text{Log}[x]$$
$$\frac{df}{dx} = 2\,x \qquad\qquad \frac{dg}{dx} = \frac{1}{x}$$

$$\frac{d(x^2\ \text{Log}[x])}{dx} = 2\,x\cdot\text{Log}[x] + x^2\cdot\frac{1}{x}$$
$$= 2\,x\cdot\text{Log}[x] + x$$

Next, we differentiate the whole expression

$$\frac{(\frac{1}{2}\,x^2\ \text{Log}[x] - \frac{1}{4}\,x^2 + c)}{dx} = x\,\text{Log}[x] + \frac{1}{2}\,x - \frac{2}{4}\,x + 0 = x\,\text{Log}[x]$$

This verifies that the indefinite integral is correct.

EXERCISE 14.27. *Compute $\int \text{Log}[x]\,dx$ using integration by parts with $u = \text{Log}[x]$ and $dv = dx$. Check your answer by differentiation.*

EXAMPLE 14.28. $\int x\,e^x\,dx$

To compute this integral, use integration by parts with

$$u = x \qquad\qquad\qquad dv = e^x\,dx$$

so

$$du = dx \qquad\qquad\qquad v = e^x$$

This gives

$$\int u \, dv = u \, v - \int v \, du$$

$$\int x \, e^x \, dx = x \, e^x - \int e^x \, dx$$

$$= x \, e^x - e^x = (x - 1) \, e^x + c$$

$$\int x \, e^x \, dx = (x - 1) \, e^x + c$$

CHECK: Differentiate using the Product Rule:

$$\frac{d(f(x) \cdot g(x))}{dx} = \frac{df}{dx} \cdot g + f \cdot \frac{dg}{dx}$$

$$f(x) = (x - 1) \qquad g(x) = e^x$$

$$\frac{df}{dx} = 1 \qquad \frac{dg}{dx} = e^x$$

$$\frac{d((x - 1) \, e^x)}{dx} = e^x + (x - 1) \, e^x$$

$$= x \, e^x$$

EXAMPLE 14.29. *A Reduction of One Integral to a Previous One*

The integral $\int x^2 \, e^x \, dx$ can be done "by parts" in two steps. First, take the parts

$$u = x^2 \qquad\qquad dv = e^x \, dx$$

so

$$du = 2 \, x \, dx \qquad\qquad v = e^x$$

This gives

$$\int u \, dv = u \, v - \int v \, du$$

$$\int x^2 \, e^x \, dx = x^2 \, e^x - 2 \int x \, e^x \, dx$$

$$= x^2 \, e^x - 2(x - 1) \, e^x + c$$

Note that the second integral was computed in the previous example, so

$$\int x^2 \, e^x \, dx = (x^2 - 2 \, x + 2) \, e^x + c$$

CHECK: Differentiate using the Product Rule:

$$\frac{d(f(x) \cdot g(x))}{dx} = \frac{df}{dx} \cdot g + f \cdot \frac{dg}{dx}$$

$$f(x) = (x^2 - 2 \, x + 2) \qquad g(x) = e^x$$

$$\frac{df}{dx} = 2 \, x - 2 \qquad \frac{dg}{dx} = e^x$$

$$\frac{d((x^2 - 2x + 2)e^x)}{dx} = (2x - 2)e^x + (x^2 - 2x + 2)e^x$$

$$= x^2 e^x$$

EXAMPLE 14.30. *Circular Parts Still Gives an Answer*

We compute the integral $\int e^{2x} \operatorname{Sin}[3x]\, dx$ by the parts

$$u = e^{2x} \qquad\qquad dv = \operatorname{Sin}[3x]\, dx$$

so

$$du = 2e^{2x}\, dx \qquad\qquad v = -\frac{1}{3}\operatorname{Cos}[3x]$$

This gives

$$\int u\, dv = u\, v - \int v\, du$$

$$\int e^{2x} \operatorname{Sin}[3x]\, dx = -\frac{1}{3}e^{2x}\operatorname{Cos}[3x] + \frac{2}{3}\int e^{2x}\operatorname{Cos}[3x]\, dx$$

Now use the parts on the second integral,

$$w = e^{2x} \qquad\qquad dz = \operatorname{Cos}[3x]\, dx$$

so

$$dw = 2e^{2x}\, dx \qquad\qquad z = \frac{1}{3}\operatorname{Sin}[3x]$$

This gives

$$\int w\, dz = w\, z - \int z\, dw$$

$$\int e^{2x} \operatorname{Cos}[3x]\, dx = \frac{1}{3}e^{2x}\operatorname{Sin}[3x] - \frac{2}{3}\int e^{2x}\operatorname{Sin}[3x]\, dx$$

Substituting this into the second integral above, we obtain

$$\int e^{2x} \operatorname{Sin}[3x]\, dx = -\frac{1}{3}e^{2x}\operatorname{Cos}[3x] + \frac{2}{3}\left[\frac{1}{3}e^{2x}\operatorname{Sin}[3x] - \frac{2}{3}\int e^{2x}\operatorname{Sin}[3x]\, dx\right]$$

$$= -\frac{1}{3}e^{2x}\operatorname{Cos}[3x] + \frac{2}{9}e^{2x}\operatorname{Sin}[3x] - \frac{4}{9}\int e^{2x}\operatorname{Sin}[3x]\, dx$$

Bringing the like integral to the left side, we obtain

$$\int e^{2x} \operatorname{Sin}[3x]\, dx + \frac{9}{4}\int e^{2x}\operatorname{Sin}[3x]\, dx = \frac{2}{9}e^{2x}\operatorname{Sin}[3x] - \frac{1}{3}e^{2x}\operatorname{Cos}[3x] + c$$

$$\frac{4+9}{4}\int e^{2x}\operatorname{Sin}[3x]\, dx = \frac{2}{9}e^{2x}\operatorname{Sin}[3x] - \frac{1}{3}e^{2x}\operatorname{Cos}[3x] + c$$

$$\int e^{2x} \operatorname{Sin}[3x]\, dx = \frac{2}{13}e^{2x}\operatorname{Sin}[3x] - \frac{3}{13}e^{2x}\operatorname{Cos}[3x] + c$$

EXERCISE 14.31. *Check the previous indefinite integral by using the Product Rule to dif-ferentiate $\frac{2}{13} e^{2x} \operatorname{Sin}[3x] - \frac{3}{13} e^{2x} \operatorname{Cos}[3x]$.*
Compute the integral $\int e^{2x} \operatorname{Sin}[3x] \, dx$ by the parts

$$u = \operatorname{Sin}[3x] \qquad\qquad\qquad dv = e^{2x} \, dx$$

so

$$du = 3 \operatorname{Cos}[3x] \, dx \qquad\qquad\qquad v = \frac{1}{2} e^{2x}$$

EXAMPLE 14.32. *A Two Step Computation of $\int (\operatorname{Cos}[x])^2 \, dx$*

This integral can also be computed without the trig identities in Example 14.18. Use the parts

$$u = \operatorname{Cos}[x] \qquad\qquad\qquad dv = \operatorname{Cos}[x] \, dx$$

so

$$du = -\operatorname{Sin}[x] \, dx \qquad\qquad\qquad v = \operatorname{Sin}[x]$$

Yielding the integration formula

$$\int (\operatorname{Cos}[x])^2 \, dx = -\operatorname{Sin}[x] \operatorname{Cos}[x] + \int (\operatorname{Sin}[x])^2 \, dx$$

$$= -\operatorname{Sin}[x] \operatorname{Cos}[x] + \int [1 - (\operatorname{Cos}[x])^2] \, dx$$

$$= -\operatorname{Sin}[x] \operatorname{Cos}[x] + \int 1 \, dx - \int (\operatorname{Cos}[x])^2 \, dx$$

so $2 \int (\operatorname{Cos}[x])^2 \, dx = x - \operatorname{Sin}[x] \operatorname{Cos}[x] + c$ and

$$\int (\operatorname{Cos}[x])^2 \, dx = \frac{x}{2} - \frac{1}{2} \operatorname{Sin}[x] \operatorname{Cos}[x] + c$$

EXERCISE 14.33. *Calculate $\int (\operatorname{Sin}[x])^2 \, dx$.*

Here are some practice problems. (Remember that you can check your work with *Mathematica*.)

EXERCISE 14.34. *Drill on Integration by Parts*

a) $\int \theta \operatorname{Cos}[\theta] \, d\theta = ?$

b) $\int \theta^2 \operatorname{Sin}[\theta] \, d\theta = ?$

c) $\int_0^{\pi/2} \theta \operatorname{Sin}[\theta] \, d\theta = ?$

d) $\int_0^{\pi/2} \theta \operatorname{Cos}[\theta] \, d\theta = ?$

e) $\int x e^{2x} \, dx = ?$

f) $\int x^2 e^{2x} \, dx = ?$

g) $\int e^{5x} \operatorname{Sin}[3x] \, dx = ?$

h) $\int e^{3x} \operatorname{Cos}[5x] \, dx = ?$

There's something indefinite about these integrals.

EXERCISE 14.35. *What is wrong with the following computation*

$$\int \frac{dx}{x} = \int \frac{x}{x^2} dx = -1 + \int \frac{dx}{x}$$

using integration by parts with $u = x$, $dv = \frac{dx}{x^2}$, $du = dx$ and $v = \frac{-1}{x}$? Subtracting $\int \frac{dx}{x}$ from both sides of the equations above yields

$$0 = -1$$

The proof of the Integration by Parts formula is actually easy.

EXERCISE 14.36. *The Product Rule in Reverse*
Use the product rule for differentiation to prove the indefinite integral form of 'Integration by Parts.' Notice that if $H(x)$ is any function, $\int dH(x) = H(x) + c$, by definition. Let $H = uv$ and show that $dH = u\,dv + v\,du$. Indefinitely integrate both sides of the dH equation,

$$u \cdot v = H(x) = \int dH = \int u\,dv + \int v\,du$$

Once you have this indefinite rule, use the Fundamental Theorem to prove the definite rule.

14.7. Combined Integration

Here are some tougher problems where you need to use more than one method at a time.

EXERCISE 14.37.

(1) $\int x \cos[x] \sin[x]\, dx = ?$
    Note $\int (\cos[\theta])^2\, d\theta$ is done above two ways.

(2) $\int \theta\, (\cos[\theta])^2\, d\theta = ?$
    Use integrals from the previous drill problems.

(3) $\int \frac{x^3}{\sqrt{x^2-1}}\, dx = ?$
    Hint: Use parts $u = x^2$ and $dv = \frac{x\,dx}{\sqrt{x^2-1}}$ and compute the dv integral.

(4) $\int \frac{1}{x^3} \sqrt{\frac{1}{x} - 1}\, dx = ?$
    Use parts $u = \frac{1}{x}$ and $dv = \frac{1}{x^2}\sqrt{\frac{1}{x} - 1}$. Compute the dv integral.

(5) $\int_0^1 \text{ArcTan}[x]\, dx = ?$
    Use parts $u = \text{ArcTan}[x]$ and $dv = dx$

(6) $\int_0^1 x \, \text{ArcTan}[x]\, dx = ?$
    Use parts and the previous exercise.

(7) $\int_0^1 \text{ArcTan}[\sqrt{x}]\, dx = ?$
    Use parts $u = \text{ArcTan}[\sqrt{x}]$ and change variables in the resulting integral.

(8) $\int_4^9 \text{Sin}[\sqrt{x}]\ dx = \ ?$
    Use parts $u = \text{Sin}[\sqrt{x}]$ and $dv = dx$, then change variables with $w = \sqrt{x}$

(9) $\int \text{Log}^2[x]\ dx = \ ?$
    Use the parts $u = \text{Log}[x]$ and $dv = \text{Log}[x]\ dx$. Calculate the dv integral.

14.8. Impossible Integrals

There are important limitations to symbolic integration that go beyond the practical difficulties of learning all the tricks. This section explains why.

If you can compute changes of variables and use Integration by Parts, you can compute a host of integrals. Traditionally, calculus courses spent more time and effort in this pursuit than we have. Perhaps this chapter is already too long if you consider that *Mathematica* (and several other symbolic computer packages) can compute all the integrals in the chapter in seconds. Integration by Parts and the change of variable and differential are important ideas for the theoretical transformation of integrals, so we tried to include just enough drill for you to learn the methods well and see many basic examples, but not so much that you felt you had to wallow in symbols.

Before the discovery of very general antidifferentiation algorithms and practical implementations of the algorithms on computers, human development of integration skills was an important part of the training of scientists and engineers. Many applications are easier to understand with explicit symbolic formulas for the integrals that arise. Early in the days of calculus, it was quite impressive that integration could be used to learn many many new formulas like the classical formulas for the area of a circle or volume of a sphere. We saw how easy it was to generalize the integration approach to the volume of a cone in the beginning of the last chapter. However, some simple looking integrals have no antiderivative what-so-ever. This is not the result of peculiar mathematical examples.

The arclength of an ellipse just means the length measured as you travel along an ellipse. In the next chapter there is an exercise for you to find integral formulas for this arclength. Early developers of calculus must have tried very hard to compute those integrals with antiderivatives, but after more than a century of trying, Liouville proved that there is no analytical expression for that antiderivative in terms of the classical functions. There are many important examples like this. The **SymbolicIntegr.ma** NoteBook shows you several of them.

The fact that the antiderivative has no expression in terms of old functions does not mean that the integral does not exist. If you type the following in *Mathematica* you will see a peculiar result,

In[1]=
   Integrate[Cos[x∧2],x]
Out[1]=
   Sqrt[$\frac{\text{Pi}}{2}$] FresnelC[Sqrt[$\frac{\text{Pi}}{2}$] x]

The innocent looking integral $\int \text{Cos}[x^2]\ dx$ is not innocent at all. The expression above is written in terms on a non-elementary function called a Fresnel integral (which arises in scientific applications.) The function $\text{Cos}[x^2]$ is perfectly smooth and well-behaved, it just doesn't have an antiderivative that can be expressed in terms of known functions.

Mathematica has many such non-elementary functions. When they arise as integrals, the numerical computation of the functions can be performed with numerical integration just as you did in Exercise 13.64.

The bottom line is this: Integrals are used to DEFINE and numerically compute important new functions in science and mathematics, even when they do not have expressions in terms of elementary functions.

CHAPTER 15

Applications of Integration

This chapter explores deeper applications of integration, especially to computation of geometric quantities. The most important part of integration is setting the integrals up and understanding when the techniques of the last chapter apply. *Mathematica* can perform the symbolic calculus of the techniques, but they must be applicable and you must give *Mathematica* the correct approximation sum or integrand.

15.1. The Infinite Sum Theorem: Duhamel's Principle

We have seen that some care must be taken in the more delicate geometric approximation problems (such as arclength) in order to be sure that a sum of infinitely many infinitesimal errors does not build up. It is NOT sufficient to have the error of each slice tend to zero. The Infinite Sum Theorem will give us a general accuracy test to apply to finding integral formulas. The secret is to measure the error on a scale of the increment 'δx.'

Additivity of the integral means that if we define a function of two variables, $I(u, v) = \int_u^v f(x) \, dx$ (for $f(x)$ continuous on $[a, b]$), then we have the additivity property

$$I(u, v) + I(v, w) = I(u, w)$$

for any $u < v < w$ in $[a, b]$. Additivity expresses a simple property and many geometric functions such as accumulated area, length, volume, or surface area have this property. This fact means that we can apply the next result to finding integral formulas of many geometric quantities.

THEOREM 15.1. *The Infinite Sum Theorem or Duhamel's Principle*
Let $B(u, v)$ satisfy $B(u, v) + B(v, w) = B(u, w)$ for $u < v < w$ in $[a, b]$. If $f(x)$ is a continuous real function on $[a, b]$ such that for any infinitesimal subinterval $[x, x+\delta x] \subseteq [a, b]$,

$$B(x, x + \delta x) = f(x) \ \delta x + \varepsilon \cdot \delta x$$

with $\varepsilon \approx 0$, that is, if the quantity $f(x) \, \delta x$ approximates an infinitesimal 'slice' of B on a scale of δx, then

$$B(a, b) = \int_a^b f(x) \, dx$$

PROOF: By repeated use of Additivity (and extension of the formula to infinitesimal increments)

$$B(a, b) = \text{Sum}[B(x, x + \delta x), \{a, b - \delta x, \delta x\}]$$
$$= \text{Sum}[f(x)\delta x, \{a, b - \delta x, \delta x\}] + \text{Sum}[\varepsilon \cdot \delta x, \{a, b - \delta x, \delta x\}]$$
$$\approx \int_a^b f(x)\, dx + \text{Sum}[\varepsilon \cdot \delta x, \{a, b - \delta x, \delta x\}]$$

and

$$\text{Sum}[\varepsilon \cdot \delta x, \{a, b - \delta x, \delta x\}] \approx 0$$

by the triangle inequality estimate we used in the proof of the first half of the Fundamental Theorem (and the Existence Theorem and the zero derivative theorem.) So the two real quantities satisfy

$$B(a, b) \approx \int_a^b f(x)\, dx$$

forcing them to be equal and proving the theorem.

EXERCISE 15.2. *Define the accumulated energy function*

$$E(t_1, t_2)$$

to be the amount of energy a household consumes from time t_1 to time t_2. Explain why $E(s, t)$ is additive. (This is just a matter of saying what the terms in the formula mean.)

EXAMPLE 15.3. *Distance is an Integral*

We return to the distance example of section 12.2 above. Suppose that $R(t)$ is any continuous rate (speed) function. By the Extreme Value Theorem 11.12, $R(t)$ has a max and a min over the interval $[t, t + \delta t]$, say

$$R_m \leq R(s) \leq R_M \qquad \text{for} \quad t \leq s \leq t + \delta t$$

The distance traveled during this time interval must lie between the extremes

$$R_m \cdot \delta t \leq \text{Actual Distance Traveled} \leq R_M \cdot \delta t$$

However, since $R(t)$ is continuous,

$$R_m \approx R(t) \approx R_M$$

so

$$\text{Actual Distance Traveled} = R(t) \cdot \delta t + \varepsilon \cdot \delta t$$

with $\varepsilon \approx 0$.

Finally, let $D(t_1, t_2)$ denote the distance traveled between the times t_1 and t_2. This is additive, because if $t_1 < t_2 < t_3$, $D(t_1, t_3) = D(t_1, t_2) + D(t_2, t_3)$, so the Infinite Sum Theorem applies to show that

$$\text{Total Distance} = \int_a^b R(t)\, dt$$

EXAMPLE 15.4. *Length Formula $L = \int_a^b \sqrt{1 + (f'[x])^2}\, dx$*

We don't want you to just "plug in" formulas, but rather to understand integral formulas and derive them for yourself. Exercise 13.16 actually asked you to derive the arclength formula for a general explicit curve $y = f(x)$. A simpler way to summarize that work is to imagine measuring the length of a segment of the curve viewed inside an infinitesimal microscope. We will show now that what we see in the microscope gives the correct answer more easily than that early exercise.

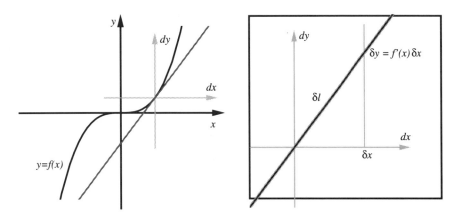

FIGURE 15.5: Microscopic View of Length

The point of re-working this example is that we can see that the formula for the length of the curve is

$$L = \int_a^b \sqrt{1 + (f'[x])^2} \; dx$$

because we know that the line we see in a microscopic view of a smooth function has equation $dy = f'[x] \, dx$, so the height of the triangle is $f'[x] \, \delta x$, for a base of δx. The Pythagorean Theorem says the length is

$$\sqrt{(\delta x)^2 + (f'[x])^2 \delta x^2} = \sqrt{1 + (f'[x])^2} \; \delta x$$

except for errors that are infinitesimal when viewed in the microscope. This is exactly what is needed for The Infinite Sum Theorem 15.1 or Duhamel's Principle,

$$\text{Length between } x \text{ and } x + \delta x = \sqrt{1 + (f'[x])^2} \; \delta x + \iota \, \delta x$$

The Theorem says that we can add the linear pieces we see in the microscopic views to obtain the full nonlinear quantity. The details of the computation were contained in Exercise 13.16.

EXAMPLE 15.6. *Parametric Length* $L = \int_a^b \sqrt{[dx(t)]^2 + [dy(t)]^2}$

The simple shortcut of finding the length integral with the view in an infinitesimal microscope also works for paramteric curves. Sometimes parametric integral formulas are better behaved than explicit formulas and sometimes curves do not even have single explicit formulas, so that we must use parametric formulas. What are parametric formulas for curves?

A circle is given by the parametric equations

$$x(\theta) = \text{Cos}[\theta]$$
$$y(\theta) = \text{Sin}[\theta]$$

These equations simply express the relationship between radian measure of angles and the corresponding (x, y)-point on the unit circle. We used this idea in Chapter 3 when we reviewed radian measure and in Chapter 5 when we computed the increments of sine and cosine and proved the differentiation formulas for sine and cosine. A tiny increment of the circle looks as follows under a powerful microscope:

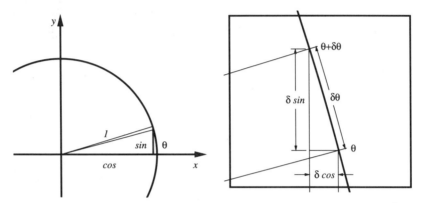

FIGURE 15.7: Increments of Sine and Cosine

The small triangle in the magnified view has:

Horizontal base $= -\delta \operatorname{Cos}[\theta] = -(\operatorname{Cos}[\theta + \delta\theta] - \operatorname{Cos}[\theta])$

Vertical side $= \delta \operatorname{Sin}[\theta] = \operatorname{Sin}[\theta + \delta\theta] - \operatorname{Sin}[\theta]$

Hypotenuse $= \delta\theta$

The length of the "hypotenuse" is $\delta\theta$, because radian measure is defined to be the length measured along the unit circle. However, the side that looks like a hypotenuse is actually a magnified circle. Clearly, the error between the length of the circular arc and the approximating straight line is small compared to $\delta\theta$, so the length of the circle is given by the integral

$$\int_0^{2\pi} d\theta$$

It is still worthwhile to see how this is related to the computation with increments.

The basic definition of derivative 5.15 and the formulas for the derivatives of sine and cosine give us:

$\delta x = \delta \operatorname{Cos}[\theta] = \operatorname{Cos}[\theta + \delta\theta] - \operatorname{Cos}[\theta] = -\operatorname{Sin}[\theta]\delta\theta + \iota_1\delta\theta$

$\delta y = \delta \operatorname{Sin}[\theta] = \operatorname{Sin}[\theta + \delta\theta] - \operatorname{Sin}[\theta] = \operatorname{Cos}[\theta]\delta\theta + \iota_2\delta\theta$

The Pythagorean Theorem says that the length of the true straight hypotenuse is the square root of the sum of the squares of the lengths of the legs,

$$
\begin{aligned}
\text{Small Hypotenuse} &= \sqrt{[\delta x]^2 + [\delta y]^2} \\
&= \sqrt{(\text{Cos}[\theta + \delta\theta] - \text{Cos}[\theta])^2 + (\text{Sin}[\theta + \delta\theta] - \text{Sin}[\theta])^2} \\
&= \sqrt{(x'[\theta]\delta\theta + \iota_1\delta\theta)^2 + (y'[\theta]\delta\theta + \iota_1\delta\theta)^2} \\
&= \sqrt{(-\text{Sin}[\theta]\delta\theta + \iota_1\delta\theta)^2 + (\text{Cos}[\theta]\delta\theta + \iota_2\delta\theta)^2} \\
&= \sqrt{(x'[\theta] + \iota_1)^2 + (y'[\theta] + \iota_1)^2} \; \delta\theta \\
&= \sqrt{(-\text{Sin}[\theta]\delta\theta + \iota_1)^2 + (\text{Cos}[\theta]\delta\theta + \iota_2)^2} \; \delta\theta \\
&= \sqrt{(x'[\theta])^2 + (y'[\theta])^2} \; \delta\theta + \iota_3\delta\theta \\
&= \sqrt{(-\text{Sin}[\theta])^2 + (\text{Cos}[\theta])^2} \; \delta\theta + \iota_3\delta\theta \\
&\approx\approx \sqrt{(x'[\theta])^2 + (y'[\theta])^2} \; \delta\theta \\
&= \sqrt{(-\text{Sin}[\theta])^2 + (\text{Cos}[\theta])^2} \; \delta\theta
\end{aligned}
$$

Since $(\text{Cos}[\theta])^2 + (\text{Sin}[\theta])^2 = 1$, we have

$$\text{Small Straight Triangular Hypotenuse} \approx\approx \delta\theta$$

This means that we may compute the length by the integral of $d\theta$.

In the integral for the length of an ellipse, the expression $\sqrt{(x')^2 + (y')^2}$ is more complicated and we can not do the final step of replacing $\sqrt{(-\text{Sin}[\theta])^2 + (\text{Cos}[\theta])^2}$ by 1 as above. Parametric equations describing an ellipse are given in the project on geometric integrals below.

EXAMPLE 15.8. *A Simple Parametric Length*

Let us consider another example, the length of the parametric curve

$$
\begin{aligned}
x &= t^2 \\
y &= t^3
\end{aligned}
$$

for $0 \le t \le 1$. We have the increments

$$
\begin{aligned}
\delta x &= x'(t)\delta t + \iota_1\delta t & \delta y &= y'(t)\delta t + \iota_2\delta t \\
&= 2t\delta t + \iota_1\delta t & &= 3t^2\,\delta t + \iota_2\delta t
\end{aligned}
$$

so that the length of the hypotenuse of the triangle we would see in an infinitesimal microscope is

$$\delta l = \sqrt{[\delta x]^2 + [\delta y]^2} = \sqrt{[2t]^2 + [3t^2]^2}\,\delta t + \iota_3\delta t$$

and the Infinite Sum Theorem says,

$$L = \int_0^1 \sqrt{[2t]^2 + [3t^2]^2}\,dt = \frac{(13)^{3/2} - 8}{27} \approx 1.4971$$

EXERCISE 15.9. *Explicit Length*

Verify the parametric computation that we have just done by using the explicit equation for arclength, $L = \int_a^b \sqrt{1 + [f'(x)]^2}\, dx$, on the curve $y = x^{3/2}$. This is the same as the parametric curve above, since $t = \sqrt{x} = x^{1/2}$, so $y = t^3 = (x^{1/2})^3 = x^{3/2}$.

EXAMPLE 15.10. *Volume of a Half Bagel*

The upper half disk of radius 1 centered at $(2,0)$ is revolved about the y axis. The equation of the semicircular boundary of the half disk is

$$y = \sqrt{1 - (x-2)^2}$$

The resulting figure looks like the top half of a sliced bagel.

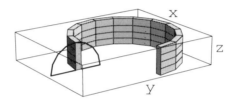

FIGURE 15.11: Half Bagel Shells

We approximate the accumulated volume between radius x and radius $x + \delta x$ by the cylindrical 'shell' with inner radius x, outer radius $x + \delta x$ and height $ht\,(x) = \sqrt{1 - (x-2)^2}$. Outer cylinder - inner cylinder $=$

$$\pi R^2 ht - \pi r^2 ht = \pi[(x + \delta x)^2 - x^2]ht$$

$$= \pi[2x\delta x + \delta x^2]ht$$

$$= \pi[2x\delta x + \delta x^2]\sqrt{1 - (x-2)^2}$$

$$= \pi 2x\delta x\sqrt{1 - (x-2)^2} + \delta x[\pi \delta x\sqrt{1 - (x-2)^2}$$

$$= \pi 2x\delta x\sqrt{1 - (x-2)^2} + \varepsilon \cdot \delta x$$

The δx^2 term can be neglected in our integral formula, by the Infinite Sum Theorem, because it produces a term of the form $\varepsilon \cdot \delta x$ with $\varepsilon \approx 0$.

The flat top on our cylinder produces another error between the accumulated vlolume from x to $x + \delta x$. If we are on the rising side of the circle, the left height, $ht\,(x) = \sqrt{1 - (x-2)^2}$ produces a shell that lies completely inside the bagel while the right height

$ht\left(x+\delta x\right)=\sqrt{1-\left(\left[x+\delta x\right]-2\right)^{2}}$ produces a shell that includes the cylindrical slice of the bagel (with its curved top.) This means

$$\pi 2x\delta x\sqrt{1-\left(x-2\right)^{2}}+\varepsilon_{1}\delta x\leq V\left(x,x+\delta x\right)\leq\pi 2x\delta x\sqrt{1-\left(\left[x+\delta x\right]-2\right)^{2}}+\varepsilon_{2}\delta x$$

On the falling part of the circle, the outside terms in this inequality are interchanged, the left height is above and the right height below. In either case

$$V\left(x,x+\delta x\right)=\pi 2x\delta x\sqrt{1-\left(\left[x+\delta x\right]-2\right)^{2}}+\varepsilon_{3}\delta x$$

with $\varepsilon_{3}\approx 0$, so we have

$$V\left(a,b\right)=2\pi\int_{1}^{3}x\sqrt{1-\left(x-2\right)^{2}}dx$$

This integral can be computed with a trig substitution,

$$x-2=\mathrm{Sin}\left(\theta\right)\qquad dx=\mathrm{Cos}\left(\theta\right)d\theta$$
$$x=1\Leftrightarrow\theta=-\pi/2\qquad x=3\Leftrightarrow\theta=\pi/2$$

so the integral becomes

$$V\left(a,b\right)=2\pi\int_{-\pi/2}^{\pi/2}\left(\mathrm{Sin}\left(\theta\right)+2\right)\sqrt{1-\mathrm{Sin}^{2}\left(\theta\right)}\,\mathrm{Cos}\left(\theta\right)d\theta$$
$$=2\pi\int_{-\pi/2}^{\pi/2}\left(\mathrm{Sin}\left(\theta\right)+2\right)\mathrm{Cos}^{2}\left(\theta\right)d\theta$$
$$=2\pi\int_{-\pi/2}^{\pi/2}\mathrm{Sin}\left(\theta\right)\mathrm{Cos}^{2}\left(\theta\right)d\theta+2\pi\int_{-\pi/2}^{\pi/2}2\mathrm{Cos}^{2}\left(\theta\right)d\theta$$
$$=2\pi\left(\frac{-1}{3}\mathrm{Cos}^{3}\left(\theta\right)\right)|_{-\pi/2}^{\pi/2}+2\pi\int_{-\pi/2}^{\pi/2}\left(1+\mathrm{Cos}\left(2\theta\right)\right)d\theta$$
$$=0+2\pi[\theta+\frac{1}{2}\mathrm{Sin}\left(2\theta\right)]|_{-\pi/2}^{\pi/2}$$
$$=2\pi^{2}$$

EXAMPLE 15.12. *Trig Substitutions and Parametric Forms*

The change of variables in the previous example can be viewed as parametric equations for the outline of the donut. A unit semicircle is given by the parametric equations

$$x=\mathrm{Sin}[\theta]$$
$$y=\mathrm{Cos}[\theta]$$

where x goes from -1 to $+1$ as θ goes from $-\pi/2$ to $+\pi/2$. You should verify that y only takes positive values in this range and that this pair traces out the top half circle of unit radius centered at zero.

Adding 2 to the value of x moves the center of the semicircle to $(2,0)$, so it forms the outline semicircle of our half bagel,

$$x = 2 + \text{Sin}[\theta]$$
$$y = \text{Cos}[\theta]$$

The integral change of variables formula for dx can be thought of as expressing the thickness of the shells in terms of θ. It is also perfectly OK to just think of the change of variables as a formal manipulation that makes the integral easier to compute.

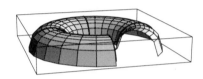

FIGURE 15.13: Area of Donut Icing

EXAMPLE 15.14. *The Icing on the Donut*

The surface of the top half of the torus is obtained by rotating the semicircle

$$y = \sqrt{1 - (x - 2)^2}$$

about the y axis. We will approximate the area by considering what happens to a tiny segment of the semicircle as we rotate that segment about the y axis. The slope of the segment matters in calculating the area of one of these 'barrel hoops.'

FIGURE 15.15: Two Barrel Hoops of Different Slant and Same Thickness

The area is approximately the total length of the generating segment, δl, times the distance through which it travels. It is rotated about the y axis, so it goes around a circle of radius x. The circumference of a circle is $C = 2\pi r = 2\pi x$ in this case. This makes the area of the hoop at x

$$\delta A = 2\pi x \, \delta l$$

We need an expression for δl in terms of the x variable so that we can form an integral.

The length of a tiny segment of the curve is approximately

$$\delta l \approx\approx \sqrt{\delta x^2 + \delta y^2}$$

with an error that is small compared to δx. Since we have $y = y[x]$ as an explicit function of x, we can write

$$\delta y \approx\approx y'[x] \, \delta x$$

so that our length becomes $\delta l \approx\approx \sqrt{\delta x^2 + (y'[x])^2 \delta x^2} = \sqrt{1 + (y'[x])^2} \, \delta x$ and our approximating area is

$$\delta A \approx\approx 2\pi x \sqrt{1 + (y'[x])^2} \, \delta x$$

Finally, by the infinite sum theorem,

$$A = 2\pi \int_1^3 x \sqrt{1 + (y'[x])^2} \, dx$$

$$= 2\pi \int_1^3 x \sqrt{1 + \frac{(x-2)^2}{1 - (x-2)^2}} \, dx$$

$$= 2\pi \int_1^3 \frac{x}{\sqrt{1 - (x-2)^2}} \, dx$$

This is a nasty integral when $x = 1$ or $x = 3$. Of course, *Mathematica* might do the integration for us, but there is a more geometric way to see how to proceed.

15.1.1. A Second Approach. This time let's represent the generating semicircle of radius 1 centered at 2 on the x axis by parametric equations:

$$x[t] = 2 + \mathrm{Sin}[t]$$
$$y[t] = \mathrm{Cos}[t] \qquad \text{for } -\pi/2 \le t \le \pi/2$$

We go back to the length portion of the area of the approximating hoops. The length of a tiny segment of the curve is approximately

$$\delta l \approx\approx \sqrt{\delta x^2 + \delta y^2}$$

but in this case, the increments satisfy

$$\delta x[t] = x[t + \delta t] - x[t] = x'[t] \, \delta t + \varepsilon \, \delta t$$
$$\approx\approx \mathrm{Cos}[t] \, \delta t$$
$$\delta y[t] = y[t + \delta t] - y[t] = y'[t] \, \delta t + \varepsilon \, \delta t$$
$$\approx\approx -\mathrm{Sin}[t] \, \delta t$$

so the length is $\delta l \approx\approx \sqrt{\delta x^2 + \delta y^2} = \sqrt{\text{Cos}^2[t] + \text{Sin}^2[t]}\ \delta t = \delta t$. We must express the increment of area in terms of t,

$$\delta A = 2\pi\,x\,\delta l = 2\pi\,x\,\delta t$$
$$= 2\pi(2 + \text{Sin}[t])\,\delta t$$
$$= 2\pi(2 + \text{Sin}[t])\,\delta t$$

and the area is given by

$$A = 2\pi \int_{-\pi/2}^{\pi/2} (2 + \text{Sin}[t])\,dt$$
$$= 2\pi \int_{-\pi/2}^{\pi/2} 2\,dt + 2\pi \int_{-\pi/2}^{\pi} \text{Sin}[t]\,dt$$
$$= 4\pi^2 + 0$$

since the integral of sine over a half period is zero.

Not only is this parametric integral for the area easy to compute, it also does not have the mathematical singularities of the cartesian form above when $x = 1$ and $x = 3$.

15.2. A Project on Geometric Integrals

This section is an important mathematical project. We have not put it in the Mathematical Background because it is so important.

Find integral formulas for the following quantities. Use the Infinite Sum Theorem to verify the correctness of your formulas. If possible, compute your integral by antidifferentiation, otherwise use *Mathematica* 's 'Integrate' or 'NIntegrate.' Each problem has the steps

(1) Slice the figure some way and find the formula for one slice.
(2) Express that formula in the form $f(x)\,dx$ or $f(y)\,dy$ or $f(\theta)\,d\theta$.
(3) Find the limits of integration.
(4) Compute using rules or numerical integration with or without *Mathematica*.

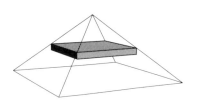 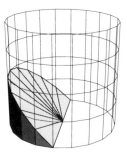

FIGURE 15.16: A Sliced Pyramid and A Round Wedge

PROBLEM 15.17. *Discs*
Find the volume of a right circular cone of base radius r and height h.

PROBLEM 15.18. *Square Slices*
Find the volume of a square pyramid with base area B and height h. Does it matter whether or not it is a right pyramid or slant pyramid? Does the base have to be square?

PROBLEM 15.19. *Triangles*
A wedge is cut from a (cylindrical) tree trunk of radius r by cutting the tree with two planes meeting on a diameter. One plane is perpendicular to the axis and the other makes an angle θ with the first. Find the volume of the wedge.

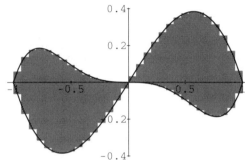

FIGURE 15.20: Area Between $y = x^5 - x^3$ and $y = x - x^3$

PROBLEM 15.21. *Area Between Curves*
Find the area of the bounded regions between the curves $y = x^5 - x^3$ and $z = x - x^3$. Notice that the curves define two regions, one with z on top and the other with y on top.

PROBLEM 15.22. *Find the area between the tangent curves of Problem 7.28.*

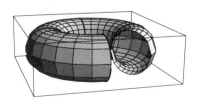

FIGURE 15.23: A Torus

PROBLEM 15.24. *Slice by Shells*
Find the volume of the solid torus (donut)

$$(x - R)^2 + y^2 \leq r^2 \qquad (0 < r < R)$$

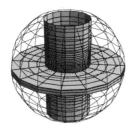
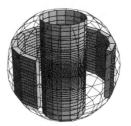

FIGURE 15.25: Slice by Washers or Slice by Shells

PROBLEM 15.26. *Slice by Washers or Shells*
A (cylindrical) hole of radius r is bored through the center of a sphere of radius R. Find the volume of the remaining part of the sphere.

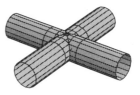

FIGURE 15.27: A Spherical Cap and Crossing Cylinders

PROBLEM 15.28. *Part of a Sphere*
Find the volume of the portion of a sphere that lies above a plane a distance c above the center of the sphere $(0 < c < r)$ the radius of the sphere.

PROBLEM 15.29. *Intersecting Cylinders*
Two circular cylinders of equal radii r intersect through their centers at right angles. Find the volume of the common part. (HINT: The intersection can actually be sliced into square cross sections.)

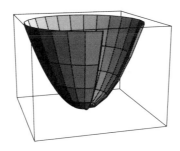

FIGURE 15.30: A Parabolic Antenna

PROBLEM 15.31. *Surface Area*
Find the surface area of the surface obtained by revolving $y = \frac{1}{2}x^2$ about the y axis.

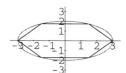   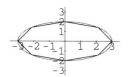   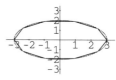

FIGURE 15.32: Equal θ Partition of The Ellipse

PROBLEM 15.33. *Arclength of an Ellipse*
An ellipse is given parametrically by the pair of equations

$$x = 3 \operatorname{Cos}(\theta)$$
$$y = 2 \operatorname{Sin}(\theta)$$

with $-\pi < \theta < \pi$.

Find an integral formula for the length of the curve by "looking in an infinitesimal microscope," as we did in the computation of the length of a parametric curve above. In a microscope we will see a right triangle with the change in x on the horizontal leg, the change in y on the vertical leg and the length along the hypotenuse. The Pythagorean theorem says that the corresponding increment of length is given by

$$\delta l = \sqrt{\delta x^2 + \delta y^2}$$

This time, the length will NOT be equal to the change in the angle, $\delta\theta$, because an ellipse is a circle that has been strecthed different amounts in the x and y directions. The changes in the coordinates are function changes

$$\delta x(\theta) = x(\theta + \delta\theta) - x(\theta)$$
$$\delta y(\theta) = y(\theta + \delta\theta) - y(\theta)$$

Use the increment approximation for these changes to express them approximately in terms of $\delta\theta$. Substitute the approximations into the Pythagorean expression above.

*Test your formula on the circle (where you know the answer) as well as the 'parametric'
equations*

$$x = \theta$$

$$y = 2\sqrt{1 - \left(\frac{\theta}{3}\right)^2}$$

(which can be compared to the explicit arclength formula $L = \int_a^b \sqrt{1 + [f'(x)]^2}\ dx$.)

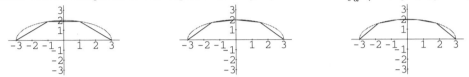

FIGURE *15.34: Equal x Partition of The Ellipse*

The *(correct) arclength formula for the ellipse cannot be computed by antidifferentiation, so you must use numerical integration such as Mathematica NIntegrate[.]. Why is the
parametric integral for the ellipse better behaved than the explicit formula?*

15.3. Other Projects

"Improper" integrals such as $\int_0^1 \frac{1}{\sqrt{x}}\ dx$, whose integrand tends to infinity and is discontinuous at $x = 0$ or $\int_1^\infty \frac{1}{x^2}\ dx$, which is integrated over an infinite interval, are studied in
the Mathematical Background and used in Chapter 18 on infinite series.

Part 3

Vector Geometry

CHAPTER 16

Basic Vector Geometry

We live in three space dimensions with constantly changing time, so most of the applications of calculus take place in more than one variable. This chapter gives the basics of 3 dimensional analytical geometry in its powerful vector formulation.

Geometrically, vectors are 'arrows,' but these arrows can be measured in various ways. The simplest way to measure vectors is with cartesian coordinates, but then the challenge becomes, "How do I calculate the geometric quantity I need from the components?" or the inverse problem, "How do I formulate the geometric condition I have analytically in terms of coordinates?" The real topic of this chapter is basic translation back and forth between pictures and formulas. You really only need to learn a few formulas, but you need to learn to use them geometrically. The next chapter will test your 'fluency.'

This is hard stuff, because once you get the idea, it is easy, yet while you are just watching you sometimes miss the important translating that is going on. At first you will understand the pictures and the formulas separately. In order to have a working knowledge of 3 dimensional mathematics, it is not sufficient to just memorize formulas. You must learn to translate between the languages of algebra and geometry. It will "pay off" in many ways.

Vectors arise in physics as (non-geometrical) forces or electrical fields and in many other areas. We will use them mathematically in total derivatives of functions of several variables and in differential equations. Our translation ability allows us to apply both geometry and analysis to all these problems.

If you learned some 3 dimensional geometry in high school, go right to the next chapter and solve exercises. If you can do those exercises, you probably don't need to read this chapter.

16.1. Cartesian Coordinates

Our first task is to review cartesian coordinates in 1, 2 and 3 dimensions. Once this is done, we will view the coordinates as (one representation of) a vector. Cartesian coordinates are simply a way to measure the location of a point on a line, in a plane or in three space. This measurement requires a reference point, 0, a scale or size of unit, and orientation, the relative position of the axes.

16.1.1. Cartesian Coordinates in 1 Dimension.

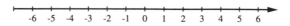

FIGURE 16.1: A Number Line

To locate a point on a line, we select one point to be the origin, select a unit scale and select a positive direction (usually to the right on a horizontal line). Any point can be located by measuring the real number distance associated with that point. Positive numbers are to the right, and negative numbers are measured to the left of the origin. This real number is called the coordinate of the geometrical point.

EXAMPLE 16.2. *Draw a number line and graph the coordinate 5.2.*

SOLUTION:

A line is shown below with origin 0, scale 1 and a rightward positive direction. Simply measure 5.2 units to the right.

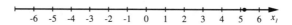

FIGURE 16.3: A Coordinate Point

EXAMPLE 16.4. *Draw a number line and graph the coordinate −3.5.*

SOLUTION:

Measure 3.5 units in the negative direction to the left.

FIGURE 16.5: A Negative Coordinate

EXAMPLE 16.6. *How far apart are −3.5 and 5.2?*

SOLUTION:

Begin at the point 3.5 units to the left of the origin and measure all the way to the point 5.2 units to the right of the origin. This is a total of $3.5 + 5.2 = 8.7$ units.

EXERCISE 16.7. *Give a general formula for the distance between a point with coordinate a and a point with coordinate b. Explain how your formula covers the various cases of positive and negative signs for a and b. (HINT: How much are each of the following: $a − b$, $b − a$, $|b − a|$?)*

Once we have decided to use positive and negative to denote right and left on our line, we can also ask:

EXAMPLE 16.8. *How is it from −3.5 to 5.2 as a directed distance? And how far is it from 5.2 to −3.5 as a directed distance?*

SOLUTION:

Beginning at the point 3.5 units to the left of the origin, we move to the right to get to the point 5.2 units to the right of the origin. Moving to the right makes the direction of the distance positive, so the answer is +8.7.

On the other hand, to move from 5.2 to   3.5 we go to the left, so the distance is negative and the answer from 5.2 to −3.5 is −8.7.

EXERCISE 16.9. *Give a general formula for the directed distance from a point with coordinate a to a point with coordinate b. Explain how your formula covers the various cases of positive and negative signs for a and b.*

16.1.2. Cartesian Coordinates in 2 Dimensions. To locate a point in a plane we place two perpendicular number lines on the plane. We call the point of intersection the origin of a cartesian coordinate system. The number lines are the axes of the system. When viewed from the usual perspective the x_1-axis extends positively to the right and the x_2-axis extends positively upward, with both axes having the same scale. This is a counterclockwise orientation from the first to the second axis. We associate a point with the pair of numbers - in order - that it takes to measure their location. If the point is X and the numbers are x_1 and x_2, we denote the point $X(x_1, x_2)$

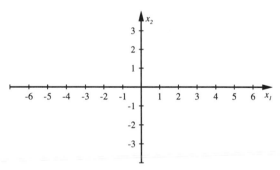

FIGURE 16.10: A Coordinate Plane

If we want to locate a point $X(x_1, x_2)$ in the plane we start at the origin and move a directed distance x_1 along the first axis. Next we move from this point along a line parallel to the second axis a directed distance x_2.

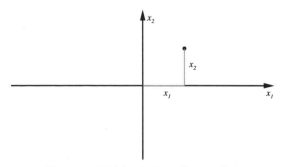

FIGURE 16.11: A Coordinate Pair

EXAMPLE 16.12. *Use a 2-dimensional Cartesian coordinate system to locate the point* $P(4, 5.3)$.

SOLUTION:

First we move 4 units to the right of the origin along the x_1-axis, then we move from this point 5.3 units up, parallel to the x_2-axis.

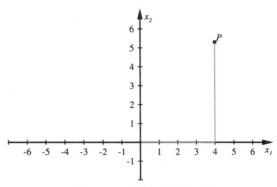

FIGURE 16.13: $P(4, 5.3)$

Given the location of a point we may find its coordinates by drawing a line through the point parallel to the x_2-axis. The directed distance on this line from its intersection with the x_1-axis to the point is its x_2 coordinate. The directed distance from the origin to the intersection of this line with the x_1-axis is the x_1 coordinate.

EXAMPLE 16.14. *Use the 2-dimensional graph below to find the coordinates of point A.*

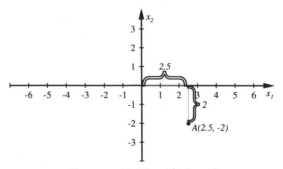

FIGURE 16.15: $A(2.5, -2)$

SOLUTION:

Drop a perpendicular to the x-axis. Notice that the distance from the origin to the base of the perpendicular is 2.5 units to the right. The distance along the perpendicular from the x-axis to the point is 2 units down. Combining these measurements, we have $A(2.5, -2)$

16.1.3. Cartesian coordinates in 3 Dimensions. To locate a point in space we place three mutually perpendicular number lines in space. Each number line is perpendicular to the other two and the lines intersect each other at the point we call the origin. The number lines are the axes of the system. When viewed from the usual perspective the x_1-axis extends positively out toward our observing position, the x_2-axis extends positively to the right and the x_3-axis extends positively upward, with all axes having the same scale. This system is right-hand oriented, that is, if you align your right thumb with the positive x_1-axis, your first finger with the positive x_2-axis, and your second finger with the x_3-axis then your second finger points in the positive x_3 direction. Your left hand would have your middle finger pointing in the negative x_3-direction.

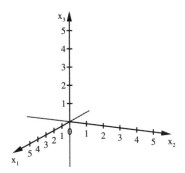

FIGURE 16.16: Space Coordinates

To locate a particular point from its three coordinates (x_1, x_2, x_3), first move a directed distance x_1 from the origin along the x_1- axis, then move a directed distance x_2 from this point along a line parallel to the x_2-axis, and finally, move from this point along a line parallel to the x_3-axis. The final location is the point $X(x_1, x_2, x_3)$.

EXAMPLE 16.17. *Locate the point in three-space with coordinates X(3,-2,4).*

SOLUTION:

Move three units out along the first axis, which is shown pointing toward us. Then move 2 units negatively along a line parallel to the second axis. This is left 2 units. Finally, move 4 units up along a line parallel to the third axis. The point **X** is at the top of the final segment shown in the figure.

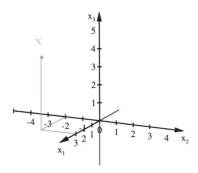

FIGURE 16.18: $X(3, -2, 4)$

EXERCISE 16.19. *Locate the points $X(-3, 2, 4)$, $Y(3, 2, -4)$ and $Z(3, 2, 4)$.*

16.2. Position Vectors

Coordinates may also be thought of as position vectors in 1, 2 and 3 dimensions. A position vector is the 'arrow' that points from the origin in a coordinate system to the point with those coordinates. This arrow gives a direction and a magnitude starting at the origin of a coordinate system, but we will compute these geometric quantities from cartesian coordinates. The necessary formulas appear below.

EXAMPLE 16.20. *Graph the 1 dimensional position vector 3.*

SOLUTION:

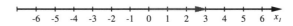

FIGURE 16.21: A 1-D Vector

EXAMPLE 16.22. *Graph the 2 dimensional position vector $\begin{bmatrix} 4 \\ -2 \end{bmatrix}$.*

SOLUTION:

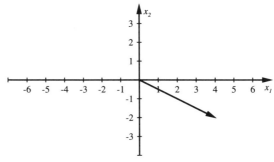

FIGURE 16.23: A 2-D Vector

EXAMPLE 16.24. *Graph the 3 dimensional position vector $\begin{bmatrix} 3 \\ -2 \\ 4 \end{bmatrix}$.*

SOLUTION:

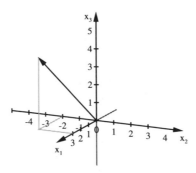

FIGURE 16.25: A 3-D Vector

A position vector will be labeled with a boldface letter, **X**, a horizontal ordered triple (or pair), $X(2,1,3)$, or a vertical ordered triple (or pair), $\begin{bmatrix} 2 \\ 1 \\ 3 \end{bmatrix}$. Notice that a point and a position vector are denoted the same way, because we want to treat them the same way whether or not we draw the arrow connecting the origin and the point.

Also notice that position vectors always start at the origin; this is very important. As far as the computations going from algebra to geometry are concerned, arrows that do not start at the origin are illegal and the formulas do not apply to illegal arrows. We'll still draw them, but figure out ways to find legal computations associated with the shifted illegal arrows.

EXERCISE 16.26. *Here is some practice with position vectors.*

(1) *Graph the following position vectors.*
a) $A(-3)$ b) $B(6)$ c) $C(5.333)$

(2) *Graph the following position vectors,*
a) $A(3,4)$ b) $B(-3,-2)$ c) $C(-2,6)$
d) $D(5,-3)$ e) $E(4.1,2.7)$ f) $F(-3.5,1)$

(3) *Graph the following position vectors,*
a) $A(2,1,5)$ b) $B(-3,4,2)$ c) $C(-2,-4,-1)$
d) $D(3,-2,-5)$ e) $E(3.25,2.1,-6.5)$ f) $F(8.3,0,7)$

(4) *Find the ordered pairs or triples designating the following position vectors.*

a)
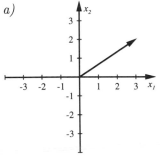

b)
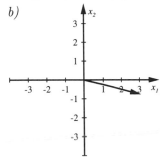

c)
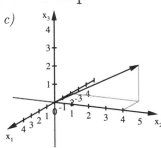

d)
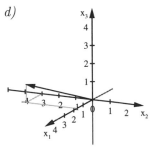

(5) *Use the floor of your classroom as a 2-dimensional cartesian coordinate system with a corner as the origin. Form a system with equal scales using the edge of the floor as the positive axes. Locate and state the coordinates of a point at the base of*
   (a) *your desk*
   (b) *the doorway*
   (c) *the instructor's desk*

(6) *In a rectangular room, choose a corner to be the origin. Label the three edges at the corner between the walls and between the walls and floor so that they form the positive axes of a right-handed coordinate system. Use equal scales along the axes and measure the coordinates of the position vector pointing to the following*
   (a) *the top of your desk*
   (b) *the light switch*
   (c) *the top of the instructor's desk*
   (d) *the corner at the ceiling diagonally opposite your origin corner.*
Locate a point for each of the following and estimate its coordinates.
   (a) *the front door of the building*
   (b) *your parent's home.*

16.3. Basic Geometry of Vectors

This section shows how to find 'magnitude and direction' from cartesian coordinates. It also gives the useful and simple projection formula and the vector cross product that is unique to three dimensions. The cross product is a determinant computation, so it is a little complicated. However, the cross product contains powerful geometric information including orientation.

16.3.1. Length of 2-D Vectors. The length of a position vector is a unique non-negative real number that gives the distance from the origin to the tip of the position vector. Using the Pythagorean Theorem we see that the length in 2 dimensions is given by $|\mathbf{X}| = \sqrt{x_1^2 + x_2^2}$

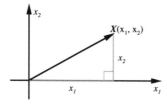

$$|\mathbf{X}| = \sqrt{x_1^2 + x_2^2}$$

For example, the length of the position vector $\begin{bmatrix} 2 \\ -3 \end{bmatrix}$ is

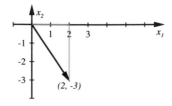

$$\sqrt{2^2 + (-3)^2}$$
$$\sqrt{4 + 9}$$
$$\sqrt{13}$$

EXERCISE 16.27. *Draw the following vectors and find their lengths.*
a) $\begin{bmatrix} 4 \\ 2 \end{bmatrix}$ b) $\begin{bmatrix} -5 \\ -2 \end{bmatrix}$ c) $\begin{bmatrix} -3 \\ 7 \end{bmatrix}$ d) $\begin{bmatrix} 6 \\ -2 \end{bmatrix}$

16.3.2. Length of 3-D Vectors. To find the length of a position vector in three dimensions we have to apply the Pythagorean Theorem twice. Consider position vector $\mathbf{X} = \begin{bmatrix} x_1 \\ x_2 \\ x_3 \end{bmatrix}$ as shown in the diagram below.

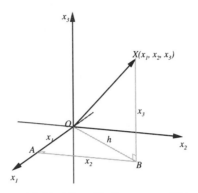

FIGURE 16.28: Two Pythagorean Triangles

First we apply the Pythagorean Theorem to find h, the length of the hypotenuse of right triangle AOB which lies in the x_1x_2-plane and has legs x_1 and x_2.

$$\begin{aligned} h^2 &= x_1^2 + x_2^2 \\ h &= \sqrt{x_1^2 + x_2^2} \end{aligned}$$

Now we can apply the Pythagorean Theorem in triangle XOB with hypotenuse the length of $\mathbf{X}$ and legs h and x_3.

$$|\mathbf{X}| = \sqrt{h^2 + x_3^2}$$

Substituting the value of h^2 from above we obtain

LENGTH FORMULA IN 3-D The length of a 3 dimensional position vector $\mathbf{X} = \begin{bmatrix} x_1 \\ x_2 \\ x_3 \end{bmatrix}$ is

$$|\mathbf{X}| = \sqrt{x_1^2 + x_2^2 + x_3^2}$$

For example, the length of position vector $\begin{bmatrix} 3 \\ -2 \\ 4 \end{bmatrix}$ is

$$\sqrt{3^2 + (-2)^2 + 4^2}$$
$$\sqrt{9 + 4 + 16}$$
$$\sqrt{29}$$

EXERCISE 16.29. *Draw the following vectors and find their lengths.*

a) $\begin{bmatrix} 4 \\ 6 \\ 5 \end{bmatrix}$ b) $\begin{bmatrix} -4 \\ 7 \\ -2 \end{bmatrix}$ c) $\begin{bmatrix} 6 \\ 8 \\ -10 \end{bmatrix}$ d) $\begin{bmatrix} -6 \\ -9 \\ -1 \end{bmatrix}$

In general, a d dimensional position vector $\mathbf{X} = \begin{bmatrix} x_1 \\ \vdots \\ x_d \end{bmatrix}$, has length

$$|\mathbf{X}| = \sqrt{x_1^2 + \ldots + x_d^2}$$

where d is any dimension (usually 2 or 3).

16.3.3. Dot Product. The dot (or inner) product does not have an immediate geometric interpretation, but is used to compute angles. It is algebraically very simple.

Consider the position vectors $\mathbf{X} = \begin{bmatrix} x_1 \\ \vdots \\ x_d \end{bmatrix}$ and $\mathbf{Y} = \begin{bmatrix} y_1 \\ \vdots \\ y_d \end{bmatrix}$, where d is any dimension. The dot product of $\mathbf{X}$ and $\mathbf{Y}$ is the scalar quantity

DOT PRODUCT

$$\langle \mathbf{X} \bullet \mathbf{Y} \rangle = x_1 y_1 + \cdots + x_d y_d$$

Notice that the length is related to the dot product by

$$|\mathbf{X}| = \sqrt{\langle \mathbf{X} \bullet \mathbf{X} \rangle} = \sqrt{x_1 x_1 + \cdots + x_d x_d}$$

EXAMPLE 16.30. *Find the dot product of the position vectors* $\begin{bmatrix} 4 \\ 3 \end{bmatrix}$ *and* $\begin{bmatrix} 7 \\ -11 \end{bmatrix}$.

SOLUTION:

$$
\begin{aligned}
\langle \begin{bmatrix} 4 \\ 3 \end{bmatrix} \bullet \begin{bmatrix} 7 \\ -11 \end{bmatrix} \rangle &= 4 \cdot 7 + 3 \cdot (-11) \\
&= 28 - 33 \\
&= -5
\end{aligned}
$$

EXAMPLE 16.31. *Find the dot product of the position vectors* $\begin{bmatrix} 2 \\ -3 \\ 5 \end{bmatrix}$ *and* $\begin{bmatrix} -4 \\ 3 \\ 6 \end{bmatrix}$.

SOLUTION:

$$
\begin{aligned}
\langle \begin{bmatrix} 2 \\ -3 \\ 5 \end{bmatrix} \bullet \begin{bmatrix} -4 \\ 3 \\ 6 \end{bmatrix} \rangle &= 2 \cdot (-4) + (-3) \cdot 3 + 5 \cdot 6 \\
&= -8 - 9 + 30 \\
&= 13
\end{aligned}
$$

EXERCISE 16.32. *Draw each pair of vectors and find their dot product*

a) $\begin{bmatrix} -3 \\ -2 \end{bmatrix}, \begin{bmatrix} -8 \\ 12 \end{bmatrix}$ b) $\begin{bmatrix} 2 \\ 1 \end{bmatrix}, \begin{bmatrix} -4 \\ 5 \end{bmatrix}$ c) $\begin{bmatrix} 7 \\ -3 \end{bmatrix}, \begin{bmatrix} -2 \\ -8 \end{bmatrix}$ d) $\begin{bmatrix} 6 \\ -9 \end{bmatrix}, \begin{bmatrix} -3 \\ 2 \end{bmatrix}$

e) $\begin{bmatrix} 9 \\ 1 \end{bmatrix}, \begin{bmatrix} 2 \\ -18 \end{bmatrix}$ f) $\begin{bmatrix} 5 \\ 2 \\ -3 \end{bmatrix}, \begin{bmatrix} 2 \\ -2 \\ 2 \end{bmatrix}$ g) $\begin{bmatrix} 6 \\ -3 \\ -3 \end{bmatrix}, \begin{bmatrix} 2 \\ 9 \\ -5 \end{bmatrix}$ h) $\begin{bmatrix} 3 \\ 5 \\ 6 \end{bmatrix}, \begin{bmatrix} 4 \\ -2 \\ 3 \end{bmatrix}$

i) $\begin{bmatrix} 5 \\ 6 \\ 3 \end{bmatrix}, \begin{bmatrix} 1 \\ 3 \\ 7 \end{bmatrix}$ j) $\begin{bmatrix} -2 \\ -3 \\ -5 \end{bmatrix}, \begin{bmatrix} 4 \\ 2 \\ 6 \end{bmatrix}$

16.3.4. Angle Between Two Vectors. Given two position vectors $\mathbf{X}$ and $\mathbf{Y}$ with $\theta = \angle\langle\mathbf{X}, \mathbf{Y}\rangle$ the angle between them, the cosine of θ can be found by the following formula:

$$\mathrm{Cos}[\theta] = \frac{\langle \mathbf{X} \bullet \mathbf{Y} \rangle}{|\mathbf{X}|\,|\mathbf{Y}|}$$

In order to find the actual angle, we need to calculate the ArcCosine of the expression on the right. ArcCosine is ugly to compute by hand, but easy for your calculator or *Mathematica*.

EXAMPLE 16.33. *If* $\mathbf{X} = \begin{bmatrix} 3 \\ 5 \\ 6 \end{bmatrix}$, $\mathbf{Y} = \begin{bmatrix} 4 \\ -2 \\ 3 \end{bmatrix}$, *and* θ *is the angle between them, then*

$$\mathrm{Cos}[\theta] = \frac{\left\langle \begin{bmatrix} 3 \\ 5 \\ 6 \end{bmatrix} \bullet \begin{bmatrix} 4 \\ -2 \\ 3 \end{bmatrix} \right\rangle}{\left| \begin{bmatrix} 3 \\ 5 \\ 6 \end{bmatrix} \right| \left| \begin{bmatrix} 4 \\ -2 \\ 3 \end{bmatrix} \right|}$$

$$= \frac{12 - 10 + 18}{\sqrt{9 + 25 + 36}\sqrt{16 + 4 + 9}}$$

$$= \frac{20}{\sqrt{2030}}$$

$$\theta = \mathrm{ArcCos}[\frac{20}{\sqrt{2030}}]$$

$$\approx 1.11085 \quad (radians)$$

EXERCISE 16.34. $\mathbf{A} = \begin{bmatrix} a_1 \\ a_2 \\ a_3 \end{bmatrix} = \begin{bmatrix} 1 \\ 3 \\ 2 \end{bmatrix}$ *and* $\mathbf{B} = \begin{bmatrix} b_1 \\ b_2 \\ b_3 \end{bmatrix} = \begin{bmatrix} -2 \\ 1 \\ 3 \end{bmatrix}$ *Compute* $\langle\mathbf{A} \bullet \mathbf{B}\rangle$, $|\mathbf{A}|$, $|\mathbf{B}|$, $\mathrm{Cos}[\angle\langle A, B\rangle]$ *and if you have a calculator, find the inverse cosine and* $\theta = \angle\langle\mathbf{A}, \mathbf{B}\rangle$.

EXERCISE 16.35. *Find the angle between each pair of vectors in Exercise* 16.32. *Compare your computations with the figures you drew.*
Which pairs of vectors are perpendicular?

16.3.5. Translation Exercises. Here is some basic practice at using geometry and algebra.

EXERCISE 16.36. *Circle*
Describe geometrically, the set of points that consists of the tips of all 2-D position vectors that have a length of five. Algebraically, this is the set of vectors $X(x_1, x_2)$ *satisfying an equation in* x_1 *and* x_2. *What is this equation?*

EXERCISE 16.37. *Sphere*
Repeat Exercise 16.36 using three-dimensional position vectors and give the equation for the three coordinates.

EXERCISE 16.38. *Dots in 2-D, Circle in 3-D*
What is the intersection of a plane and a sphere? What is the intersection of a line and a circle?

(1) *Find and draw on the same coordinate system all 2-D position vectors of the form* $\begin{bmatrix} x_1 \\ 7 \end{bmatrix}$ *that also have length 25. The first condition is the equation* $x_2 = 7$*. What does it mean geometrically? What is the condition that the unknown vector* $X(x_1, x_2)$ *has length 25 in terms of algebra? What is the combined algebraic condition? (What are the simultaneous equations?)*

(2) *Find and draw on the same coordinate system all 3-D position vectors of the form* $\begin{bmatrix} x_1 \\ 7 \\ x_3 \end{bmatrix}$ *that also have length 25. The first condition is the equation* $x_2 = 7$*. What does it mean geometrically? What is the condition that the vectors have length 25 in terms of algebra? What is the geometric meaning of the equation* $x_1^2 + x_2^2 + x_3^2 = 625$*? What is the combined algebraic condition* $x_2 = 7$ *and* $x_1^2 + x_2^2 + x_3^2 = 625$*? What is the geometric set?*

(3) *Find and draw on the same coordinate system all 3-D position vectors of the form* $\begin{bmatrix} x_1 \\ x_2 \\ 7 \end{bmatrix}$*. What does it mean geometrically? What is the condition that the unknown vector* $X(x_1, x_2, x_3)$ *satisfies* $x_1^2 + x_2^2 = 625$*? (Careful,* x_3 *is not restricted.) What is the combined algebraic condition* $x_3 = 7$ *and* $x_1^2 + x_2^2 = 625$*? What is the geometric set? Do these vectors all have the same length?*

16.3.6. Perpendicular Vector Projection. Vector projection may be visualized as one position vector casting a "shadow" on the other position vector under special lighting conditions.

Let **X** and **Y** be position vectors (2-dimensional or 3-dimensional). We are wish to compute the position vector formed by the "shadow" of **Y** cast onto **X** if the source of light is perpendicular to position vector **X** and is contained by the plane in which both **X** and **Y** lie.

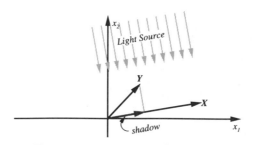

FIGURE 16.39: Vector Projection

Notice from the diagram that the "shadow" is just a scalar multiple of $\mathbf{X}$. Algebraically the projection of $\mathbf{Y}$ onto $\mathbf{X}$, denoted by $\text{Proj}_{\mathbf{X}}(\mathbf{Y})$, is

$$\text{Proj}_{\mathbf{X}}(\mathbf{Y}) = \frac{\langle \mathbf{Y} \bullet \mathbf{X} \rangle}{\langle \mathbf{X} \bullet \mathbf{X} \rangle} \mathbf{X}$$

Since the quotient of dot products is a scalar (or number), $c = \frac{\langle \mathbf{Y} \bullet \mathbf{X} \rangle}{\langle \mathbf{X} \bullet \mathbf{X} \rangle}$, the $\text{Proj}_{\mathbf{X}}(\mathbf{Y})$ becomes $c\mathbf{X}$. Position vector projection is a special scalar multiplication. We will see below that scalar multiplication just produces a stretch, compression or reversal of a vector. The "shadow" lies along the direction of $\mathbf{X}$.

You can derive this formula using trigonometry and the dot product formula for the cosine of an angle. But whether you derive it or simply memorize it, this formula is worth remembering because it can be computed without square roots or ArcCosines. In other words, if you can reduce a geometric problem to computing a perpendicular projection, you have an easy computation.

EXAMPLE 16.40. *Given* $\mathbf{X} = \mathbf{X}(-5, 2)$ *and* $\mathbf{Y} = \mathbf{Y}(-2, 6)$ *calculate and sketch a diagram of the projection of* $\mathbf{Y}$ *onto* $\mathbf{X}$.

SOLUTION:

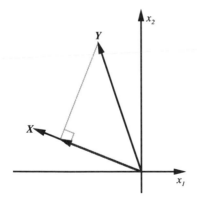

$$\text{Proj}_{\mathbf{X}}(\mathbf{Y}) = \frac{\begin{bmatrix} -5 \\ 2 \end{bmatrix} \bullet \begin{bmatrix} -2 \\ 6 \end{bmatrix}}{\begin{bmatrix} -5 \\ 2 \end{bmatrix} \bullet \begin{bmatrix} -5 \\ 2 \end{bmatrix}} \begin{bmatrix} -5 \\ 2 \end{bmatrix}$$

$$= \frac{22}{29} \begin{bmatrix} -5 \\ 2 \end{bmatrix}$$

$$= \begin{bmatrix} -\frac{110}{29} \\ \frac{44}{29} \end{bmatrix}$$

EXAMPLE 16.41. *Given* $\mathbf{R} = \mathbf{R}(4, 6, -1)$ *and* $\mathbf{S} = \mathbf{S}(-2, 3, 1)$ *calculate and sketch a diagram of the projection of* $\mathbf{S}$ *onto* $\mathbf{R}$.

SOLUTION:

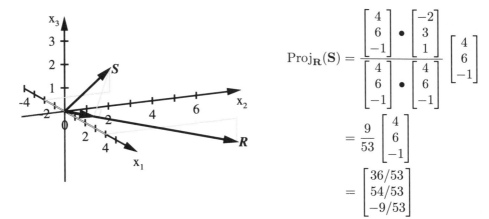

$$\text{Proj}_{\mathbf{R}}(\mathbf{S}) = \frac{\begin{bmatrix} 4 \\ 6 \\ -1 \end{bmatrix} \bullet \begin{bmatrix} -2 \\ 3 \\ 1 \end{bmatrix}}{\begin{bmatrix} 4 \\ 6 \\ -1 \end{bmatrix} \bullet \begin{bmatrix} 4 \\ 6 \\ -1 \end{bmatrix}} \begin{bmatrix} 4 \\ 6 \\ -1 \end{bmatrix}$$

$$= \frac{9}{53} \begin{bmatrix} 4 \\ 6 \\ -1 \end{bmatrix}$$

$$= \begin{bmatrix} 36/53 \\ 54/53 \\ -9/53 \end{bmatrix}$$

16.3.7. Perpendicularity.

EXERCISE 16.42. *Give the simplest possible algebraic condition you can use to test whether two vectors are perpendicular. HINT: What is $\text{Cos}[\frac{\pi}{2}]$? For example, are* $\begin{bmatrix} 1 \\ -3 \\ 2 \end{bmatrix}$ *and* $\begin{bmatrix} 2 \\ 2 \\ 1 \end{bmatrix}$ *perpendicular? Are* $\begin{bmatrix} 1 \\ -3 \\ 2 \end{bmatrix}$ *and* $\begin{bmatrix} 1 \\ 2 \\ 1 \end{bmatrix}$ *? What are the angles between them? Does the denominator matter?*

YOUR CONDITION FOR PERPENDICULARITY

$$\mathbf{A} \quad \text{is perpendicular to} \quad \mathbf{B} \quad \Leftrightarrow?$$

16.3.8. Cross Product in 3 Dimensions. The cross product is a powerful geometric tool in 3 dimensions. Before you read the computation, think how you might find a vector perpendicular to two given vectors such as $\begin{bmatrix} 1 \\ 3 \\ 1 \end{bmatrix}$ and $\begin{bmatrix} -2 \\ -3 \\ 1 \end{bmatrix}$. Sketch a figure and convince yourself that there is a whole line of possibilities. Also convince yourself that the computation is not obvious.

ALGEBRAIC CROSS PRODUCT: The 3-dimensional vector cross product is given by the following symbolic determinant computation

$$\mathbf{Z} = \mathbf{X} \times \mathbf{Y} = \det \begin{vmatrix} i & j & k \\ x_1 & x_2 & x_3 \\ y_1 & y_2 & y_3 \end{vmatrix} = \begin{bmatrix} x_2 y_3 - x_3 y_2 \\ x_3 y_1 - x_1 y_3 \\ x_1 y_2 - x_2 y_1 \end{bmatrix}$$

GEOMETRIC CROSS PRODUCT: The cross product is a new vector with the following properties:

(1) $\mathbf{Z}$ is perpendicular to both $\mathbf{X}$ and $\mathbf{Y}$
(2) The length of $\mathbf{Z}$ equals the area of the parallelogram spanned by the edges of $\mathbf{X}$ and $\mathbf{Y}$.

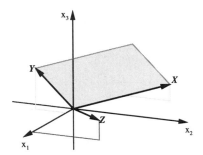

FIGURE 16.43: $\mathbf{Z} = \mathbf{X} \times \mathbf{Y}$

(3) The triple $\mathbf{X}$, $\mathbf{Y}$, $\mathbf{Z}$ forms a hight-hand frame of vectors, that is, you can point your right thumb in the direction of $\mathbf{X}$, your first finger in the direction of $\mathbf{Y}$ and your second finger in the direction of $\mathbf{Z}$. If you try this with your left hand, your second finger will point opposite the direction of $\mathbf{Z}$.

EXAMPLE 16.44. *Compute the cross product* $\mathbf{Z} = \mathbf{X} \times \mathbf{Y}$,

$$\mathbf{Z} = \begin{bmatrix} -8 \\ 9 \\ 2 \end{bmatrix} \times \begin{bmatrix} 4 \\ -3 \\ 5 \end{bmatrix}$$

SOLUTION:

$$\mathbf{Z} = \mathbf{X} \times \mathbf{Y}$$

$$= \det \begin{vmatrix} i & j & k \\ x_1 & x_2 & x_3 \\ y_1 & y_2 & y_3 \end{vmatrix} = \det \begin{vmatrix} i & j & k \\ -8 & 10 & 2 \\ 2 & -4 & 5 \end{vmatrix}$$

$$= \begin{bmatrix} x_2 y_3 - x_3 y_2 \\ x_3 y_1 - x_1 y_3 \\ x_1 y_2 - x_2 y_1 \end{bmatrix} = \begin{bmatrix} 10 \cdot 5 - 2 \cdot (-4) \\ 2 \cdot 2 - (-8) \cdot 5 \\ (-8) \cdot (-4) - 10 \cdot 2 \end{bmatrix}$$

$$= \begin{bmatrix} 50 + 8 \\ 4 + 40 \\ 32 - 20 \end{bmatrix} = \begin{bmatrix} 58 \\ 44 \\ 12 \end{bmatrix}$$

EXERCISE 16.45. *Compute the vector (cross) product* $\mathbf{M} = \mathbf{P} \times \mathbf{D}$, *where* $\mathbf{P} = \begin{bmatrix} p_1 \\ p_2 \\ p_3 \end{bmatrix} = \begin{bmatrix} 1 \\ 3 \\ 1 \end{bmatrix}$ *and* $\mathbf{D} = \begin{bmatrix} d_1 \\ d_2 \\ d_3 \end{bmatrix} = \begin{bmatrix} -2 \\ -3 \\ 1 \end{bmatrix}$. *Find a vector perpendicular to BOTH* $\mathbf{P}$ *and* $\mathbf{D}$. *Sketch the plane that contains* 0, $\mathbf{P}$ *and* $\mathbf{D}$. *Sketch* $\mathbf{M}$.

EXERCISE 16.46. *Areas and Angles*
The area of a parallelogram with edges $\mathbf{X}$ *and* $\mathbf{Y}$ *of lengths* $|\mathbf{X}|$ *and* $|\mathbf{Y}|$ *can be computed by simple trigonometry. In the figure below, a 'height' or segment perpendicular to the side of* $\mathbf{X}$ *is shown.*

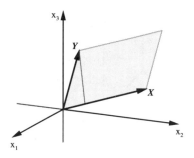

FIGURE 16.47: Area by Trig

1) Show that the length of the 'height' segment is $h = |\mathbf{Y}| \operatorname{Sin}[\theta]$, where θ is the angle between $\mathbf{X}$ and $\mathbf{Y}$.

2) Show that the area of the parallelogram is $|\mathbf{X}| \cdot |\mathbf{Y}| \operatorname{Sin}[\theta]$. Use trigonometry to find the height $h = |\mathbf{Y}| \operatorname{Sin}[\theta]$ and base $b = |\mathbf{X}|$ of the parallelogram.

3) We also know that the area of the parallelogram is $|\mathbf{X} \times \mathbf{Y}|$ from the geometric definition of cross product. Show that

$$\operatorname{Sin}[\theta] = \frac{|\mathbf{X} \times \mathbf{Y}|}{|\mathbf{X}| \cdot |\mathbf{Y}|}$$

4) Which is an easier computation, finding the angle between $X(-6, 8, 2)$ and $Y(2, 3, 8)$ by the dot product formula for cosine or using the cross product formula above for sine? (In both cases, use your calculator to find the arcsine or arccosine.)

16.4. The Geometry of Vector Addition

Analytical geometry from computer graphics packages to high-tech theory is built on simple formulas. The ideas are very fundamental, but vector addition is a little peculiar. It is algebraically easy, but to add vectors geometrically, we slide one parallel to itself until its tail is at the tip of the other. The translated vector is illegal, but it points to the legal sum vector that starts at the origin. This is the key to handling illegal arrows and understanding more advanced analytical geometry.

16.4.1. Vector Sum in 2-D: Tips-to-Tails Add Like Components. Suppose we wish to add two 2-dimensional position vectors $\mathbf{R} = R(r_1, r_2)$ and $\mathbf{S} = S(s_1, s_2)$.

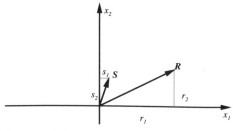

FIGURE 16.48: Vectors and Components

The idea of position vector addition is to translate $\mathbf{S}$ to the tip of the position vector $\mathbf{R}$. The arrow parallel to $\mathbf{S}$ starting at the tip of $\mathbf{R}$ is illegal, but it can be measured by putting parallel coordinate axes at the tip of $\mathbf{R}$ and using the coordinates of $\mathbf{S}$ in the translated coordinate system.

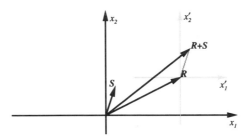

FIGURE 16.49: Tips to Tails Vector Sum

The point at the tip of the oriented displacement parallel to $\mathbf{S}$ is the position vector for the sum $\mathbf{R} + \mathbf{S}$.

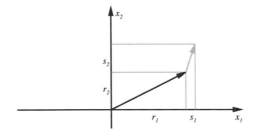

FIGURE 16.50: Sum Components

Let us see why this is called "addition." From the diagram the x_1-coordinate of the position vector $\mathbf{R} + \mathbf{S}$ is $r_1 + s_1$, the sum of the x_1-coordinate of $\mathbf{R}$ and the x_1-coordinate of $\mathbf{S}$. Similarly the x_2 coordinate of $\mathbf{R} + \mathbf{S}$ is $r_2 + s_2$, the sum of the x_2-coordinate of $\mathbf{R}$ and the x_2-coordinate of $\mathbf{S}$. Algebraically the position vector sum in two dimensions is:

$$\mathbf{R} + \mathbf{S} = \begin{bmatrix} r_1 \\ r_2 \end{bmatrix} + \begin{bmatrix} s_1 \\ s_2 \end{bmatrix} = \begin{bmatrix} r_1 + s_1 \\ r_2 + s_2 \end{bmatrix}$$

We could also move from the origin to the tip of position vector $\mathbf{S}$. Then move to the tip of an oriented displacement parallel to $\mathbf{R}$ starting at the tip of $\mathbf{S}$.

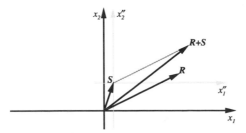

FIGURE 16.51: Tips to Tails The Other Way

The point at the tip of this oriented displacement is the position vector for the sum $\mathbf{S} + \mathbf{R}$.

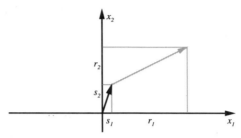

FIGURE 16.52: Sum Components

From the diagram the x_1 coordinate of the position vector $\mathbf{S} + \mathbf{R}$ is $s_1 + r_1$ and the x_2 coordinate is $s_2 + r_2$. Algebraically

$$\mathbf{S} + \mathbf{R} = \begin{bmatrix} s_1 \\ s_2 \end{bmatrix} + \begin{bmatrix} r_1 \\ r_2 \end{bmatrix} = \begin{bmatrix} s_1 + r_1 \\ s_2 + r_2 \end{bmatrix} = \begin{bmatrix} r_1 + s_1 \\ r_2 + s_2 \end{bmatrix}$$

So $\mathbf{R} + \mathbf{S} = \mathbf{S} + \mathbf{R}$ and we see that it does not matter from which position vector we start.

Notice the sum of two position vectors is another position vector with its tail legally fixed at the origin.

If we combine both oriented displacements in one diagram,

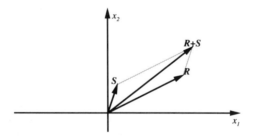

FIGURE 16.53: Parallelogram Vector Addition

we can see that the position vectors $\mathbf{R}$ and $\mathbf{S}$ form two adjacent sides of a parallelogram with one vertex at the origin. The sum $\mathbf{R} + \mathbf{S}$ is the directed diagonal from the origin of this parallelogram. This is the geometrical way to see that $\mathbf{R} + \mathbf{S} = \mathbf{S} + \mathbf{R}$.

EXAMPLE 16.54. *Given*

$$\mathbf{X} = \begin{bmatrix} 3 \\ 1 \end{bmatrix} \qquad and \qquad \mathbf{Y} = \begin{bmatrix} -5 \\ 6 \end{bmatrix}$$

geometrically and algebraically find $\mathbf{X} + \mathbf{Y}$.

SOLUTION, GEOMETRIC PART: Draw the position vectors $\begin{bmatrix} 3 \\ 1 \end{bmatrix}$ and $\begin{bmatrix} -5 \\ 6 \end{bmatrix}$. Next draw an oriented displacement from the tip of $\begin{bmatrix} 3 \\ 1 \end{bmatrix}$ parallel to $\begin{bmatrix} -5 \\ 6 \end{bmatrix}$. You are now at the point (-2,7) draw the position vector to that point from the origin and it is the position vector $\mathbf{X} + \mathbf{Y}$.

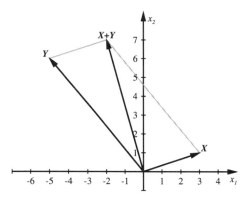

FIGURE 16.55: Another Vector Sum

SOLUTION, ALGEBRAIC PART:

$$\mathbf{X} + \mathbf{Y} = \begin{bmatrix} 3 \\ 1 \end{bmatrix} + \begin{bmatrix} -5 \\ 6 \end{bmatrix} = \begin{bmatrix} 3 + (-5) \\ 1 + 6 \end{bmatrix} = \begin{bmatrix} -2 \\ 7 \end{bmatrix}$$

EXAMPLE 16.56. *Given*

$$\mathbf{X} = \begin{bmatrix} 2 \\ -1 \end{bmatrix} \qquad and \qquad \mathbf{Y} = \begin{bmatrix} -3 \\ -4 \end{bmatrix}$$

geometrically and algebraically find $\mathbf{Y} + \mathbf{X}$.

GEOMETRIC SOLUTION: Draw the position vectors $\begin{bmatrix} 2 \\ -1 \end{bmatrix}$ and $\begin{bmatrix} -3 \\ -4 \end{bmatrix}$. Draw and oriented displacement from the tip of $\begin{bmatrix} -3 \\ -4 \end{bmatrix}$ parallel to $\begin{bmatrix} 2 \\ -1 \end{bmatrix}$. Draw a position vector from the origin to this point $(-1, -5)$ and this vector is $\mathbf{X} + \mathbf{Y}$.

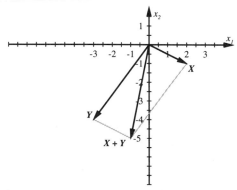

FIGURE 16.57: $X(2, -1) + Y(-3, -4) = Z(-1, -5)$

ALGEBRAIC SOLUTION:

$$\mathbf{Y} + \mathbf{X} = \begin{bmatrix} -3 \\ -4 \end{bmatrix} + \begin{bmatrix} 2 \\ -1 \end{bmatrix} = \begin{bmatrix} -3 + 2 \\ -4 + (-1) \end{bmatrix} = \begin{bmatrix} -1 \\ -5 \end{bmatrix}$$

16.4.2. Vector Sum in 3-D: Tips-to-Tails Add Like Components. We will work through an example of adding the vectors

$$\mathbf{R} = R(r_1, r_2, r_3) = R(-6, 8, 2) \qquad \text{and} \qquad \mathbf{S} = S(s_1, s_2, s_3) = S(2, -3, 3)$$

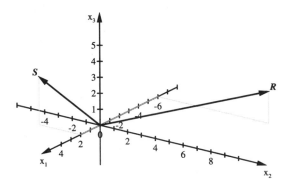

FIGURE 16.58: Vectors $\mathbf{R}$ and $\mathbf{S}$

Let us form the geometric sum $\mathbf{R} + \mathbf{S}$. We move from the origin to the tip of position vector $\mathbf{R}$ first. Then move to the tip of an oriented displacement parallel to $\mathbf{S}$ starting at the tip of $\mathbf{R}$.

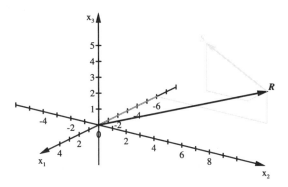

FIGURE 16.59: $\mathbf{S}$ Drawn Illegally at Tip of $\mathbf{R}$

The point at the tip of this oriented displacement is the position vector for the sum $\mathbf{R}+\mathbf{S}$. From the diagram the x_1 coordinate of the position vector $\mathbf{R}+\mathbf{S}$ is r_1+s_1, the x_2 coordinate is $r_2 + s_2$, and the x_3 coordinate is $r_3 + s_3$. Algebraically

$$\mathbf{R} + \mathbf{S} = \begin{bmatrix} r_1 \\ r_2 \\ r_3 \end{bmatrix} + \begin{bmatrix} s_1 \\ s_2 \\ s_3 \end{bmatrix} = \begin{bmatrix} r_1 + s_1 \\ r_2 + s_2 \\ r_3 + s_3 \end{bmatrix} = \begin{bmatrix} s_1 + r_1 \\ s_2 + r_2 \\ s_3 + r_3 \end{bmatrix}$$

Now, suppose we wish to add the two 3-dimensional position vectors in the other order. Geometrically, to form $\mathbf{S} + \mathbf{R}$, we translate $\mathbf{R}$ so that its tail begins at the tip of $\mathbf{S}$. This oriented displacement or illegal arrow parallel to $\mathbf{R}$ can be measured from the components of $\mathbf{R}$ by using a translated coordinate system with its origin at the tip of $\mathbf{S}$.

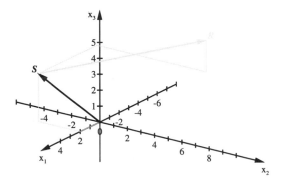

FIGURE 16.60: **R** Drawn Illegally at Tip of **S**

The point at the tip of the oriented displacement is the position vector for the sum $\mathbf{S} + \mathbf{R}$.

The net movement on the x_1 axis for the position vector $\mathbf{S} + \mathbf{R}$ is $s_1 + r_1$, the sum of the x_1 coordinate of $\mathbf{S}$ and the x_1 coordinate of $\mathbf{R}$. Similarly the x_2 coordinate of $\mathbf{S} + \mathbf{R}$ is $s_2 + r_2$ and the x_3 coordinate of $\mathbf{S} + \mathbf{R}$ is $s_3 + r_3$. Algebraically the position vector sum in three dimensions is

$$\mathbf{S} + \mathbf{R} = \begin{bmatrix} s_1 \\ s_2 \\ s_3 \end{bmatrix} + \begin{bmatrix} r_1 \\ r_2 \\ r_3 \end{bmatrix} = \begin{bmatrix} s_1 + r_1 \\ s_2 + r_2 \\ s_3 + r_3 \end{bmatrix} = \begin{bmatrix} -4 \\ 5 \\ 5 \end{bmatrix}$$

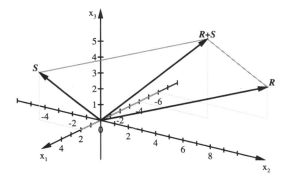

FIGURE 16.61: The Vector Sum $\mathbf{R} + \mathbf{S}$

Again $\mathbf{S} + \mathbf{R} = \mathbf{R} + \mathbf{S}$ and we see that it does not matter from which position vector we begin. Geometrically, we can see this from a plot of both additions. The vector sum is the diagonal through the origin of a parallelogram with $\mathbf{R}$ on one edge and $\mathbf{S}$ on another.

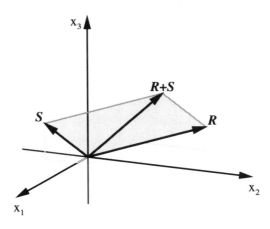

FIGURE 16.62: 3-D Vector Parallelogram Sum

EXAMPLE 16.63. *Given*

$$\mathbf{X} = \begin{bmatrix} 5 \\ 2 \\ 3 \end{bmatrix} \qquad and \qquad \mathbf{Y} = \begin{bmatrix} -3 \\ 4 \\ 2 \end{bmatrix}$$

find $\mathbf{X} + \mathbf{Y}$ *with geometry and with algebra.*

GEOMETRIC SOLUTION: Draw the position vectors $\begin{bmatrix} 5 \\ 2 \\ 3 \end{bmatrix}$ and $\begin{bmatrix} -3 \\ 4 \\ 2 \end{bmatrix}$. Draw an oriented

displacement from the tip of $\begin{bmatrix} -3 \\ 4 \\ 2 \end{bmatrix}$ parallel to $\begin{bmatrix} 5 \\ 2 \\ 3 \end{bmatrix}$. Draw a position vector from the origin

to this point $P(2, 6, 5)$. This position vector is $\mathbf{X} + \mathbf{Y}$.

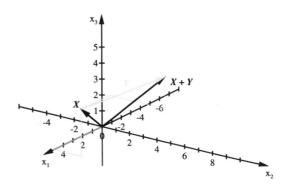

FIGURE 16.64: The Vector Sum $\mathbf{X} + \mathbf{Y}$

ALGEBRAIC SOLUTION:

$$\mathbf{X} + \mathbf{Y} = \begin{bmatrix} 5 \\ 2 \\ 3 \end{bmatrix} + \begin{bmatrix} -3 \\ 4 \\ 2 \end{bmatrix} = \begin{bmatrix} 5 + (-3) \\ 2 + 4 \\ 3 + 2 \end{bmatrix} = \begin{bmatrix} 2 \\ 6 \\ 5 \end{bmatrix}$$

EXERCISE 16.65. *For each problem use geometry and algebra to find the sum of the two position vectors.*

(1) $\mathbf{S} = \begin{bmatrix} 3 \\ 5 \end{bmatrix}$ $\mathbf{T} = \begin{bmatrix} 2 \\ 1 \end{bmatrix}$ (2) $\mathbf{R} = \begin{bmatrix} -3 \\ 4 \end{bmatrix}$ $\mathbf{S} = \begin{bmatrix} -2 \\ -5 \end{bmatrix}$

(3) $\mathbf{X} = \begin{bmatrix} 5 \\ 3 \\ 4 \end{bmatrix}$ $\mathbf{Y} = \begin{bmatrix} 1 \\ 6 \\ 2 \end{bmatrix}$ (4) $\mathbf{Y} = \begin{bmatrix} -3 \\ -4 \end{bmatrix}$ $\mathbf{Z} = \begin{bmatrix} 4 \\ -2 \end{bmatrix}$

(5) $\mathbf{S} = \begin{bmatrix} 4 \\ 6 \\ 2 \end{bmatrix}$ $\mathbf{W} = \begin{bmatrix} -5 \\ 2 \\ 3 \end{bmatrix}$ (6) $\mathbf{X} = \begin{bmatrix} -2 \\ -4 \\ 2 \end{bmatrix}$ $\mathbf{Y} = \begin{bmatrix} 2 \\ 6 \\ 3 \end{bmatrix}$

(7) Using a diagram, explain that the sum $\mathbf{X} + \mathbf{Y}$ of two 3-dimensional position vectors $\mathbf{X}$ and $\mathbf{Y}$ is the directed diagonal of a parallelogram with $\mathbf{X}$ and $\mathbf{Y}$ as two adjacent sides.

16.5. The Geometry of Scalar Multiplication

16.5.1. Scalar Multiplication in 2-D.

The algebraic definition of multiplication of a number (or scalar) c times a vector $\mathbf{X} = \begin{bmatrix} x_1 \\ x_2 \end{bmatrix}$ is to simply multiply each component by the scalar,

$$c\,\mathbf{X} = c \begin{bmatrix} x_1 \\ x_2 \end{bmatrix} = \begin{bmatrix} c \cdot x_1 \\ c \cdot x_2 \end{bmatrix}$$

Geometrically, a scalar multiple of one vector points along the same line through the origin, but is stretched, compressed or reversed (and possibly also stretched or compressed.)

When we multiply a position vector by a scalar we get another position vector. This differs greatly from the dot product when you "multiply" two position vectors and obtain a scalar answer.

We want to examine scalar multiplication geometrically. Let's begin by looking at the position vectors $\mathbf{X} = \begin{bmatrix} 5 \\ 2 \end{bmatrix}$ and $\mathbf{Y} = \begin{bmatrix} 10 \\ 4 \end{bmatrix}$

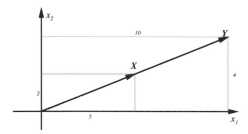

FIGURE 16.66: $\mathbf{X}$ and $\mathbf{Y}$

Notice that position vector $\mathbf{Y}$ has twice the length of position vector $\mathbf{X}$.

(1) $|\mathbf{Y}| = \sqrt{10^2 + 4^2} = 2\sqrt{5^2 + 2^2}$ $|\mathbf{X}| = \sqrt{5^2 + 2^2}$

The position vector $\mathbf{Y} = \begin{bmatrix} 10 \\ 4 \end{bmatrix}$ could be rewritten as $\begin{bmatrix} 2 \cdot 5 \\ 2 \cdot 2 \end{bmatrix}$ or $2\begin{bmatrix} 5 \\ 2 \end{bmatrix}$, but that means $\mathbf{Y} = 2\mathbf{X}$. This is consistent with our observation about the length of $\mathbf{Y}$ compared to the length of $\mathbf{X}$.

Let us test our observations with another example. Examine the position vectors $\mathbf{T} = \begin{bmatrix} 1 \\ 2 \end{bmatrix}$ and $\mathbf{U} = \begin{bmatrix} 3 \\ 6 \end{bmatrix}$.

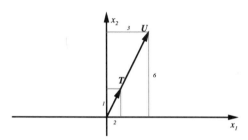

FIGURE 16.67: $\mathbf{T}$ and $\mathbf{U}$

This time position vector $\mathbf{U}$ has three times the length of position vector $\mathbf{T}$. Position vector $\mathbf{U} = \begin{bmatrix} 3 \\ 6 \end{bmatrix}$ can be rewritten as $\begin{bmatrix} 3 \cdot 1 \\ 3 \cdot 2 \end{bmatrix}$ or $3\begin{bmatrix} 1 \\ 2 \end{bmatrix}$ and it is true that $\mathbf{U} = 3\mathbf{T}$.

When a position vector $\mathbf{X}$ is multiplied by a number (a scalar) $c > 1$, the result is another position vector obtained by multiplying each element of the original position vector by the scalar. This multiplication stretches the position vector into a new, longer position vector and the length formula says the length of $\mathbf{Y}$ is c times the length of $\mathbf{X}$ when $\mathbf{Y} = c\mathbf{X}$,

$$|\mathbf{Y}| = \left| \begin{bmatrix} c\,x_1 \\ c\,x_2 \end{bmatrix} \right| = \sqrt{(c\,x_1)^2 + (c\,x_2)^2}$$

$$= \sqrt{c^2 x_1^2 + c^2 x_2^2}$$

$$= c\sqrt{x_1^2 + x_2^2}$$

$$c\,|\mathbf{X}| = c\sqrt{x_1^2 + x_2^2}$$

What is the geometric situation when $0 < c < 1$? The above algebraic computation still works, but what does it show? Let's examine the position vectors $\mathbf{S} = \begin{bmatrix} 8 \\ 2 \end{bmatrix}$ and $\mathbf{R} = \begin{bmatrix} 4 \\ 1 \end{bmatrix}$.

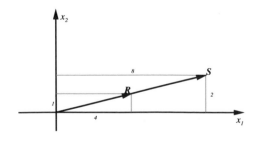

FIGURE 16.68: **R** and **S**

Here we observe the length of **R** is one half the length of **S**. So $\mathbf{R} = \begin{bmatrix} 4 \\ 1 \end{bmatrix}$ may be rewritten

$\begin{bmatrix} \frac{1}{2} \cdot 8 \\ \frac{1}{2} \cdot 2 \end{bmatrix}$ or $\frac{1}{2} \begin{bmatrix} 8 \\ 2 \end{bmatrix}$ and that is $\frac{1}{2}\mathbf{S}$ and $\mathbf{R} = \frac{1}{2}\mathbf{S}$. In this case the position vector **S** shrinks to a new position vector **R** when multiplied by the scalar $\frac{1}{2}$.

EXERCISE 16.69. *Draw a diagram of the scalar multiplication $c\mathbf{X}$ if $\mathbf{X} = \begin{bmatrix} 5 \\ 4 \end{bmatrix}$ and $c = 1$ using the above arguments as examples. Draw a $\mathbf{X}$ in the case $a = 0$.*

Consider the algebraic proof that scalar multiplication changes length by the factor c,

$$|\mathbf{Y}| = \left| \begin{bmatrix} c\,x_1 \\ c\,x_2 \end{bmatrix} \right| = \sqrt{(c\,x_1)^2 + (c\,x_2)^2}$$

$$= c\sqrt{x_1^2 + x_2^2}$$

$$c\,|\mathbf{X}| = c\sqrt{x_1^2 + x_2^2}$$

This computation is not correct if $c < 0$, because $\sqrt{c^2} \neq c$ in that case. Now compare the position vectors $\mathbf{R} = \begin{bmatrix} 4 \\ 3 \end{bmatrix}$ and $\mathbf{S} = \begin{bmatrix} -8 \\ -6 \end{bmatrix}$.

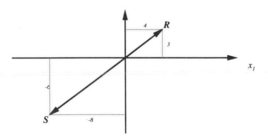

FIGURE 16.70: **R** and **S**

Here position vector **S** has twice the length of position vector **R** but is oriented in exactly the opposite direction. So $\mathbf{S} = \begin{bmatrix} -8 \\ -6 \end{bmatrix}$ or $\begin{bmatrix} -2 \cdot 4 \\ -2 \cdot 3 \end{bmatrix}$ or $-2 \begin{bmatrix} 4 \\ 3 \end{bmatrix}$ but that is $-2\mathbf{R}$. So $\mathbf{S} = -2\mathbf{R}$.

EXERCISE 16.71. *Draw a diagram of the scalar multiplication of $-\frac{1}{3}\mathbf{U}$ if $\mathbf{U} = \begin{bmatrix} -3 \\ 6 \end{bmatrix}$. Is the resulting position vector in the quadrant you expected? Is it shorter or longer than position vector $\mathbf{U}$?*

There are four basic geometric cases of scalar multiplication for a scalar c and a position vector **X**, where $c\,\mathbf{X} = c \begin{bmatrix} x_1 \\ x_2 \end{bmatrix} = \begin{bmatrix} c \cdot x_1 \\ c \cdot x_2 \end{bmatrix}$.

16.5.2. Scalar Multiplication in 3-D. The algebraic definition of multiplication of a number (or scalar) c times a vector $\mathbf{X} = \begin{bmatrix} x_1 \\ x_2 \\ x_3 \end{bmatrix}$ is to simply multiply each component by the scalar,

$$c\,\mathbf{X} = c \begin{bmatrix} x_1 \\ x_2 \\ x_3 \end{bmatrix} = \begin{bmatrix} c \cdot x_1 \\ c \cdot x_2 \\ c \cdot x_3 \end{bmatrix}$$

In three dimensions the geometric scalar multiple examples will be similar to the ones in 2-D, corresponding to stretching, compressing, or reversing the vector.

Again we have the length computation for positive c,

$$|\mathbf{Y}| = \left| \begin{bmatrix} c\,x_1 \\ c\,x_2\,c\,x_3 \end{bmatrix} \right| = \sqrt{(c\,x_1)^2 + (c\,x_2)^2 + (c\,x_3)^2}$$

$$= \sqrt{c^2 x_1^2 + c^2 x_2^2 + c^2 x_3^2}$$

$$= c\,\sqrt{x_1^2 + x_2^2 + x_3^2}$$

$$c\,|\mathbf{X}| = c\,\sqrt{x_1^2 + x_2^2 + x_3^2}$$

This says that the length of a positive scalar multiple is the multiple of the original length.

If c is negative, we get $|c\,\mathbf{X}| = |c|\,|\mathbf{X}|$, but not $|c\,\mathbf{X}| = c\,|\mathbf{X}|$. The direction of the vector is reversed, but the reversed vector still has a positive length.

Examine the position vector $\mathbf{X} = \begin{bmatrix} 1 \\ 4 \\ 2 \end{bmatrix}$ and the position vector $\mathbf{Y} = \begin{bmatrix} 3 \\ 12 \\ 6 \end{bmatrix}$.

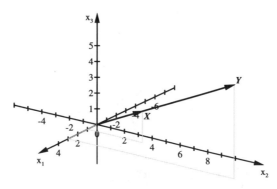

FIGURE 16.72: $\mathbf{X}$ and $\mathbf{Y}$

EXERCISE 16.73. *Compute the length of* $\mathbf{Y}$ *and the length of* $\mathbf{X}$*. They differ by a factor of three.*

Position vector $\mathbf{Y} = \begin{bmatrix} 3 \\ 12 \\ 6 \end{bmatrix}$ can be rewritten as $\begin{bmatrix} 3 \cdot 1 \\ 3 \cdot 4 \\ 3 \cdot 2 \end{bmatrix}$ or $3 \begin{bmatrix} 1 \\ 4 \\ 2 \end{bmatrix}$. By the algebraic defini-

tion above, that is $\mathbf{Y} = 3\mathbf{X}$.

Now examine the position vector $\mathbf{R} = \begin{bmatrix} 3 \\ 9 \\ 6 \end{bmatrix}$ and the position vector $\mathbf{S} = \begin{bmatrix} 1 \\ 3 \\ 2 \end{bmatrix}$.

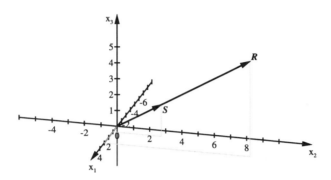

FIGURE 16.74: $\mathbf{R}$ and $\mathbf{S}$

Position vector $\mathbf{S}$ is one-third the length of position vector $\mathbf{R}$. $\mathbf{S} = \begin{bmatrix} 1 \\ 3 \\ 2 \end{bmatrix}$ can be rewritten as

$\begin{bmatrix} \frac{1}{3} \cdot 3 \\ \frac{1}{3} \cdot 9 \\ \frac{1}{3} \cdot 6 \end{bmatrix}$ or $\frac{1}{3} \begin{bmatrix} 3 \\ 9 \\ 6 \end{bmatrix}$ but that is $\frac{1}{3}\mathbf{R}$.

A similar argument will work with position vectors $\mathbf{X} = \begin{bmatrix} 2 \\ 4 \\ 6 \end{bmatrix}$ and $\mathbf{Y} = \begin{bmatrix} -1 \\ -2 \\ -3 \end{bmatrix}$.

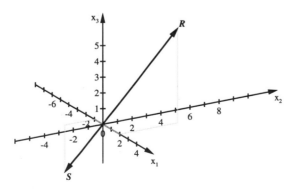

FIGURE 16.75: $\mathbf{X}$ and $\mathbf{Y}$

Notice that position vector $\mathbf{Y}$ is one half the length of position vector $\mathbf{X}$ but in exactly the opposite direction. $\mathbf{Y} = \begin{bmatrix} -1 \\ -2 \\ -3 \end{bmatrix}$ can be rewritten as $\begin{bmatrix} -\frac{1}{2} \cdot 2 \\ -\frac{1}{2} \cdot 4 \\ -\frac{1}{2} \cdot 6 \end{bmatrix}$ or $-\frac{1}{2}\begin{bmatrix} 2 \\ 4 \\ 6 \end{bmatrix}$ but that is $-\frac{1}{2}\mathbf{X}$.

There are four basic cases for the geometric interpretation of scalar multiplication as follows. These are the same in two and three dimensions, so we illustrate them in 3-D. STRETCHING: If $c > 1$ then $c\,\mathbf{X}$ is a position vector in the *same* direction as $\mathbf{X}$ but of longer length, in fact, the length is c times longer.

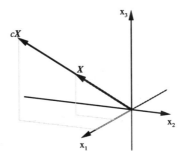

FIGURE 16.76: Stretching by $c > 1$

If $c = 1$ the position vector remains unchanged since $c\mathbf{X} = 1\mathbf{X} = \mathbf{X}$

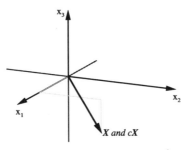

FIGURE 16.77: No Change

COMPRESSING: If $0 < c < 1$, then $c\,\mathbf{X}$ is a position vector in the same direction as $\mathbf{X}$, but of shorter length.

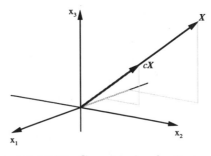

FIGURE 16.78: Compressing by $0 < c < 1$

REVERSING: If $c < 0$ then $c\mathbf{X}$ is a position vector in the *opposite* direction as $\mathbf{X}$ and its length is longer or shorter depending on the absolute value of c, $|c|$. We can reduce this to the case of scalar multiplication by -1, followed by a stretch or compression of $|c|$.

$c = -1$

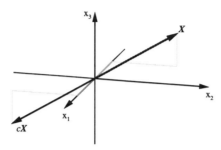

FIGURE 16.79: Reversal of A Vector

EXAMPLE 16.80. *Given* $\mathbf{R} = \mathbf{R}(-3, 2)$ *and* $c = 3$ *calculate and sketch* $c\mathbf{R}$.

SOLUTION:

$$c\mathbf{R} = 3 \begin{bmatrix} -3 \\ 2 \end{bmatrix} = \begin{bmatrix} 3 \cdot -3 \\ 3 \cdot 2 \end{bmatrix} = \begin{bmatrix} -9 \\ 6 \end{bmatrix}$$

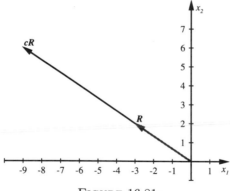

FIGURE 16.81:

EXAMPLE 16.82. *Given* $\mathbf{S} = \mathbf{S}(-5, 8, 2)$ *and* $c = 1/4$ *calculate and sketch* $c\mathbf{S}$.

SOLUTION:

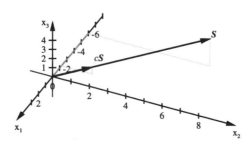

FIGURE 16.83:

EXERCISE 16.84. *For each problem use both geometry in a sketch and algebra in a calculation to find the scalar multiplication using the position vector and scalar given.*

(1) $\mathbf{S} = \begin{bmatrix} 6 \\ 4 \end{bmatrix}$ $c = 2$ (2) $\mathbf{T} = \begin{bmatrix} -3 \\ 1 \end{bmatrix}$ $c = 3$

(3) $\mathbf{U} = \begin{bmatrix} 8 \\ 5 \\ 2 \end{bmatrix}$ $c = \dfrac{1}{2}$ (4) $\mathbf{V} = \begin{bmatrix} 12 \\ -3 \\ 9 \end{bmatrix}$ $c = -\dfrac{1}{3}$

(5) $\mathbf{W} = \begin{bmatrix} -5 \\ -2 \end{bmatrix}$ $c = -2$ (6) $\mathbf{X} = \begin{bmatrix} -4 \\ -6 \\ 4 \end{bmatrix}$ $c = 1$

(7) $\mathbf{Y} = \begin{bmatrix} 5 \\ 3 \end{bmatrix}$ $c = \dfrac{2}{3}$ (8) $\mathbf{Z} = \begin{bmatrix} -6 \\ -3 \\ -8 \end{bmatrix}$ $c = -1$

EXAMPLE 16.85. *Unit Vectors*

Scalar multiplication can be used to find a vector of unit length that points in the same direction as a given vector. For example, the vector

$$\mathbf{Z} = \begin{bmatrix} -6 \\ -3 \\ -8 \end{bmatrix} \qquad \text{has length} \qquad |\mathbf{Z}| = \sqrt{6^2 + 3^2 + 8^2} = \sqrt{109}$$

We need to compress the vector by the factor $1/\sqrt{109}$ in order to make it unit length,

$$\mathbf{Z}_u = \frac{1}{\sqrt{109}} \cdot \begin{bmatrix} -6 \\ -3 \\ -8 \end{bmatrix} = \begin{bmatrix} \frac{-6}{\sqrt{109}} \\ \frac{-3}{\sqrt{109}} \\ \frac{-8}{\sqrt{109}} \end{bmatrix} \approx - \begin{bmatrix} 0.575 \\ 0.287 \\ 0.794 \end{bmatrix}$$

EXERCISE 16.86. *Calculate a unit length vector in the same direction as each of the vectors in the previous exercise. Add these vectors to the sketches for that problem. Verify that each of your vectors has unit length.*

EXERCISE 16.87. *Describe geometrically, the set of points that consists of the tips of all position vectors of the form* $\begin{bmatrix} 2c \\ 3c \end{bmatrix}$, *where c takes on all real number values. Sketch several values, $c = 1$, $c = 3/2$, $c = 2$, $c = 5/2$, $c = 1/2$, $c = -1/2$, $c = -1$, $c = -3/2 \cdots$ until you see the general pattern.*

EXERCISE 16.88. *Repeat Exercise 16.87 using the position vector* $\begin{bmatrix} 3c \\ 5c \\ 8c \end{bmatrix}$.

16.6. Vector Difference and Oriented Displacements

A position vector **X** in 2-space or in 3-space can be visualized geometrically by an arrow from the origin to the point **X**. It is essential that we always treat vectors as starting from the origin in our algebraic computations. This arrow can be thought of as an oriented displacement from the origin to the point **X** as shown below.

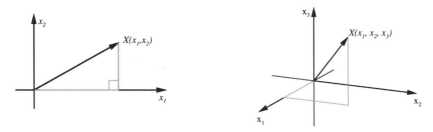

FIGURE 16.89: Two-D and Three-D Vectors

We also want to consider oriented displacements (or arrows) from any one point to another. We must be careful to not treat these the same as position vectors. These vectors are sometimes called 'free vectors.' We prefer to call them oriented displacements or illegal arrows. Consider any two points **S** and **T** in a 2-dimensional cartesian coordinate system. The oriented displacement from **S** to **T** can be visualized geometrically by an arrow from **S** to **T** as shown below.

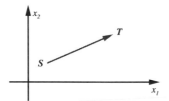

FIGURE 16.90: A 2-D Displacement

This oriented displacement from **S** to **T** cannot be described using a position vector, because the tail of this arrow would not lie at the origin. If we first move the x_1x_2-coordinate axes such that the translated $x_1'x_2'$-coordinate axes are parallel to the corresponding x_1x_2-coordinate axes and point **S** is at the origin of the $x_1'x_2'$-coordinate system as shown below, then the components of the new arrow are a position vector in the new system. However, we do not want to mix coordinate systems.

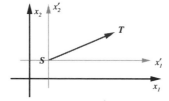

FIGURE 16.91: New Coordinates

An ordinary position vector is associated with the oriented displacement from $\mathbf{S}$ to $\mathbf{T}$ because of the geometric 'tips-to-tails' interpretation of vector sum. If an ordinary unknown position vector $\mathbf{X}$ is added to $\mathbf{S}$ and the sum $\mathbf{S} + \mathbf{X}$ points legally from the origin to the tip of $\mathbf{T}$, then the 'tips-to-tails' law of vector addition says that the arrow with its tail at $\mathbf{S}$ the same length and parallel to $\mathbf{X}$ connects the tip of $\mathbf{S}$ and the tip of $\mathbf{T}$.

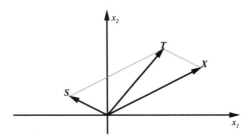

FIGURE 16.92: An Unknown Parallel to the Displacement

Algebraically, we have $\mathbf{S} + \mathbf{X} = \mathbf{T}$, so if both $\mathbf{S}$ and $\mathbf{T}$ are known, we solve

$$\mathbf{X} = \mathbf{T} - \mathbf{S}$$

This is an important observation: The (legal) position vector parallel to the (illegal) arrow pointing from the tip of $\mathbf{S}$ to the tip of $\mathbf{T}$ is given by the vector difference.

EXAMPLE 16.93. *Find the position vector $\left[\begin{smallmatrix} x_1 \\ x_2 \end{smallmatrix}\right]$ describing the oriented displacement from $S(2,3)$ to $T(5,9)$.*

SOLUTION: We begin by plotting the points $\mathbf{S}$ and $\mathbf{T}$ and drawing the arrow from $\mathbf{S}$ to $\mathbf{T}$.

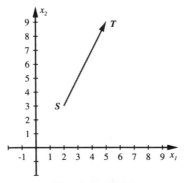

FIGURE 16.94:

Next we translate the axes, so we can see the derivation of the formula $\mathbf{X} = \mathbf{T} - \mathbf{S}$ again.

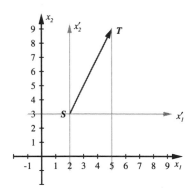

FIGURE 16.95:

From the diagram we see that the coordinates of $T(5,9)$ in the $x_1' x_2'$-coordinate system are $X(3,6)$ and hence the vector $\left[\begin{smallmatrix} 3 \\ 6 \end{smallmatrix}\right]$ is the desired position vector. Algebraically, $\left[\begin{smallmatrix} 5 \\ 9 \end{smallmatrix}\right] - \left[\begin{smallmatrix} 2 \\ 3 \end{smallmatrix}\right] = \left[\begin{smallmatrix} 3 \\ 6 \end{smallmatrix}\right]$.

EXAMPLE 16.96.   *Given the position vectors* $\mathbf{X} = \mathbf{X}(5,3)$ *and* $\mathbf{Y} = \mathbf{Y}(2,-4)$ *find and sketch a diagram of* $\mathbf{Y} - \mathbf{X}$.

SOLUTION:

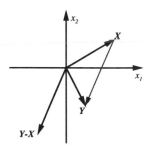

$$\mathbf{Y} - \mathbf{X} = \begin{bmatrix} 2 \\ -4 \end{bmatrix} - \begin{bmatrix} 5 \\ 3 \end{bmatrix} = \begin{bmatrix} -3 \\ -7 \end{bmatrix}$$

EXERCISE 16.97.   *For each of the following, use a diagram to find the position vector that describes the oriented displacement from* $\mathbf{S}$ *to* $\mathbf{T}$. *Compare your answer with the difference,* $\mathbf{S} - \mathbf{T}$.

a) $S(4,5); T(6,9)$ b) $S(3,7); T(2,4)$
c) $S(-4,6); T(3,-5)$ d) $S(-3,-4); T(5,-4)$

EXERCISE 16.98.   *Finish - Start*
In the diagram below an illegal arrow or oriented displacement from the end of $\mathbf{R}$ *to the end of* $\mathbf{S}$ *is show. Sketch the parallel legal position vector and give an algebraic formula for it.*

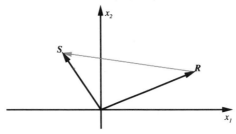

FIGURE 16.99:

An oriented displacement from point **S** to point **T** in a 3-dimensional cartesian coordinate system can also be described using an unknown position vector with the tips-to-tails law of vector addition. The procedure is similar to that used in 2-dimensions.

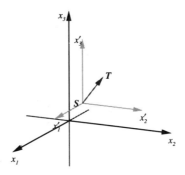

FIGURE 16.100: A 3-D Displacement and New Coordinates

To derive the formula for the position vector $\mathbf{X} = \begin{bmatrix} x_1 \\ x_2 \\ x_3 \end{bmatrix}$ describing the oriented displacement from $S(2, 3, 4)$ to $T(5, 7, 9)$ we begin by plotting dots at **S** and **T** and drawing the (illegal) arrow from **S** to **T** as shown below.

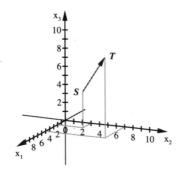

FIGURE 16.101:

If we translate the axes, we see the following figure with **X** plotted in the new coordinate system. This is also the tips-to-tails interpretation of $\mathbf{S} + \mathbf{X} = \mathbf{T}$, so $\mathbf{X} = \mathbf{T} - \mathbf{S}$.

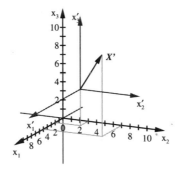

FIGURE 16.102: **X** in New Coordinates

From the diagram the coordinates of **X** in the $x'_1x'_2x'_3$-coordinate system are $(3, 4, 5)$ and hence $\begin{bmatrix} 3 \\ 4 \\ 5 \end{bmatrix}$ is the desired position vector. Verify this by subtraction.

EXAMPLE 16.103. *Given the position vector* **R** $=$ **R**$(-3, 4, 1)$ *and* **S** $=$ **S**$(2, 3, -2)$ *find and sketch a diagram of* **S** $-$ **R**.

SOLUTION:

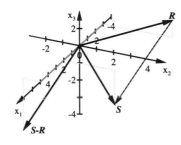

$$ \mathbf{S} - \mathbf{R} = \begin{bmatrix} 2 \\ 3 \\ -2 \end{bmatrix} - \begin{bmatrix} -3 \\ 4 \\ 1 \end{bmatrix} = \begin{bmatrix} 5 \\ -1 \\ -3 \end{bmatrix} $$

3-D EXERCISES:

EXERCISE 16.104. *For each of the following, use a diagram to find the position vector that describes the oriented displacement from* **S** *to* **T**. *Verify your results by subtraction.*
a) $S(2, 5, 3)$; $T(7, 10, 12)$ b) $S(13, 10, 9)$; $T(4, 6, 2)$ c) $S(4, -5, 8)$; $T(-2, -6, 4)$

CHAPTER 17

Analytical Vector Geometry

This chapter develops and tests your skill at using position vectors to formulate geometric problems analytically. The formulas summarized in the first section allow you to translate geometric ideas into algebra and algebraic ideas into geometry. You need to memorize the formulas along with their geometric meaning. The parametric line and implicit plane will be explained in detail in sections below. The other formulas were explained in detail in the previous chapter.

17.1. A Lexicon of Geometry and Algebra

This section is a the "translation dictionary" that will allow you to go back and forth between geometry and algebra. You should memorize the algebraic computation of each subsection and draw a diagram to which it corresponds. Once you can translate these ten "words" that have counterparts in both the language of algebra and the language of geometry, all you need to do three dimensional analytical geometry is put your problem in a form where your lexicon applies, make the translation and proceed in the other language.

Recall that our formulas are for "position vectors" with their tail at the origin. When we want to draw an arrow with its tail at another point, which is "an illegal vector," we use the geometric meaning of vector addition to find the corresponding true position vector. Sometimes the vector difference is more convenient in making computations with illegal vectors. This depends on which vectors are known and which you need to compute.

17.1.1. Scalar Multiplication and Stretching. The scalar product $c\mathbf{X}$ is a vector that stretches, shrinks or reverses $\mathbf{X}$, but does not change its direction.

$$c\,\mathbf{X} = c \begin{bmatrix} x_1 \\ x_2 \\ x_3 \end{bmatrix} = \begin{bmatrix} c \cdot x_1 \\ c \cdot x_2 \\ c \cdot x_3 \end{bmatrix}$$

17.1.2. Vector Addition and Tips-to-Tails. The sum $\mathbf{X} + \mathbf{Y}$ is the legal position vector you reach by drawing $\mathbf{Y}$ illegally with its tail at the tip of $\mathbf{X}$.

$$\mathbf{X} + \mathbf{Y} = \begin{bmatrix} x_1 \\ x_2 \\ x_3 \end{bmatrix} + \begin{bmatrix} y_1 \\ y_2 \\ y_3 \end{bmatrix} = \begin{bmatrix} x_1 + y_1 \\ x_2 + y_2 \\ x_3 + y_3 \end{bmatrix}$$

17.1.3. Vector Difference and Displacement. The legal vector parallel to the illegal arrow pointing from the tip of **B** to the tip of **A** is **A** − **B**.

$$\mathbf{A} - \mathbf{B} = \begin{bmatrix} a_1 \\ a_2 \\ a_3 \end{bmatrix} - \begin{bmatrix} b_1 \\ b_2 \\ b_3 \end{bmatrix} = \begin{bmatrix} a_1 - b_1 \\ a_2 - b_2 \\ a_3 - b_3 \end{bmatrix}$$

17.1.4. Length.

$$|\mathbf{X}| = \sqrt{x_1^2 + x_2^2 + x_3^2}$$

17.1.5. Angle.

$$\mathrm{Cos}[\theta] = \frac{\langle \mathbf{X} \bullet \mathbf{Y} \rangle}{|\mathbf{X}|\,|\mathbf{Y}|}$$
$$= \frac{x_1 \cdot y_1 + x_2 \cdot y_2 + x_3 \cdot y_3}{\sqrt{x_1^2 + x_2^2 + x_3^2} \cdot \sqrt{y_1^2 + y_2^2 + y_3^2}}$$

17.1.6. Perpendicularity. Nonzero vectors **X** and **Y** are perpendicular if and only if

$$\langle \mathbf{X} \bullet \mathbf{Y} \rangle = 0$$
$$x_1 \cdot y_1 + x_2 \cdot y_2 + x_3 \cdot y_3 = 0$$

17.1.7. Projection. The vector projection of **Y** in the direction of **X** is given by

$$\mathrm{Proj}_{\mathbf{X}}(\mathbf{Y}) = \frac{\langle \mathbf{Y} \bullet \mathbf{X} \rangle}{\langle \mathbf{X} \bullet \mathbf{X} \rangle}\, \mathbf{X}$$

$$= \frac{x_1 \cdot y_1 + x_2 \cdot y_2 + x_3 \cdot y_3}{x_1 \cdot x_1 + x_2 \cdot x_2 + x_3 \cdot x_3} \begin{bmatrix} x_1 \\ x_2 \\ x_3 \end{bmatrix}$$

17.1.8. Cross Product. The 3-D cross product is given by

$$\mathbf{Z} = \mathbf{X} \times \mathbf{Y} = \det \begin{vmatrix} i & j & k \\ x_1 & x_2 & x_3 \\ y_1 & y_2 & y_3 \end{vmatrix} = \begin{bmatrix} x_2 y_3 - x_3 y_2 \\ x_3 y_1 - x_1 y_3 \\ x_1 y_2 - x_2 y_1 \end{bmatrix}$$

and **Z** is the vector with the properties that

(1) **Z** is perpendicular to both **X** and **Y**.
(2) The length of **Z** equals the area of the parallelogram spanned by the edges of **X** and **Y**.
(3) The frame **X**, **Y**, **Z** is right-handed.

17.1.9. Parametric Line. The line in the direction of the vector $\mathbf{D}$ passing through the tip of the vector $\mathbf{P}$ is the set of all vectors $\mathbf{X}$ of the form $\mathbf{X} = \mathbf{P} + t\mathbf{D}$, for some real number t. You need to also be able to recognize this vector equation in its classical form

$$
\begin{aligned}
x_1 &= p_1 + d_1\,t \\
x_2 &= p_2 + d_2\,t \\
x_3 &= p_3 + d_3\,t
\end{aligned}
\qquad \text{for example} \qquad
\begin{aligned}
x &= 2 - 5\,t \\
y &= -3 + 2\,t \\
x &= 1 + 7\,t
\end{aligned}
$$

You should understand that the vector equation describes a line because it is a combination of vector addition and scalar multiplication, as explained in Section 17.2 below. You must work Exercise 17.4.

17.1.10. Implicit Plane (or 2-D Implicit Line). Given known vectors $\mathbf{N}$ and $\mathbf{R}$, the set of all (unknown) vectors $\mathbf{X}$ satisfying

$$ \langle \mathbf{N} \bullet (\mathbf{X} - \mathbf{R}) \rangle = 0 $$

is the plane perpendicular to $\mathbf{N}$ passing through the tip of $\mathbf{R}$. (In 2-D, this is the line perpendicular to $\mathbf{N}$ through $\mathbf{R}$.) You should be able to recognize the vectors if you are given the equation in a classical form like

$$ n_1\,x_1 + n_2\,x_2 + n_3\,x_3 = k $$

for example,

$$ x - 3\,y + 5\,z = 2 $$

This is explained in Section 17.5 below, and is based on geometric vector difference and the condition for perpendicularity. You must work Exercise 17.26.

Not every algebraic computation has a geometric interpretation and not every geometric construction corresponds to a single simple formula in our lexicon. For example, if we want a vector of unit length in the same direction as a given vector $\mathbf{X}$, we use the length formula below to compute its length. Then we use scalar multiplication to find the unit vector by stretching or shrinking the original vector by an appropriate amount to make its length 1. It is not necessary to have a separate formula for this idea if you understand both formulas separately.

EXERCISE 17.1. *Custom uses of the length formula*

Compute the length of the vectors

$$ \mathbf{Y} = \begin{bmatrix} 4 \\ -4 \\ 2 \end{bmatrix} \qquad and \qquad \mathbf{Z} = \begin{bmatrix} 2 \\ -2 \\ 1 \end{bmatrix} $$

Find a scalar c so that $\mathbf{Z} = c\mathbf{Y}$.

Give your own personal formula for the unit length vector that points in the same direction as a given general vector, $\mathbf{X} = \begin{bmatrix} x_1 \\ x_2 \\ x_3 \end{bmatrix}$. *It is not essential that you add this formula to your personal lexicon, but you may if you wish. It is essential that you can find the unit vector. For example, test your formula on*

$$\begin{bmatrix} \frac{2}{3} \\ -\frac{1}{3} \\ \frac{2}{3} \end{bmatrix} \qquad \text{the unit vector in the direction of} \qquad \begin{bmatrix} 2 \\ -1 \\ 2 \end{bmatrix}$$

PROBLEM 17.2. *The Geometric Half of the Lexicon*
Draw the geometric figure that corresponds to each of the formulas in the ten subsections above.

17.2. The Vector Parametric Line

This section shows you how a line can be described by a parametric vector equation. It simply uses vector sums and scalar multiples, so we begin with a reminder of their geometric meaning..

GENERAL INSTRUCTIONS: Plot (legal) position vectors as black arrows (with their tails at the origin, 0). Plot illegal (free) vectors which have their tails at other places as dotted red arrows.

EXERCISE 17.3. *Warm Up*

$$\mathbf{A} = \begin{bmatrix} a_1 \\ a_2 \\ a_3 \end{bmatrix} = \begin{bmatrix} 1 \\ 3 \\ 2 \end{bmatrix} \text{ and } \mathbf{B} = \begin{bmatrix} b_1 \\ b_2 \\ b_3 \end{bmatrix} = \begin{bmatrix} -2 \\ 1 \\ 3 \end{bmatrix} \quad Plot\ A,\ B,\ \tfrac{1}{2}A,\ 3B$$

1) Draw the illegal arrow parallel to $3B$, but with its tail at the tip of $\frac{1}{2}A$. Draw the illegal arrow parallel to $\frac{1}{2}A$, but with it's tail at the tip of $3B$. Compute $3B + \frac{1}{2}A$ and plot it as a legal vector.
2) Draw the illegal arrow from the tip of B to the tip of A. Compute $A - B$. Plot $A - B$ as a legal vector.

The next exercise uses the lexicon entries for scalar multiplication and vector addition to derive the parametric equation of a line.

PROBLEM 17.4. *A Line through $\mathbf{P}$ in the direction of $\mathbf{D}$*

$$\mathbf{P} = \begin{bmatrix} p_1 \\ p_2 \\ p_3 \end{bmatrix} = \begin{bmatrix} 1 \\ 3 \\ 1 \end{bmatrix} \qquad and \qquad \mathbf{D} = \begin{bmatrix} d_1 \\ d_2 \\ d_3 \end{bmatrix} = \begin{bmatrix} -2 \\ -3 \\ 1 \end{bmatrix}$$

a) Start a new sheet of paper. Let $t = 1$.
   a.1) Draw a dot at the tip of the legal vector $\mathbf{P}$.
   a.2) Draw the vector $t\mathbf{D}$ as a legal vector and as a dotted illegal vector with its tail at the tip of $\mathbf{P}$.
   a.3) Compute $\mathbf{P} + t\mathbf{D}$ and plot it as a legal vector.
b) Start a new sheet of paper. Let $t = \frac{1}{2}$.
   b.1) Draw a dot at the tip of the legal vector $\mathbf{P}$.
   b.2) Draw the vector $t\mathbf{D}$ as a legal vector and as a dotted illegal vector with its tail at the tip of $\mathbf{P}$.
   b.3) Compute $\mathbf{P} + t\mathbf{D}$ and plot it as a legal vector.
c) Start a new sheet of paper. Let $t = 3$.
   c.1) Draw a dot at the tip of the legal vector $\mathbf{P}$.

c.2) Draw the vector tD as a legal vector and as a dotted illegal vector with its tail at the tip of **P***.*

c.3) Compute **P** + t**D** *and plot it as a legal vector.*

d) Start a new sheet of paper. Let t = −1.

d.1) Draw a dot at the tip of the legal vector **P***.*

d.2) Draw the vector tD as a legal vector and as a dolled illegal vector with its tail at the tip of **P***.*

d.3) Compute **P** + t**D** *and plot it as a legal vector.*

e) What is the set of tips of (legal) position vectors of the form X = **P** + t**D***, for all real numbers t?*

f) Explain why the coordinate equations
$$\begin{bmatrix} x_1 \\ x_2 \\ x_3 \end{bmatrix} = \begin{bmatrix} 1 \\ 3 \\ 1 \end{bmatrix} + t \begin{bmatrix} -2 \\ -3 \\ 1 \end{bmatrix} \text{ or}$$

$$x = 1 - 2t$$
$$y = 3 - 3t$$
$$z = 1 + t$$

describe a parametric line that goes through the point $\mathbf{P} = \begin{bmatrix} 1 \\ 3 \\ 1 \end{bmatrix}$ *and points parallel to the*

vector $\mathbf{D} = \begin{bmatrix} -2 \\ -3 \\ 1 \end{bmatrix}$.

*g) Draw this line by hand and check your work with the Mathematica NoteBook **ParametricLine.ma**.*

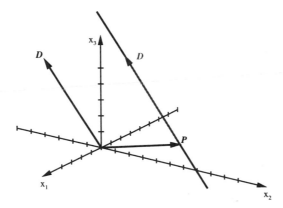

FIGURE 17.5: A Line through **P** in direction **D**

EXERCISE 17.6. *Classical Parametric Equations*
The parametric equations

$$x = 4 - 3t$$
$$y = 1 + t$$
$$z = 2t$$

define a line. Use the vector form

$$\mathbf{X} = \mathbf{P} + t\mathbf{D}$$

to plot it. What are $\mathbf{P}$ and $\mathbf{D}$? Why does the vector form help to make the plot?

Find two non-parallel vectors $\mathbf{M}$ and $\mathbf{N}$ both perpendicular to the direction of this line.

Not every geometric fact needs to have a formula. We know geometrically that "two points determine a line," but that does not mean we need to enter a formula corresponding to that fact in our lexicon. Instead, we can use the vector difference to find a direction and then use the parametric line to determine the line algebraically. The geometric fact that two points determine a line means that we can use the lexicon to find an algebraic representation of the line.

EXERCISE 17.7. *Two Points Determine a Line*
Find the parametric equations of a line through the points

$$\mathbf{Q} = \begin{bmatrix} 2 \\ 3 \\ 4 \end{bmatrix} \quad \textit{and} \quad \mathbf{R} = \begin{bmatrix} 1 \\ -2 \\ 3 \end{bmatrix}$$

First, sketch both vectors.

Next, use the geometric meaning of the vector difference to find a direction vector for the line. Sketch the illegal vector pointing from $\mathbf{Q}$ to $\mathbf{R}$ and compute its legal counterpart, $\mathbf{D}$.

Finally, use one point and your direction vector to give equations for the line in the form $\mathbf{X} = \mathbf{P} + t\mathbf{D}$.

There is more than one way to solve the previous exercise of determining the parametric equation of a line through two points. The vector from $\mathbf{Q}$ to $\mathbf{R}$ is one direction vector for the line and the vector from $\mathbf{R}$ to $\mathbf{Q}$ is another. Of course, these two vectors are simply the negative of one another, but they are different. They both point along the line, but in opposite directions. If $\mathbf{D}$ is one vector in the direction of the line, any scalar multiple $c\mathbf{D}$ also points along the line. In addition, any point $\mathbf{P}$ that lies on the line can be used in the parametric form $\mathbf{X} = \mathbf{P} + t\mathbf{D}$.

EXERCISE 17.8. *Non-Uniqueness of Parametric Equations*
Show that both of the following sets of parametric equations go through both of the points
$$\mathbf{R} = \begin{bmatrix} 2 \\ 3 \\ 4 \end{bmatrix} \text{ and } \mathbf{Q} = \begin{bmatrix} 1 \\ -2 \\ 3 \end{bmatrix}.$$

$x = 1 + t$	$x = 2 - 2s$
$y = -2 + 5t$	$y = 3 - 10s$
$z = 3 + t$	$z = 4 - 2s$

In other words, find values of $t = t_R$, t_Q, $s = s_R$ and s_Q so that each of the general vectors of the equations $X(x, y, z)$ match both $\mathbf{R}$ and $\mathbf{Q}$.

Sometimes we need to use the lexicon indirectly. We may want to find a certain formula, but have no lexicon entry for it, but rather have a lexicon entry for a closely associated vector. We compute the associated vector and use that to find the one we really want. The next exercise shows you how this works in finding the distance from the origin to a line.

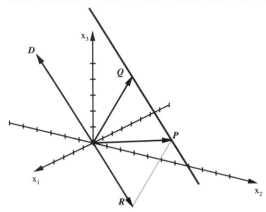

FIGURE 17.9: Distance to a Line

EXERCISE 17.10. *Parallel and Perpendicular Components*
This exercise shows you how to use various parts of the algebra- geometry lexicon to find the distance from the origin to a line.
1) Let $\mathbf{P} = \begin{bmatrix} p_1 \\ p_2 \\ p_3 \end{bmatrix} = \begin{bmatrix} 1 \\ 3 \\ 1 \end{bmatrix}$ *and* $\mathbf{D} = \begin{bmatrix} d_1 \\ d_2 \\ d_3 \end{bmatrix} = \begin{bmatrix} -2 \\ -3 \\ 1 \end{bmatrix}$. *Sketch the line that points parallel to the direction of the arrow associated with* $\mathbf{D}$ *and passes through the tip of* $\mathbf{P}$. *This is the set of legal position vectors of the form* $\mathbf{X} = \mathbf{P} + t\mathbf{D}$, *a parametric line.*
2) Compute the vector $\mathbf{R}$ = *the projection of* $\mathbf{P}$ *in the direction of* $\mathbf{D}$. *Sketch your result.*
3) Compute $\mathbf{Q} = \mathbf{P} - \mathbf{R}$ *and add a sketch of* $\mathbf{Q}$ *as a legal position vector with its tail at the origin. (Hint: Use the geometric interpretation of vector difference to find an illegal version of* $\mathbf{Q}$.)
4) Sketch $\mathbf{R}$ *drawn as an illegal vector with its tail at the tip of* $\mathbf{Q}$. *Where does the tip of the illegal vector point? Why?*
5) PROVE (by computation) that $\mathbf{Q}$ *is perpendicular to* $\mathbf{D}$. *HINT: It is easy to do this by computation.*

$$\mathbf{Q} \bullet \mathbf{D} = (\mathbf{P} - \mathbf{R}) \bullet \mathbf{D}$$
$$= \mathbf{P} \bullet \mathbf{D} - \mathbf{R} \bullet \mathbf{D}$$
$$= \mathbf{P} \bullet \mathbf{D} - \frac{\mathbf{P} \bullet \mathbf{D}}{\mathbf{D} \bullet \mathbf{D}} \mathbf{D} \bullet \mathbf{D}$$

6) Use your figure to show that the distance from the origin to the line is $|\mathbf{Q}|$. *Explain.*

PROBLEM 17.11. *A Custom Program for Distance*
Given a general line in terms of the parametric equation $\mathbf{X} = \mathbf{P} + t\mathbf{D}$ *and a point* $\mathbf{Q}$ *anywhere in space, use vector projection to find the distance from* $\mathbf{Q}$ *to the line. Rather than give a single formula, you may want to outline a procedure of several steps. Ideally, you should be able to use your procedure to write a brief Mathematica program to compute the distance.*

17.3. Radian Measure and Parametric Curves

This section recalls the definition of radian measure, sine and cosine, and uses them together with vector operations to move in circles, loops and spirals. More information on sine and cosine as parametric equations for a circle is contained in the Mathematical Background Chapter on High School Review with *Mathematica*.

The radian measure of an angle θ is a length measured along a unit circle. Place the circle with its center at the apex of the angle. The two sides of the angle intercept the circle and the radian measure of the angle is the distance between the sides. For our connection with the sine and cosine, we take one side of the angle to be the positive x axis as shown next.

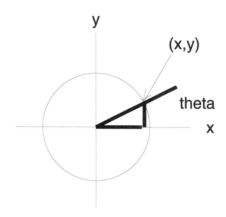

FIGURE 17.12: Radian Measure

The point where the other side of the angle meets the unit circle has coordinates

$$x = \text{Cos}[\theta]$$
$$y = \text{Sin}[\theta]$$

These equations follow from SOH-CAH-TOA. The right triangle with one leg on the x axis and a vertical leg up to the intersection point has a hypotenuse of length 1, since that hypotenuse is a radius of the unit circle. Thus,

$$\text{Cos}[\theta] = \frac{\text{adjacent}}{\text{hypotenuse}} = \frac{x}{1}$$
$$\text{Sin}[\theta] = \frac{\text{opposite}}{\text{hypotenuse}} = \frac{y}{1}$$

A vector view of the pair of equations is that the vector

$$\mathbf{X}(\theta) = \begin{bmatrix} \text{Cos}[\theta] \\ \text{Sin}[\theta] \end{bmatrix}$$

points to the unit circle at the angle θ, a distance of θ measured along the circle. This is a vector parametric equation for a circle. In other words, if we plotted all vectors of this form, without plotting θ, we would fill the unit circle.

EXERCISE 17.13. *Parametric Circles*

(1) *Use scalar multiplication to show that the parametric vector equation of a circle of radius r centered at the origin is*

$$\mathbf{X}(\theta) = r \begin{bmatrix} \mathrm{Cos}[\theta] \\ \mathrm{Sin}[\theta] \end{bmatrix}$$

(2) *Use vector addition to show that the parametric vector equation of a circle of radius r centered at the point* $\begin{bmatrix} a \\ b \end{bmatrix}$ *is*

$$\mathbf{X}(\theta) = \begin{bmatrix} r\,\mathrm{Cos}[\theta] + a \\ r\,\mathrm{Sin}[\theta] + b \end{bmatrix}$$

but not

$$\mathbf{X}(\theta) = r \begin{bmatrix} \mathrm{Cos}[\theta] + a \\ \mathrm{Sin}[\theta] + b \end{bmatrix}$$

(3) *Verify that you derivations are correct by making Mathematica animations of a point traveling around these circles using the NoteBook **Circles.ma**.*

17.3.1. Sums of Circles: Cycloids. A small wheel rolls around a large fixed wheel without slipping. What path does a point on the edge of the small wheel trace out as it goes? You can answer this with vectors and an understanding of radian measure. The Spirograph toy makes such plots. The next figure is a sample path.

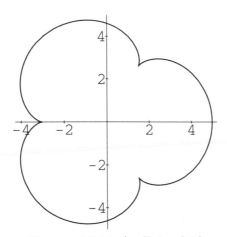

FIGURE 17.14: An Epicycloid

This is not trivial - vectors help - and here is a special case to show you why. Suppose the large wheel has radius 3 and the small wheel has radius 1. How many turns must the small wheel make while traveling once around the large wheel? This is a famous question because it appeared on a national college entrance test as a multiple choice question. The reason it is famous is that the correct answer was not among the choices.

Here is the obvious solution. The big wheel has three times the circumference of the small wheel (you can use $C = 2\pi r$), so the small one must mark off three revolutions to cover the same distance and travel all the way around the large one. Obvious, but wrong. The small wheel turns four times. Run the *Mathematica* animation in **EpiCyAnimate.ma** to see for yourself.

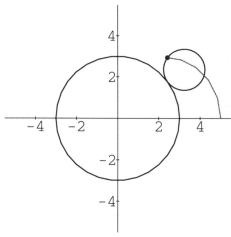

FIGURE 17.15: **EpiCyAnimate.ma**

We will use vectors to analyse the motion of a point on the little wheel. This will give us parametric equations for the whole path, but also show us why the small wheel turns 4 times in the 3 to 1 ratio case. You will need to know that the length of a circular segment of angle θ and radius r is

$$\text{arc length} = r\,\theta$$

For example, the length all the way around a circle of radius r is what we get when we go an angle $\theta = 2\pi$, so $C = 2\pi\, r$.

We begin with one vector $\mathbf{C}(\theta)$ pointing to the center of the small wheel. We know from the sine-cosine discussion above that

$$\mathbf{C}(\theta) = \begin{bmatrix} c_1 \\ c_2 \end{bmatrix} = \begin{bmatrix} R\,\text{Cos}[\theta] \\ R\,\text{Sin}[\theta] \end{bmatrix}$$

if θ measures the angle from the x-axis and $R = r_b + r_s$, the sum of the radii of the large and small wheels.

We also have a second vector $\mathbf{W}(\omega)$ pointing from the center of the small wheel to the tracing point on its rim. Again using sine-cosine, we have

$$\mathbf{W}(\omega) = \begin{bmatrix} w_1 \\ w_2 \end{bmatrix} = \begin{bmatrix} r_s\,\text{Cos}[\omega] \\ r_s\,\text{Sin}[\omega] \end{bmatrix}$$

where ω is the angle measured around the small wheel starting from the horizontal and r_s is the radius of the small wheel.

The vector $\mathbf{W}(\omega)$ is illegal; it does not start at the origin. The legal interpretation is simply that

$$\mathbf{X} = \mathbf{C} + \mathbf{W}$$

is the arrow from the origin to the reference point on the small wheel. This may be written in components as

$$x_1 = c_1 + w_1$$
$$x_2 = c_2 + w_2$$

where the components of $\mathbf{C}$ and $\mathbf{W}$ are given by the formulas above.

The two vectors start in the position shown next with both pointing to the right of their centers.

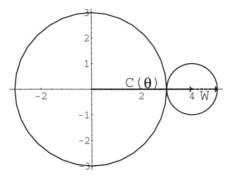

FIGURE 17.16: Starting Position

After a small increase in θ, the vectors look like the next figure.

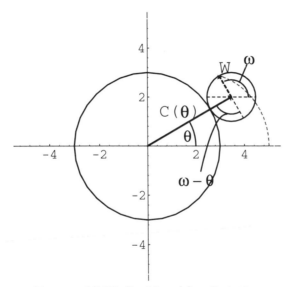

FIGURE 17.17: Position After Rotation

The length of portion along the big wheel where contact has been made is $r_b\,\theta$. The portion contacted along the small wheel is $r_s\,(\omega - \theta)$, because the contact point lags behind the original horizontal reference line. "No slipping" means that equal lengths are marked off along both wheels, so

$$r_b\,\theta = r_s\,(\omega - \theta)$$
$$r_s\,\omega = (r_b + r_s)\,\theta$$
$$\omega = \frac{r_b + r_s}{r_s}\,\theta$$

Consider the case mentioned above where $r_b = 3$ and $r_s = 1$. When $\theta = 2\pi$, so that the small wheel has gone all the way around the large wheel, we have $\omega = \frac{3+1}{1}\,2\pi = 4 \cdot 2\pi$. The small wheel turns 4 times a full turn.

The *Mathematica* NoteBook **Cycloid.ma** contains the formulas for the epicycloid we have just derived and uses them to plot the curve. We want you to extend the idea of using parametric circles and vector addition together in the next exercise.

EXERCISE 17.18. *The Cycloid, The Hypocycloid*

(1) *A wheel of radius $r_s = 1$ rolls to the right along a straight line at speed v without slipping. The vector pointing to the center of the wheel has the form*

$$\mathbf{C}(t) = \begin{bmatrix} c_1 \\ c_2 \end{bmatrix} = \begin{bmatrix} v \cdot t \\ ? \end{bmatrix}$$

where t is the time.

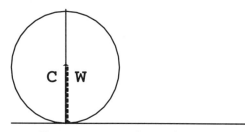

FIGURE *17.19: Start of a Cycloid*

The vector pointing from the center of the wheel to a reference point on the wheel has the form

$$\mathbf{W}(\omega) = \begin{bmatrix} w_1 \\ w_2 \end{bmatrix} = \begin{bmatrix} -r \, \mathrm{Sin}[\omega] \\ -r \, \mathrm{Cos}[\omega] \end{bmatrix}$$

In what position does the wheel begin? If $\omega = 0$ when $t = 0$?

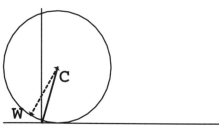

FIGURE *17.20: After a Small Change*

The distance moved along the line at time t is $v \cdot t$. How much is the distance marked along the wheel in terms of ω? Equate these, assuming no slipping, and find ω in terms of t.

*Modify the Mathematica NoteBook **Cycloid.ma** to plot this curve, known as a cycloid.*

(2) *A small wheel of radius r_s rolls around the inside of a large fixed wheel of radius r_b. Use vectors to find parametric equations for the path of a point on the tip of the small wheel and plot the path using Mathematica. Three main things change*

compared with the epicycloid above: First, the small wheel rolls in the opposite direction as its center. You could account for this by taking

$$\mathbf{W}(\omega) = r_s \cdot \begin{bmatrix} \mathrm{Cos}[-\omega] \\ \mathrm{Sin}[-\omega] \end{bmatrix}$$

Why does a negative angle correspond to clockwise rotation? Second, the distance from the center of the big wheel to the center of the small wheel is no longer the sum of the radii. Third, the equation that says the same lengths are contacted along the large and small wheel, or no slipping occurs, is different. This time the opposite rotation causes angles to add in the length equation along the small wheel.

(3) *If the large wheel has radius 3 and the small wheel has radius 1, how many turns does the small wheel make when going around the large wheel once for the epicycloid? For the hypocycloid? (What is ω when $\theta = 2\pi$?)*

(4) *(Optional) What happens if you use the epicycloid program and equations with $r = -1$ or a negative r_s in general? Try this with Mathematica and then show that the curve really is a hypocycloid, but corresponds to drawing the curve with a Spirograph starting the small wheel in a different position.*

An integral computation of the area under one loop of the cycloid is taken up in the Mathematical Background Chapter on Integration Drill.

EXERCISE 17.21. *The Helicopter*
This is a simple three dimensional exercise like the cycloid problem. It is simple because there is not a "no slipping" equation.

The rotor on a helicopter has radius 10 feet and rotates 40 times per minute. The helicopter flies along a straight line at an elevation of 500 feet at 50 miles per hour due East. The rotor turns in a horizontal plane (as the helicopter is coasting). Give parametric equations for the path of a point on the tip of the rotor with respect to a point on the ground.

17.4. Parametric Tangents and Velocity Vectors

Suppose that we have a parametric graph like the epicycloid

$$x = 4 \, \mathrm{Cos}[\theta] + \mathrm{Cos}[4\,\theta]$$
$$y = 4 \, \mathrm{Sin}[\theta] + \mathrm{Sin}[4\,\theta]$$

and want to find the tangent line to the graph at a point like $\theta = \pi/6$, where $4\theta = 2\pi/3$, so

$$x = 4\,\frac{\sqrt{3}}{2} - \frac{1}{2} \approx 2.9641$$

$$y = 4\,\frac{1}{2} + \frac{\sqrt{3}}{2} \approx 2.86603$$

We place parallel local coordinates (dx, dy) centered at this (x, y) point, $(x[\theta], y[\theta])$. The differentials of the original parametric equations

$$dx = (-4 \, \mathrm{Sin}[\theta] - 4 \, \mathrm{Sin}[4\theta])d\theta$$
$$dy = (4 \, \mathrm{Cos}[\theta] + 4 \, \mathrm{Cos}[4\theta])d\theta$$

at this specific $\theta = \pi/6$ give the parametric equations

$$dx = -(2 + 2\sqrt{3})d\theta \approx -5.46 \, d\theta$$
$$dy = (2\sqrt{3} - 2)d\theta \approx 1.46 \, d\theta$$

These equations are parametric equations for the tangent at $\pi/6$. They describe a line through the (dx, dy) origin at $(x, y) = (x[\pi/6], y[\pi/6])$ pointing in the approximate direction

$$\begin{bmatrix} -5.46 \\ 1.46 \end{bmatrix}$$

because the vector form of the parametric line is

$$d\mathbf{X} = \mathbf{P} + d\theta \cdot \mathbf{D}$$

with components

$$\begin{bmatrix} dx \\ dy \end{bmatrix} = \begin{bmatrix} 0 \\ 0 \end{bmatrix} + d\theta \begin{bmatrix} -5.46 \\ 1.46 \end{bmatrix}$$

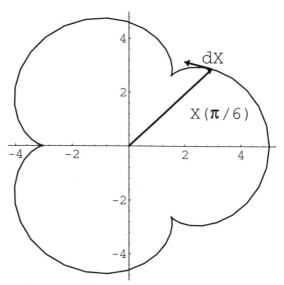

FIGURE 17.22: The Tangent at $\theta = \pi/6$

Once we know the tangent vector, we can sketch the tangent simply by putting the pen at the (x, y) point of tangency and sketching the line through this vector (drawn with its tail at the tangency point.)

17.4.1. Parametric Tangents. In general, the family of differentials of equations defining parametric curves define the parametric tangent line in local coordinates (where t is considered fixed):

Curve in (x,y)-coordinates	Tangent in (dx,dy)-coordinates at (x_0, y_0)
$x = x[t]$	$dx = x'[t_0]\,dt$
$y = y[t]$	$dy = y'[t_0]\,dt$
$z = z[t]$	$dz = z'[t_0]\,dt$

$$\text{Tangent Vector} \;=\; \begin{bmatrix} x'[t_0] \\ y'[t_0] \\ z'[t_0] \end{bmatrix}$$

EXERCISE 17.23. *Sketch the parametric curve and its tangent at $t = \pi/6$*

Curve in (x,y)-coordinates	Tangent in (dx,dy)-coordinates at $(\sqrt{3}/2, 1/2)$
$x = x[t] = \mathrm{Cos}[t]$	$dx = x'[t]\,dt$
$y = y[t] = \mathrm{Sin}[t]$	$dy = y'[t]\,dt$

Simply compute the proper vector and sketch it with its tail at the point of tangency.

17.4.2. Velocity Vectors. The parametric tangent vector can also be viewed as a velocity in the case where the parameter is time t and the equations represent position (x, y, z). The reason that the parametric tangent is velocity comes from the increment equations defining derivatives with respect to t as in Definition 5.15 or Definition 5.20. A change in position over a small time increment is

$$\frac{1}{\delta t}(\mathbf{X}[t + \delta t] - \mathbf{X}[t]) = \begin{bmatrix} (x[t+\delta t] - x[t])/\delta t \\ (y[t+\delta t] - y[t])/\delta t \\ (z[t+\delta t] - z[t])/\delta t \end{bmatrix} = \begin{bmatrix} x'[t] \\ y'[t] \\ z'[t] \end{bmatrix} + \begin{bmatrix} \varepsilon_x \\ \varepsilon_y \\ \varepsilon_z \end{bmatrix} \approx \begin{bmatrix} x'[t] \\ y'[t] \\ z'[t] \end{bmatrix}$$

PROBLEM 17.24. *A simple case of velocity vectors is the vector equation* $\mathbf{X} = \mathbf{P} + t\mathbf{D}$, *such as*

$$x = 1 - 2\,t$$
$$y = 3 - 1\,t$$
$$z = 1 + 2\,t$$

Suppose you are in space moving along a path described by these equations for time, t in seconds, where x, y, and z are measured in miles. Make sketches as you do the following computations.

Find your position at time zero, $t = 0$, and $\mathbf{X}_0 = ?$

Find your position one time unit later, $t = 1$, and $\mathbf{X}_1 = ?$

Find the vector displacement that you moved during this second. You should think of this as given in units of miles per second.

Find the length of the vector displacement. This represents your speed. Why?

Compute the velocity vector $\mathbf{X}'(t) = \begin{bmatrix} x'(t) \\ y'(t) \\ z'(t) \end{bmatrix}$. *Why is this the same as your displacement in one second?*

A klingon starship moves along the path

$$x = 1 - 4t$$
$$y = 3 - 2t$$
$$z = 1 + 4t$$

with the same time units, t. How fast is it going? When are you together? Why won't you ever be together again?

17.5. The Implicit Equation of a Plane

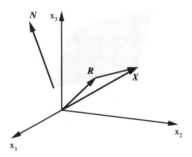

FIGURE 17.25: A Plane Normal to **N** through **R**

This section begins with the fundamental exercise that helps you find the geometric form of the equation of a plane. That equation says in algebraic terms: 'The plane is the set of vectors whose displacements from a certain point **R** are perpendicular to a certain normal vector **N**.'

PROBLEM 17.26. *The Implicit Equation of a Plane*
*We are given vectors **R** on a plane and **N** perpendicular (or normal) to it. We wish to find an equation in the unknown* $\mathbf{X} = \begin{bmatrix} x_1 \\ x_2 \\ x_3 \end{bmatrix}$ *that describes the plane. That is, so that the tips of all the **X** that satisfy the equation lie on the plane and completely fill it.*

Take the example $\mathbf{N} = \begin{bmatrix} -6 \\ -4 \\ 3 \end{bmatrix}$ *and* $\mathbf{R} = \begin{bmatrix} -6 \\ 8 \\ 2 \end{bmatrix}$. *Sketch the plane through the tip of **R** that is perpendicular to **N**.*

*Draw a generic unknown vector **X** whose tip lies somewhere on the plane.*

*Draw the illegal arrow that points from the tip of **R** to the tip of **X**. Show that the legal vector parallel to this arrow is* **X** − **R**.

*Write the statement "**X** − **R** is perpendicular to **N**" as an algebraic formula.*

Show that the equation of this specific example is

$$\mathbf{N} \bullet (\mathbf{X} - \mathbf{R}) = 0$$

$$\begin{bmatrix} -6 \\ -4 \\ 3 \end{bmatrix} \bullet \begin{bmatrix} x_1 + 6 \\ x_2 - 8 \\ x_3 - 2 \end{bmatrix} = 0$$

$$6\,x_1 + 4\,x_2 - 3\,x_3 = -10$$

Geometrically, three points determine a plane. The next exercise shows you how to use the lexicon to find the algebraic equation of the plane the three points determine. Solution of the exercise is an algebraic proof that three points really do determine a plane, but it is a specific proof, because it gives you the specific equation.

PROBLEM 17.27. *Three points Determine a Plane*
1) Plot the vectors

$$\mathbf{R} = \begin{bmatrix} r_1 \\ r_2 \\ r_3 \end{bmatrix} = \begin{bmatrix} 3 \\ 0 \\ 0 \end{bmatrix}, \qquad \mathbf{S} = \begin{bmatrix} s_1 \\ s_2 \\ s_3 \end{bmatrix} = \begin{bmatrix} 0 \\ 2 \\ 0 \end{bmatrix} \quad and \quad \mathbf{T} = \begin{bmatrix} t_1 \\ t_2 \\ t_3 \end{bmatrix} = \begin{bmatrix} 0 \\ 0 \\ 1 \end{bmatrix}$$

Also plot the illegal arrows $\mathbf{E}$ *from the tip of* $\mathbf{R}$ *to the tip of* $\mathbf{S}$ *and the illegal arrow* $\mathbf{F}$ *from* $\mathbf{R}$ *to* $\mathbf{T}$.
2) Compute the legal vectors $\mathbf{E} = \mathbf{S} - \mathbf{R}$ *and* $\mathbf{F} = \mathbf{T} - \mathbf{R}$. *These are parallel to the arrows drawn in part 1 by our lexicon entry for vector difference.*
3) Compute and sketch a vector $\mathbf{N}$ *that is perpendicular to both* $\mathbf{E}$ *and* $\mathbf{F}$.
4) Calculate $\mathbf{N} \bullet \mathbf{R}$, $\mathbf{N} \bullet \mathbf{S}$, $\mathbf{N} \bullet \mathbf{T}$. *Explain algebraically why* $\mathbf{R}$, $\mathbf{S}$ *and* $\mathbf{T}$ *are not perpendicular to* $\mathbf{N}$. *Do they appear to be perpendicular on your plot?*
5) The equation $\mathbf{N} \bullet (\mathbf{X} - \mathbf{R}) = 0$ *in the unknown vector* $\mathbf{X}$ *says:* $\mathbf{X}$ *such that the displacement from* $\mathbf{R}$ *to* $\mathbf{X}$ *is perpendicular to* $\mathbf{N}$. *Sketch this in a figure and explain why this is the equation for the plane determined by the tips of the three vectors* $\mathbf{R}$, $\mathbf{S}$ *and* $\mathbf{T}$.
7) The equations $\mathbf{N} \bullet (\mathbf{X} - \mathbf{R}) = 0$, $\mathbf{N} \bullet (\mathbf{X} - \mathbf{S}) = 0$ *and* $\mathbf{N} \bullet (\mathbf{X} - \mathbf{T}) = 0$ *in the unknown vector* $\mathbf{X}$ *are all equations for this plane, because each is an equation of a plane through one of* $\mathbf{R}$, $\mathbf{S}$ *or* $\mathbf{T}$ *and perpendicular to the same normal* $\mathbf{N}$. *In this form, the equations are different, but write them explicitly in components and simplify them to the form*

$$a\,x + b\,y + c\,z = k$$

Are their simplified forms different? Why?

You need to be able to recognize the geometry contained in old-fashioned equations.

EXERCISE 17.28. *The Classical Form of a Plane*
Consider parameters a, b, c *and* k *and unknowns* x, y, z. *The graph of the general linear equation in 3 variables* $ax + by + cz = k$ *is a plane. Write this equation in the vector form* $\mathbf{N} \bullet \mathbf{X} = k$.
   In terms of the given letters, what is the vector $\mathbf{N}$? *What is the unknown vector* $\mathbf{X}$?
   Take the special case $x + 3y + 2z = 6$. *What is the vector normal to this plane?*

Find the three intercepts of this plane and the axes. (HINT: An unknown vector of the

form $\mathbf{X} = \begin{bmatrix} x \\ 0 \\ 0 \end{bmatrix}$ *lies on the x axis.) Sketch the plane and its (illegal) normal vector with its*

tail somewhere on the plane.

How could you put this equation in the form $\langle \mathbf{N} \bullet (\mathbf{X} - \mathbf{R}) \rangle = 0$?

Now combine your knowledge of lines and planes.

EXERCISE 17.29. *Find an equation of the line that contains the point* $\begin{bmatrix} 1 \\ 3 \\ 2 \end{bmatrix}$ *and is perpendicular to the plane* $2x + 3y - 4z = 5$.

Vector projection is a very useful algebraic computation. In order to use it effectively, often we must compute an auxiliary vector.

EXERCISE 17.30. *The Distance from the Origin to* $\langle \mathbf{N} \bullet \mathbf{X} \rangle = k$
Suppose we are given a plane in the form $\langle \mathbf{N} \bullet \mathbf{X} \rangle = ax + by + cz = k$. *Our first step asks you to find a vector* $\mathbf{M}$ *parallel to* $\mathbf{N}$ *that lies on the plane. We want a general formula that you could use to write a Mathematica program, but a sample equation is*

$$2x - y + 2z = 6 \qquad \Leftrightarrow \qquad \begin{bmatrix} 2 \\ -1 \\ 2 \end{bmatrix} \bullet \begin{bmatrix} x \\ y \\ z \end{bmatrix} = 6$$

1) If $\mathbf{M}$ *is parallel to* $\mathbf{N}$, *then it must be a scalar multiple of* $\mathbf{N}$. *Why? Say,* $\mathbf{M} = h\,\mathbf{N}$. *Solve for* h *in terms of* a, b, c, *and* k. *Show that your component equations are equivalent to*

$$h = \frac{k}{\langle \mathbf{N} \bullet \mathbf{N} \rangle}$$

In particular, find the value of h *for the sample above.*
2) Use $\mathbf{M}$ *to find a formula for the distance from the origin to the plane. (Hint: Draw the figure of* $\mathbf{M}$ *pointing from the origin to the plane. Why is it perpendicular to the plane?)*
3) The equation $\langle \mathbf{N} \bullet \mathbf{X} \rangle = ax + by + cz = k$ *is not unique, because we can multiply both sides by any non-zero constant without changing the solutions. For example, the following equations all describe the same plane:*

$$2x - y + 2z = 6 \quad \Leftrightarrow \quad \begin{bmatrix} 2 \\ -1 \\ 2 \end{bmatrix} \bullet \begin{bmatrix} x \\ y \\ z \end{bmatrix} = 6$$

$$\Leftrightarrow$$

$$\frac{2}{3}x - \frac{1}{3}y + \frac{2}{3}z = 2 \quad \Leftrightarrow \quad \begin{bmatrix} \frac{2}{3} \\ -\frac{1}{3} \\ \frac{2}{3} \end{bmatrix} \bullet \begin{bmatrix} x \\ y \\ z \end{bmatrix} = 2$$

Apply your formula from parts (1) and (2) above to finding the distance from the origin to this plane using each of the formulas for the plane.

4) The unit length vector in the same direction as $\mathbf{N}$ *is* $\mathbf{U} = \frac{1}{|\mathbf{N}|} \mathbf{N}$. *Show that* $\ell = \frac{k}{|\mathbf{N}|}$
makes the following equations equivalent:

$$\langle \mathbf{N} \bullet \mathbf{X} \rangle = k \qquad \Leftrightarrow \qquad \langle \mathbf{U} \bullet \mathbf{X} \rangle = \ell$$

5) Prove that the distance from the origin to the plane $\langle \mathbf{N} \bullet \mathbf{X} \rangle = k$ *is* $\frac{k}{|\mathbf{N}|}$. *(Hint: Which multiple of* $\mathbf{U}$ *lies on the equivalent plane* $\langle \mathbf{U} \bullet \mathbf{X} \rangle = \ell$?)

PROBLEM 17.31. *The Distance from a Point to a Plane*
Given a plane in the form $\langle \mathbf{N} \bullet \mathbf{X} \rangle = k$ *and a point* $\mathbf{Q}$ *anywhere in space, use vector projection to describe a procedure to compute the distance from* $\mathbf{Q}$ *to the plane.*

17.6. Wrap-up Exercises

EXERCISE 17.32. *Write the vector A as a linear combination* $\mathbf{A} = b\mathbf{B} + c\mathbf{C}$, *that is, find*
b and c. $\mathbf{A} = \begin{bmatrix} 2 \\ -1 \\ -1 \end{bmatrix}$, $\mathbf{B} = \begin{bmatrix} 0 \\ 2 \\ 1 \end{bmatrix}$, $\mathbf{C} = \begin{bmatrix} -10 \\ 3 \\ 4 \end{bmatrix}$ *Draw the parallelogram with a vertex at the*
origin, one side the arrow $b\mathbf{B}$ *and another side the arrow* $c\mathbf{C}$. *Show the vector* $\mathbf{A}$ *on your figure.*

EXERCISE 17.33. *Sketch and find the area a of the triangle with vertices*

$$\mathbf{A} = \begin{bmatrix} 1 \\ 1 \\ 1 \end{bmatrix} \quad , \quad \mathbf{B} = \begin{bmatrix} 2 \\ 3 \\ 5 \end{bmatrix} \quad , \quad \mathbf{C} = \begin{bmatrix} -1 \\ 3 \\ 1 \end{bmatrix}$$

EXERCISE 17.34. *Show that the tips of the vectors* $\mathbf{A} = \begin{bmatrix} 3 \\ 2 \\ 0 \end{bmatrix}$, $\mathbf{B} = \begin{bmatrix} 1 \\ 1 \\ -1 \end{bmatrix}$, $\mathbf{C} = \begin{bmatrix} 5 \\ 3 \\ 1 \end{bmatrix}$ *are colinear. Find an equation for the line on which they lie and show that all three satisfy the equation.*

EXERCISE 17.35. *Show that the tips of the vectors* $\mathbf{A} = \begin{bmatrix} 0 \\ 2 \\ 1 \end{bmatrix}$, $\mathbf{B} = \begin{bmatrix} 1 \\ -1 \\ -1 \end{bmatrix}$, $\mathbf{C} = \begin{bmatrix} 13 \\ -1 \\ 5 \end{bmatrix}$,
$\mathbf{D} = \begin{bmatrix} 14 \\ 2 \\ 8 \end{bmatrix}$ *are coplanar. Find the equation of the plane of three and show that all four satisfy that equation.*

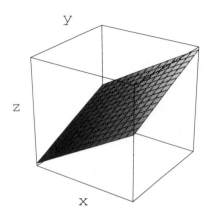

y

z

x

FIGURE 17.36: $z = 3x + 2y$

EXERCISE 17.37. *Normal and Tangential Components*

1) Suppose we have a plane given by

$$z = Ax + By$$

For example, consider $A = 3$ and $B = 2$. Give a 3-D vector $\mathbf{N}$ normal (or perpendicular) to the plane with an upward z-component.

2) Also suppose that a force $\mathbf{F}$ acts on the plane. For example, consider $\mathbf{F} = \begin{bmatrix} 1 \\ 2 \\ 3 \end{bmatrix}$. Compute the projection of $\mathbf{F}$ in the direction of the normal $\mathbf{N}$.

$$\mathbf{F}_N = \mathrm{Proj_{\mathbf{N}}}(\mathbf{F})$$

What is the magnitude of the force acting against the plane?
3) Sketch $\mathbf{N}$, $\mathbf{F}$, $\mathbf{F}_N$ and $\mathbf{F} - \mathbf{F}_N$ on the figure above. How much force acts along or tangent to the plane?

EXERCISE 17.38. *Find the angle at which the line*

$$x = 1 - t$$
$$y = 2t - 3$$
$$z = 5$$

intersects the plane $z = 3x + 2y + 7$

EXERCISE 17.39. *Show that the lines below never intersect by trying to solve for values of the parameters that would make x, y, and z equal.*

$$\begin{aligned} x &= 1 - t \\ y &= 2t - 3 \\ z &= 5 \end{aligned} \qquad\qquad \begin{aligned} x &= s - 1 \\ y &= 3 - 2s \\ z &= 1 + s \end{aligned}$$

Find the distance between these lines.

The flight deck of an aircraft carrier is inclined 10° off the axis of the ship. If you proceed with velocity vector $\mathbf{V}$ in still air, the wind will not blow straight down the flight deck, because it gives the apparent wind of $-\mathbf{V}$. However, if there is a real wind blowing with velocity $\mathbf{W}$, then we can set a course so that the wind does blow straight down the flight deck.

EXERCISE 17.40. *Find the heading (unit vector in the direction of your velocity $\mathbf{V}$) and speed ($|\mathbf{V}|$) to proceed in a wind blowing at velocity $\mathbf{W}$ in order for the apparent wind to blow down the flight deck at 30 knots. What conditions does $\mathbf{W}$ have to satisfy if the maximum speed of the ship is 30 knots?*

The parameter in a parametric line can represent time. Two airplanes flying along intersecting lines only collide if the point of intersection occurs at the same time.

EXERCISE 17.41. *Airplane $\mathbf{X}$ leaves the runway position $(0,0,0)$ at time $t = 0$ with a velocity of $U(300, 450, 15)$. What is the vector parametric equation in terms of t for the line along which it travels? $\mathbf{X} =$?*

Airplane $\mathbf{Y}$, flying nearby, is at position $P(100, 0, 5)$ at time $t = 0$ when the first plane takes off. It is traveling with a velocity of $V(0, 500, 0)$. What is the vector parametric equation in terms of t for the line along which it travels? $\mathbf{Y} =$?

Show that the paths of $\mathbf{X}$ and $\mathbf{Y}$ cross, but that they do not collide.

Airplane $\mathbf{Z}$, flying nearby, is at position $Q(40, 44, 2)$ at time $t = 0$ when the first plane takes off. It is traveling with a velocity of $W(0, 120, 0)$. What is the vector parametric equation in terms of t for the line along which it travels? $\mathbf{Z} =$?

Show that $\mathbf{X}$ and $\mathbf{Z}$ collide if they both maintain their courses.

CHAPTER 18

Linear Functions and Graphs in Several Variables

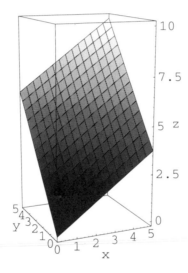

FIGURE 18.1: $z = \frac{3}{5}x + \frac{4}{5}y$

This chapter looks at plotting in 3-D and analyses the case of linear functions in detail. The main idea of differential calculus is that smooth functions microscopically 'look' like linear functions, so we need to understand the linear case before we proceed with calculus of several variables. Vectors help us understand the geometry of plotting in 3-D, but we need to be able to use different formulations. The basic linear function of two variables is given by the explicit equation

$$z = A\,x + B\,y + h$$

or $z = L(x,y) = A\,x + B\,y + h$, for constants A, B and h. For example, if $A = \frac{3}{5}$, $B = \frac{4}{5}$ and $h = 0$, then $z = \frac{3}{5}x + \frac{4}{5}y$. The explicit plot of $z = L(x,y)$ is a plane in 3-space.

We already know a geometric formula for a plane in 3-D,

$$\langle \mathbf{N} \bullet \mathbf{X} \rangle = k$$

where $\mathbf{N} = \begin{bmatrix} a \\ b \\ c \end{bmatrix}$ is the (fixed) normal vector to the plane and $\mathbf{X} = \begin{bmatrix} x \\ y \\ z \end{bmatrix}$ is the unknown (variable) vector of coordinates. (Exercise 17.30 shows that $k = |N| \times$[distance from the origin to the plane.

In the example above $\mathbf{N} = \begin{bmatrix} 2 \\ -1 \\ 1 \end{bmatrix}$ is the vector normal to the plane. (Sketch $\mathbf{N}$ on the graph.)

We may write the vector equation $\langle \mathbf{N} \bullet \mathbf{X} \rangle = k$ in components

$$a\,x + b\,y + c\,z = k$$

Simple algebra gives us the connection between the parameters. If we start with the explicit formula, then $a = -A$, $b = -B$, $c = 1$ and $k = h$ gives the parameters for the geometric form. Notice that it is easy to plot a plane if we know its normal and one point on it.

$$z = A \cdot x + B \cdot y + h \quad \Leftrightarrow \quad \langle \begin{bmatrix} -A \\ -B \\ 1 \end{bmatrix} \bullet \begin{bmatrix} x \\ y \\ z \end{bmatrix} \rangle = h \quad \Leftrightarrow \quad -A \cdot x - B \cdot y + z = h$$

$$z = -\frac{a}{c} x - \frac{b}{c} \cdot y + \frac{k}{c} \quad \Leftrightarrow \quad \langle \begin{bmatrix} a \\ b \\ c \end{bmatrix} \bullet \begin{bmatrix} x \\ y \\ z \end{bmatrix} \rangle = k \quad \Leftrightarrow \quad a \cdot x + b \cdot y + c \cdot z = h$$

If we start with the geometric form, knowing a, b, c and k, then we solve for z and obtain $A = -\frac{a}{c}$, $B = -\frac{b}{c}$ and $h = \frac{k}{c}$.

Which formula is best? Neither one. The vector equation gives us some useful geometric information, but the explicit equation is useful in other contexts and the parameters A and B have direct geometric interpretations in contour plots.

There are three major types of graphs: explicit, implicit and parametric. We studied parametric lines and curves in the last chapter, and while parametric surfaces are very useful (especially in computer graphics), we will not include them in this chapter. Explicit graphs are of the form $z = f(x)$ or $z = f(x, y)$ with all variables plotted and are probably the most familiar to you. Implicit or contour graphs are seldom used in one variable, but are often used in two. The contour graph of the height of a mountain is called a topographical map. If $z = h(x, y)$ gives the height z above a base plane point (x, y), then the contour map is a plane filled with curves of constant height

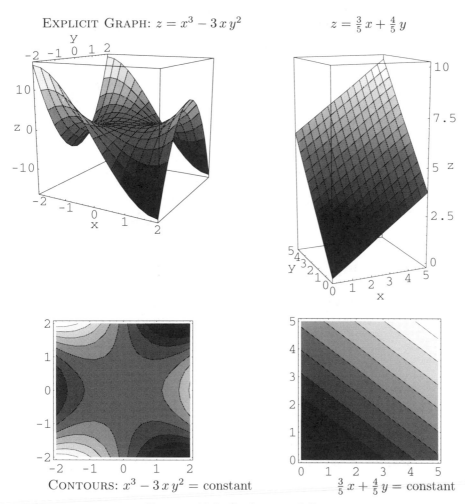

EXPLICIT GRAPH: $z = x^3 - 3\,x\,y^2$ $z = \frac{3}{5}\,x + \frac{4}{5}\,y$

CONTOURS: $x^3 - 3\,x\,y^2 = $ constant $\frac{3}{5}\,x + \frac{4}{5}\,y = $ constant

FIGURE 18.2: Surfaces and Contours

One of the main goals of the chapter is to connect the parameters in $z = A\,x + B\,y + h$ with the contour plot of the linear function. The 2-D vector

$$\mathbf{G} = \begin{bmatrix} A \\ B \end{bmatrix}$$

points in the x-y-direction of fastest increase of $z = A\,x + B\,y + h$ and we may plot $\mathbf{G}$ in the contour plot or base plane of the 3-D plot. This is called the gradient vector. We must learn to not confuse it with the normal to the explicit plane in 3 dimensions

$$\mathbf{N} = \begin{bmatrix} -A \\ -B \\ +1 \end{bmatrix}$$

18.1. Vertical Slices and Chickenwire Plots

This section asks you to make some hand-drawn graphs - linear, quadratic and cubic. Once you understand what these graphs mean, *Mathematica* will be immensely helpful in graphing nonlinear graphs. It is also very helpful in visualizing as you will see in the NoteBook **SurfaceFlyBy.ma**.

EXAMPLE 18.3. *Vertical Slices of* $z = x^2 + \frac{1}{2}y^2$

We begin with the function $z = f(x,y) = x^2 + \frac{1}{2}y^2$ finding the vertical slices parallel to the $x - z$-plane. If we fix the value of y, say $y = -2$, then the function becomes $z = f(x,-2) = x^2 + 2$. This is a simple parabola and its graph is shown in the first figure below plotted from $x = -2$ to $x = 3$. The plot is drawn on the vertical (x,z)-plane where $y = -2$.

Next, we fix $y = -1$ and plot $z = f(x,-1) = x^2 + \frac{1}{2}$. This is also a parabola with the same basic shape as the first, but translated down by $\frac{3}{2}$. It is shown in the second figure below plotted in the vertical plane where $y = -1$ and x and z vary.

The next 4 figures below show the plots $z = x^2$ on the plane $y = 0$, $z = x^2 + \frac{1}{2}$ on the plane $y = 1$, $z = x^2 + 2$ on the plane $y = 2$, and $z = x^2 + \frac{9}{2}$ on the plane $y = 3$.

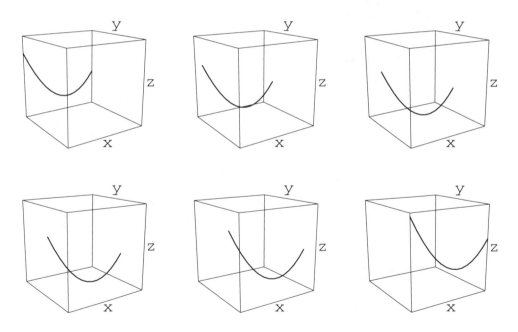

FIGURE 18.4: $z = f(x,a)$,　　$a = -2, -1, 0, 1, 2, 3$

We continue with the function $z = f(x,y) = x^2 + \frac{1}{2}y^2$ considering vertical slices parallel to the $y - z$-plane. If we fix the value of x, say $x = -2$, then the function becomes $z = f(-2,y) = 4 + \frac{1}{2}y^2$. This is a simple parabola and its graph is shown in the first figure below plotted from $y = -2$ to $y = 3$. The plot is drawn on the vertical (y,z)-plane where $x = -2$.

Next, we plot $z = f(-1, y) = 1 + \frac{1}{2}y^2$. This is also a parabola with the same basic shape as the first, but translated down by 3. It is shown in the second figure below plotted in the vertical plane where $x = -1$ and y and z vary.

The next 4 figures below show the plots $z = \frac{1}{2}y^2$ on the plane $y = 0$, $z = 1 + \frac{1}{2}y^2$ on the plane $x = 1$, $z = 4 + \frac{1}{2}y^2$ on the plane $x = 2$, and $z = 9 + \frac{1}{2}y^2$ on the plane $x = 3$.

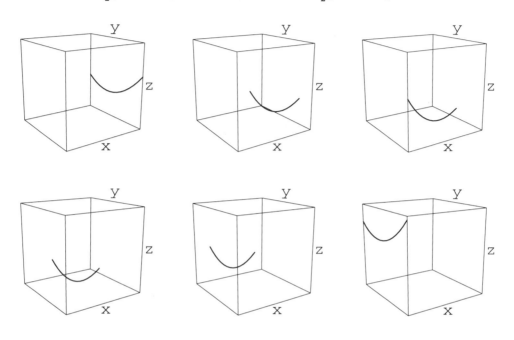

FIGURE 18.5: $z = f(a, y)$,　　$a = -2, -1, 0, 1, 2, 3$

When all these curves are shown together and the segments between them are shaded in, we see the surface as follows.

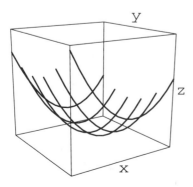 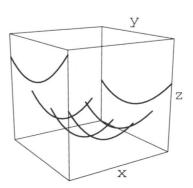

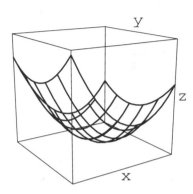 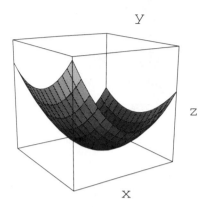

FIGURE 18.6: $z = x^2 + \frac{1}{2}y^2$

EXERCISE 18.7. *Explicit Surfaces*
For each of the functions $z = f(x, y)$ below with their Mathematica graphs:

1) Plot the curves $z = f(x, -2)$, $z = f(x, -1)$, $z = f(x, 0)$, $z = f(x, 1)$, $z = f(x, 2)$, $z = f(x, 3)$.

2) Plot the curves $z = f(-2, y)$, $z = f(-1, y)$, $z = f(0, y)$, $z = f(1, y)$, $z = f(2, y)$, $z = f(3, y)$.

3) Combine your plots from above on a single 3-D graph, plotting $z = f(x, 2)$ in the vertical plane $y = 2$, and plotting $z = f(-1, y)$ in the vertical plane $x = -1$, etc.

*4) Modify the **SurfaceSlices.ma** NoteBook to check your hand sketches.*

*5) Run the Mathematica NoteBook **SurfaceFlyBy.ma** and modify it to fly by each of these surfaces.*

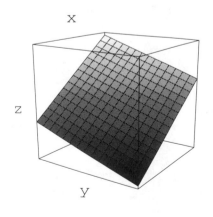 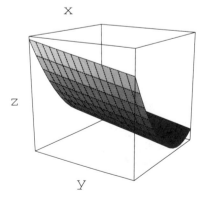

FIGURE 18.8: $z = -x - \frac{1}{2}y$ $z = x^2 - \frac{1}{2}y$

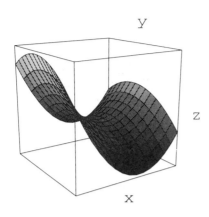 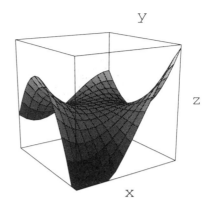

FIGURE 18.9: $z = x^2 - \frac{1}{2}y^2$ $\qquad\qquad$ $z = x^3 - 3\,x\,y^2$

EXERCISE 18.10. *Consider the formula*

$$T = M \operatorname{Sin}\left[\pi \frac{t}{6}\right] e^{-d/20}$$

Let $M = 15$ at first, then later show the parameter M on your plots. (T is temperature in Celsius, t time in years and d depth in meters for seasonably fluctuating sea water.)

(1) *Plot T vs. t treating d as a parameter. How do the graphs change as d increases? As the depth increases, do seasonal fluctuations become bigger or smaller?*
(2) *Plot T vs. d treating t as a parameter.*
(3) *Use Mathematica's Plot3D[T, $\{t, 0, 24\}$, $\{d, 0, 60\}$] to see both families at once.*

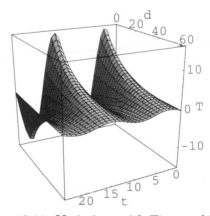

FIGURE 18.11: Variations with Time and Depth

EXERCISE 18.12. *Here are some additional functions to plot using Mathematica's Plot3D[.]. You will recognize them if you think about the curves you get by holding x fixed and varying y or holding y fixed and varying x.*

a) $\quad z = x^2 + y^2$ $\qquad\qquad\qquad$ b) $\quad z = (x/5)^2 + (y/3)^2$

c) $z = x\,y^3$ d) $z = x^2\,y^3$

e) $z = \text{Sin}[x] + \text{Cos}[y]$ f) $z = x \cdot \text{Cos}(2\pi y)$

g) $z = e^{-(x^2+y^2)}$ h) $z = e^{-(x^2+y^2)}\,\text{Sin}[x^2 + y^2]$

   To make a Mathematica plot simply type a Mathematica program like this:

f[x_ ,y_] := x∧2 + y∧2;
Plot3D[f[x,y], { x, -1, 1} ,{ y, -1, 1}];

18.2. Horizontal Slices and Contour Graphs

EXAMPLE 18.13. *Horizontal Slices of $z = x^2 + \frac{1}{2}y^2$*

   If we slice horizontally, we hold the value of z fixed. Consider the example above $z = x^2 + \frac{1}{2}y^2$. If we let $z = 0$ we have the equation $0 = x^2 + \frac{1}{2}y^2$. This has only the solution $(x,y) = (0,0)$. This is the single point at the very bottom of the surface.

   If we let $z = 1$, the intersection of the plane with equation $z = 1$ and the surface $z = x^2 + \frac{1}{2}y^2$ has the equation $1 = x^2 + \frac{1}{2}y^2$. This is an ellipse.

   Similarly, the intersections of the planes $z = $ constant, are larger ellipses of similar shapes. The contour plot of the function $z = x^2 + \frac{1}{2}y^2$ is the family of these curves.

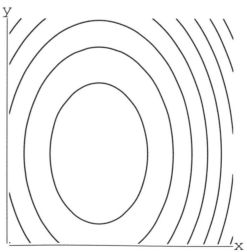

FIGURE 18.14: $z = x^2 + \frac{1}{2}\,y^2$

EXERCISE 18.15. *Contour Graphs*
Find the equations of the contours in the following examples for the z-values $z = -1$, $z = 0$, $z = 1$, $z = 2$. The plots are shown for $-2 \le x, y \le 3$. (Even though Mathematica puts lines at the left and bottom.)

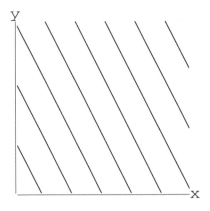
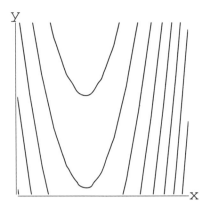

FIGURE 18.16: $z = -x - \frac{1}{2}y$ $z = x^2 - \frac{1}{2}y$

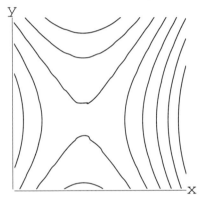
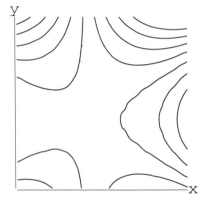

FIGURE 18.17: $z = x^2 - \frac{1}{2}y^2$ $z = x^3 - 3xy^2$

EXERCISE 18.18. *Make ContourPlots of the functions from Exercise* ?? *and compare them to your surface plots.*

To make a contour plot type a Mathematica program like this:

f[x_ ,y_] := x∧2 + y∧2;
ContourPlot[f[x,y], { x, -1, 1} ,{ y, -1, 1}];

18.3. *Mathematica* Plots

The functions $z = f(x, y)$ or $z = f(\mathbf{X})$ given below are based on simple formulas, yet graphing them by hand would be difficult because it is not at all easy to see what slices look like when x or y is fixed. *Mathematica* makes graphing them easy.

a) $z = xy^2 + x^3y$

b) $z = \mathrm{Sin}^3(x) + \mathrm{Cos}^4(y)$

c) $z = \mathrm{Sin}(xy)\,\mathrm{Cos}(xy)$

d) $z = \sqrt{1 - x^2 - y^2} \cdot \mathrm{Cos}(2\pi xy)$

e) $z = \sqrt{x^2 + \mathrm{Sin}(y^2)}$

f) $z = \frac{1}{\sqrt{x^2+y^2}}$

g) $z = (2x^2 + 3y^2)e^{-(x^2+y^2)}$

h) $z = 2xye^{-(x^2/2+y^2/8)}$

i) $z = \text{ArcTan}[x^2 y^3]$ j) $z = \text{Log}[x^2 + y^4]$

EXERCISE 18.19. *Use the Mathematica NoteBook* **ExplicitSurfaces.ma** *and the Mathematica NoteBook* **ContourPlots.ma** *to plot all of the functions above. Compare the surface plots with their topographical maps. (We have hidden one wrong contour plot in the Note-Book to keep you on your toes.)*

Save this work, because we will return to these examples in the next chapter.

18.4. Linear Functions and Gradient Vectors

This section helps you discover convenient formulas to move back and forth between the geometry and algebra of the contour plot of a linear function and the geometry and algebra of the explicit graph of the same linear function. These geometric formulas are used in the next chapter to study nonlinear functions. Remember, calculus lets you compute what you would see in a powerful microscope focused on a nonlinear function, by giving you its local linear approximation through rules of differentiation. In order to "see," you need to be able to use some linear geometric computations. It helps if you devise formulas that work from the most easily computed quantities. The main ideas are in the next exercise.

You should try to devise a simple geometric way to go from values for A, B and h to the graph of the line $A\,x + B\,y = h$. At first you will want to re-write things in the form $y = m\,x + b$, but it turns out to be better to work more directly from the vector $\mathbf{G} = \begin{bmatrix} A \\ B \end{bmatrix}$. The main advantage is that it is easy to plot the vector. The main complication is in relating h to this perpendicular vector.

PROBLEM 18.20. *Contour Line Density*
The purpose of this exercise is to have you work out the formulas needed to relate the steepness of an explicit linear graph $z = A\,x + B\,y$ and the density of the lines in its contour plot $A\,x + B\,y = h$. The secret of the whole exercise is in relating the length of the vector $G(A, B)$ to the steepness and density.
Consider the family of lines in 2-D given by

$$A\,x + B\,y = h; \qquad for \quad h = 0, \pm 1, \pm 2, \cdots$$

We want you to develop a systematic geometric approach to this kind of plot, but you can use any methods you like in the first two parts. Check your later vector formulas on the first two special cases at each step of the exercise.
1) Plot the family of lines on a sheet of graph paper if $(A, B) = (\sqrt{3}, 1)$. Plot the vector $\mathbf{G} = \begin{bmatrix} A \\ B \end{bmatrix}$ on top of your family of lines. The next part of the exercise asks you to do the same thing for a shorter parallel vector. Remember the title of the problem.
2) Plot the family of lines on a sheet of graph paper if $(A, B) = (\frac{\sqrt{3}}{2}, \frac{1}{2})$ Plot the vector $\mathbf{G} = \begin{bmatrix} A \\ B \end{bmatrix}$ on top of your family of lines.

Note $\text{Cos}[\pi/6] = \sqrt{3}/2$ and $\text{Sin}[\pi/6] = 1/2$, or the vectors from parts (1) and (2) are both inclined 30° above the x axis. The contour lines from parts (1) and (2) are both perpendicular *to this direction, but the lines are closer together in one of them. Which has closer contour lines (for $h = 0, \pm1, \pm2, \cdots$), part (1) or part (2)?*

Now consider the general geometric case of the family of lines $Ax + By = h$ for $h = 0, \pm1, \pm2, \cdots$ with unknown, but fixed A and B.

3) Plot the vector $\mathbf{G} = \begin{bmatrix} A \\ B \end{bmatrix}$ and the line through the origin perpendicular to $\mathbf{G}$. Show that this is the line $\langle \mathbf{G} \bullet \mathbf{X} \rangle = Ax + By = 0$. (Hint: Use your geometric-algebraic lexicon.) Compare this procedure with the two test cases you did at the start of the exercise.

4) Show that the 2-vector $\mathbf{G}$ is perpendicular to all the lines. Notice that

$$\langle \mathbf{G} \bullet \mathbf{X} \rangle = Ax + By = h; \qquad h = 0, \pm1, \pm2, \cdots$$

5) What are all the vectors perpendicular to the lines $Ax + By = h$?

6) What is the length of $\mathbf{G}$? (Compute the value in your two test cases.)

7) Which unknown vector of the form $\mathbf{X} = t\mathbf{G}$ for some real t lies on the line $\mathbf{G} \bullet \mathbf{X} = 1$? You should treat the components of $\mathbf{G}$ as known parameters and solve for t. Sketch this vector $\mathbf{X}$ on your two test cases for $\mathbf{G}$ (parts (1) and (2) above).

8) Show that the distance between the lines $\langle \mathbf{G} \bullet \mathbf{X} \rangle = 0$ and $\langle \mathbf{G} \bullet \mathbf{X} \rangle = +1$ is $1/|\mathbf{G}|$.

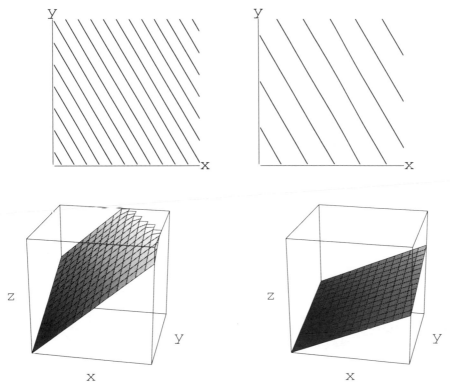

FIGURE 18.21: $z = \sqrt{3}\,x + y$ and $z = \sqrt{3}\,x/2 + y/2$

In general, the distance between the line $\langle \mathbf{G} \bullet \mathbf{X} \rangle = k_1$ and the second line $\langle \mathbf{G} \bullet \mathbf{X} \rangle = k_2$ is $|k_2 - k_1|/|\mathbf{G}|$. Roughly speaking, 'The longer $\mathbf{G}$, the steeper the explicit graph and the denser the contour lines...' Write a verbal description of the connection between the closeness of the lines $A\,x + B\,y = h$; $h = 0, \pm 1, \pm 2, \cdots$ and the length of $\mathbf{G}$. The next problem will help you explain the connection between the steepness of the explicit graph $z = A\,x + B\,y$ and the length of $\mathbf{G}$.

PROBLEM 18.22. *Steepness and* $|\mathbf{G}|$
An explicit plane through the origin in 3-D has an equation of the form

$$z = A\,x + B\,y$$

1) For example, $z = \sqrt{3}\,x + y$. Sketch this plane by x-y-slices, holding x fixed and varying y, holding y fixed and varying x, but plotting everything in 3-D. How do A and B appear on your plot?

2) It is easier to plot by re-writing the equation in the form $\mathbf{N} \bullet \mathbf{X} = 0$, where $\mathbf{N} = \begin{bmatrix} -A \\ -B \\ 1 \end{bmatrix}$.

Show that this is equivalent to the original equation. What is the geometric meaning of $\mathbf{N} \bullet \mathbf{X} = 0$? Sketch the 3-D vector $\mathbf{N}$ on your 3-D figure for this case. (Hint: Use your geometric-algebraic lexicon.)
3) Repeat the exercise for the equation $z = \sqrt{3}\,x/2 + y/2$, that is, sketch the plane perpendicular to $\begin{bmatrix} -\frac{\sqrt{3}}{2} \\ -\frac{1}{2} \\ 1 \end{bmatrix}$ through the origin. Do you need to plot slices?

4) As the 2-D vector $\mathbf{G} = \begin{bmatrix} A \\ B \end{bmatrix}$ gets larger, what happens to the 3-D vector $\mathbf{N} = \begin{bmatrix} -A \\ -B \\ 1 \end{bmatrix}$?

For example, plot $\begin{bmatrix} -A \\ -B \\ 1 \end{bmatrix}$ and the plane perpendicular to it through the origin in case $A =$ $B = 1/10$ and in the case $A = B = 10$.
How is the steepness of the plane related to the length of $\mathbf{G}$?

A contour map of an explicit surface $z = f(x, y)$ is the planar sketch of the curves that cut the surface at various fixed values of $z = k$, a constant. In the case of a linear function, $z = A\,x + B\,y$, the contours are parallel straight lines. The questions then are:

  (1) In which direction does the family of lines point?
  (2) How far apart are the lines for a given spacing of the constant values of z?

The analogous questions about the 3-D plane which you will answer in the next Exercise are:

  (1) In which x-y-direction does the plane slope up most steeply?
  (2) How steep is the plane in this direction?

PROBLEM 18.23. *Use the general results of the two Problems 18.20 and 18.22 to write a brief essay explaining how you answer the two general questions preceding this problem. 'Given the linear function $z = A\,x + B\,y$, the contour map is the family of lines perpendicular to ... The distance between the contour at height $z = k_2$ and $z = k_1$ is ... This means that in order to increase z by 1 unit in the most efficient way, we move ... units in the direction of ...' Hint: The most efficient direction is $\frac{1}{|\mathbf{G}|}\mathbf{G}$. Why?*

18.5. Explicit, Implicit and Parametric Graphs

This section reviews what we already know about the three main kinds of formulas for graphs with brief graphical reminders. We shall see that whenever we have an explicit formula, the differential is an explicit formula for the associated tangent. Whenever we have an implicit formula, the differential is an implicit formula for the tangent and whenever we have a parametric formula, the differential is a parametric formula for the tangent. In other words, calculus always gives the tangent formula as the same type of graph, explicit, implicit or parametric.

We pause to review cases that we have already seen. When we use calculus, there is a symbolic step between the specific tangent line and the nonlinear formula. In addition, if we wish to express the tangent in x, y-coordinates, then there is another step for that. The main idea, however, is that the differential is the formula for the tangent.

Type	Equation Nonlinear in x, y	Tangent Linear in dx, dy
Explicit 2-D	$y = f[x]$	$dy = m\,dx$
	$y = \sqrt{25 - x^2}$	$dy = -(4/3)dx$
Implicit 2-D	$F[x, y] = c$	$a\,dx + b\,dy = 0$
	$x^2 + y^2 = 25$	$4\,dx + 3\,dy = 0$
Parametric 2-D	$x = x[t]$	$dx = a\,dt$
	$y = y[t]$	$dy = b\,dt$
	$x = 5\,\mathrm{Cos}[t]$	$dx = -3\,dt$
	$y = 5\,\mathrm{Sin}[t]$	$dy = 4\,dt$

18.5.1. Explicit Graphs and Tangents - 2D. The graph of an explicit function is the set of (x, y) coordinates that satisfy $y = f[x]$, for example, $y = \sqrt{25 - x^2}$ shown next. Its tangent is the explicit line $dy = m\,dx$ where $m = f'(x) = -x/\sqrt{25 - x^2}$ and x is considered fixed in the (dx, dy)-coordinates. For example, this is shown for $x = 4$, where $y = 3$ and $-x/\sqrt{25 - x^2} = -4/3$

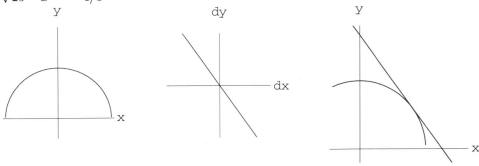

FIGURE 18.24: $y = \sqrt{25 - x^2}$ and $dy = -4\,dx/3 \Leftrightarrow y - 3 = -4(x - 4)/3$

EXERCISE 18.25. *Use Mathematica to plot the curve* $y = \frac{3}{4}\sqrt{25 - x^2}$. *Find the symbolic differential* $dy = f'[x]\, dx$ *and use it to find the equation of the tangent to this curve when* $x = 4$. *Finally, use Mathematica to plot both curves.*

18.5.2. Explicit Graphs and Tangents - 3D. The graph of an explicit function is the set of (x, y, z) coordinates that satisfy $z = f[x, y]$, for example, $z = \operatorname{Cos}[xy]$ shown next. In the next chapter we will see how to use rules of differentiation in order to find the values of A and B that make the plane tangent to a given nonlinear graph at a specified point. The plane is given by the explicit linear equation in two variables, $dz = A\, dx + B\, dy$.

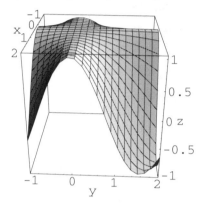   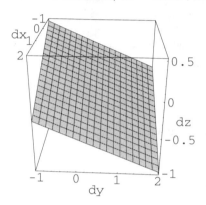

FIGURE 18.26: $y = \operatorname{Cos}[xy]$ and $dz = -0.265\, dx - 0.229\, dy$

18.5.3. Implicit Equations. An equation like the formula for the circle of radius 5,

$$x^2 + y^2 = 5^2$$

does not give an explicit way to compute y from x, but it is implicit in the formula that only certain values of y work for a given x. In general, an implicit formula $F(x, y) = c$ is associated with the set of (x, y)-coordinates that satisfy the equation.

An implicit linear equation in the two variables dx and dy has the form

$$a\, dx + b\, dy = k$$

for constants a, b, and k.

Implicit differentiation of the circle example gives

$$x^2 + y^2 = 25 \qquad \Leftrightarrow \qquad 2x\, dx + 2y\, dy = 0$$

When $(x, y) = (4, 3)$, we have the particular implicit tangent $4\, dx + 3\, dy = 0$ with normal vector $(3, 4)$. In other words, the tangent is perpendicular to the radius vector. (Which is clear geometrically without calculus.) If we write this in terms of the original (x, y)-coordinates, we replace dx by $(x - 4)$ and replace dy by $(y - 3)$, obtaining the line $4 \cdot (x - 4) + 3 \cdot (y - 3) = 0$. In vector notation, this is the line perpendicular to $(4, 3)$ through the point $(4, 3)$.

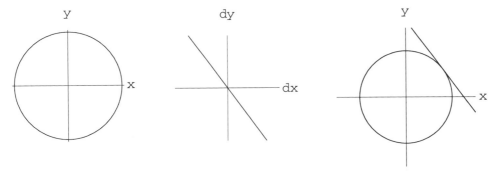

FIGURE 18.27: $x^2 + y^2 = 5$ and $4\,dx + 3\,dy = 0$

In the next chapter, we will use the observation that the vector (A, B) is perpendicular to the tangent line $A\,dx + B\,dy = 0$ to understand the gradient. The contour plot of a function $f[x, y]$ is a collection of 2-D implicit curves, $f[x, y] = c$, for constants c.

EXERCISE 18.28. *Use Mathematica to plot the curve* $\frac{x^2}{16} + \frac{y^2}{9} = \frac{25}{16}$. *Find the symbolic differential* $A\,dx + B\,dy = 0$ *and use it to find the equation of the tangent to this curve when* $x = 4$. *Finally, use Mathematica to plot both curves.*

18.5.4. Parametric Curves. The most important parametric equations in mathematics are the equations for the unit circle in terms of radian measure (as discussed in the Mathematical Background Chapter on High School Review with *Mathematica*). A variation on that is the circle of radius 5 given by the following equations, shown with their symbolic differentials and the differentials at a specific point.

$$x = 5\,\text{Cos}[t] \qquad dx = -5\,\text{Sin}[t] \qquad dx = -3\,dt$$
$$y = 5\,\text{Sin}[t] \qquad dy = 5\,\text{Cos}[t] \qquad dy = 4\,dt$$

If we take $t = \text{ArcTan}[3/4]$, then $\text{Sin}[t] = 3/5$ and $\text{Cos}[t] = 4/5$, so $x = 4$, $y = 3$, $dx = -3\,dt$, and $dy = 4\,dt$. The specific differentials are parametric equations for the tangent line to the circle at $(4, 3) = (x, y)$. They say in vector terms that the direction of the tangent line is the vector $(-3, 4)$. This can be viewed as the velocity vector if t represents time and x and y are positions. If we want to express the tangent in the same (x, y)-coordinates, we would replace dx by $(x - 4)$ and dy by $(y - 3)$. The parametric vector equations for the line in direction $(-3, 4)$ through the point $(4, 3)$ are $x = 4 - 3\,dt$ and $dy = 3 + 4\,dt$.

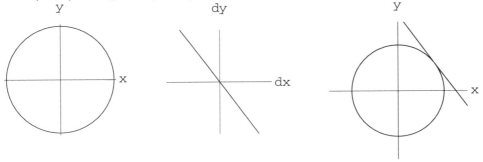

FIGURE 18.29:
$$x = 5\,\text{Cos}[t], \quad y = 5\,\text{Sin}[t], \quad dx = -3dt, \quad dy = 4dt, \quad x = -3dt + 4, \quad y = 4dt + 3$$

It would not be necessary to replace the parameter dt, since it is not plotted.

EXERCISE 18.30. *Use Mathematica to plot the parametric curve* $x = 5\,\mathrm{Cos}[t]$, $y = \frac{15}{4}\,\mathrm{Sin}[t]$. *Find the symbolic differentials* $dx = a\,dt$ *and* $dy = b\,dt$ *and use them to find the equation of the tangent to this curve when* $t = \mathrm{ArcTan}[4/3]$. *Finally, use Mathematica to plot both curves.*

Part 4

Differentiation in Several Variables

CHAPTER 19

Differentiation of Functions of Several Variables

This chapter shows how to use 'microscopic views' of nonlinear functions of two variables to analyse various aspects of their behavior. The Increment Principle says that small enough views appear to be linear, so small nonlinear changes are approximately linear. The new thing is that the linear function approximating the nonlinear one is itself a function of two variables.

For functions of one variable, we can summarize the graphical and symbolic theorems, or rules, of differentiation as follows:

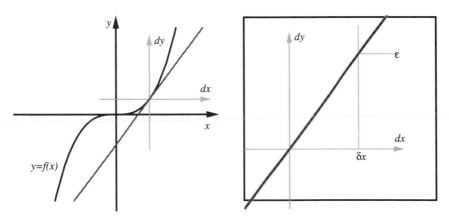

FIGURE 19.1: $dy = m \cdot dx$ in an Infinitesimal Microscope

THEOREM 19.2. *Summary of 1 Variable Differentiation Results*
The graph of the linear function given by

$$dy = m \ dx$$

in local (dx, dy) - coordinates is the tangent line (at x, where x is considered fixed) to the explicit nonlinear graph

$$y = f(x)$$

provided $m = f'(x)$ *and* $f'(x)$ *can be computed by the rules yielding a formula valid in an interval around* x.

The condition of (uniform) "tangency" is expressed by the microscopic error formula

$$f(x + \delta x) = f(x) + f'(x) \cdot \delta x + \varepsilon \cdot \delta x$$

where the error is small, $\varepsilon \approx 0$, *whenever the perturbation is small,* $\delta x \approx 0$. *This approximation applies when* x *is a finite number away from the boundary* $h \not\approx x \not\approx k$ *where the rules are valid,* $h < x < k$. *The approximation means that a microscopic view of a tiny piece of the graph* $y = f(x)$ *looks the same as the linear graph* $dy = m \cdot dx$.

We studied the geometric meaning of ε in terms of an ideal microscope of infinite power. The graphical idea can be reviewed on the *Mathematica* NoteBooks called **Microscope.ma** or **Zoom.ma**. Those NoteBooks animate the microscope, showing you successive magnifications of a graph. You can modify the function $f(x)$ and press $< Enter >$ to see an animation of the function of your choice.

19.1. Definition of Partial and Total Derivatives

This section presents the definition of the derivative with respect to a two dimensional vector independent variable (which may also be thought of as two scalar independent variables). The 3-D parts of the *Mathematica* NoteBook **Microscope.ma** presents animations of both explicit graphs and contour graphs with their magnifications. You can also modify the function $f(x, y)$ in the 3-D parts of the microscope notebook to view the function of your choice. Please try it; it's easy and the 'movies' are great.

Partial derivatives $\frac{\partial f}{\partial x}$ and $\frac{\partial f}{\partial y}$ are calculated with the same rules we used before by treating the other variable as a constant (or parameter.) We will discuss the details of computing partial derivatives below.

DEFINITION 19.3. *Total and Partial Derivatives in 2 Variables*
The linear function in local (dx, dy, dz) *- coordinates given by*

$$dz = A\ dx + B\ dy$$

is the graph of a plane through $(dx, dy, dz) = (0, 0, 0)$. *This plane is tangent to the surface graph of the nonlinear function*

$$z = f(x, y)$$

at a fixed (x, y, z) *at the origin of the parallel* (dx, dy, dz) *coordinates, provided* $A = \frac{\partial f}{\partial x}(x, y)$, $B = \frac{\partial f}{\partial y}(x, y)$, *and the partial derivatives can be computed by the rules yielding formulas valid in a rectangle around* (x, y).

The condition of "tangency" is a special approximation expressed by the "microscopic" error formula

$$f(x + \delta x, y + \delta y) = f(x, y) + \frac{\partial f}{\partial x}\delta x + \frac{\partial f}{\partial y}\delta y + \varepsilon \sqrt{\delta x^2 + \delta y^2}$$

where the error $\varepsilon \approx 0$ *is small whenever the perturbations are small,* $\delta x \approx 0$ *and* $\delta y \approx 0$. *Geometrically, this approximation just means that the surface and its tangent plane are indistinguishable when viewed in an infinitesimal microscope.*

The microscopic magnification makes the small term $\sqrt{\delta x^2 + \delta y^2}$ appear unit size and the condition $\varepsilon \approx 0$ means that the difference between the linear and nonlinear graphs appears small even after magnification. (The expression $\sqrt{\delta x^2 + \delta y^2}$ is just the length of the perturbation vector $\delta \mathbf{X} = (\delta x, \delta y)$ which is magnified to unit length.)

This approximation is valid when x and y are finite hyperreal numbers away from the boundary of validity, say $h \not\approx x \not\approx k$ and $m \not\approx y \not\approx n$ when the partial differentiation rules are valid on $h < x < k$ and $m < y < n$. Functions satisfying this condition on a region of the plane are called "locally linear" or smooth.

INTUITIVE INCREMENT APPROXIMATION

Roughly speaking, the change in a smooth function f is approximately given by the linear function in (dx, dy) (where x and y are considered fixed)

$$`f(x + dx, y + dy) - f(x, y) \simeq A \ dx + B \ dy'$$

The error is small even compared to the size of the small vector perturbation (dx, dy), with size $\sqrt{dx^2 + dy^2} = \left\| \begin{bmatrix} dx \\ dy \end{bmatrix} \right\|$.

The total differential and 3-D microscope approximation can also be written in terms of vectors as follows:

$$dz = \langle \mathbf{G} \bullet dX \rangle$$

is the total differential of the function

$$z = f(\mathbf{X})$$

of the vector variable $\mathbf{X} = \begin{bmatrix} x \\ y \end{bmatrix}$, with the tangency condition (microscopic approximation):

$$f(\mathbf{X} + \delta \mathbf{X}) = f(X) + \langle \mathbf{G} \bullet \delta \mathbf{X} \rangle + \varepsilon \, |\delta \mathbf{X}|$$

where $\mathbf{G} = \nabla f(X) = \begin{bmatrix} \frac{\partial f}{\partial x}(x, y) \\ \frac{\partial f}{\partial y}(x, y) \end{bmatrix}$, $X = \begin{bmatrix} x \\ y \end{bmatrix}$ and $\delta X = \begin{bmatrix} \delta x \\ \delta y \end{bmatrix}$. The vector ∇f is pronounced 'nabla f' (or sometimes 'grad f') and is called the gradient of f at $\mathbf{X}$. In this notation, we have the total differential given by the dot product of the gradient and the perturbation vector,

$$dz = \langle \nabla f(\mathbf{X}) \bullet d\mathbf{X} \rangle = \langle \begin{bmatrix} \frac{\partial f}{\partial x}(x, y) \\ \frac{\partial f}{\partial y}(x, y) \end{bmatrix} \bullet \begin{bmatrix} dx \\ dy \end{bmatrix} \rangle$$

This is the linear part of the microscope approximation. It is linear in the perturbation $d\mathbf{X}$ where $\mathbf{X}$ is fixed.

19.2. Geometric Interpretation of the Total Derivative

The vector form of the microscope approximation helps us to interpret the error term in the manner of the 2-D microscope. The function $f(\mathbf{X})$ is smooth if and only if infinitesimal vector views of its surface graph appear indistinguishable from linear planes, that is, $\varepsilon \approx 0$ in the microscopic approximation:

$$f(\mathbf{X} + \delta \mathbf{X}) = f(X) + \langle \mathbf{G} \bullet \delta \mathbf{X} \rangle + \varepsilon \, |\delta \mathbf{X}|$$

when $|\delta \mathbf{X}| \approx 0$.

If we view the explicit graph $z = f(\mathbf{X})$ with the center of our microscope above $\mathbf{X}$ and make the magnification $1/|\delta\mathbf{X}|$, then the point above $f(\mathbf{X} + \delta\mathbf{X})$ appears to be in the same location as the point above $\delta\mathbf{X}$ on the tangent plane. The tangent plane position is computed by $dz = \nabla f(\mathbf{X}) \bullet \delta\mathbf{X}$, while the distance on the nonlinear surface from the (dx, dy, dz) origin is computed by $f(\mathbf{X} + \delta\mathbf{X}) - f(\mathbf{X})$. The difference between these two is $f(\mathbf{X} + \delta\mathbf{X}) - f(\mathbf{X}) - \nabla f(\mathbf{X}) \bullet \delta\mathbf{X} = \varepsilon |\delta\mathbf{X}|$. Magnification by $1/|\delta\mathbf{X}|$ means that we see ε, so if $\varepsilon \approx 0$, the plane and surface appear to coincide under magnification.

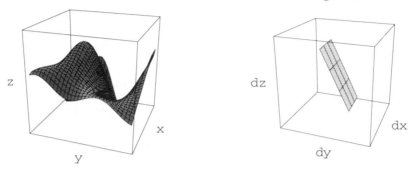

FIGURE 19.4: A Magnified Portion of a Surface

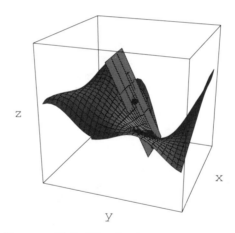

FIGURE 19.5: The Surface and Tangent

It is clear why we call this local linearity - nearby, or locally, things appear linear. We will also formulate the geometric meaning of infinite magnifications of the contour graphs later in this section and in the Microscope NoteBook.

The gradient vector, $\nabla f(\mathbf{X})$ tells us the direction to move in the $\mathbf{X} = \begin{bmatrix} x \\ y \end{bmatrix}$ -plane in order to make f increase fastest. This is only the geometrical interpretation; physically, it can be interpreted as the direction opposite heat flow, as an electrical or magnetic field, and in many other ways. Its vector nature makes it quite useful.

19.3. Partial differentiation examples

The partial derivatives, $\frac{\partial f}{\partial x}$ and $\frac{\partial f}{\partial y}$ are computed using the rules we learned in one variable by treating y as a constant or parameter when differentiating with respect to x and treating

x as a parameter when differentiating with respect to y. Technically, this is an easy extension of what we have learned and the major difficulty that arises is in understanding where the formulas are valid. This difficulty is especially apparent in computations that involve the chain rule.

You should review the rules of differentiation, especially the chain rule and product rule. Now you will have to use the rules for several different letters.

EXAMPLE 19.6. $z = f(x, y) = x^2 + y^2/3$

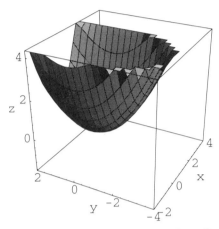

FIGURE 19.7: $z = f(x, y) = x^2 + y^2/3$

Then

$$\frac{\partial f}{\partial x}(x, y) = 2x$$

just treating the $3y^2$ term as a constant when differentiating with respect to x. Also

$$\frac{\partial f}{\partial y}(x, y) = 2y/3 = \frac{2}{3}y$$

treating the x^2 term as a constant when differentiating with respect to y. Putting these computations together, we obtain the total differential

$$dz = \frac{\partial f}{\partial x}\,dx + \frac{\partial f}{\partial y}\,dy = 2x\,dx + \frac{2}{3}y\,dy$$

Now consider the graph near $(1, -1)$ and a magnified view. The total differential for the magnified view is

$$dz = 2\,dx - \frac{2}{3}\,dy$$

FIGURE 19.8: Zooming toward $(1, -1)$ on the Surface $z = x^2 + 3\,y^2$

EXAMPLE 19.9. $z = f(x, y) = x^2 y^3$

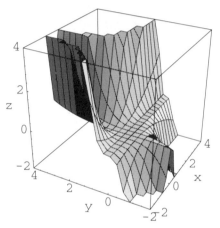

FIGURE 19.10: $z = f(x, y) = x^2 y^3$

Then

$$\frac{\partial f}{\partial x}(x, y) = 2xy^3$$

just treating the y^3 term as a constant when differentiating with respect to x. Also

$$\frac{\partial f}{\partial y}(x, y) = 3x^2 y^2$$

treating the x^2 term as a constant when differentiating with respect to y. Putting these computations together, we obtain the total differential

$$dz = \frac{\partial f}{\partial x}\,dx + \frac{\partial f}{\partial y}\,dy = 2xy^3\,dx + 3x^2 y^2\,dy$$

Now consider the graph near $(1, 1)$ and a magnified view. The total differential for the magnified view is

$$dz = 2\,dx + 3\,dy$$

FIGURE 19.11: Zooming toward $(1,1)$ on the Surface $z = x^2 y^3$

EXAMPLE 19.12. $z = f(x,y) = \text{Cos}(xy)$

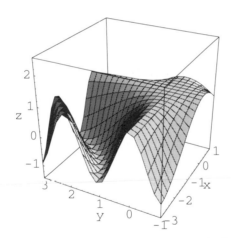

FIGURE 19.13: $z = f(x,y) = \text{Cos}[xy]$

Then

$$\frac{\partial f}{\partial x}(x,y) = -y \cdot \text{Sin}(xy)$$

treating the y term as a constant when differentiating with respect to x. This is the chain rule, $z = \text{Cos}(u)$ and $u = xy$, so $\frac{\partial z}{\partial u} = -\text{Sin}(u)$ and $\frac{\partial u}{\partial x} = y$. The chain rule yields

$$\frac{\partial z}{\partial x} = \frac{\partial z}{\partial u} \cdot \frac{\partial u}{\partial x} = -\text{Sin}(u) \cdot y = -y \cdot \text{Sin}(xy)$$

Also

$$\frac{\partial f}{\partial y}(x,y) = -x \cdot \text{Sin}(xy)$$

treating the x term as a constant when differentiating with respect to y. Putting these computations together, we obtain the total differential

$$dz = \frac{\partial f}{\partial x} \ dx + \frac{\partial f}{\partial y} \ dy = -y \cdot \text{Sin}(xy) \ dx - x \cdot \text{Sin}(xy) \ dy$$

Now consider the graph near $(-1, 1)$ and a magnified view. The total differential for the magnified view is

$$dz = \text{Sin}[1] \ dx - \text{Sin}[1] \ dy$$
$$\approx 0.8415 \ dx - 0.8415 \ dy$$

FIGURE 19.14: Zooming toward $(-1, 1)$ on the Surface $z = \text{Cos}(xy)$

EXAMPLE 19.15. $z = f(x, y) = \sqrt{y^2 - x^3}$

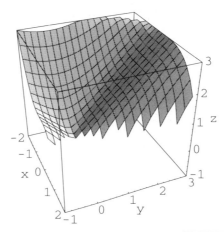

FIGURE 19.16: $z = f(x, y) = \sqrt{y^2 - x^3}$

The function $f(x, y)$ is defined when $y^2 - x^3 \geq 0$. We find the partial derivatives by using the chain rule.

$$z = u^{\frac{1}{2}} \qquad\qquad u = y^2 - x^3$$

$$\frac{\partial z}{\partial u} = \frac{1}{2} u^{-\frac{1}{2}} \qquad\qquad \frac{\partial u}{\partial x} = -3x^2$$

and

$$\frac{\partial u}{\partial y} = 2y$$

so

$$\frac{\partial z}{\partial x} = \frac{\partial z}{\partial u}\frac{\partial u}{\partial x} = \frac{1}{2}u^{-\frac{1}{2}} \cdot -3x^2 \qquad\qquad \frac{\partial z}{\partial y} = \frac{\partial z}{\partial u}\frac{\partial u}{\partial y} = \frac{1}{2}u^{-\frac{1}{2}} \cdot 2y$$

$$\frac{\partial z}{\partial x} = \frac{-3x^2}{2\sqrt{y^2 - x^3}} \qquad\qquad \frac{\partial z}{\partial y} = \frac{y}{\sqrt{y^2 - x^3}}$$

This makes the total differential

$$dz = \frac{\partial f}{\partial x}\,dx + \frac{\partial f}{\partial y}\,dy = \frac{-3x^2}{2\sqrt{y^2 - x^3}}\,dx + \frac{y}{\sqrt{y^2 - x^3}}\,dy$$

but this formula is not valid on the curve $y^2 = x^3$ nor in the region where the original function was undefined, $y^2 < x^3$. This is the region in the first quadrant below the explicit curve $y = x^{3/2}$ plus all of the fourth quadrant. (You should sketch it in the x-y-plane of the 3-D picture.)

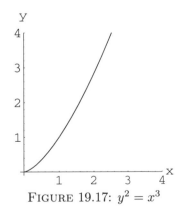

FIGURE 19.17: $y^2 = x^3$

Now consider the graph near $(0, 1)$ and a magnified view. The total differential for the magnified view is

$$dz = 0\ dx + 1\ dy$$

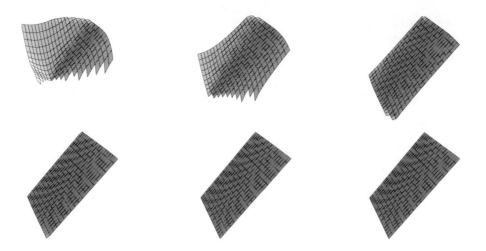

FIGURE 19.18: Zooming toward $(0,1)$ on the Surface $z = \sqrt{y^2 - x^3}$

EXAMPLE 19.19. $z = \mathrm{Tan}[(x^2 + y^2)]$

We may express the function as

$$f(x,y) = \mathrm{Tan}[(x^2 + y^2)] = \frac{\mathrm{Sin}[(x^2 + y^2)]}{\mathrm{Cos}[(x^2 + y^2)]} = \mathrm{Sin}[(x^2 + y^2)] \cdot (\mathrm{Cos}[(x^2 + y^2)])^{-1}$$

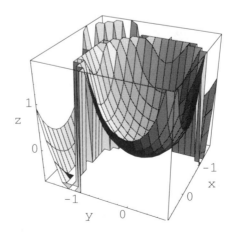

FIGURE 19.20: $z = f(x,y) = \mathrm{Tan}[(x^2 + y^2)]$

This function is undefined if cosine is zero, in other words on the circles $x^2 + y^2 = \pi\frac{1}{2}$, $= \pi\frac{3}{2}$, $= \pi\frac{5}{2}$, ... The partial derivative computations require another use of the chain rule along with the product rule as follows.

$$z = \mathrm{Sin}(u)[\mathrm{Cos}(u)]^{-1} \qquad\qquad u = (x^2 + y^2)$$

so

$$\frac{\partial z}{\partial u} = \mathrm{Cos}(u)[\mathrm{Cos}(u)]^{-1} + \mathrm{Sin}(u) \cdot (-1)[\mathrm{Cos}(u)]^{-2} \cdot (-\mathrm{Sin}(u))$$

$$= \frac{\mathrm{Cos}^2(u) + \mathrm{Sin}^2(u)}{\mathrm{Cos}^2(u)} = \frac{1}{\mathrm{Cos}^2(u)}$$

and

$$\frac{\partial u}{\partial x} = 2x \qquad\qquad \frac{\partial u}{\partial y} = 2y$$

so

$$\frac{\partial z}{\partial x} = \frac{\partial z}{\partial u}\frac{\partial u}{\partial x} = \frac{2x}{\mathrm{Cos}^2(u)} \qquad\qquad \frac{\partial u}{\partial y} = \frac{\partial z}{\partial u}\frac{\partial u}{\partial y} = \frac{2y}{\mathrm{Cos}^2(u)}$$

$$\frac{\partial z}{\partial x} = \frac{2x}{\mathrm{Cos}^2[(x^2 + y^2)]} \qquad\qquad \frac{\partial u}{\partial y} = \frac{2y}{\mathrm{Cos}^2[(x^2 + y^2)]}$$

The total derivative is thus

$$dz = \frac{\partial f}{\partial x}\,dx + \frac{\partial f}{\partial y}\,dy = \frac{2x}{\mathrm{Cos}^2[(x^2 + y^2)]}\,dx + \frac{2y}{\mathrm{Cos}^2[(x^2 + y^2)]}\,dy$$

which is valid for the same inputs as the function, $x^2 + y^2 \neq \pi(\frac{1}{2} + k)$. (Note that part of the computation above shows $\frac{d\,\mathrm{Tan}(u)}{du} = \frac{1}{\mathrm{Cos}^2(u)}$. If you already knew that, you could have shortened the calculation.)

Now consider the graph near $(-1/2, 1/2)$ and a magnified view. The total differential for the magnified view is

$$dz \approx -1.298\,dx + 1.298\,dy$$

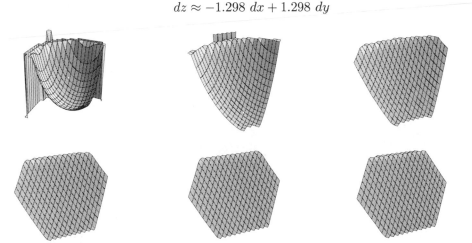

FIGURE 19.21: Zooming toward $(-1/2, 1/2)$ on the Surface $z = \mathrm{Tan}[x^2 + y^2]$

EXERCISE 19.22. *Tangent to a Sphere*
It is geometrically clear that the tangent plane to a sphere is the plane perpendicular to a radius vector ending at the point of tangency. One tangent plane is shown below. We would like you to verify that calculus gives this result. To "see" this, you must use your ability to translate geometry into algebra and algebra into geometry.

The simplest implicit equation of a sphere of (fixed) radius r is

$$x^2 + y^2 + z^2 = r^2$$

because the vectors $\mathbf{X} = \begin{bmatrix} x \\ y \\ z \end{bmatrix}$ *of length r are those satisfying* $|\mathbf{X}| = r$ *or*

$$\sqrt{x^2 + y^2 + z^2} = r$$

Squaring both sides gives the simpler equation.

The vector $\mathbf{X}$ *is the radius vector pointing from the origin to a point on the sphere.*

1) Show that the vector $\mathbf{X} = \begin{bmatrix} 2 \\ 1 \\ 3 \end{bmatrix}$ *lies on the sphere of radius* $\sqrt{14}$ *and sketch.*

A general radius vector $\mathbf{X} = \begin{bmatrix} x \\ y \\ z \end{bmatrix}$ *may be considered as fixed somewhere on the sphere, while we consider other vectors in the parallel local (dx, dy, dz) coordinate system centered at the tip of* $\mathbf{X}$*. We might call an unknown vector in the local system* $d\mathbf{X} = \begin{bmatrix} dx \\ dy \\ dz \end{bmatrix}$.

2) Sketch the (dx, dy, dz) axes at the tip of your radius vector from part (1) above.
3) What is the geo-algebraic condition for $\mathbf{X}$ *and* $d\mathbf{X}$ *to be perpendicular? Show that this is the equation*

$$x \cdot dx + y \cdot dy + z \cdot dz = 0$$

when written in components. In particular, which vectors $d\mathbf{X}$ *are perpendicular to* $\mathbf{X} = \begin{bmatrix} 2 \\ 1 \\ 3 \end{bmatrix}$

on the sphere of radius $\sqrt{14}$*? (Hint:* $d\mathbf{X} = \begin{bmatrix} 2 \\ 2 \\ -2 \end{bmatrix}$ *is perpendicular, but* $\begin{bmatrix} -2 \\ 2 \\ 2 \end{bmatrix}$ *is not.) Add a sketch of the perpendicular illegal vector* $d\mathbf{X} = \begin{bmatrix} 2 \\ 2 \\ -2 \end{bmatrix}$ *starting at the tip of* $\mathbf{X}$*.*

If we want to use explicit equations, we need to solve for z and use the equation

$$z = \sqrt{r^2 - x^2 - y^2}$$

for the top half of the sphere of radius r. (Later, we will use the implicit equation.)
4) Use the chain rule to show that

$$\frac{\partial z}{\partial x} = \frac{-x}{\sqrt{r^2 - x^2 - y^2}} = -\frac{x}{z} \quad \text{and} \quad \frac{\partial z}{\partial x} = \frac{-y}{\sqrt{r^2 - x^2 - y^2}} = -\frac{y}{z}$$

so that the total differential of the explicit function the Surface $z = z[x, y]$ above is

$$dz = -\frac{x}{z} \cdot dx - \frac{y}{z} \cdot dy$$

which is algebraically equivalent to the equation for perpendicularity

$$x \cdot dx + y \cdot dy + z \cdot dz = 0$$

when $z > 0$.

5) The partial derivatives

$$\frac{\partial z}{\partial x} = \frac{-x}{\sqrt{r^2 - x^2 - y^2}} \qquad and \qquad \frac{\partial z}{\partial x} = \frac{-y}{\sqrt{r^2 - x^2 - y^2}}$$

are undefined on the circle $x^2 + y^2 = r^2$. Why? Sketch this circle on your figure.

6) What happens to the tangent to the sphere along the circle $x^2 + y^2 = r^2$? Is there no tangent or is there something else going on mathematically that makes the partial derivatives undefined?

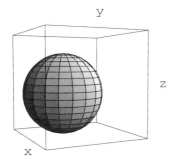
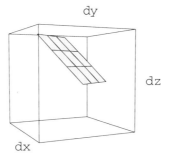

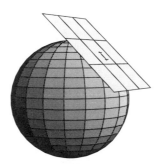

FIGURE 19.23: The Tangent Plane to a Sphere

EXERCISE 19.24. *The functions $z = f(x, y)$ or $z = f(\mathbf{X})$ are given below. In each case, compute $\frac{\partial f}{\partial x}$, $\frac{\partial f}{\partial y}$, $\nabla f(\mathbf{X})$ and the total differential. For which values of the vector $\mathbf{X}$ are your answers valid?*

a) $z = xy^2 + x^3y$

b) $z = \text{Sin}^3(x) + \text{Cos}^4(y)$

c) $z = \text{Sin}(xy)\,\text{Cos}(xy)$

d) $z = \sqrt{1 - x^2 - y^2} \cdot \text{Cos}(2\pi xy)$

e) $z = \sqrt{x^2 + \text{Sin}(y^2)}$

f) $z = \dfrac{1}{\sqrt{x^2 + y^2}}$

g) $z = (2x^2 + 3y^2)e^{-(x^2 + y^2)}$

h) $z = 2xye^{-(x^2/2 + y^2/8)}$

i) $z = \text{ArcTan}[x^2 y^3]$

j) $z = \text{Log}[x^2 + y^4]$

EXERCISE 19.25. *For each of the functions above, find one X or (x, y) point where the graph has a horizontal tangent plane, if there is one. Use your Mathematica homework from Exercise 18.19 to help see where to look for the solution of the appropriate condition on the partial derivatives you computed above.*

*Run the 3-D portions of the **Microscope.ma** NoteBook. It has a complicated built-in function and the microscopic animations in explicit surfaces as well as contour plots.*

*Modify the function in the **Microscope.ma** NoteBook to show microscopic views of some of the functions of the previous exercise. We especially recommend that you magnify $z = \sqrt{x^2 + \text{Sin}(y^2)}$ at $(x, y) = (0, 0)$ and explain what you see.*

19.4. Applications of the Total Differential Approximation

The microscope equation in two variables is

$$f(x + dx, y + dy) = f(x, y) + \frac{\partial f}{\partial x}\, dx + \frac{\partial f}{\partial y}\, dy + \varepsilon \left| \begin{bmatrix} dx \\ dy \end{bmatrix} \right|$$

This section applies this approximation to the estimation of errors in measurement.

19.4.1. Application to Gas Volume.

EXAMPLE 19.26.

This example uses the differential to estimate the error in measuring the amount of gas in a container. We use the ideal gas law

$$PV = nRT$$

where P is the pressure in dynes/cm^2, V is the volume (in cm^3) taken up by the gas, T is the absolute temperature in degrees Kelvin, n is the amount of gas in moles and R is the constant 8.3136×10^7 (for these units). At a fixed temperature $T = 294°K = 21°C$, the amount of gas is

$$n = n(P, V) = \frac{1}{RT} PV = kPV$$

where $k = 4.0913 \times 10^{-11}$.

We store the gas in a cylindrical container of measured radius 10 cm and height 1500 cm at ten times atmospheric pressure. (One atmosphere $= 1.01325 \times 10^6$ dynes/cm^2, so our measured pressure is 1.0133×10^7.) Nominally, the number of moles of gas is

$$n = kPV = 4.0913 \times 10^{-11} \times 1.0133 \times 10^7 \times V$$

where
$$V = \pi r^2 h = 3.1416 \times 10^2 \times 1500$$

making
$$n = 195.36$$

If our measurements of P, r and h are each accurate to at least 1%, what is the possible error in our computation of the amount of gas? We will use the approximation

$$n(P + dP, V + dV) - n(P, V) \simeq dn = \frac{\partial n}{\partial P}dP + \frac{\partial n}{\partial V}dV$$

obtained from the microscope equation. The differential of n is

$$dn = k(VdP + PdV)$$
$$= 4.0913 \times 10^{-11}(4.7124 \times 10^5 dP + 1.0133 \times 10^7 dV)$$

at the measured pressure and volume, while since $V = \pi r^2 h$

$$dV = 2\pi rh dr + \pi r^2 dh = 9.4248 \times 10^4 dr + 314.16 \ dh$$

One percent errors mean, $|dP| \leq 1.0133 \times 10^5$, $|dr| \leq .1$ and $|dh| \leq 15$, so the largest

$$dV = 9.4248 \times 10^4 \times .1 + 314.16 \times 15 = 14137.$$

and

$$dn = 4.0913 \times 10^{-11}(4.7124 \times 10^5 \times 1.0133 \times 10^5 + 1.0133 \times 10^7 \times 1.4137 \times 10^4)$$
$$= 7.8144$$

or a 4 % error.

PROBLEM 19.27. *You are interested in the accuracy of your speedometer and perform the following experiment on an isolated stretch of flat straight Interstate highway. You drive at constant speed with your speedometer reading 60 mph crossing between two consecutive mile markers in 57 seconds. We know "distance equals rate times time" or $D = RT$, so $R = D/T$ when time is in hours and distance is in miles.*
a) Express R in miles per hour as a function of t in seconds and D in miles.
b) Compute the total differential $dR = \frac{\partial R}{\partial t}dt + \frac{\partial R}{\partial D}dD$ symbolically and for the case where $D = 1$ and $t = 60$ corresponding to exactly 60 mph.
c) Use the approximation from part (b) to estimate the error in your speedometer (when you travel 1 mile in 57 seconds, with the speedometer reading 60 mph, so $dt = -3$ and $dD = 0$.) Also use your calculator (or Mathematica) to compute the error exactly (to the accuracy of your calculator).
d) Suppose that there is as much as $\frac{1}{4}$ of one percent error in the placement of the mile markers and as much as 1 percent error in your time measurement of $t = 57$. What is the approximate absolute range of error of your speedometer?
e) What other inaccuracies effect this experiment?
f) What do all these parts together mean???

We can re-write the differential for the amount of gas above in a 'percentage form.' The change in the amount (in moles) is dn and the amount itself is n, so the fraction dn/n is the relative change, or $\frac{dn}{n} \times 100\%$ is the percent change. Similarly, dV/V and dP/P are the relative changes in volume and pressure. Algebra yields a simple formula,

$$dn = \frac{1}{RT}(V\,dP + P\,dV)$$

$$\frac{dn}{n} = \frac{RT}{PV}\frac{1}{RT}(V\,dP + P\,dV)$$

$$\frac{dn}{n} = \frac{dP}{P} + \frac{dV}{V}$$

To be sure of signs, perhaps we should summarize with the error inequality that says the relative error in amount is no more than the sum of the relative errors in pressure and volume,

$$|\frac{dn}{n}| \le |\frac{dP}{P}| + |\frac{dV}{V}|$$

The relative volume error becomes

$$dV = 2\pi r\,,h\,dr + \pi r^2\,dh$$

$$\frac{dV}{V} = \frac{2\pi r h}{\pi r^2 h}\,dr + \frac{\pi r^2}{\pi r^2 h}\,dh$$

$$\frac{dV}{V} = 2\frac{dr}{r} + \frac{dh}{h}$$

and we can see our 4% error in the formula

$$\frac{dn}{n} = \frac{dP}{P} + \frac{dV}{V}$$

$$\frac{dn}{n} = \frac{dP}{P} + 2\frac{dr}{r} + \frac{dh}{h}$$

when each single relative error can be as much as 1% and of any sign.

EXERCISE 19.28. *Small oscillations of a pendulum may be approximated by a linear differential equation, yielding the equation for the time period of a complete oscillation, T, in terms of the length, L, and the gravitational constant, g.*

$$T = 2\pi\sqrt{\frac{L}{g}}$$

Use this formula to find a relative error formula between these quantities,

$$\frac{dT}{T} = \frac{1}{2}\left(\frac{dL}{L} - \frac{dg}{g}\right)$$

(There is a project on the real nonlinear pendulum in the Scientific Projects where you can test the accuracy of these formulas.)

We use a pendulum to measure the gravitational constant on the moon. A 1.0 meter pendulum swings through a full oscillation in 5.0 seconds. What is the nominal value of g?

If the length and time period are both accurate to two significant figures, how accurate is your measurement of the moon's gravitational constant?

19.5. The Meaning of the Gradient Vector

Now we turn our attention to a direct interpretation of the gradient vector $\nabla f(X)$. The first idea is to continue to use the approximation

$$f(\mathbf{X} + d\mathbf{X}) - f(\mathbf{X}) \simeq dz = \langle \nabla f(\mathbf{X}) \bullet d\mathbf{X} \rangle$$

but now to consider what happens as we vary the perturbation vector $d\mathbf{X}$. We see the gradient arising directly in the study of heat flow. This kind of use of gradient vectors arises in many parts of science.

Since we are studying the nonlinear function $z = f(\mathbf{X})$ by examining the small-scale behavior of its "best local linear approximation," we first review the linear facts that we learned in Exercise 18.20. The reader is urged to sketch a rough figure illustrating both the explicit graph and contour graph of a linear function and showing the results mentioned in the next theorem. The contour graph can lie in the (dx, dy) - plane below the explicit plane to aid in the descriptive nature of your sketch. Show the 2-D gradient vector $\mathbf{G}$ on your figure.

THEOREM 19.29. *Summary of the Linear Gradient*
The linear equation

$$dz = A\ dx + B\ dy$$

has a plane explicit graph in local $\begin{bmatrix} dx \\ dy \\ dz \end{bmatrix}$ *- coordinates through* $\begin{bmatrix} dx \\ dy \\ dz \end{bmatrix} = \begin{bmatrix} 0 \\ 0 \\ 0 \end{bmatrix}$ *and perpendicular*

to $\begin{bmatrix} -A \\ -B \\ +1 \end{bmatrix}$. *This same equation has a contour graph consisting of lines in local* $\begin{bmatrix} dx \\ dy \end{bmatrix}$ *-*

coordinates that are perpendicular to the gradient vector $\mathbf{G} = \begin{bmatrix} A \\ B \end{bmatrix}$. $\mathbf{G}$ *points towards lines with higher values of $dz = k$. The rate of change of dz with respect to linear movement in this direction is* $|\mathbf{G}| = \sqrt{A^2 + B^2}$.

The contour lines are spaced so that the (dx, dy)-distance between the lines for $dz = k_1$ and $dz = k_2$ is

$$\frac{|k_1 - k_2|}{|\mathbf{G}|} = \frac{|k_1 - k_2|}{\sqrt{A^2 + B^2}}$$

If dz changes by 1, the input distance is $1/|\mathbf{G}|$, so output/input $= |\mathbf{G}|$. Big gradient vectors produce close contour lines and fast increase in the gradient direction.

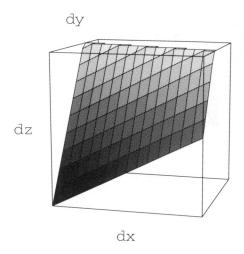

FIGURE 19.30: $z = x + 3y$

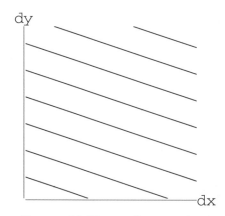

FIGURE 19.31: $x + 3y =$ constant

In the *Mathematica* Microscope NoteBook, we have seen that small magnified parts of either the explicit or contour graph of $z = f(\mathbf{X})$ are indistinguishable from the corresponding explicit or contour graph of the associated linear approximation $dz = \nabla f(\mathbf{X}) \bullet d\mathbf{X}$. The result above when applied to the total differential equation says that $\nabla f(\mathbf{X})$ points in the direction of fastest increase on the contour graph.

Another way to see this is the formula for the angle between two vectors.

$$\text{Cos}(\theta) = \frac{\mathbf{A} \bullet \mathbf{B}}{|\mathbf{A}| \cdot |\mathbf{B}|} \qquad \text{or} \qquad \mathbf{A} \bullet \mathbf{B} = |\mathbf{A}| \cdot |\mathbf{B}|\, \text{Cos}(\theta)$$

We think of moving in various directions $d\mathbf{X}$, but all of the same magnitude, $|d\mathbf{X}|$. In the linear case these perturbations could all be unit size, but they must be small to treat $f(\mathbf{X} + d\mathbf{X}) - f(\mathbf{X})$ as near dz relative to $|d\mathbf{X}|$. It is clear that $\text{Cos}(\theta)$ lies between -1 and +1, while $\nabla f(\mathbf{X})$ and $|d\mathbf{X}|$ are fixed, so

$$\nabla f(\mathbf{X}) \bullet d\mathbf{X} = |\nabla f(\mathbf{X})| \cdot |d\mathbf{X}| \, \mathrm{Cos}(\theta)$$

is largest when $\theta = 0$ [Cos(0) = 1], the dot product is zero when $\theta = \frac{\pi}{2}$ or $\theta = -\frac{\pi}{2}$, and the dot product is smallest when $\theta = \pi$ [Cos(π) = −1]. This has a simple meaning: Locally near X, $f(\mathbf{X})$ increases fastest in the direction of $\nabla f(\mathbf{X})$, is unchanged (up to linear approximation) in the directions perpendicular to $\nabla f(\mathbf{X})$, and decreases fastest in the direction $-\nabla f(\mathbf{X})$.

EXERCISE 19.32.

1) *Suppose we make a small change in* $\mathbf{X}$ *parallel to* $\mathbf{G} = \nabla f(\mathbf{X})$. *Show that the rate of change of* $z = f(\mathbf{X})$ *is approximately* $|\nabla f(\mathbf{X})|$.

2) *Suppose we make a small change in* $\mathbf{X}$ *parallel to* $\begin{bmatrix} -\frac{\partial f}{\partial y}(x,y) \\ \frac{\partial f}{\partial x}(x,y) \end{bmatrix}$. *Show that the rate of change of* $z = f(\mathbf{X})$ *is approximately zero.*

19.5.1. Heat Flow and the Gradient ∇f of a nonlinear $f(\mathbf{X})$.

EXAMPLE 19.33.

Heat flows from hot to cold, not just hot to cold, but hottest to coldest. This means it moves in the direction that makes the temperature function decrease at the fastest rate. This is exactly the description of the gradient we just gave - heat moves opposite the gradient.

Let $T(x,y)$ be the temperature at a point in the (x,y)-plane. Heat will flow in the direction of the vector $-\nabla T(\mathbf{X})$.

For example, suppose we have a heated square plate described by coordinates $0 \le x \le 1$ and $0 \le y \le 1$. We heat the plate so that the initial temperature distribution is given by

$$T(x,y) = xy(1-x)(1-y)$$

Notice that $T = 0$ all around the edge of the square plate and $T > 0$ inside the plate. Suppose we hold the edges at zero temperature. Heat will flow toward the edges to cool the interior. It flows in the direction of $-\nabla T(\mathbf{X})$. At $(x,y) = (\frac{1}{4}, \frac{1}{3})$ we have

$$\nabla T(\mathbf{X}) = \begin{bmatrix} \frac{\partial T}{\partial x} \\ \frac{\partial T}{\partial y} \end{bmatrix} = \begin{bmatrix} (1-2x)y(1-y) \\ (1-2y)x(1-x) \end{bmatrix}$$

$$\nabla T(\begin{bmatrix} \frac{1}{4} \\ \frac{1}{3} \end{bmatrix}) = \begin{bmatrix} \frac{1}{9} \\ \frac{1}{16} \end{bmatrix}$$

so heat flows in the opposite direction, $-\begin{bmatrix} \frac{1}{9} \\ \frac{1}{16} \end{bmatrix}$. We know that a microscopic view of the contour graph is the same as the contour graph of the linear equation

$$\frac{dx}{9} + \frac{dy}{16} = \text{constant}$$

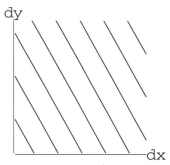

FIGURE 19.34: View at $(\frac{1}{4}, \frac{1}{3})$

If we want this direction as an angle, rather than a vector, we could use the formula

$$\mathrm{Tan}(\theta) = \frac{y}{x}$$

so $\theta = \mathrm{ArcTan}(9/16) - \pi \approx -2.6292$(radians $\approx -151°$). Actually, a better way to give a direction is with a unit length vector in the same direction. Our gradient has length $\sqrt{\frac{1}{9}^2 + \frac{1}{16}^2} = \frac{\sqrt{337}}{144}$. The vector

$$\frac{-144}{\sqrt{337}} \begin{bmatrix} \frac{1}{9} \\ \frac{1}{16} \end{bmatrix} = \begin{bmatrix} \mathrm{Cos}(\theta) \\ \mathrm{Sin}(\theta) \end{bmatrix}$$

is the unit vector for the same angle θ. This is "better" in the sense that it generalizes more easily to three (and more) dimensions. Any of the answers is sufficient, the vector, the unit vector in the same direction, or the angle θ. A contour plot of this temperature distribution is as follows. The contours here are points of equal temperature (isotherms).

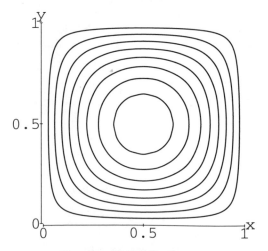

FIGURE 19.35: Isotherms

Y

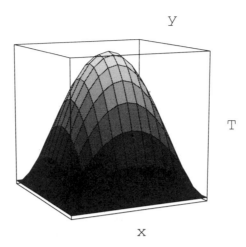

T

X

FIGURE 19.36: Explicit Graph of T

It is natural to ask now for the equations of the curves along which heat flows. We have the increments toward higher temperature proportional to the gradient,

$$dx = k\,y(1-y)(1-2\,x)$$
$$dy = k\,x(1-x)(1-2\,y)$$

so the slope of the low to high temperature curves is

$$\frac{dy}{dx} = \frac{x(1-x)(1-2\,y)}{y(1-y)(1-2\,x)}$$

We separate like variables and obtain

$$\frac{y(1-y)}{1-2\,y}\,dy = \frac{x(1-x)}{1-2\,x}\,dx$$

We can find antiderivatives of both sides by using long division and standard tricks from Chapter 12. However, *Mathematica* gives us

$$\int \frac{y(1-y)}{1-2\,y}\,dy = \int \frac{x(1-x)}{1-2\,x}\,dx$$
$$\frac{y^2}{4} - \frac{y}{4} - \frac{1}{8}\,\mathrm{Log}[2\,y-1] = \frac{x^2}{4} - \frac{x}{4} - \frac{1}{8}\,\mathrm{Log}[2\,x-1] + c$$

for an unknown constant c. We don't know a simple way to solve for y, but we can simply make the contour plot of

$$\frac{y^2}{4} - \frac{y}{4} - \frac{1}{8}\,\mathrm{Log}[2\,y-1] - (\frac{x^2}{4} - \frac{x}{4} - \frac{1}{8}\,\mathrm{Log}[2\,x-1])$$

being careful to avoid the converging flow lines at $(x,y) = (0.5, 0.5)$. This plot shows the (x,y)-curves at various values of the constant c. The graph of the flow lines and the graph of the isotherms are shown next.

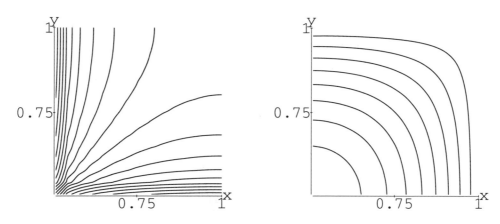

FIGURE 19.37: Flow lines and Isotherms

EXERCISE 19.38. *Integration Review*
Show that $\int \frac{x(1-x)}{1-2x} \, dx = \frac{x^2}{4} - \frac{x}{4} - \frac{1}{8}\text{Log}[2x-1] + c$ *with the following hints.*

    Use long division to show that $\frac{x(1-x)}{1-2x} = \frac{x}{2} - \frac{1}{4} - \frac{1}{4(2x-1)}$.

    Use a change of variables, $u = 4(2x-1)$, $du = 8\,dx$ *to show that*

$$\int \frac{1}{4(2x-1)} \, dx = \frac{1}{8}\text{Log}[2x-1]$$

We conclude the section with an exercise about following the steepest path down the mountain.

PROBLEM 19.39. *The Perfecto Ski Bowl has perfectly manicured slopes in the shape*

$$a = x^2 + 2y^2$$

where a is the altitude, x measures east and y measures north for $0 \le x, y \le 10$, all measured in leagues.

a) Sketch a contour map of Perfecto and sketch the path of a downhill racer who starts at $(x, y) = (5, 10)$ always pointing her skis straight down the steepest direction. (Draw some gradient vectors and sketch in a path. Check your work with a Mathematica ContourPlot.)

b) Write the specific vector equation

$$\begin{bmatrix} dx \\ dy \end{bmatrix} = -k\nabla a$$

for some unknown constant k and the specific gradient of a. What does this equation represent?

c) Solve the equation above for the unknown constant k obtaining $-k = \frac{dx}{2x} = \frac{dy}{4y}$, so

$$\frac{dy}{dx} = \frac{2y}{x}$$

d) Show that the curves $y = cx^2$ satisfy the differential equation in part c and sketch these curves on your contour map for various values of c. (Mathematica could make a ContourPlot of x^2/y or y/x^2, but the singularities cause technical problems.) Which value of c passes thru $(5,10)$ in particular?

19.6. Review Exercises

19.6.1. Inside the Microscope. Calculus lets us "see" inside a powerful microscope without actually magnifying the nonlinear graph. We know that a smooth function "looks" like its tangent at high magnification and that rules of calculus allow us to compute the equation of the tangent. In several variables, we can view the tangent as a plane touching an explicit surface or as a linear contour map approximating a small piece of a contour graph.

EXERCISE 19.40.

1) *Sketch a set of (x, y, z) axes and plot the point* $P(1, 1, 2) = \mathbf{P} = \begin{bmatrix} 1 \\ 1 \\ 2 \end{bmatrix}$. *Let x, y, and z run from -1 to 3.*

2) *The point* $\mathbf{P}$ *lies on the explicit surface $z = x^2 + y^3$. Verify this.*

3) *Add a pair of (dx, dy, dz)-axes centered at the the (x, y, z)-point $P(1, 1, 2)$. How are these axes related to the (x, y, z) axes?*

4) *Use rules of calculus to show that*

$$z = x^2 + y^3 \qquad \Rightarrow \qquad dz = 2x \, dx + 3y^2 \, dy$$

5) *Substitute $(x, y) = (1, 1)$ into your differential to show that*

$$dz = 2 \, dx + 3 \, dy$$

at the (x, y, z)-point $(1, 1, 2)$ or (dx, dy, dz)-point $(0, 0, 0)$.

6) *Plot the explicit plane $dz = 2 \, dx + 3 \, dy$ on your (dx, dy, dz)-axes.*

7) *What would you see if you looked at the graph of $z = x^2 + y^3$ under a very powerful microscope?*

8) *Use the Mathematica NoteBook **Microscope.ma** to plot the function, its differential and make an animation of a microscope zooming in on the explicit graph at the (x, y, z)-point $(1, 1, 2)$.*

9) *Explain how the Differential cell of the **Microscope.ma** NoteBook is actually solving parts (2) - (7) of this exercise.*

10) *Repeat this whole exercise using contour plots.*

The next problem reviews the three ways to define curves and the associated ways to define their tangents.

EXERCISE 19.41. *Implicit, Explicit and Parametric Tangents*
Consider the same ellipse shown below and given by the three types of equations:
Implicit:

$$\left(\frac{x}{3}\right)^2 + \left(\frac{y}{4}\right)^2 = 1$$

Explicit:

$$y = 4\sqrt{1 - \left(\frac{x}{3}\right)^2} \qquad or \qquad y = -4\sqrt{1 - \left(\frac{x}{3}\right)^2}$$

Parametric:

$$x = 3 \, \mathrm{Cos}[t]$$
$$y = 4 \, \mathrm{Sin}[t]$$

1) Calculate the differential for the ellipse in each of the forms.
Implicit:

$$? \, dx + ? \, dy = 0$$

Explicit:

$$dy = ? \, dx \qquad or \qquad dy = ? \, dx$$

Parametric:

$$dx = ? \, dt$$
$$dy = ? \, dt$$

2) Use each of the forms to graph the tangent line to the ellipse at the point where $x = 3/2$ and $y = -2\sqrt{3}$, $(t = -\pi/6)$.

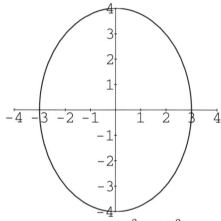

FIGURE 19.42: $\left(\frac{x}{3}\right)^2 + \left(\frac{y}{4}\right)^2 = 1$

19.6.2. Approximation by Differentials. The next problem concerns the approximation of a function of two variables and it uses the formulas

$$\frac{de^u}{du} = e^u \qquad \& \qquad \frac{d \, \text{Log}[v]}{dv} = \frac{1}{v}$$

To differentiate $z = x^y$, we first re-write it using $x = e^{\text{Log}[x]}$, so $z = [e^{\text{Log}[x]}]^y = e^{y \, \text{Log}[x]}$. Now use the chain rule.

EXERCISE 19.43. *Continuity of Exponentiation*
We are interested in the accuracy of the computation of e^π when we use the rational approximations $e \approx 2.718$ and $\pi \approx 3.142$. We let $x + dx = e$ and $y + dy = \pi$. The decimal approximations are x and y. We know $|dx| < 0.001$ and $|dy| < 0.001$. Let

$$z = x^y$$

and compute the total differential $dz = ?$. Use this to estimate the maximum error in the computation of e^π as $2.718^{3.142}$?

19.6.3. Basic Vectors.

EXERCISE 19.44. *Make your own 'lexicon' of geometric-algebraic translations or use the one you made in Chapter 17. This should be your personal list of fundamental facts that you refer to when translating between algebra and geometry. Test the usefulness of your lexicon by consciously using it when you solve the review problems.*

Here is an exercise to use to test your lexicon.

EXERCISE 19.45. *Lines and Planes*

1) A line through the point $\begin{bmatrix} 1 \\ 2 \\ 3 \end{bmatrix}$ *intersects the plane* $3x + 2y + z = 10$ *at right angles.*

Give equations for this line.

2) At what point does this line intersect the plane? Give your reason.

3) Is the vector $\begin{bmatrix} 1 \\ 2 \\ 3 \end{bmatrix}$ *perpendicular to the line? Give your reason.*

4) Is the vector $\begin{bmatrix} 1 \\ 2 \\ 3 \end{bmatrix}$ *perpendicular to the plane? Give your reason.*

Now use calculus and geometry together.

EXERCISE 19.46. *Curves and Surfaces*

1) The curve given parametrically by

$$x = t$$
$$\mathbf{X}(t): \qquad y = 2\sqrt{t}$$
$$z = e^{-t}$$

intersects the explicit surface

$$z = e^{x-y} \operatorname{Cos}[\pi x \, y]$$

at the point $\mathbf{P} = P(1, 2, 1/e)$. *Verify this.*

2) The velocity vector tangent to the curve at the time of intersection is $\begin{bmatrix} 1 \\ 1 \\ -1/e \end{bmatrix}$. *Verify this.*

3) The equation of the plane tangent to the surface at the point of intersection is $dz = \frac{1}{e}\,dx - \frac{1}{e}\,dy$. *Verify this and show that the vector* $\begin{bmatrix} 1/e \\ -1/e \\ -1 \end{bmatrix}$ *is perpendicular to the surface at this point.*

4) Find the angle that the curve makes with the surface at this point of intersection.

Here is a more dynamic test for your lexicon.

EXERCISE 19.47. *Skeet Shooting*
A new skeet range opens near your house with helium filled, gas propelled clay targets that travel with constant velocity vectors (along straight lines with fixed speed). You decide to test your skill and stand at the origin of a coordinate system which has x pointing south, y pointing east and z pointing up. Distances are measured in feet and times in seconds. Draw figures for all the following descriptions and questions.

(1) *A target thrower is located at the tip of the vector* $\mathbf{T} = \begin{bmatrix} -20 \\ 15 \\ 10 \end{bmatrix}$. *How far away is it from you? Label the distance on your figure.*

(2) *The target velocity vector is* $\mathbf{V} = \begin{bmatrix} 8 \\ 1 \\ -1 \end{bmatrix}$. *How fast is it going?*

(3) *At what time t does the target hit the ground* $(\{\mathbf{X} = \begin{bmatrix} x \\ y \\ z \end{bmatrix} : z = 0\})$?

(4) *If your eye is 5 feet above the ground, what is the angle between the horizontal and your line of sight as a function of time?*

The next exercise combines calculus and geometry to follow a curved trajectory.

EXERCISE 19.48. *A Curved Trajectory*
An old-fashioned 20th century clay target with a chip on one side comes out of the thrower and traverses the path given by

$$\mathbf{X}(t): \quad \begin{aligned} x &= \mathrm{Log}[1 + 8t] - 20 \\ y &= \mathrm{Sin}[t] + 15 \\ z &= 10 - \frac{5}{2} t^2 \end{aligned}$$

1) When does this target hit the ground?

2) How far from you is it when it hits the ground?

3) Show that the target leaves the thrower at $t = 0$ with velocity $\begin{bmatrix} 8 \\ 1 \\ 0 \end{bmatrix}$

4) How fast is it going when it hits the ground?

Now use projection of vectors to compute your tendency to slide down a slippery slope.

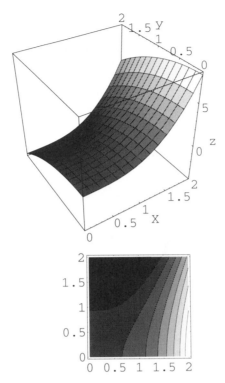

FIGURE 19.49: $z = x^3 - y^2$ and $x^3 - y^2 = $ constants

EXERCISE 19.50. *In this problem we apply a force to a slippery slope and want to know how much acts perpendicular to the slope and how much acts tangent to it. The surface has equation* $z = x^3 - y^2$. *Draw figures to go with each part of the problem.*

(1) *What 2-D vector* **G** *is perpendicular to the level curve* $0 = x^3 - y^2$ *at the point*
$$\mathbf{X} = \begin{bmatrix} x \\ y \end{bmatrix} = \begin{bmatrix} 1 \\ 1 \end{bmatrix}?$$

(2) *What 3-D vector* **N** *with upward z-component is perpendicular to the surface* $z = x^3 - y^2$ *at the point* $\mathbf{X} = \begin{bmatrix} x \\ y \\ z \end{bmatrix} = \begin{bmatrix} 1 \\ 1 \\ 0 \end{bmatrix}?$

(3) *A force* $\mathbf{F} = \begin{bmatrix} 1 \\ 2 \\ 3 \end{bmatrix}$ *is pushing up on the surface at the point* $\begin{bmatrix} x \\ y \\ z \end{bmatrix} = \begin{bmatrix} 1 \\ 1 \\ 0 \end{bmatrix}$. *What is the vector projection of* **F** *in the direction of the normal to the surface at this point? Call this normal force vector* $\mathbf{F}_n$.

(4) *Describe geometrically how the force* $\mathbf{F} - \mathbf{F}_n$ *acts at the point* $\begin{bmatrix} x \\ y \\ z \end{bmatrix} = \begin{bmatrix} 1 \\ 1 \\ 0 \end{bmatrix}$.

CHAPTER 20

Maxima and Minima in Several Variables

The differential approximation or "microscope equation" in two variables is

$$f(x + dx, y + dy) = f(x, y) + \frac{\partial f}{\partial x} dx + \frac{\partial f}{\partial y} dy + \varepsilon \left| \begin{bmatrix} dx \\ dy \end{bmatrix} \right|$$

This chapter applies the approximation to optimization in 2 independent variables. The main idea again is to replace the change in the nonlinear function corresponding to a perturbation vector $d\mathbf{X}$ by its differential

$$f(x + dx, y + dy) - f(x, y) \simeq \frac{\partial f}{\partial x} dx + \frac{\partial f}{\partial y} dy = dz$$

The differential is linear in the perturbation variables $d\mathbf{X} = (dx, dy)$. That is, when we hold $\mathbf{X}$ fixed, the partials are constant and dz has the form $dz = A dx + B dy$. Replacing the nonlinear change by the linear differential is equivalent to "looking in a powerful microscope."

When we look in a microscope, if the linear approximation has a nondegenerate graph and we can move in any direction, at least a little, then we can increase or decrease the linear function. In other words, we can not be at either a maximum or a minimum. This is the proof of the interior critical point theorem, stated below. It is an "interior" result because if we are at a boundary where we can not move in some directions, then we might not be able to make the quantity larger or smaller.

The critical point result can also be understood in terms of contour graphs. The contour graph of dz consists of the family of lines in (dx, dy)-space obtained by setting dz equal to various constants. That linear contour graph is a local approximation to the nonlinear contour graph of $z = f(\mathbf{X})$. We saw this in the **Microscope.ma** NoteBook.

If k, A and B are constants, when does the equation

$$k = A dx + B dy$$

define a line in (dx, dy) - space? You might say, 'It is always a line perpendicular to the vector $\mathbf{G} = \begin{bmatrix} A \\ B \end{bmatrix}$ and a distance $\frac{|k|}{|\mathbf{G}|}$ from the origin.' This is true unless $A = B = 0$. Vector $\mathbf{G} = 0$ is the only "bad" case. [Note that the cases where one of $A = 0$ or $B = 0$ correspond to a vertical line $dx = k/A$ $(A \neq 0, B = 0)$ or a horizontal line $dy = k/B$ $(B \neq 0, A = 0)$]. In one sense, $\mathbf{G} = 0$ is the most important case, so we deal with it first. Then we deal with the local picture of the contour graph in the case where $\nabla f \neq 0$.

20.1. Zero Gradients and Horizontal Tangent Planes

The case $A = B = 0$ corresponds to a horizontal plane for the explicit graph of $dz = A dx + B dy$. A horizontal plane has only one contour level. If we were to plot its contour map, it would either be blank, or all black, depending on whether or not we plotted that contour level. This is highly degenerate, but very interesting to our local analysis.

When $A = \frac{\partial f}{\partial x}(\mathbf{X})$ and $B = \frac{\partial f}{\partial y}(\mathbf{X})$ both are zero, the tangent plane to the nonlinear explicit graph $z = f(\mathbf{X})$ is horizontal. It is also a degenerate point which has no contour line on the linear contour graph. Algebraically it is a point where the best local linear approximation is zero - the function doesn't change "to first order."

If the vector $\mathbf{G}$ is given by

$$\mathbf{G} = \begin{bmatrix} A \\ B \end{bmatrix} = \nabla f(\mathbf{X}) = \begin{bmatrix} \frac{\partial f}{\partial x}(x,y) \\ \frac{\partial f}{\partial y}(x,y) \end{bmatrix} \neq 0$$

we know that f increases fastest for small perturbations in the direction $d\mathbf{X} = c\nabla f(\mathbf{X})$ $(c > 0)$. We know f decreases fastest in the opposite direction, $d\mathbf{X} = -c\nabla f(\mathbf{X})$ $(c > 0)$. Therefore, if we are allowed to change $\mathbf{X}$ in these directions (by at least a small amount), the point $\mathbf{X}$ where $\nabla f(\mathbf{X}) \neq 0$ cannot be a max nor min. This is the gradient vector proof of the next theorem.

THEOREM 20.1. *Interior Critical Points in 2 - D*
If $z = f(x,y)$ is defined on a plane set of (x,y)-points D and $f(x_0, y_0)$ is an extremum of f for D where there is at least a tiny rectangle about (x_0, y_0) inside D, $\{(x,y) : |x - x_0| < \eta, |y - y_0| < \theta\} \subset D$, then

$$\nabla f(X_0) = \begin{bmatrix} \frac{\partial f}{\partial x}(x_0, y_0) \\ \frac{\partial f}{\partial y}(x_0, y_0) \end{bmatrix} = \begin{bmatrix} 0 \\ 0 \end{bmatrix}$$

(We use the logical contrapositive, $[H \Rightarrow C] \equiv [\neg C \Rightarrow \neg H]$ in our statement of the result. We have shown that if $\nabla f(X) \neq 0$ $(\neg C)$ and we can move in any direction, then we are not at a max nor min $(\neg H)$. This is logically equivalent to the implication $H \Rightarrow C$, which is the way we state the theorem. See Section 11.1)

This is a negative result because it only rules out interior points $\mathbf{X}$ where $\nabla f(\mathbf{X}) \neq 0$ from being maxima or minima. It does not say where maxima or minima are located, nor even whether there are any. This is completely analogous to the critical point theorem in 1 variable, but now the interior point condition is more difficult and the zero condition is a simultaneous solution to a pair of equations.

Here are some examples and their graphs $z = x^2 \pm y^2$, $z = 3x - x^3 - 3xy^2$, $z = 6xy^2 - 2x^3 - 3x^4$. Notice that the horizontal tangents need not be maxima nor minima.

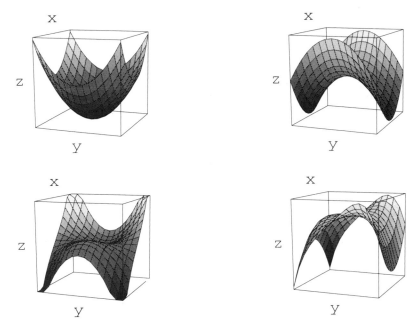

The next result is a "positive" result; it says a max and min exist. The critical point theorem tells us where not to look, but the Extreme Value Theorem says that the thing we are looking for is there. Hence we look at the places that are left, critical points and non-interior points.

THEOREM 20.2. *Extreme Values in 2 - D*
If $z = f(\mathbf{X})$ is a smooth function defined on a "compact" region D of the (x, y)-plane, then there are vectors $\mathbf{X}_m$ and $\mathbf{X}_M$ so that for all $\mathbf{X}$ in D

$$f(\mathbf{X}_m) \leq f(\mathbf{X}) \leq f(\mathbf{X}_M)$$

The big question is: What does "compact" mean for a region in the plane? In one variable, the analogous hypothesis was that we were seeking a max and min over an interval that was bounded (or did not extend to infinity in either direction) and that contained its end points. In other words $[a, b] = \{x : a \leq x \leq b\}$ for two real numbers a and b. We need two conditions in order to have a compact region in the plane. First, we need a bound on the total size of the vectors $\mathbf{X}$ in the region. Second, we need to 'include the boundary' of the region. Technically, this means that if $\{\mathbf{X}_k\}$ is a convergent sequence of vectors in the region D, then the limit point $\mathbf{X}_\infty = \lim_{k \to \infty} \mathbf{X}_k$ lies in D.

In practice, the problems we consider ask for maxima and minima over sets like triangles, circular sectors, quadrants. These regions are defined by inequalities and their non-interior, or boundary, points are defined by equations. The inside of a triangle together with its edges is compact, because sequences inside tend to points inside or on the edge and because the triangle is 'finite in extent.' In other words, we can find an upper limit on the size of all vectors that lie inside the triangle.

A quadrant is not compact, even with its boundary edges, because we can find arbitrarily large vectors in the quadrant - it extends to infinity in some directions. The function

$f(\mathbf{X}) = |\mathbf{X}|^2 = x^2 + y^2$ is smooth, but has no maximum on any region with no bound on the size of its points.

The plane less a single point is not compact for two reasons. First, the region extends to infinity as you move farther and farther from zero. Second, a sequence of vectors tending toward the excluded point converges, but not to a point in the region. Notice that the function $f(\mathbf{X}) = 1/|\mathbf{X}|$ is smooth, but has no maximum on the plane less the origin.

EXERCISE 20.3. *Sketch the set of points*

$$D = \{\mathbf{X} = \begin{bmatrix} x \\ y \end{bmatrix} : x \le 0, y \ge 0, x^2 + y^2 \le 36\}$$

Explain why it is compact. What is an upper bound on the size of points in D? Why does a convergent sequence of points from D tend to a point already included in D?
   Sketch the set of points

$$D = \{\mathbf{X} = \begin{bmatrix} x \\ y \end{bmatrix} : x > 0, y < 0, x^2 + y^2 > 36\}$$

Explain why it is NOT compact by finding one sequence of points in the region whose lengths tend to infinity and another sequence of points in the region whose limit exists, but is not in the region.

The combined effect of the two theorems is a plan of attack:

20.1.1. Two Variable Optimization on Compact Regions. To find the extrema of a smooth function $z = f(x, y)$ on a closed and bounded or compact subset of the (x, y)-plane:
a) Sketch the region D, finding the interior and boundary.
b) Compute $\frac{\partial f}{\partial x}$ and $\frac{\partial f}{\partial y}$ and then find all simultaneous solutions in x and y to the pair of equations

$$\frac{\partial f}{\partial x}(x, y) = 0 \qquad \text{and} \qquad \frac{\partial f}{\partial y}(x, y) = 0$$

such that (x, y) lies inside D. These are the interior critical points where the gradient vector is zero,

$$\nabla f(\mathbf{X}) = 0$$

c) Find the max and min of $f(\mathbf{X})$ on the boundary of D. (This is often a new 1 variable max-min problem.)
d) Find the largest and smallest value of $f(\mathbf{X})$ amongst the points from (b) and (c).

Parts (b) and (c) of this procedure look pretty scary, but we can use *Mathematica* to help us solve simultaneous equations. The important thing is to have a clear idea of what we want to do, rather than worrying about how we are going to do it.

EXERCISE 20.4. *A Dog Saddle ?*
*Use Mathematica to plot the surface of $z = 4xy(x^2 - y^2)$ for $|x|, |y| \le 2$. You could even use the NoteBook **FlyBySurfaces.ma** Use Mathematica to check your computation of $\partial z/\partial x$ and $\partial z/\partial y$ as well as to solve the simultaneous equations*

$$Solve[\{\partial z/\partial x == 0, \partial z/\partial y == 0\}]$$

and see that the only flat point on the surface is at $(0,0)$.

Where are the highest and lowest points on the dog saddle - just visually?
There are 4 depressions on this 'saddle' for a dog's legs, but would she be comfortable?

We begin studying max-min theory with some simple examples where we do not need *Mathematica*'s help with the equations. The first example can be graphed and approximately maximized by visual inspection. The bothersome negative "volumes" that spoil our graphs are the key to making this optimization a compact region problem.

EXAMPLE 20.5.

A Post Office regulation says that a package must have the sum of the length plus the girth less than or equal to 108 inches. What is the maximum volume for a rectangular box meeting these requirements? The volume of a rectangular solid is $V = lwh$, where l is the length, w is the width, and h is the height (which we will measure in inches). The postal regulation says

$$2(w + h) + l \leq 108$$

and since more is bigger, we will have $2(w + h) + l = 108$ at maximal volume. This means we can eliminate a variable and make this a two dimensional independent variable problem:

$$\text{Maximize } V = wh(108 - 2(w + h))$$

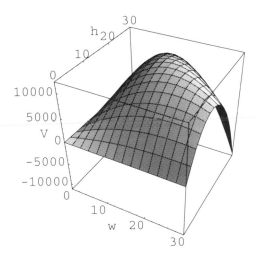

FIGURE 20.6: $V = wh(108 - 2(w + h))$

What is the region of the independent variables (w, h)? Clearly, $w \geq 0$ and $h \geq 0$. If $2(w+h) = 108$, then there is no more room for length, so our compact region is the triangle:

$$D = \{(w, h) : 0 \leq w, 0 \leq h \ \& \ 2(w + h) \leq 108\}$$

A sketch of the line $2(w + h) = 108$ or $w + h = 54$ shows that D is the triangle in the first quadrant of the (w, h)-plane below the line perpendicular to $(1, 1)$ and through the points $(54, 0)$ and $(0, 54)$.

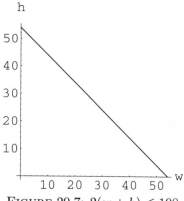

FIGURE 20.7: $2(w + h) \leq 108$

Our complete mathematical problem is:

$$\text{Maximize } [V = wh(108 - 2(w + h)) : 0 \leq w, 0 \leq h \ \& \ 2(w + h) \leq 108]$$

Step a) This is the triangle sketched above. The boundary consists of:

The segment of the w axis, $h = 0$ & $0 \leq w \leq 54 = \frac{108}{2}$

The segment of the h axis, $w = 0$ & $0 \leq h \leq 54 = \frac{108}{2}$

The line segment $w + h = 54$, for $0 \leq w, h \leq 54$

Step b)

$$\frac{\partial V}{\partial w} = h(108 - 4w - 2h) \quad \text{and} \quad \frac{\partial V}{\partial h} = w(108 - 4h - 2w)$$

So we must find all interior solutions to the pair of equations

$$h(108 - 4w - 2h) = 0 \qquad \text{and} \qquad w(108 - 2w - 4h) = 0$$

This is done in the *Mathematica* NoteBook **Maxmin.ma**, but is simple enough to do ourselves. One solution is $w = h = 0$. If $h = 0$ but $w \neq 0$, then the bottom equation gives us the condition

$$(108 - 2w) = 0$$

so $w = 54$. Similarly, if $w = 0$ but $h \neq 0$, then $h = 54$. We have found three solutions: $(w, h) = (0, 0), (54, 0)$ & $(0, 54)$. None of these is interior to D, because they are the vertices of the triangle bounding D.

The final solution (and only interior one) is when $h \neq 0$ and $w \neq 0$, so we can divide the original equations by these variables obtaining the simultaneous linear equations

$$4w + 2h = 108 \qquad \text{and} \qquad 2w + 4h = 108$$

dividing the first by -2, we obtain

$$-2w - h = -54 \qquad \text{and} \qquad 2w + 4h = 108$$

adding, we obtain

$$3h = 54, \quad \text{or} \quad h = 18$$

back substituting, we obtain

$$2w + 4 \times 18 = 108, \text{ so } w = 18$$

and our only interior solution is

$$(w, h) = (18, 18)$$

It is tempting to jump to the correct conclusion that the maximum volume is attained when $w = 18$, $h = 18$ and $l = (108 - 2(18 + 18)) = 36$, so $V = 11664$. But remember that the mathematical theory looks for both the max and the min. Steps (c) and (d) of our max-min procedure are easy in this case and they complete the reasoning.

Step c) The max and min of $f(X)$ on the boundary is easy to compute. On the edge of the triangle, $h = 0$ & $0 \leq w \leq 54 = \frac{108}{2}$, $V = 0$. On the edge of the triangle, $w = 0$ & $0 \leq h \leq 54 = \frac{108}{2}$, $V = 0$ again. On the edge of the triangle, $w + h = 54$, for $0 \leq w, h \leq 54$, $l = 0$, so again $V = 0$. The max and min of zero is zero.

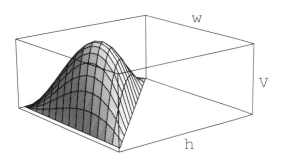

FIGURE 20.8: V plotted only over D

Step d) The two values we must compare for V are $V = 0$ and $V = 11664$. Clearly the max is 11664 and the min is 0.

EXAMPLE 20.9.

Let's change the previous mathematical problem to

$$\text{Maximize } [V = wh(108 - 2(w + h)) : 0 \leq w, 0 \leq h \text{ \& } 2w + h \leq 50]$$

so that the boundary conditions are no longer trivial. The critical point equations

$$h(108 - 4w - 2h) = 0 \qquad \text{and} \qquad w(108 - 2w - 4h) = 0$$

still only have the solutions (0,0), (54,0), (0,54), (18,18). But now only (0,0) lies in D and that point is on the boundary.

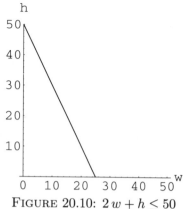

FIGURE 20.10: $2w + h \leq 50$

We also still have $V = 0$ on the w axis and on the h axis, so the boundary we need to check is

$$2w + h = 50 \text{ or } h = 50 - 2w$$

Substituting this in the formula for V we obtain

$$V_b = (50 - 2w)w(108 - 2w - 2(50 - 2w)) = (50 - 2w)w(8 + 2w)$$

along $h = 50 - 2w$, for $\quad 0 \leq w \leq 25$.

Finding extrema on this part of the boundary of D is the mathematical problem

$$\text{Extremize } [V_b = 400w + 84w^2 - 4w^3 \ : \ 0 \leq w \leq 25]$$

At the end points of this 1 variable extremum problem, we have $V_b(0) = 0$ and $V_b(25) = 0$. In the interior of the restricted problem, we have critical points when the derivative,

$$\frac{dV_b}{dw} = 400 + 168w - 12w^2$$

is zero. This only occurs when $w = 7 \pm \sqrt{7^2 + 200/6} \approx 16.0738$ (for $0 \leq w \leq 25$) where $V_b(16.0738) \approx 11520.6$. This is the maximum on the boundary and since there are no critical points inside this new region, this is also the over-all global maximum. The minimum is still zero.

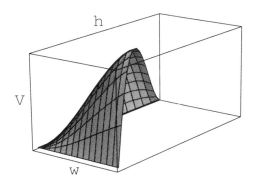

h

V

w

FIGURE 20.11: V plotted over $h + 2\,w \le 50$

EXERCISE 20.12. *Find the max and min*

$$z = x^2 - y^2 - 2x + 2y + 3 \qquad 0 \le x \le 2 \qquad\qquad 0 \le y \le 2x$$
$$z = \operatorname{Sin}[x] + \operatorname{Cos}[y] \qquad 0 \le x \le \pi \qquad\qquad 0 \le y \le \pi$$
$$z = \operatorname{Sin}[x]\operatorname{Cos}[y] \qquad 0 \le x \le \pi \qquad\qquad 0 \le y \le \pi$$
$$z = 4 - x^2 - y^2 \qquad\qquad -\sqrt{1 - x^2} \le y \le \sqrt{1 - x^2}$$

EXAMPLE 20.13.

We might as well flog the poor Post Office example to death by considering the quarter circle problem

$$\text{Extremize } [V = wh(108 - 2w - 2h) \ : \ 0 \le w \,, 0 \le h \ \& \ w^2 + h^2 \le 625]$$

This is the same function, but a new outer boundary, the circle of radius 25 centered at $(0,0)$. Since the vector $(18, 18)$ has length $\sqrt{18^2 + 18^2} \approx 25.456$, there are again no interior critical points. V is still 0 on the axes and the new problem only requires us to extremize on the quarter circle boundary.

$$\text{Extremize } [V = wh(108 - 2w - 2h) \ : \ w^2 + h^2 = 25^2 \ \& \ 0 \le w, h]$$

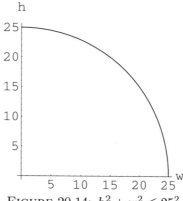

$$\text{FIGURE 20.14: } h^2 + w^2 \leq 25^2$$

Rather than solve for h in terms of w, we will use implicit differentiation to compute the critical points $\frac{dV}{dw} = 0$.

$$V = 108wh - 2w^2h - 2wh^2 \qquad\qquad : \qquad w^2 + h^2 = 25^2$$

$$dV = (108h - 4wh - 2h^2)dw + (108w - 2w^2 - 4wh)dh \qquad : \qquad wdw + hdh = 0$$

because the total differential of V is

$$dV = 108hdw + 108wdh - 4whdw - 2w^2dh - 2h^2dw - 4whdh$$

and the differential of the constraint equation is

$$2wdw + 2hdh = 0$$

Solving the constraint equation for dh gives $dh = -\frac{w}{h}dw$ which we substitute into the expression for dV yielding

$$dV = [(108h - 4wh - 2h^2) - \frac{w}{h}(108w - 2w^2 - 4wh)]dw$$

so

$$\frac{dV}{dw} = \frac{108h^2 - 4wh^2 - 2h^3 - 108w^2 + 2w^3 + 4w^2h}{h}$$

Mathematica finds the simultaneous solutions to the equations

$$108h^2 - 4wh^2 - 2h^3 - 108w^2 + 2w^3 + 4w^2h = 0 \qquad \text{and} \qquad w^2 + h^2 = 625$$

to be $(w, h) \approx (17.6777, 17.6777)$ plus 3 real solutions outside the region and 2 complex solutions. At this critical point solution, we have $V \approx 11652.9$ This is the max on the circular boundary and also on the whole region, because $V = 0$ on the rest of the boundary (including the ends of the quarter circle).

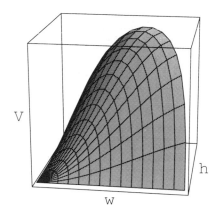

FIGURE 20.15: V plotted over $h^2 + w^2 \leq 25^2$

The moral of the various parts of this example is that the boundary extrema are more difficult in 2 dimensions, in fact, they become difficult 1 dimensional extremization problems. All these variations had compact regions, D.

The quarter circle boundary max could be done with explicit formulas or with parametric equations. The implicit method is the one we want to stress, but you can compare it to the others in the next exercise.

EXERCISE 20.16. *In the Example 20.13 above, we used implicit differentiation to find the critical point on the circular boundary. Here are two alternative methods to find the boundary extrema,*

$$\text{Extremize } [V = wh(108 - 2w - 2h) \ : \ 0 \leq w \,, 0 \leq h \ \& \ w^2 + h^2 = 625]$$

Explicit Equation
  Solve $w^2 + h^2 = 625$ for $h = \sqrt{625 - w^2}$ and substitute this into the formula for V,

$$V = w\sqrt{625 - w^2}(108 - 2w - 2\sqrt{625 - w^2})$$

Now solve the one variable max-min problem

$$\text{Extremize } [V = w\sqrt{625 - w^2}(108 - 2w - 2\sqrt{625 - w^2}) \ : \ 0 \leq w \leq 25]$$

Parametric Equation
  The vector parametric equation for the quarter circle is

$$\begin{bmatrix} w \\ h \end{bmatrix} = \begin{bmatrix} 25\,\text{Cos}[t] \\ 25\,\text{Sin}[t] \end{bmatrix} \qquad \text{for} \ \ 0 \leq t \leq \pi/2$$

Substitute this into the formula for V,

$$V = 25\,\text{Cos}[t]\ 25\,\text{Sin}[t](108 - 2 \cdot 25\,\text{Cos}[t] - 2 \cdot 25\,\text{Sin}[t])$$

Now solve the one variable max-min problem

$$\text{Extremize } [V = 625 \, \text{Cos}[t] \, \text{Sin}[t](108 - 50(\text{Cos}[t] + \text{Sin}[t]) \;:\; 0 \le t \le \frac{\pi}{2}]$$

Which approach is easiest? What else would you have to do if we asked for the max and min on the whole circle $w^2 + h^2 = 625$?

20.2. Implicit Differentiation (Again)

Implicit differentiation often simplifies problems with constraints. It can also simplify direct calculations. This brief section reviews the method in 2 variables and extends it to 3. Implicit differentiation just amounts to treating all variables equally at the differentiation stage.

In Chapter 7 we computed the tangent to a circle by implicit differentiation as in the previous example.

$$w^2 + h^2 = 25^2$$
$$2 \, w \, dw + 2 \, h \, dh = 0$$
$$\text{so} \qquad \frac{dh}{dw} = -\frac{w}{h}$$

The formula $w \, dw + h \, dh = 0$ even has the advantage that we may think of either variable as the independent variable. When $w = 25$ we have $h = 0$. The circle is smooth, but the tangent is vertical, $25 \, dw + 0 \, dh = 0$ or $dw = 0$. This is simply the local variable equation for the dh-axis. The explicit formula $h = \sqrt{25^2 - w^2}$ with derivative $\frac{dh}{dw} = -w/\sqrt{25^2 - w^2}$ is undefined at $w = 25$. We want to make an important observation on the type of equations that we are working with.

The equation $w^2 + h^2 = 25^2$ is an implicit equation in w and h. When we consider w and h fixed and located somewhere on the circle, the equation in dw and dh, $w \, dw + h \, dh = 0$ is an implicit equation for a line. For example, $(20, 15)$ lies on the circle. The implicit equation of its tangent at this point is shown in the next figure.

The implicit equation of the line is easy to interpret geometrically, if we remember that $\mathbf{X}$ and $d\mathbf{X}$ are perpendicular when $\langle \mathbf{X} \bullet d\mathbf{X} \rangle = 0$. The implicit tangent equation becomes the statement that the vectors are perpendicular,

$$w \, dw + h \, dh = \langle \begin{bmatrix} w \\ h \end{bmatrix} \bullet \begin{bmatrix} dw \\ dh \end{bmatrix} \rangle = 0$$

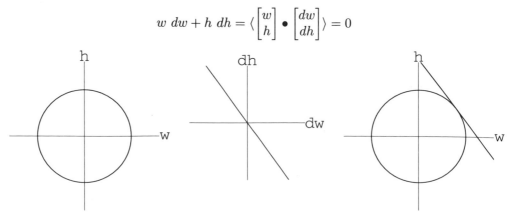

FIGURE 20.17: $20 \, dw + 15 \, dh = 0$

Implicit Curve	Implicit Tangent Line
$w^2 + h^2 = 25^2$	$w\,dw + h\,dh = 0$

EXAMPLE 20.18. *Implicit Tangent to a Sphere*

In the Exercise 19.22 of the last chapter, we calculated the tangent plane to a sphere by explicit differentiation. The implicit equation of the sphere centered at zero with radius r is

$$x^2 + y^2 + z^2 = r^2$$

(For example, we could take $r = \sqrt{14}$, so that $X(2, 1, 3)$ lies on the sphere.) Differentiating both sides with respect to each variable (including z) yields:

$$\frac{\partial(x^2 + y^2 + z^2)}{\partial x} = 2\,x, \qquad \frac{\partial(x^2 + y^2 + z^2)}{\partial y} = 2\,y, \qquad \frac{\partial(x^2 + y^2 + z^2)}{\partial z} = 2\,z$$

and $\frac{\partial r^2}{\partial -} = 0$ since r is constant. Now we write the total differential in all the variables,

$$\frac{\partial(x^2 + y^2 + z^2)}{\partial x}\,dx + \frac{\partial(x^2 + y^2 + z^2)}{\partial y}\,dy + \frac{\partial(x^2 + y^2 + z^2)}{\partial z}\,dz = 0$$

$$2\,x\,dx + 2\,y\,dy + 2\,z\,dz = 0$$

$$x\,dx + y\,dy + z\,dz = 0$$

$$\left\langle \begin{bmatrix} x \\ y \\ z \end{bmatrix} \bullet \begin{bmatrix} dx \\ dy \\ dz \end{bmatrix} \right\rangle = 0$$

which says that $\mathbf{X}$ and $d\mathbf{X}$ are perpendicular.

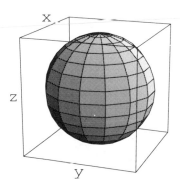

FIGURE 20.19: $x^2 + y^2 + z^2 = r^2$

Implicit Surface	Implicit Tangent Plane
$x^2 + y^2 + z^2 = r^2$	$x\,dx + y\,dy + z\,dz = 0$

EXERCISE 20.20. *Tangent to an Ellipsoid*

The ellipsoid $3\,x^2 + 2\,y^2 + z^2 = 20$ *is shown below in an example. Use implicit differentiation to show that the tangent plane to the ellipsoid at the point* $X(x, y, z)$ *is perpendicular to the vector*

$$\mathbf{N} = \begin{bmatrix} 3\,x \\ 2\,y \\ z \end{bmatrix}$$

Sketch the tangent plane at the points

$$\mathbf{X} = \begin{bmatrix} 1 \\ 1 \\ \sqrt{15} \end{bmatrix} \approx \begin{bmatrix} 1 \\ 1 \\ 3.87 \end{bmatrix} \qquad and \qquad \mathbf{X} = \begin{bmatrix} 2 \\ 2 \\ 0 \end{bmatrix}$$

In general, if we start with an implicit relationship between x, y and z, with a function of all three equal to a constant, c, we differentiate both sides of the equation with respect to all the variables and write the total differential as an equation:

Implicit Surface $\qquad\qquad$ Implicit Tangent Plane

$$f(x, y, z) = c \qquad\qquad \frac{\partial f}{\partial x}\,dx + \frac{\partial f}{\partial y}\,dy + \frac{\partial f}{\partial z}\,dz = 0$$

When we use this equation for the differential, the x, y and z must satisfy the original implicit relation, $f(x, y, z) = c$. We will use this in constrained max-min problems below.

20.3. Extrema Over Noncompact Regions

The next example asks for a minimum of a function that is not defined on a bounded region. The Extreme Value Theorem does not apply, so there are no guarantees. There is a min, but no max. You should be able to explain why there is no maximum area.

EXAMPLE 20.21.

We wish to make a rectangular box with an open top of unit volume (1 m^3). To minimize cost, we wish to minimize the combined surface area of the sum of the three sides and bottom. Let x be the width of the box, y be the length, and z be the height (all in m.) Unit volume means $V = lwh = xyz = 1$, or $z = \frac{1}{xy}$. The area of the bottom is xy, one side xz, another side yz, and there are two of each type of side, thus the total area

$$A = xy + 2xz + 2yz = xy + 2\frac{xz}{xy} + 2\frac{yz}{xy} = xy + \frac{2}{y} + \frac{2}{x}$$

and we have the mathematical problem

$$\text{Minimize } [A = xy + \frac{2}{y} + \frac{2}{x} \; : \; x > 0 \, , \; y > 0]$$

The region for minimization is neither closed nor bounded, so there is no mathematical guarantee that we will find the min. Intuitively, it is clear that there is a min, however, and it is also clear that there is no max. (Can you show this? Remember the math professor who tried farming in Exercise 11.24?)

The partials are

$$\frac{\partial A}{\partial x} = y - \frac{2}{x^2} \qquad\qquad \frac{\partial A}{\partial y} = x - \frac{2}{y^2}$$

and the equations

$$y = \frac{2}{x^2} \qquad\qquad x = \frac{2}{y^2}$$

have the simultaneous solution

$$x = 2^{\frac{1}{3}} = y$$

so

$$A = 2^{\frac{2}{3}} + 2^{\frac{5}{3}} \approx 4.7622$$

EXAMPLE 20.22.

Find the point on the surface $z = x^2 + 2y^2$ that is closest to $(3, 2, 1)$.

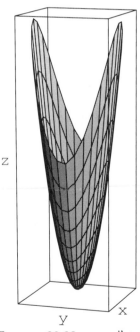

FIGURE 20.23: $z = x^2 + 2y^2$

The vector from $(3, 2, 1)$ to a general (x, y, z) is

$$\begin{bmatrix} x - 3 \\ y - 2 \\ z - 1 \end{bmatrix}$$

and its length is $\sqrt{(x-3)^2 + (y-2)^2 + (z-1)^2}$. When (x, y, z) lies on the surface $z = x^2 + 2y^2$, the distance becomes $\sqrt{(x-3)^2 + (y-2)^2 + (x^2 + 2y^2 - 1)^2}$, so this is the distance from (3,2,1) to a general point on the surface. We are asking to minimize this function over all values of x and y. This is not a bounded region and a sketch of the surface (drawn by *Mathematica*) will show that there is no maximum distance. (Sketch the vector displacement on the figure yourself now.) No maximum also follows easily from the formula: if $x = H$ is infinite (even if y is small), the distance is infinite.

Since the square function is increasing for positive values, it is easiest to minimize the squared distance

$$\text{Mimimize } [D = (x-3)^2 + (y-2)^2 + (x^2 + 2y^2 - 1)^2 \; : \; \text{all } x, y \,]$$

The total differential is

$$dD = 2(x-3)dx + 2(y-2)dy + 2(x^2 + 2y^2 - 1)(2xdx + 4ydy)$$

$$dD = 2[\{(x-3) + 2x(x^2 + 2y^2 - 1)\}dx + \{(y-2) + 4y(x^2 + 2y^2 - 1)\}dy]$$

(as our reader should verify using small steps with the chain rule.) We use *Mathematica* to solve the simultaneous equations

$$(x-3) + 2x(x^2 + 2y^2 - 1) = 0 \qquad \text{and} \qquad (y-2) + 4y(x^2 + 2y^2 - 1) = 0$$

obtaining the solution $x \approx 1.15728$, $y \approx 0.477944$ and 4 complex solutions. This makes the minimum distance

$$\sqrt{(1.155728 - 3)^2 + (0.477944 - 2)^2 + (1.155728^2 + 2 \times 0.477944^2 - 1)^2} \approx 2.51915$$

It is clear that there is no maximum distance. A simple way to show this is to take a huge number for both $x = H$ and $y = H$ (or technically, let H be an infinite hyperreal number.) Then the squared distance is also huge (or infinite).

$$(H-3)^2 + (H-2)^2 + (H^2 + 2H^2 - 1)^2 > (H-3)^2 + (H-3)^2 + (H^2 - 3)^2 > H$$

We can use max-min theory to show that our critical point is a minimum. We concoct an artificial bounded max-min problem where the max lies on the boundary of our concocted region and the min on the boundary is larger than our interior critical point. A sphere of huge radius H centered at $(3, 2, 1)$ intersects our paraboloid at squared distance H^2. The intersection is the curve satisfying both of the equations

$$z = x^2 + 2y^2 \qquad \text{and} \qquad (x-3)^2 + (y-2)^2 + (z-1)^2 = H^2$$

So we consider the problem

$$\text{Extremize } [D \; : (x-3)^2 + (y-2)^2 + (x^2 + 2y^2 - 1)^2 \leq H^2]$$

There is only one critical point interior to the region and the squared distance to the boundary is always the huge H^2. This shows that the interior point is minimal for the artificially restricted problem. It is also minimal for the whole problem, because points outside the concocted region are farther away.

The drawback to this solution is that the region given by $(x-3)^2 + (y-2)^2 + (x^2 + 2y^2 - 1)^2 \leq H^2$ can not be solved and sketched so easily. Consider instead the simpler circular region of huge radius H given by

$$x^2 + y^2 \leq H^2$$

It is intuitively clear that the nearest point to $(3, 2, 1)$ on the boundary $x^2 + y^2 = H^2$ is still a huge distance from the point $(3, 2, 1)$. Let's estimate it. The boundary minimum is more than the squared horizontal distance $(x - 3)^2 + (y - 2)^2$ which occurs along the line through $(3, 2)$ in the x-y-plane, $(3K)^2 + (2K)^2 = H^2$, so $K = H/\sqrt{13}$. This makes the minimal distance on this boundary more than the infinite amount $(K - 3)^2 + (K - 2)^2$. The Extreme Value Theorem applies to the problem

$$\text{Extremize } [D \; : (x - 3)^2 + (y - 2)^2 \leq H^2]$$

which has its minimum at the interior critical point.

The next example is a geometric distance max - min problem that is over a compact region, but, as is typical of closed and bounded surfaces, the computations are best done implicitly. The implicit computations hide the closed bounded nature to some extent.

EXAMPLE 20.24.

Find the points on the ellipsoid

$$3x^2 + 2y^2 + z^2 = 20$$

that are nearest and farthest from the point $(3, 2, 1)$.

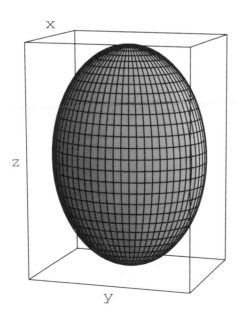

FIGURE 20.25: $3x^2 + 2y^2 + z^2 = 20$

The vector displacement from $(3, 2, 1)$ to (x, y, z) is the same as in the previous Example, but we cannot simply solve for z and substitute into the distance formula. Instead, we use the squared distance D with a general (x, y, z) and remember the constraint:

$$D = (x - 3)^2 + (y - 2)^2 + (z - 1)^2 \qquad\qquad 3x^2 + 2y^2 + z^2 = 20$$

$$dD = 2(x - 3)dx + 2(y - 2)dy + 2(z - 1)dz \qquad 6xdx + 4ydy + 2zdz = 0$$

and substituting

$$dz = -\frac{3x}{z}dx - \frac{2y}{z}dy$$

into the expression for dD we have

$$dD = \frac{2z(x - 3) - 3x(2(z - 1))}{z}dx + \frac{2z(y - 2) - 2y(2(z - 1))}{z}dy$$

Mathematica finds the solutions to the simultaneous equations

$$2z(x - 3) - 6x(z - 1) = 0$$

$$2z(y - 2) - 4y(z - 1) = 0$$

$$3x^2 + 2y^2 + z^2 = 20$$

$(x, y, z) \approx (-1.06259, -1.29127, -3.64389)$ and $(x, y, z) \approx (2.17044, 1.59387, 0.886994)$ and 4 complex solutions. These points yield the distances:

$$\sqrt{D(-1.06259, -1.29127, -3.64389)} \approx 6.99305$$

$$\sqrt{D(2.17044, 1.59387, 0.886994)} \approx 0.930528$$

20.3.1. Finding Critical Points with Constraints. To find the critical points of $w = f(x, y, z)$ when z is constrained by $g(x, y, z) = c$ a constant:

(1) Compute $dw = \frac{\partial f}{\partial x}dx + \frac{\partial f}{\partial y}dy + \frac{\partial f}{\partial z}dz$ and $\frac{\partial g}{\partial x}dx + \frac{\partial g}{\partial y}dy + \frac{\partial g}{\partial z}dz = 0$

(2) Solve the constraint differential $\frac{\partial g}{\partial x}dx + \frac{\partial g}{\partial y}dy + \frac{\partial g}{\partial z}dz = 0$ for dz

(3) Substitute this expression for dz into the expression for dw from part 1. Simplifying yields

$$dw = [\text{mess1 with } x, y, z]dx + [\text{mess 2 with } x, y, z]dy$$

$$= \frac{\partial w}{\partial x}dx + \frac{\partial w}{\partial y}dy$$

mess $1 = \frac{\partial w}{\partial x}$ and mess $2 = \frac{\partial w}{\partial y}$, but they are a mess in the sense that they have a dependent variable z in the expression. We cannot solve the equations $\frac{\partial w}{\partial x} = 0$ and $\frac{\partial w}{\partial y} = 0$ in this form because of the extra variable.

(4) Solve the simultaneous equations

$$\frac{\partial w}{\partial x} = 0$$

$$\frac{\partial w}{\partial y} = 0$$

$$g(x, y, z) = c$$

These are your critical values.

EXAMPLE 20.26. *Max-min of Volume with Fixed Surface Area*

Implicit equations are very powerful, but can sometimes be difficult to analyse. We state the following symbolically and solve it before we explain a geometric way to look at it.

$$\text{Extremize } [w = xyz : xy + yz + zx = 1, x \geq 0, y \geq 0, z \geq 0]$$

Let's use the previous procedure.

$$dw = y z\, dx + x z\, dy + x y\, dz \qquad : \qquad (y + z)dx + (x + z)dy + (x + y)dz = 0$$
$$: \qquad (x + y)dz = -(y + z)dx - (x + z)dy$$
$$: \qquad dz = -\frac{y + z}{x + y}\, dx - \frac{x + z}{x + y}\, dy$$

Now substitute the constrained value of dz in the dw-equation:

$$dw = y z\, dx + x z\, dy + x\, y\left(-\frac{y + z}{x + y}\, dx - \frac{x + z}{x + y}\, dy\right)$$
$$= (y z - x y\frac{y + z}{x + y})dx + (x z - x y\frac{x + z}{x + y})dy$$
$$= \frac{y z(x + y) - x y(y + z)}{x + y}\, dx + \frac{x z(x + y) - x y(x + z)}{x + y}\, dy$$
$$= \frac{\partial w}{\partial x}\, dx + \frac{\partial w}{\partial y}\, dy$$

Solution of the simultaneous equations, $\frac{\partial w}{\partial x} = 0$, $\frac{\partial w}{\partial y} = 0$ with the constraint is given by the *Mathematica* command

```
Solve[{y z (x + y) == x y (y + z),
x z (x + y) == x y (x + z),
x y + y z + x z == 1}]
```

Yielding only the one positive solution $x = y = z = 1/\sqrt{3}$, where $w = 1/(3\sqrt{3})$. Hummm. Is this the max or the min?

Along the boundaries $x = 0$ or $y = 0$, $w = x y z = 0$, so $w = 1/(3\sqrt{3})$ is not the minimum. Is it the maximum?

Some theory can help us. Consider the problem restricted to a huge rectangle.

$$\text{Extremize } [w = xyz : xy + yz + zx = 1, 0 \leq x \leq H, 0 \leq y \leq H, 0 \leq z]$$

This problem still does not satisfy the hypotheses of the Extreme Value Theorem, but the reason is a disguised. The constraint $xy + yz + zx = 1$ can not be satisfied if both $x = 0$ and $y = 0$. More generally, $z = (1 - x\, y)/(x + y)$ is undefined on the line $x + y = 0$ or $y = -x$. This line removes the point $(0, 0)$. Cut out an infinitesimal corner near zero,

$$\text{Extremize } [w = xyz : xy + yz + zx = 1, \text{square with corner removed shown below}]$$

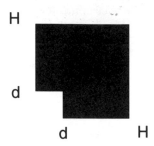

FIGURE 20.27: Region

$$[\delta \leq x \leq H \quad \& \quad 0 \leq y \leq H] \qquad \text{or} \qquad [\delta \leq y \leq H \quad \& \quad 0 \leq x \leq H]$$

This problem satisfies the hypotheses of the Extreme Value Theorem, so we know that it has a max and a min. The only interior critical point is $x = y = z = 1/\sqrt{3}$, so we examine the edges.

We know that $w = 0$ on the edges $x = 0$ and $y = 0$.

If $x = \delta \approx 0$ and $0 \leq y \leq \delta$, then $0 \leq xy \leq \delta^2$, $\delta \leq x + y$ and $z = (1 - xy)/(x + y) \leq 1/\delta$, so $w = xyz \leq \delta^2/\delta = \delta \approx 0$. In other words, w is small on the edges of the small corner.

Intuitively, if $x = H$ is huge, y and z must be very small so that $xy + yz + zx = 1$. The question is: How big is $w = xyz$, when $x = H$ is infinite? This requires some estimation:

$$xy + yz + zx = 1$$
$$Hy + Hz + yz = 1$$

Since Hy and Hz are positive and no more than 1, both y and z must be infinitesimal. This also makes

$$w = Hyz \approx 0$$

because Hy is at most 1 and finite times infinitesimal is infinitesimal. (Alternately, let $H = 10^6$, a million. Then Hy and Hz are no more than 1, so neither y not z is more than one one-millionth. This makes Hyz less than a millionth.)

Similarly, x and z must be small if $y = H$ and the computation shows that $w \approx 0$.

What have we shown? The extrema of $w = xyz$ must be either at $x = y = z = 1/\sqrt{3}$ or along the boundary where $w \approx 0$. There MUST be both a max and a min and the value along the boundary is only infinitesimal, so

$$w = xyz \qquad \text{is maximized for} \qquad x = y = z = \frac{1}{\sqrt{3}}$$

when x and y are in the cut off square and $xy + yz + zx = 1$. Outside this set, w is even smaller than it is on the large edges, so w is maximized for all positive x and y at the values $x = y = z = 1/\sqrt{3}$.

There is a geometric way to understand this problem. Consider a rectangular box with width x, length y, and height z. The area of the faces are xy, yz, and zx, and there are two of each of these, in other words,

$$\text{The surface area of the box } = 2(xy + yz + zx)$$

so our constraint $xy + yz + zx = 1$ means that the box has a fixed area of $1/2$. We also have

$$\text{The volume of the box} = w = xyz$$

and our question is

$$\text{Extremize [The volume of a box : The area is fixed]}$$

We have shown that a cube has maximum volume for an area of $1/2$. The huge edge argument amounts to showing that when one side is very long, the volume of the box is small, because you must keep the other two edges so small in order to hold the area fixed. Perhaps you can explain the geometry problem more intuitively, but there is a point to the max-min theory that we illustrate with the next example.

20.3.2. The Sailboat with No Fast or Slow Path. This is L. C. Young's example of a max-min problem with no max and no min. We wish to sail a sailboat up stream and against the wind on a river. We can sail at $45°$ and make fast progress, but we must 'come about' at the shore and sail on the opposite $45°$ tack. Fine, this works; we make progress up stream.

However, the current in the river is strongest near the middle and slowest near the shore. We would make faster progress if we 'came about' before we reached the strong current in the middle, say, we sailed on a third of the river near the bank. Conversely, we would make slower progress if we stayed in the middle third of the river sailing against the fast current. There is no end to this. It would be even faster to sail on the ninth of the river nearest the shore, and slower to sail in the ninth nearest the fastest current. The fastest path is to sail right against the shore and 'come about' infinitely often. The slowest is to stay at the strongest current and 'come about' infinitely often. Unfortunately, you can't 'come about' infinitely often and this problem has no max and no min.

EXERCISE 20.28. *Find the max and min subject to the constraints given.*

$$w = z(y - x) \qquad : x + y + z = 1, \qquad 0 \le x \le 1, \qquad 0 \le y \le 1$$
$$w = xyz \qquad : x + y + z = 1, \qquad 0 \le x \le 1, \qquad 0 \le y \le 1$$
$$w = xy + yz + zx \qquad : xyz = 1, \qquad 0 \le x, y, z, \qquad \text{is there a max? a min?}$$

EXERCISE 20.29. *Find the minimum surface area of a rectangular solid of fixed volume V.*

Interpret your solution as the inequality

$$V^2 \le \left(\frac{A}{6}\right)^3$$

For the volume, V, of any rectangular solid of area A.

The maximal volume of a rectangular solid of fixed surface area A follows from this general inequality. Why?

EXERCISE 20.30. *Find the point on the surface $z = x^2 - y^2/3$ nearest to $\mathbf{P} = \begin{bmatrix} 1 \\ -2 \\ 3 \end{bmatrix}$*

Use the Maxmin.ma NoteBook to help you solve the next problem.

EXERCISE 20.31. *A surface is given implicitly by the equation*

$$x^2 + 2y^2 + z^2 = 88$$

Find the point on the surface that is closest to

$$\mathbf{P} = \begin{bmatrix} 2 \\ -1 \\ -3 \end{bmatrix}$$

Is there a point farthest from $\mathbf{P}$ *?*

EXERCISE 20.32. *When is the speeding target of Exercise* 19.48 *closest to you?*

EXERCISE 20.33. *Find the volume of the largest rectangular solid with its faces parallel to the coordinate planes which can be inscribed in the ellipsoid* $(\frac{x}{2})^2 + (\frac{y}{3})^2 + (z)^2 = 1$. *(HINT: Try the sphere first, if you're having trouble.)*

We know that the square maximizes the area of all rectangles with a fixed perimeter. Which triangle maximizes the area of all triangles of fixed perimeter? Heron's formula says the area $A = \sqrt{\frac{p}{2}(\frac{p}{2} - a)(\frac{p}{2} - b)(\frac{p}{2} - c)}$ for a triangle with sides a, b, c. We know $p = a+b+c$ and all three of a, b, and c must be positive to make a triangle. In addition, if c is the long side, $c \geq a + b$.

EXERCISE 20.34. *Find the triangle of maximum area amongst those with fixed perimeter.*

Hints for the triangle problem: Eliminate the variable c, using $c = p - a - b$. Since p is fixed, the area $A = A(a, b)$ is a function of two variables once we substitute the expression for c into Heron's formula above. Show that the geometry requires $0 < a$, $0 < b$ and $\frac{1}{2}p < a + b < p$ for a maximum. Sketch this (a, b) region.

20.4. Projects on Max - min

The Mathematical Background project on least squares fit of a straight line to data uses max-min in two variables to find the fit. There is also a section on geometric distance problems that can be phrased as minimization problems.

Part 5

Differential Equations

CHAPTER 21

Continuous Dynamical Systems

This chapter begins the systematic study of "differential equations." We have already seen many examples where differential equations play a key role in describing change: the S-I-R epidemic, the serious law of cooling introduced in the silly canary story, Bugs Bunny's and Galileo's laws of gravity, air resistance for a falling object, a mathematical definition for irrational exponents. We did not name the chapter "differential equations," because our study will not be a study of an equation and tricks to 'solve' it, but rather of the dynamic 'movement' produced by the differential equation.

A differential equation describes 'how things change,' and if we know where we start, we should be able to predict where we go - and how fast. The 'systems' we want to study are families of initial value problems given by a differential equation and a starting position.

One good analogy for a mathematical dynamical system is the 'flow' on the surface of a smoothly moving river. The differential equation corresponds to the velocity vector at each point on the surface, while the collection of paths of all the water particles constitutes the solution flow. In this chapter we will find numerical, graphical and symbolic descriptions of mathematical flows determined by a differential equation and its initial conditions. This will shed new light even on the previous examples like the S-I-R model where we have seen the solutions, but not analyzed them geometrically. Geometry is a powerful qualitative tool, answering the 'where we go' questions. Numerics and symbolics answer the quantitative 'how fast' questions.

The most important thing is to learn to 'speak' the language of change by expressing how variables change one another. We will explicitly solve linear equations and a few nonlinear ones, but let the computer find the solution flows the rest of the time. This lets us concentrate on the most important issues:

(1) What are the assumptions that go into describing change mathematically?
(2) What dynamic movement does the change law produce?

21.1. One Dimensional Continuous Initial Value Problems

'The instantaneous rate of growth of algae is equal to the amount present.' Each bit of algae itself produces new algae, so the more you have, the faster the total increases. Let x denote the amount of algae (say in grams) in the whole lake. Let t denote the elapsed time (in hours) measured from some starting time. The growth statement above is the differential equation,

$$\frac{dx}{dt} = x$$

because the instantaneous rate of growth of x is $\frac{dx}{dt}$ and that equals x. (Notice that this simple differential equation has x on both sides of the equation, unlike the differentiation formulas from earlier in the course. For example, the derivative formula for a power in these variables looks like: 'If $x = 5\,t^3$, then $\frac{dx}{dt} = 15\,t^2$, with only the independent variable t on the right hand side.)

How much algae will be present after 3 hours? We can't be expected to answer this question with only this information. Why? Because we don't know how much algae we started with. For example, if we start with no algae, we never get any, while if we start with lots, we get much more.

If we also give the initial information,

$$x(0) = 57$$
$$\frac{dx}{dt} = x$$

we do expect to be able to predict how much we will have in three hours or any future time. How to make the prediction is another matter. (The specific answer to this question is completed in the exercises below.) The important thing is that we only expect specific future predictions to a 'change question' posed as:

(1) Where does the quantity x start.
(2) How does x change with respect to t (a continuous variable like time.)

DEFINITION 21.1. *One Dimensional Initial Value Problem*
The mathematical form of a one dimensional initial value problem at starting time t_0 is

WHERE WE START:	$x(t_0) = a$	*for some constant a*
HOW WE CHANGE:	$\dfrac{dx}{dt} = f(x,t)$	*for a function of two variables, $f(\cdot,\cdot)$*

Technically, we seek a **function** $x(t)$ defined for $t \geq t_0$ with $x(t_0) = a$ and derivative $x'(t) = \frac{dx}{dt} = f(x(t), t)$. Notice that the variable x stands for an unknown function in the differential equation, but for just a scalar variable in the expression $f(x, t)$.

The solution function is NOT the function $f(x, t)$. That function is the "growth law." In the case of the algae growth model above, $f(x, t) = x$ and f does not depend on t.

EXERCISE 21.2. *The Natural Exponential*
What is the text's official mathematical definition of $x = e^t$?

(Hint: Look it up in the chapter on exponentials and change variables.)

EXERCISE 21.3. *Exponential Growth*
Show that the function $x(t) = 7\,e^t$ satisfies:

$$x(0) = 7$$
$$\frac{dx}{dt}(t) = x(t) \qquad or \qquad \frac{dx}{dt} = x$$

In other words, $x(t) = 7 e^t$ satisfies this initial value problem.
   Find the solution to another initial value problem

$$x(0) = 57$$
$$\frac{dx}{dt} = x$$

Find the solution to the general initial value problem

$$x(0) = x_0 \qquad x_0 \text{ a constant}$$
$$\frac{dx}{dt} = x$$

The way we want to study differential equations is to ask questions like the following. For the algae growth model

$$x(0) = x_0$$
$$\frac{dx}{dt} = x$$

which starts with an amount x_0, how long does it take to grow a metric ton? In other words, how does this time depend on the starting amount? Mathematically, we want to find t_1 (depending on x_0) such that $x(t_1) = 1,000,000$.

EXERCISE 21.4. *Show that the algae growth model*

$$x(0) = 57$$
$$\frac{dx}{dt} = x$$

predicts $x(t_1) = 1,000,000$ when $t_1 = \text{Log}[\frac{10^6}{57}] = 6 \, \text{Log}[10] - \text{Log}[57] \approx 9.77246$
   Show that the general algae growth model

$$x(0) = x_0$$
$$\frac{dx}{dt} = r\,x$$

predicts $x(t_1) = 1,000,000$ when $t_1 = \text{Log}[\frac{10^6}{x_0}]/r$.

A somewhat more realistic algae growth model has the 'logistic' form

$$x(0) = x_0$$
$$\frac{dx}{dt} = x(1 - \frac{x}{3457.6})$$

This differential equation has the function $f(x,t) = x(1 - \frac{x}{3457.6})$ for the form in Definition 21.1. Notice that $f(x,t)$ does not depend on t.
   Let's see what this growth law says in English (which isn't very good at describing change, but is more familiar). This model will never produce a metric ton of algae and that will be clear once we understand simply *what* the growth law says.

Since we are interested only in solutions when $x > 0$, we divide both sides by x obtaining the per capita growth law

$$\frac{1}{x}\frac{dx}{dt} = 1 - \frac{x}{3457.6}$$

The left hand side of this equation is the per capita rate of growth, that is, the rate of adding new x per unit x - new algae babies per algae parent.

Notice that the per capita growth in our first model is

$$\frac{1}{x}\frac{dx}{dt} = 1$$

This is the growth law that says each individual produces at a certain rate no matter how large the population becomes. The per capita growth rate in the logistic law changes with x.

So how do we express this logistic growth law in English? 'The per capita growth of algae is equal to one minus the fraction of x over 3457.6.' This is literally correct, but misses the meaning. It is like a word by word translation from German to English; all the words are there, but the meaning is lost in the jumble.

We can understand this growth law by describing how it works qualitatively. If $x \approx 0$, we have

$$\frac{1}{x}\frac{dx}{dt} = 1 - \frac{x}{3457.6} \approx 1$$

which is approximately our original model, $\frac{dx}{dt} = x$. However, as x grows larger, $1 - \frac{x}{3457.6}$ decreases until $x = 3457.6$ where the right hand side equals zero. At zero growth, x doesn't change. If $x > 3457.6$, the right hand side is negative, and this would make the per capita growth negative. Negative growth is the same thing as decline or decrease, so if x starts at a value larger than 3457.6, it decreases toward this value. All this shows:

$$(\frac{dx}{dt} < 0 \quad \text{if} \quad x < 0)$$

$$\frac{dx}{dt} = 0 \quad \text{if} \quad x = 0$$

$$\frac{dx}{dt} > 0 \quad \text{if} \quad 0 < x < 3457.6$$

$$\frac{dx}{dt} = 0 \quad \text{if} \quad x = 3457.6$$

$$\frac{dx}{dt} < 0 \quad \text{if} \quad x > 3457.6$$

The best translation of this information is in the language of geometry. If we mark an x-line with arrows that point in the direction of the change in x, the above information is simply the figure

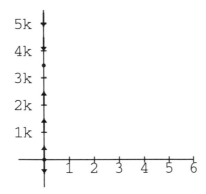

FIGURE 21.5: Directions of Algae Growth by Value

From this geometric translation, we can see 'where things go,' namely, all the initial value problems with $x_0 > 0$, tend toward a limit of 3457.6. Convince yourself of this! It's easy. In fact, it is easier to use the direction line of this equation than it is to use the explicit solution. An exercise will convince you of this.

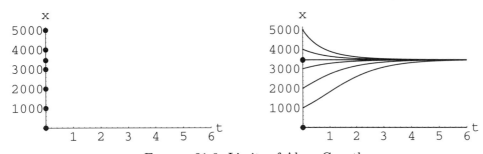

FIGURE 21.6: Limits of Algae Growth

EXERCISE 21.7. *Directions are Easier than Formulas*
The explicit solution to the initial value problem

$$x(0) = x_0$$

$$\frac{dx}{dt} = x(1 - \frac{x}{3457.6})$$

is given by the formula

$$x(t) = \frac{k\,e^t}{1 + \frac{k}{3457.6}e^t}$$

where $k = \frac{x_0}{1 - x_0/3457.6}$. Verify that this is the solution. (Ugh, just differentiating and showing $\frac{dx}{dt} = x(1 - x/c)$ is messy.) If $x_0 > 0$, show that $\lim_{t\to\infty} x(t) = 3457.6$.

You can use *Mathematica* as follows:
In[1]
```
x[t_ ] := k Exp[t]/(1 + k Exp[t]/c);
x[t]
```

In[2]

y[t_] := Evaluate[D[x[t] , t]];
y[t]

In[3]
  y[t] - x[t] (1 - x[t]/c)

In[4]
  Simplify[%]
The formula in the previous exercise can answer quantitative questions like 'how fast' does the solution get somewhere, but the qualitative question of what limit the solution approaches is easier to see simply from a direction field.

EXERCISE 21.8. *Stagnant Points*
Show that the solution to the initial value problem

$$x(0) = x_0$$

$$\frac{dx}{dt} = x(1 - \frac{x}{3457.6})$$

for $x_0 = 0$ is $x(t) = 0$ and for $x_0 = 3457.6$ is $x(t) = 3457.6$. Why are these the only constant solutions? Is it easier to answer this question using the differential equation or the explicit formula in the previous exercise? (What if you first had to find the explicit solution yourself?)

DEFINITION 21.9. *Autonomous Differential Equation*
A differential equation of the form

$$\frac{dx}{dt} = f(x)$$

where the function f does NOT depend on the independent variable, t, is called autonomous.

An autonomous equation has a growth law that does not depend on time. This is essential in the analysis of growth by just direction of change, because otherwise the arrows on our x-line would move in time.

EXERCISE 21.10. *Show that the function $x(t) = c\,e^{t^2}$, for a constant c, is a solution to the differential equation $\frac{dx}{dt} = 2\,t\,x$.*
  What is the initial value $x(0) =$?
  Is this an autonomous differential equation?

The logistic equation above is autonomous, because $\frac{dx}{dt} = x(1 - \frac{x}{3457.6})$ has $f(x,t) = x(1 - \frac{x}{3457.6})$ which does not depend on t.

DEFINITION 21.11. *Direction Line*
An x-line with arrows pointing in the direction of x-change corresponding to an autonomous differential equation is called a one dimensional direction field or direction line.

Points on the direction line with zero derivative are stagnant, that is, if you start there, the flow never carries you off. A more official name for this is "equilibrium point."

DEFINITION 21.12. *Equilibria, Attractors and Repellers*
A point x_e on a direction line for the differential equation $\frac{dx}{dt} = f(x)$ with $f(x_e) = 0$ is called an equilibrium point. An equilibrium point is called a local attractor if the direction arrows point toward it on both sides. An equilibrium point is called a repeller if the direction arrows point away from it on both sides.

It is possible for an equilibrium point to be neither an attractor nor a repeller. If the direction field points toward the equilibrium point on one side and away on the other, then the equilibrium point is a 'neither.'

EXERCISE 21.13. *Show that the equilibrium point of*

$$\frac{dx}{dt} = x^2$$

is neither an attractor nor a repeller. (Which point is the equilibrium?)
Find and classify all the equilibrium points of the differential equation

$$\frac{dx}{dt} = \mathrm{Sin}[x]$$

(This equation does not arise naturally, but has a simple direction line. A similar looking equation describes the pendulum.)

It is not very hard to find out if an equilibrium point is an attractor or repeller in one dimension. In two dimensions, later in the chapter, we will see that this is much harder. There we will use derivatives to find a criterion. The analogous idea works in one dimension. See if you can figure it out; it is quite simple.

PROBLEM 21.14. *Consider an autonomous differential equation $\frac{dx}{dt} = f(x)$ with an equilibrium point x_e with $f(x_e) = 0$. (The two previous examples have $f(x) = x^2$ and $f(x) = \mathrm{Sin}[x]$.) Give a condition on the value of $f'[x_e]$ that will tell if x_e is an attractor or repeller. What are the specific numerical values of $f'[x_e]$ for the equilibria in the previous exercise?*

The following exercises cast the examples of this section in a little more useful form. We want you to understand what the growth laws say in terms of parameters and to get some practice with the new terminology.

EXERCISE 21.15. *General Exponential Growth*
If a population has a constant per capita fertility rate, independent of size, then its growth is governed by the law

$$\frac{dx}{dt} = r\,x \qquad \text{for some constant } r$$

Explain. Show that the function $x(t) = x_0\,e^{r\,t}$ has $x(0) = x_0$ and satisfies this differential equation. Draw the direction line and show that, mathematically, zero is the only equilibrium point and that it is a repeller (though the negative values are biologically meaningless).

Let's consider a practical example of exponential growth.

EXAMPLE 21.16. *Exponential Mice*

In the spring, the growth of the mouse population in a field near my house is governed by the law

$$\frac{dx}{dt} = r\,x \qquad \text{for some constant } r$$

where x represents the number of female mice per acre and t is time in days. We want to understand the meaning of the per-capita growth rate parameter, r. Monday morning before work, there were $1,000$ (female) mice per acre and 8 hours later there were $1,100$.

EXERCISE 21.17. *On the average at this rate, how many babies is each mother having per day? (Hint: The change in mice is? The change in days is? The ratio is? There are how many mice? The ratio is? Keep track of units.) Since this is a fraction, re-express this rate as the average number of days between each mother mouse's birth of a baby.*

The strict analytical solution to the initial value problem $x(0) = 1000$ and $\frac{dx}{dt} = r\,x$ is

$$x(t) = 1000\,e^{r\,t}$$

Eight hours later is $t = 1/3$ days, and we know $x(1/3) = 1000\,e^{r/3}$, so

$$1100 = 1000\,e^{r/3}$$
$$e^{r/3} = 1.1$$
$$\text{Log}[e^{r/3}] = \text{Log}[1.1]$$
$$\frac{r}{3} \approx 0.0953102$$
$$r \approx 0.285931$$

EXERCISE 21.18. *What is the instantaneous rate at which each female mouse is having babies? (Hint: What is the expression for the rate of increase in mice per day? How many females are there - that is, what variable tells you this? What is the ratio?)*

Explain the similarity of your answer to this question and the answer to the previous question. (As a rough approximation, we could say $r \approx 0.285931 \sim 0.333 = 1/3$. And $1/0.285931 \sim 3$.)

Radioactive substances decay at a rate proportional to the amount of the substance present. When a radioactive substance decays, it changes to something else, hence is gone. Roughly speaking, each atom has an independent probability of emitting a particle, so the more you have, the more decay you see.

EXERCISE 21.19. *Exponential Decay*
Let x equal the amount of a radioactive substance (in grams) and let t denote the time in years. What expression about x represents decay? What does $-\frac{dx}{dt}$ represent? Express the statement, "The rate of decay is proportional to the amount present." as a differential equation. Show that the function $x(t) = x_0\,e^{-r\,t}$ has $x(0) = x_0$ and satisfies your differential equation for the correct choice of a constant r. How is this constant referred to in the original statement?

Draw the direction line and show that, mathematically, zero is the only equilibrium point and that it is an attractor (though the negative values are physically meaningless).

A typical natural environment can only support so many individuals of a particular species. This is referred to as the carrying capacity of the environment. The logistic algae model above had a carrying capacity of 3457.6.

EXERCISE 21.20. *Geometric Analysis of Logistic Growth*
A general "logistic growth" model is the initial value problem

$$x(0) = x_0$$
$$\frac{dx}{dt} = r\, x \left(1 - \frac{x}{c}\right)$$

for positive constants r and c. Draw a direction line corresponding to this family of initial value problems and explain why all positive solutions tend to the limit x = c. The constant c is called the "carrying capacity" because of this result. It represents the number of individuals that the 'environment' can support.

How does the constant r effect the solutions? (Consider cases where r is very big and very small, but positive.) Does r effect the direction line?

How could you measure the constants r and c in the field?

Show that, mathematically, x = 0 and x = c are the only equilibrium points and that one is a repeller (though the negative directions are biologically meaningless) and the other an attractor. Attraction to the equilibrium x = c is important biologically. Why? (What if the population gets temporarily depressed below c by disease?)

Logistic-like equations arise in other contexts. For example, the Scientific Project on S-I-S diseases and endemic limits studies the system of equations

$$\frac{ds}{dt} = b\, i - a\, s\, i$$
$$\frac{di}{dt} = a\, s\, i - b\, i$$

where t is time in days, s is the fraction of the population that is susceptible (or well) and i is the fraction of the population that is infectious (or sick). We know (in this context) that the sum of the infective and susceptible fractions is one, $s + i = 1$. Substituting $i = 1 - s$ into the first equation gives

$$\frac{ds}{dt} = (b - a\, s)(1 - s)$$
$$= b(1 - \frac{a}{b} s)(1 - s)$$
$$= b(1 - c\, s)(1 - s)$$

where $\frac{a}{b} = c$, the contact number (a constant related to the infectiousness of the disease. The constant b is one over the number of days a person is infectious, another disease-specific constant.). This is just a logistic equation with the origin shifted.

PROBLEM 21.21. *S-I-S Diseases and the Logistic Equation*
Show that the susceptible fraction values $s_1 = 1$ and $s_e = 1/c$ are the only equilibria of

$$\frac{ds}{dt} = b(1 - c\, s)(1 - s)$$

Consider the case $c > 1$ (say, $c = 1.5$) and show that $s_e = 1/c$ is an attractor. In this case show that the disease is endemic, with the limit of the susceptible fraction $\lim_{t \to \infty} s(t) = s_e$, no matter what initial value $0 < s(0) < 1$ we begin with. This is the 'endemic limit.' What does this mean about the long-term behavior of the disease in a population?

Consider the case $0 < c < 1$ (say, $c = .75$). Show that $s_1 = 1$ is an attractor. Prove that if $0 < s(0) \leq 1$, then $\lim_{t \to \infty} s(t) = 1$. What does this mean about the long-term behavior of the disease in a population?

The contact constant c has an intuitive interpretation as 'the number of people you contact sufficiently closely during your whole illness to transmit the disease (if they are well).' Why does $c > 1$ result in a persistent infection of the population and $c < 1$ not do so?

In each of the two cases above, how does the parameter b affect the solutions?

Responses to loudness of sounds, brightness of lights and other broad range phenomena are often nonlinear in order to accommodate the range. The 'Weber-Fechner law' even proposes a specific nonlinear psychological response law. These things are awkward to state in English, but really not very hard to understand once you express them mathematically. Here is an example:

EXERCISE 21.22. *Sometimes Diff E Q s are better than English*
The "Richter scale" of earthquake magnitudes is defined by the statements:

(1) *The lowest intensity setting of a seismograph is magnitude 0 at an energy level of 0.01754 kilowatt-hours of energy.*
(2) *The rate of change of magnitude with respect to energy intensity is inversely proportional to intensity (or, in other words, is proportional to the reciprocal of intensity).*

Identify the independent and dependent variables in the above statements and give them variable names. Write the two statements using your variables as a mathematical initial value problem. Show that the solution to your initial value problem is given by the natural logarithm with some appropriate constants. You need more information to determine these constants.

Here are some actual data:

Energy Intensity millions of kw-h	Magnitude Richter scale	Earthquake
8.8	5.8	*Armenia 1988*
197	6.7	*California 1971*
279	6.8	*Armenia 1988*
$7. \times 10^4$	8.4	*Alaska 1964*
1.4×10^5	8.6	*San Francisco 1906*
3.94×10^5	8.9	*Japan 1933*

Find the Richter magnitude as a function of the energy intensity for each of these. (They are not all exactly the same.) Use your formula from the 1964 Alaska quake (or an average) to compute the magnitude of an earthquake half as intense as the 1964 Alaska quake.

21.2. Euler's Method

Euler's approximate solution of an initial value problem uses where we are and how we change to move ahead a small step. Once ahead, we know where we are and can compute how we change and move ahead another step.

Our basic numerical method for solving initial value problems is quite simple provided you remember the increment equation (or microscope approximation) from Definition 5.15:

$$x(t + \delta t) = x(t) + \frac{dx}{dt}(t)\, \delta t + \varepsilon \delta t$$

where $\varepsilon \approx 0$ whenever $\delta t \approx 0$. Even though we do not know a formula for the unknown function $x(t)$ in an initial value problem

$$x(0) = x_0$$
$$\frac{dx}{dt} = f(x,t)$$

we do know $x(0) = x_0$ and thus can compute $\frac{dx}{dt}(0) = f(x(0),0)$. Substituting what we know into the right hand side of the increment equation above, we obtain new information on the left

$$x(0 + \delta t) = x(0) + f(x(0),0)\, \delta t + \varepsilon_1 \delta t$$
$$x(0 + \delta t) \approx x(0) + f(x(0),0)\, \delta t$$

The approximation $x(\delta t) = x(0) + f(x(0),0)\, \delta t$ produces an error that is small compared to δt, specifically, $x_{true}(t_1) = x(t_1) + \varepsilon_1 \delta t$, where $t_1 = 1 \times \delta t$. If we substitute this approximate value in the right hand side of the increment equation again, we obtain

$$x(t_1 + \delta t) \approx x(t_1) + f(x(t_1),t_1)\, \delta t + \varepsilon \delta t$$
$$x(t_1 + \delta t) \approx x(t_1) + f(x(t_1),t_1)\, \delta t$$

Assuming that $f(x,t)$ is continuous, we have $\theta = [f(x(t_1),t_1) - f(x(t_1) + \iota,t_1)] \approx 0$, for $\iota \approx 0$. This means that the error between the actual $x(t_1)$ and the approximation above is at most $\varepsilon_1 \delta t$ plus the next microscope error $\varepsilon \, \delta t$ and an error of the form $\theta \, \delta t$. Calculating $x(t_2) = x(2 \times \delta t)$ by the formula $x(t_2) = x(t_1) + f(x(t_1),t_1)\, \delta t$ produces a total error of $(\varepsilon_1 + \varepsilon + \theta)\, \delta t$. We let $\varepsilon_2 = \varepsilon + \theta \approx 0$, so we have $x_{true}(t_2) = x(t_2) + \varepsilon_1 \delta t + \varepsilon_2 \delta t$ with $\varepsilon_i \approx 0$. The important thing to observe is that the errors accumulate or add up, but that they do have a factor of δt.

Roughly speaking, if we keep substituting the newly approximated values into the increment equation we obtain an approximation satisfying

$$x_{true}(t_n) = x(t_n) + \varepsilon_1 \delta t + \varepsilon_2 \delta t + \cdots + \varepsilon_n \delta t$$

and the total error at time $t_n = n \times \delta t$ satisfies

$$|\varepsilon_1 \delta t + \varepsilon_2 \delta t + \cdots + \varepsilon_n \delta t| \leq \text{Max}[\varepsilon_i](\delta t + \cdots + \delta t) = \varepsilon_p \times n \times \delta t = \varepsilon_p \times t_n$$

for some particular $\varepsilon_p \approx 0$. This procedure of computing new values and substituting them in the increment equation is known as Euler's method of approximating the solution to an initial value problem.

If we summarize Euler's method, we use the approximation $x(t+dt) - x(t) \approx dx$ and compute $dx = f(x,t)\,dt$. To help remember this, we will write our differential equations using differentials (instead of derivatives),

$$x(t_0) = x_0$$
$$dx = f(x,t)\,dt$$

Euler's method is easy to compute using *Mathematica* lists. We begin with the list $\{t_0, x_0\}$ and add $\{t_{n+1}, x_{n+1}\}$ to the list using the computations:

$$x_1 = x_0 + f(x_0, t_0)\,dt \qquad\qquad t_1 = t_0 + dt$$
$$x_2 = x_1 + f(x_1, t_1)\,dt \qquad\qquad t_2 = t_2 + dt$$
$$x_3 = x_2 + f(x_2, t_2)\,dt \qquad\qquad t_3 = t_2 + dt$$
$$x_4 = x_3 + f(x_3, t_3)\,dt \qquad\qquad \cdots$$

where $dt \approx 0$ is a small number. The 'rule' to compute the next value of x can be given as a vector function of a vector input, $\mathbf{X}_{n+1} = \mathbf{G}[\mathbf{X}_n]$,

$$\begin{bmatrix} t_{n+1} \\ x_{n+1} \end{bmatrix} = \begin{bmatrix} t_n + dt \\ x_n + f(x_n, t_n)\cdot dt \end{bmatrix} = \mathbf{G}\begin{bmatrix} t_n \\ x_n \end{bmatrix} = \mathbf{G}[\mathbf{X}_n]$$

In this notation the successive approximations are $\{t_1, x_1\} = \mathbf{G}[\{t_0, x_0\}]$, $\{t_2, x_2\} = \mathbf{G}[\{t_1, x_1\}] = \mathbf{G}[\mathbf{G}[\{t_0, x_0\}]]$, $\{t_3, x_3\} = \mathbf{G}[\{t_2, x_2\}] = \mathbf{G}[\mathbf{G}[\mathbf{G}[\{t_0, x_0\}]]]$ and so forth.

Our solution is the list

$$\{\{t_0, x_0\}, \mathbf{G}[\{t_0, x_0\}], \mathbf{G}[\mathbf{G}[\{t_0, x_0\}]], \mathbf{G}[\mathbf{G}[\mathbf{G}[\{t_0, x_0\}]]], \cdots\}$$

which can be generated with the NestList[.] command.

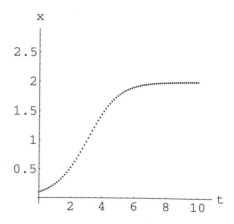

FIGURE 21.23: Euler Solution of $dx = x(1 - x/2)\,dt$

EXERCISE 21.24. *Computer and Exact Logistic Growth*
Use Euler's method to solve the initial value problem

$$x(0) = 1 \quad\quad \& \quad\quad dx = x\,(1 - x/2)dt$$

Type the following Mathematica programs and run the final one with dt = N[1/2], 1/4, 1/16, 1/265. Compare your approximations to the exact solution given in Example 21.32 below.

First, (with a clear kernel)
f[x_,t_] := x (1 - x/2);
G[{ t_, x_}] := { t + dt, x + f[x,t] dt} ;
solnList = NestList[G, { 0,1} ,3]

Entering this computation should give you a symbolic answer with *dt* treated as an unknown. NestList[.] makes the first three iterations of the function *G*.

Second, add the specific value of *dt* = 1/9 and iterate the function *G* 9 times,
f[x_,t_] := x (1 - x/2);
G[{ t_, x_}] := { t + dt, x + f[x,t] dt} ;
dt = 1/9
solnList = NestList[G, { 0,1} ,9]

Finally, plot your approximation up to time 7, iterating 70 times with a time step of 0.1,
f[x_,t_] := x (1 - x/2);
G[{ t_, x_}] := { t + dt, x + f[x,t] dt} ;
dt = 0.1
solnList = NestList[G, { 0,1} , 70];
plot1 = ListPlot[solnList];
y[t_] := 2 Exp[t]/(1 + Exp[t]);
plot 2 = Plot[y[t], { t, 0, 7}];
Show[plot1 , plot2]

EXERCISE 21.25. *Computer Analysis of Logistic Growth*
Modify the **EulerApprox.ma** *NoteBook to compute solutions to the logistic growth model $dx = x(1 - x/c)\,dt$ with $c = 2$. Compute Euler solutions for the initial conditions $x_0 = 0.1$, $x_0 = 0.5$, $x_0 = 1.0$, $x_0 = 1.5$, $x_0 = 2.5$ and $x_0 = 3.0$ and run the animations of successively better and better approximations.*

Recall that a discrete dynamical system (from Chapter 12) was of the form

$$x(0) = x_0$$
$$x(t + 1) = x(t) + F[x(t)]$$

and generates a solution $x(0)$, $x(1)$, $x(2) \cdots$. Euler's method is a discrete dynamical system (when the equation is autonomous), but the time moves in steps of size of δt instead of step size 1,

$$x(0) = x_0$$
$$x(t + \delta t) = x(t) + f[x(t)]\,\delta t$$

and generates a solution $x(0)$, $x(1 \times \delta t)$, $x(2 \times \delta t) \cdots$. At an equilibrium point for our continuous dynamical system we have $f(x_e) = 0$, so the corresponding discrete system has $F[x_e] = f(x_e)\,\delta t = 0$ as well.

EXERCISE 21.26. *Discrete Stability in Euler's Method*
Apply the criterion for the equilibrium x_e to be an attractor in the discrete system of Euler's method at a fixed step size. (You may need to look the condition up in the chapter on discrete dynamical systems.) How does that translate into stability when $F[x] = f(x)\,\delta t$?

What happens to your condition in the limit as $\delta t \to 0$? Is this condition sufficient to show that the continuous system $dx = f(x)dt$ has an attractor at x_e?

21.3. Some Theory

The orientation of this course, and especially this chapter, is to use calculus to describe change in concrete and scientific contexts. So why do we need any theory? We know where we start and how we change. Can't we just use Euler's method to see where we go - and how fast? First of all, if we did write down a mathematical model that happened not to have any solution, it wouldn't be very useful in understanding science and the world. Euler's method might turn out 'approximations' that aren't converging toward anything. That is not a very common occurrence; most models based on initial value problems have solutions. However, uniqueness of the solution turns out to be problematic. Scientifically, when we say a model describes a situation we mean that the mathematical solution is THE solution, not one of several possible behaviors. If our model mathematically has two or more solutions, then it does not determine the scientific outcome. Mathematically, in the non-unique case Euler's method might oscillate between two solutions. We need a theorem that says we have solutions and that they determine the behavior completely. This will tell us that the mathematical formulation of our problem is complete; it does determine THE outcome.

How could a single initial value problem have two outcomes? Unfortunately, the answer is, "It's easy." Consider

$$x(0) = 0$$
$$dx = 3\,x^{\frac{2}{3}}\,dt$$

The function $x_1(t) = 0$ is a solution. The function $x_2(t) = t^3$ also is a solution. We can even piece solutions together along the t axis. For example, the function

$$x(t) = \begin{cases} (t + \frac{3}{2})^3, & \text{if } t < -\frac{3}{2} \\ 0, & \text{if } -\frac{3}{2} \le t \le \frac{1}{2} \\ (t - \frac{1}{2})^3, & \text{if } t > \frac{1}{2} \end{cases}$$

is a solution (shown below with flow lines). Having the dependent variable x in the expression for $\frac{dx}{dt}$ makes it quite different from the explicit formulas $\frac{dx}{dt} = f(t)$ with which we are most familiar.

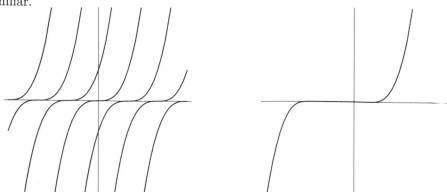

FIGURE 21.27: Tangent Solution Curves of $dx = 3\,x^{\frac{2}{3}}\,dt$

Fortunately, there is a simple criterion to use to find out if an initial value problem has unique solutions. The theorem says there is a solution and there is only one.

THEOREM 21.28. *Existence & Uniqueness for I. V. P.s*
Suppose that the functions $f(x,t)$ and $\frac{\partial f}{\partial x}(x,t)$ are continuous in a rectangle around (x_0, t_0).
Then the initial value problem

$$x(t_0) = x_0$$
$$dx = f(x,t)\,dt$$

has a unique solution $x(t)$ defined for some small time interval $t_0 - \Delta < t < t_0 + \Delta$. In this case, Euler's Method converges to $x(t)$ on closed subintervals $[t_0, t_1]$, for $t_1 < t_0 + \Delta$.

This theorem is easy enough to use: Simply calculate enough partials to be sure that the functions are continuous. (We know that smooth implies continuous.)

EXERCISE 21.29. *Show that the hypotheses of the Existence and Uniqueness Theorem do not apply to the function $f(x,t) = 3\,x^{\frac{2}{3}}$ at certain initial values. Is the partial derivative $\frac{\partial f}{\partial x}$ continuous? Show that the functions $x_1(t) = 0$ and $x_2(t) = t^3$ are both solutions with initial value 0. Are there other solutions?*
Show that the hypotheses of the Existence and Uniqueness theorem do apply the the exponential growth and decay laws $\frac{dx}{dt} = r\,x$, for a positive or negative constant r.
Show that the hypotheses of the Existence and Uniqueness theorem do apply the the logistic growth laws $\frac{dx}{dt} = r\,x\,(1 - x/c)$, for constants r and c.

The solutions of an initial value problem can "explode to infinity" in finite time. This means that the Δ in the Existence and Uniqueness theorem may be limited.

EXERCISE 21.30. *Explosion in Finite Time*
Show that the function $x(t) = \frac{1}{\frac{1}{x_0} - t}$ is a solution to the initial value problem

$$x(0) = x_0$$
$$dx = x^2\,dt$$

Also show that the function $f(x,t) = x^2$ has continuous partial derivatives, so satisfies the hypotheses of the Existence and Uniqueness Theorem. At what time does x 'escape to infinity'? What is the maximum Δ in the Existence and Uniqueness Theorem in this case?

Uniqueness has important consequences even in terms of sketching solutions. For example, the equilibrium point $x_e = c$ in the logistic equation $dx = r\,x\,(1-x/c)\,dt$ is an attractor, but while positive solutions all tend toward $x = c$, they can not get there in finite time. Why? Because $x_c(t) = c$ for all t, $-\infty < t < \infty$, is a solution. Another solution can not satisfy $x_1(t_0) = c$ if $x(t)$ is not identically c, because both $x_c(t)$ and $x_1(t)$ would satisfy the differential equation and the initial condition $x(t_0) = c$.

21.4. Separation of Variables

There are many many tricks for finding solutions of differential equations. We are only interested in two main ones:

(1) Guessing that the answer might be an exponential.
(2) 'Separation of Variables.'

The first is used in linear equations, while the second works if we can write the differential equation with all the $x's$ on one side and all the $t's$ on the other. This is the form

$$g(x)\,dx = h(t)\,dt$$

To solve this differential equation integrate both sides and solve for x. Keep the constants of indefinite integration so your expressions represent families of solutions to the differential equation without initial conditions. It doesn't always work, but it so simple that it is helpful to know the idea.

This method is conceptually backwards. First we find formulas for solutions with integration constants in them, then we use 'where we start.' The formulas represent 'how we change.' Don't think about the method, just do it.

EXAMPLE 21.31. *Separation of Variables*

We begin with the differential equation

$$\frac{dx}{dt} = x \cdot t$$

which we re-write as

$$\frac{dx}{x} = t\,dt$$

Now indefinitely integrate both sides

$$\int \frac{dx}{x} = \int t\,dt$$

$$\text{Log}[x] = \frac{1}{2}t^2 + c$$

$$e^{\text{Log}[x]} = e^{\frac{1}{2}t^2 + c} = e^{\frac{1}{2}t^2}e^c$$

$$x = C\,e^{\frac{1}{2}t^2}$$

Note that we used the exponential functional identity $e^{\alpha+\beta} = e^{\alpha} \cdot e^{\beta}$. Also, since e^c is a constant, we may as well just introduce a simpler name $C = e^c$. If we want $x(0) = 3$, we let $C = 3$. The function $x(t) = 3e^{\frac{1}{2}t^2}$ is the unique solution to that initial value problem by the Existence and Uniqueness Theorem.

EXAMPLE 21.32. *Integration and Solution Troubles*

The steps in the separation of variables can become technical. We could have trouble finding indefinite integrals, though *Mathematica* might help there. We could also have trouble solving for x. Here is an example where things still work out.

$$dx = r\, x(1 - \frac{x}{c})\, dt$$

$$\frac{dx}{x(1 - \frac{x}{c})} = r\, dt$$

$$\int \frac{dx}{x(1 - \frac{x}{c})} = \int r\, dt = r\, t + k \qquad \text{for a constant } k$$

But how do we compute $\int \frac{dx}{x(1-\frac{x}{c})}$?

A trick from high school algebra says that we can write 'partial fractions' as follows. Write the combined denominator in terms of separate denominators with unknown constants a and b in the numerator,

$$\frac{1}{x(1 - \frac{x}{c})} = \frac{a}{x} + \frac{b}{1 - \frac{x}{c}}$$

$$= \frac{a(1 - \frac{x}{c})}{x(1 - \frac{x}{c})} + \frac{b\,x}{x(1 - \frac{x}{c})}$$

$$= \frac{a - \frac{a}{c}x + b\,x}{x(1 - \frac{x}{c})}$$

After expanding in terms of these unknowns, the numerators have to match, so $a + (b - a/c)\,x = 1$. Hence $a = 1$ and $b = \frac{1}{c}$, making

$$\int \frac{dx}{x(1 - \frac{x}{c})} = \int \frac{1}{x} + \frac{1}{c}\frac{1}{1 - \frac{x}{c}}\, dx$$

$$= \int \frac{1}{x}\, dx + \int \frac{1}{c}\frac{1}{1 - \frac{x}{c}}\, dx$$

$$= \text{Log}[x] - \int \frac{1}{u}\, du \qquad \text{where } u = 1 - \frac{x}{c}$$

$$= \text{Log}[x] - \text{Log}[1 - \frac{x}{c}]$$

$$= \text{Log}[\frac{x}{1 - \frac{x}{c}}]$$

Therefore our separation of variables yields

$$\text{Log}[\frac{x}{1 - \frac{x}{c}}] = r\, t + k$$

$$e^{\text{Log}[\frac{x}{1 - \frac{x}{c}}]} = e^{r\, t + k} = e^k\, e^{r\, t}$$

$$\frac{x}{1 - \frac{x}{c}} = K\, e^{r\, t}$$

Now we solve for x,

$$x = \left(1 - \frac{x}{c}\right) K e^{rt}$$

$$x + \frac{K}{c} e^{rt} x = K e^{rt}$$

$$x\left(1 + \frac{K}{c} e^{rt}\right) = K e^{rt}$$

$$x = \frac{K e^{rt}}{1 + \frac{K}{c} e^{rt}}$$

Whew!

Here is an outline of the method of

21.4.1. Separation of Variables.

(1) Given a differential equation $dx = f(x,t)\, dt$, separate the variables in the form

$$g(x)\, dx = h(t)\, dt$$

(2) Indefinitely integrate both sides, keeping the 'constant of integration' k

$$\int g(x)\, dx = \int h(t)\, dt$$

$$G(x) = H(t) + k$$

(3) Solve the equation $G(x) = H(t) + k$ for x.

If you can't do any one of these steps, use another method.

EXERCISE 21.33. *Practice with Partial Fraction Integration*
Solve for constants a and b so that

$$\frac{1}{(x-3)(x-2)} = \frac{a}{x-3} + \frac{b}{x-2}$$

If these constants are known, use properties of the integral to prove

$$\int \frac{1}{(x-3)(x-2)}\, dx = a \int \frac{1}{x-3}\, dx + b \int \frac{1}{x-2}\, dx$$

Combine the first two parts and use change of variables to show that

$$\int \frac{1}{(x-3)(x-2)}\, dx = \text{Log}[\frac{x-3}{x-2}] + C$$

You can check your work with *Mathematica* as follows:
Integrate[1/((x - 3)(x - 2)) , x]
Now apply the method to an interesting application.

PROBLEM 21.34. *Partial Fraction Integration and the S-I-S Model*
Use separation of variables to find symbolic solutions of the S-I-S differential equation,

$$\frac{ds}{dt} = b(1 - c\,s)(1 - s)$$

Here is some help. Let c be a constant and s be a variable. Show that

$$\frac{1}{(1 - s)(1 - c\,s)} = a\left[\frac{c}{1 - c\,s} - \frac{1}{1 - s}\right]$$

for $a = 1/(c - 1)$. Use this to re-write the S-I-S differential equation in the form

$$\left[\frac{c}{1 - c\,s} - \frac{1}{1 - s}\right] ds = r\,dt, \qquad where\ r = b(c - 1)$$

Show that

$$\int \frac{c}{1 - c\,s}\,ds = -\operatorname{Log}[1 - c\,s] + k_1 \qquad and \qquad -\int \frac{1}{1 - s}\,ds = \operatorname{Log}[1 - s] + k_2$$

so

$$\int \left[\frac{c}{1 - c\,s} - \frac{1}{1 - s}\right] ds = ? + k_3$$

for constants k_j.
   Consider the separated differential equation and show that

$$\operatorname{Log}[\frac{1 - s}{1 - c\,s}] = r\,t + k$$

$$\frac{1 - s}{1 - c\,s} = e^{r\,t + k} = e^k \cdot e^{r\,t} = K\,e^{r\,t}$$

where $r = b(c - 1)$ and K is constant.
   Solve the equation

$$\frac{1 - s}{1 - c\,s} = E$$

for s in terms of c and E and use your solution to show that the solution to the S-I-S equation has the form

$$s[t] = \frac{1 - k\,e^{-r\,t}}{c - k\,e^{-r\,t}}$$

   Now, solve the initial value problem by finding K so that

$$s_0 = \frac{1 - k}{c - k}$$

(Hint: $k = (c\,s_0 - 1)/(s_0 - 1)$, you can use Mathematica .)
   Finally, use your analytical solution to prove that

$$\lim_{t \to \infty} s[t] = \frac{1}{c}, \qquad for\ 0 < s_0 < 1, \qquad when\ c > 1$$

$$\lim_{t \to \infty} s[t] = 1, \qquad for\ 0 < s_0 < 1, \qquad when\ 0 < c < 1$$

(Don't forget to compute the sign of r in each case.)

Compare the solution to the previous problem with the geometric solution of the "endemic limit" in Exercise 21.21.

EXERCISE 21.35. *Symbolic Solutions and a Check*
Use separation of variables to solve the initial value problems:

(1) $x(0) = 1$ $dx = x\,t\,dt$
(2) $x(0) = 0$ $dx = e^{-t-x}\,dt$
(3) $x(1) = 0$ $dx = \frac{1}{t}\,dt$
(4) $x(0) = 0$ $dx = 2t\,dt$
(5) $x(0) = 2$ $dx = x^2\,dt$
(6) $x(0) = 0$ $dx = 2x\,dt$
(7) $x(0) = 0$ $dx = 3\,x^{2/3}\,dt$ *Which solution is Euler's?*

*Use the Mathematica NoteBook **EULER&exact.ma** along with your solution to compare the Euler approximations to the exact solution of part (2). (This comparison is already done for you for part (1) and part (3) as well as for the logistic equation.)*

You will notice that Euler's method is not very accurate. In some examples, the accumulating errors make small enough increment computations infeasible, either because there are too many computations, or because machine arithmetic errors enter into the accuracy of the computations.

You will also notice computational difficulties in the growth of solutions. Of course, true solutions can 'explode' to infinity in finite time, and in that case, Euler's method will grow very large. However, we know that exponential functions grow very rapidly (see the NoteBook **ExpGth.ma**) and *Mathematica* cannot compute numerical values like 10^{30000}. You could easily encounter this kind of growth by trying to solve a growing differential equation too long.

We have prepared several packages based on "Runge-Kutta" methods of solution. The idea behind these methods is to use a higher degree local fit. In other words, to take not just the slope of the function into account, but its curvature, etc. as well. We also have 'truncated' or 'bounded' the solutions in these methods. If the numerical solutions get bigger than 1,000,000, we simply stop the growth. This may be wrong mathematically, but causes no problems in solutions that remain below 1,000,000.

EXERCISE 21.36. *Accurate Solutions to Differential Equations*
*Use the Mathematica NoteBook **AccDEsol.ma** to solve one of the differential equations above where Euler's method seemed rather inaccurate. Use a bigger step size with the accurate method, for example, if you used $dt = 0.05$ with Euler's method, try $dt = 0.1$ with the accurate method.*

21.5. The Geometry of Autonomous Equations in Two Dimensions

The conceptual idea of an initial value problem extends easily to vector equations,

$$\mathbf{X}(0) = \mathbf{X}_0$$
$$d\mathbf{X} = \mathbf{F}(\mathbf{X}, t)\,dt$$

which in 2 dimensional components looks like:

$$\begin{bmatrix} x(0) \\ y(0) \end{bmatrix} = \begin{bmatrix} x_0 \\ y_0 \end{bmatrix}$$

$$\begin{bmatrix} dx \\ dy \end{bmatrix} = \begin{bmatrix} f(x,y,t) \\ g(x,y,t) \end{bmatrix} dt$$

or as given in the next result without brackets. The vector formula $d\mathbf{X} = \mathbf{F}(\mathbf{X},t)\, dt$ should be thought of as the equation telling you an increment $d\mathbf{X} \approx \mathbf{X}(t+dt) - \mathbf{X}(t)$ that you would move in an 'instant' dt from the given information t, $x(t)$ and $y(t)$. This is simply Euler's method in vectors.

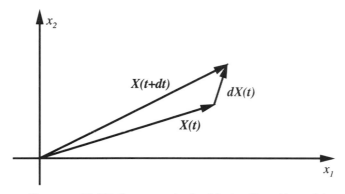

FIGURE 21.37: Increment of a Vector Function of t

The basic theory is similar, but the geometry of solutions is much richer. We state the 2-D versions of the theory first and then examine some mathematical examples. Use of the phase plane analysis is taken up in the examples below where the whole story is presented:

   (1) What are the assumptions that go into describing change mathematically?
   (2) How does the phase plane or flow describe the dynamic movement?

THEOREM 21.38. *Vector Existence & Uniqueness for I. V. P.s*
Suppose that the functions $f(x,y,t)$, $g(x,y,t)$, $\frac{\partial f}{\partial x}(x,y,t)$, $\frac{\partial f}{\partial y}(x,y,t)$, $\frac{\partial g}{\partial x}(x,y,t)$ and $\frac{\partial g}{\partial y}(x,y,t)$ are continuous in a box around (x_0, y_0, t_0). Then the initial value problem

$$x(0) = x_0$$
$$y(0) = y_0$$
$$dx = f(x,y,t)\, dt$$
$$dy = g(x,y,t)\, dt$$

has a unique vector solution with components $x(t)$, $y(t)$ defined for some small time interval $t_0 - \Delta < t < t_0 + \Delta$. In this case, Euler's Method converges to $x(t)$, $y(t)$ on closed subintervals $[t_0, t_1]$, for $t_1 < t_0 + \Delta$.

Geometric analysis of solutions by direction of change only makes sense when the direction of change does not vary with time.

DEFINITION 21.39. *Autonomous Differential Equation*
A vector differential equation of the form

$$\frac{d\mathbf{X}}{dt} = \mathbf{F}(\mathbf{X})$$

where the function $\mathbf{F}$ *does NOT depend on the independent variable, t, is called autonomous. In components this becomes*

$$dx = f(x, y) \, dt$$
$$dy = g(x, y) \, dt$$

where f and g depend on x and y, but not on t.

An autonomous equation has a growth law that does not depend on time. This is essential in the analysis of growth by just direction of change, because otherwise the arrows on our x-y-plane would move in time at the same x-y position.

DEFINITION 21.40. *Direction Fields*
An x-y-plane with arrows pointing in the direction of x-y-change corresponding to an autonomous differential equation is called a two dimensional direction field.

The solution curves $x(t)$, $y(t)$, plotted parametrically (without t) pass through the direction field tangent to the arrows at every point.

Points on the direction field with both component derivatives zero are "in equilibrium," that is, if you start there, the flow never carries you off; the constant function $(x(t), y(t)) = (x_e, y_e)$ for all t is a solution.

DEFINITION 21.41. *Equilibria, Attractors and Repellers*
A point (x_e, y_e) *on a direction field for the vector differential equation*

$$dx = f(x, y) \, dt$$
$$dy = g(x, y) \, dt$$

with $f(x_e, y_e) = g(x_e, y_e) = 0$ *is called an equilibrium point. An equilibrium point is called a local attractor if solutions that start from nearby initial conditions tend toward the point. An equilibrium point is called a repeller if solutions that start from nearby initial conditions tend away from the point.*

An equilibrium point does not have to be either an attractor or a repeller; some nearby points may move in while others move out.

EXAMPLE 21.42. *Competition*

Our first geometric analysis will be of the equations

$$dx = r\,x(1 - \frac{x}{7} - \frac{y}{7})\,dt$$
$$dy = s\,y(1 - \frac{x}{10} - \frac{y}{5})\,dt$$

for positive constants r and s. We can think of these equations as describing an interaction between two species. Notice that if $y(t) = 0$ for any time, then $dy = 0$, so y remains zero for all future time. In this case, the first equation becomes $dx = r\,x(1 - x/7)\,dt$, which is a logistic growth equation like the ones we studied above. The points $(0,0)$ and $(7,0)$ are equilibria, with $(0,0)$ repelling along the x-axis and $(7,0)$ attracting along the x-axis. Similarly, if $x = 0$, then $dx = 0$, so x doesn't change and the equation $dy = s\,y(1 - y/5)\,dt$ is another logistic growth equation with $(0,0)$ repelling along the y-axis and $(0,5)$ attracting along the y-axis.

The important new ingredient we need to study is the interaction (or coupling) between x and y. First, the change is strictly vertical if x doesn't change or $dx = 0$. This is the equation

$$r\,x(1 - \frac{x}{7} - \frac{y}{7}) = 0$$

We have already discussed the solution $x = 0$. This is when we are on the y-axis and have the dynamics of $dy = s\,y(1 - y/5)\,dt$. The other solution is the line

$$(1 - \frac{x}{7} - \frac{y}{7}) = 0$$

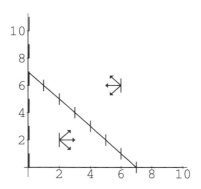

FIGURE 21.43: Vertical
$$dx = r\,x(1 - \tfrac{x}{7} - \tfrac{y}{7})\,dt = 0$$

To the left of this line in the first quadrant (where x and y are positive), the value of dx is positive, so the vector

$$dX = \begin{bmatrix} dx \\ dy \end{bmatrix}$$

points 'easterly.' It may point due east, northeast, or southeast, but it cannot have a westerly component.

To the right of the line $(1 - \frac{x}{7} - \frac{y}{7}) = 0$, the value of dx is negative, so the vector dX points 'westerly.'

The change is strictly horizontal if y doesn't change or $dy = 0$. This is the equation

$$s\,y(1 - \frac{x}{10} - \frac{y}{5}) = 0$$

We have already discussed the solution $y = 0$. This is when we are on the x-axis and have the dynamics of $dx = r\,x(1 - x/7)\,dt$. The other solution is the line

$$(1 - \frac{x}{10} - \frac{y}{5}) = 0$$

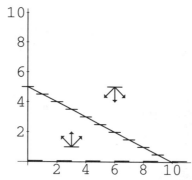

FIGURE 21.44:   Horizontal

$$dy = s\,y(1 - \tfrac{x}{10} - \tfrac{y}{5})\,dt = 0$$

In the first quadrant below this line, the value of dy is positive, so the vector $d\mathbf{X}$ points 'northerly.'

Above the line $(1 - \tfrac{x}{10} - \tfrac{y}{5}) = 0$, the value of dy is negative and the vector $d\mathbf{X}$ points 'southerly.'

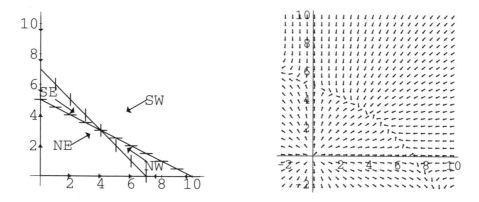

FIGURE 21.45: Compass Directions & Direction Field

for
$$dx = r\,x(1 - \tfrac{x}{7} - \tfrac{y}{7})\,dt$$
$$dy = s\,y(1 - \tfrac{x}{10} - \tfrac{y}{5})\,dt$$

A more detailed computer calculation of directions for this example is shown in the next figure. You can draw direction fields with the *Mathematica* NoteBook **DirField.ma**. Either of these figures can be used to sketch solutions of this vector differential equation. Put your pencil down on the compass direction chart and move in the directions indicated or on the computer drawn direction field and move keeping your motion tangent to the little arrows.

The exact solution curves $(x(t), y(t))$, drawn parametrically, must pass through points in the x-y plane so that their tangents point in the directions of the direction field. The next figure shows the last frame of a *Mathematica* animation of many solutions dynamically moving under the change law described by these differential equations. It was created with the NoteBook **Flow2D.ma**. The heavy dots are the current positions of points in the solution.

Notice that the point of intersection of the lines

$$1 - \frac{x}{7} - \frac{y}{7} = 0$$
$$1 - \frac{x}{10} - \frac{y}{5} = 0$$

is $(x, y) = (4, 3)$, so that there are equilibrium points at $(4, 3)$, $(7, 0)$, $(0, 5)$ and $(0, 0)$. You can simply look at the solution flow to see that $(4, 3)$ is a local attractor, that $(0, 0)$ is a local repeller and the other two attract in some directions and repel in others. We called this example 'competition' because, while the two 'species' x and y each tend to diminish the other (with the terms $\frac{x}{10}$ and $\frac{y}{7}$), they do balance this competition at the point $(4, 3)$.

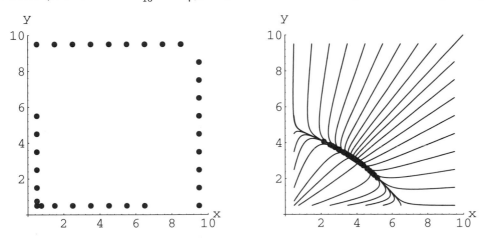

FIGURE 21.46: Solution Flow at $t = 0$ and $t = 10$

$$dx = r\,x(1 - \tfrac{x}{7} - \tfrac{y}{7})\,dt$$
$$dy = s\,y(1 - \tfrac{x}{10} - \tfrac{y}{5})\,dt$$

Our next example has very similar looking equations, but very different dynamics.

EXAMPLE 21.47. *Fierce Competition*

This dynamical system is

$$dx = r\,x(1 - \frac{x}{10} - \frac{y}{5})\,dt$$
$$dy = s\,y(1 - \frac{x}{7} - \frac{y}{7})\,dt$$

Notice that if $y(t) = 0$ for any time, then $dy = 0$, so y remains zero for all future time. In this case, the first equation becomes $dx = r\,x(1 - x/10)\,dt$, which is a logistic growth equation like the ones we studied above. The points $(0, 0)$ and $(10, 0)$ are equilibria, with $(0, 0)$ repelling along the x-axis and $(10, 0)$ attracting along the x-axis. Similarly, if $x = 0$, then $dx = 0$, so x doesn't change and the equation $dy = s\,y(1 - y/7)\,dt$ is another logistic growth equation with $(0, 0)$ repelling along the y-axis and $(0, 7)$ attracting along the y-axis.

The change vector $d\mathbf{X}$ is strictly vertical if x doesn't change or $dx = 0$. This is the equation

$$r\,x(1 - \frac{x}{10} - \frac{y}{5}) = 0$$

We have already discussed the solution $x = 0$. The other solution is the line

$$(1 - \frac{x}{10} - \frac{y}{5}) = 0$$

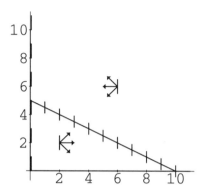

FIGURE 21.48: Vertical

$$dx = r\,x(1 - \frac{x}{10} - \frac{y}{5})\,dt = 0$$

To the left of this line in the first quadrant (where x and y are positive), the value of $d\mathbf{X}$ points 'easterly.' It may point due east, northeast, or southeast, but it cannot have a westerly component.

To the right of the line $(1 - \frac{x}{10} - \frac{y}{5}) = 0$, the value of dx is negative, so the vector $d\mathbf{X}$ points 'westerly.'

The change is strictly horizontal if y doesn't change or $dy = 0$. This is the equation

$$s\,y(1 - \frac{x}{7} - \frac{y}{7}) = 0$$

We have already discussed the solution $y = 0$. This is when we are on the x-axis and have the dynamics of $dx = r\,x(1 - x/10)\,dt$. The other solution is the line

$$(1 - \frac{x}{7} - \frac{y}{7}) = 0$$

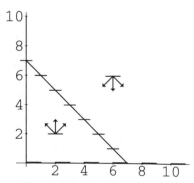

FIGURE 21.49: Horizontal

$$dy = s\,y(1 - \tfrac{x}{7} - \tfrac{y}{7})\,dt = 0$$

In the first quadrant below this line, the value of dy is positive, so the vector $d\mathbf{X}$ points 'northerly.'

Above the line $(1 - \tfrac{x}{10} - \tfrac{y}{5}) = 0$, the value of dy is negative and the vector $d\mathbf{X}$ points 'southerly.'

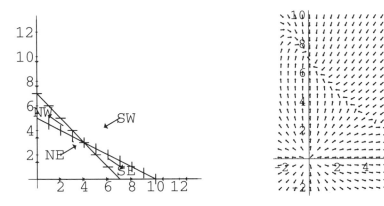

FIGURE 21.50: Compass Directions & Direction Field

for
$$dx = r\,x(1 - \tfrac{x}{10} - \tfrac{y}{5})\,dt$$
$$dy = s\,y(1 - \tfrac{x}{7} - \tfrac{y}{7})\,dt$$

A more detailed computer calculation of directions is shown in the next figure. This is the direction field for this example. Use either the compass direction chart or the direction field to sketch several solutions of the differential equations now.

The solution curves $(x(t), y(t))$, drawn parametrically, must pass through points in the x-y plane so that their tangents point in the directions of the direction field.

Notice that the point of intersection of the lines

$$1 - \frac{x}{7} - \frac{y}{7} = 0$$
$$1 - \frac{x}{10} - \frac{y}{5} = 0$$

is still $(x,y) = (4,3)$, so that there are equilibrium points at $(4,3)$, $(10,0)$, $(0,7)$ and $(0,0)$. This time the central equilibrium $(4,3)$ is not attracting. Most solutions that begin in the first quadrant end up either at $(10,0)$ or $(0,7)$. This is ferocious competition where either x or y wins, killing their whole competing species.

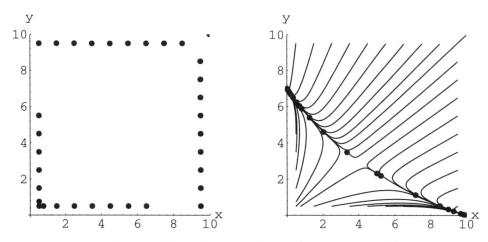

FIGURE 21.51: Solution Flow at $t = 0$ and $t = 10$

$$dx = r\, x(1 - \tfrac{x}{10} - \tfrac{y}{5})\, dt$$
$$dy = s\, y(1 - \tfrac{x}{7} - \tfrac{y}{7})\, dt$$

21.5.1. Flows vs. Explicit Solutions. Even the static last frame of the flow animation for these equations shows a 'big picture' that is hard to get from individual solutions. The (vector) solution starting at $(2, 2)$ passes near the central equilibrium, but then goes on toward the $(0, 7)$ equilibrium. As explicit solutions they look like:

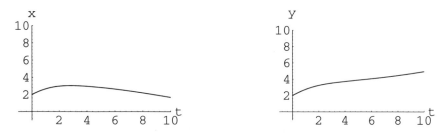

FIGURE 21.52: Explicit Solutions Starting at (2,2)

The solution starting at $(4, 2)$ tends to the $(10, 0)$ equilibrium and look explicitly as follows:

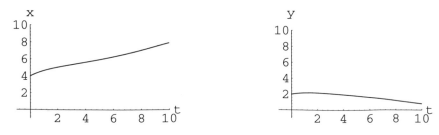

FIGURE 21.53: Explicit Solutions Starting at (4,2)

The solution starting at $(4, 6)$ tends to the $(0, 7)$ equilibrium and look explicitly as follows:

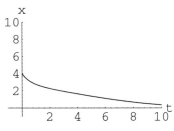
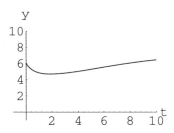

FIGURE 21.54: Explicit Solutions Starting at (4,6)

The 'phase plane' of complete parametric curves gives us a simple geometric way to see how the whole dynamical system works. The static picture loses the 'how fast' information, but the *Mathematica* flow animation even captures that. The next section will show what phase plane analysis adds to your understanding of the S-I-R epidemic model.

EXERCISE 21.55. *Sketching Explicit Solutions*
Use the solution flow figure to sketch the explicit solution $(x(t), y(t))$ to these equations that starts at $(x_0, y_0) = (4, 8)$. Notice that $x(t)$ is decreasing, while $y(t)$ first decreases and then increases again.

EXERCISE 21.56. *Direction Fields & Flows*
Geometrically analyze the differential equations

$$dx = 2x(1 - \frac{x}{4} - \frac{y}{4}) \, dt$$
$$dy = y(1 - \frac{x}{3} - \frac{y}{6}) \, dt$$

*Do the calculations both by hand as in the above discussion and check your work with the example NoteBook **CowSheep.ma**.*

Now apply your knowledge to an application.

PROBLEM 21.57. *Predator-Prey Interaction*
The Lotka-Volterra equations describe a simple predator-prey system. The equations are:

$$dx = (p - m\,y)x \, dt$$
$$dy = (-q + n\,x)y \, dt$$

One of these variables represents the density of rabbits, and the other represents the density of foxes in an environment with unlimited space and grass. Analyze the cases where one species is absent ($x = 0$ and $y = 0$) as well as how one species effects the other. Which variable is foxes and which rabbits?

Sketch the lines where the vector field is horizontal. Which population has leveled off along these lines?

Sketch the lines where the vector field is vertical. Which population has leveled off along these lines?

Make a compass heading chart and a direction field and use them to hand sketch some parametric solutions of the equation.

*Check your work with the example NoteBook **FoxRabbit.ma** when $p = 0.1$, $q = 0.04$, $m = 0.005$, $n = 0.00004$.*

After watching the flow animation you should sketch the solutions as functions of time, x vs. t and y vs. t.

*Use **AccDEsol.ma** to solve the equations with the initial conditions (1000,5), (1000,20). Compare these solutions to your sketches.*

21.6. Flow Analysis of S-I-R Epidemics

Recall the differential equations from Chapter 2 describing an S-I-R epidemic:

$$\frac{ds}{dt} = -a\,s\,i$$

(S-I-R DE's)

$$\frac{di}{dt} = a\,s\,i - b\,i$$

$r = 1 - s - i$. Where the variables are:

t	=	time measured in days continuously from t=0 at the start of the epidemic
s	=	the fraction of the population that is susceptible
i	=	the fraction of the population that is infected
r	=	the fraction of the population that is removed

with n equal to the (fixed) size of the total population.

The parameter b is the average daily rate of recovery,

$$b = 1/(\text{the number of days infectious})$$

The parameter a is the rate at which members of the population mix in a manner sufficiently close to transmit the particular disease. When the contact involves a susceptible and infected person, this results in a new infection. This means that the number of new infected people per day is a times the number of infectious people ($i \times n$) times the fraction of contacted people who are still susceptible, $s = \frac{\text{number susceptible}}{n}$:

$$\text{number of new infected per day} = a\,s\,i \times n$$

This means that the rate of decrease of s is $a\,s\,i$, since we divide by n to get the fractional variable, s.

The ratio

$$c = \frac{a}{b}$$

is called the "contact number" and intuitively represents the average number of contacts each infected has over the course of the disease. c is the parameter that can be measured. Typical values of a and b for rubella are:

$$b = 1/11$$
$$c = 6.8$$
$$a = b\,c = 6.8/11$$

You found explicit computer solutions to these differential equations in Chapter 2 and you could re-compute these with **AccDEsol.ma** if they're lost. However, the explicit solutions are not the best way to understand some aspects of the epidemic model. Recall that c is measured by solving a messy equation for its smallest root. Why is this? More fundamentally (but closely related), what is a formula for

$$\lim_{t \to \infty} s[t] =?$$

Or, at least, how does the number of people who escape infection during the epidemic depend on the number of people who are susceptible at the start of the epidemic? These questions become easy to understand with a hand-drawn sketch of the flow of the S-I-R equations.

PROBLEM 21.58. *Escaping Infection for Individuals and the Herd*
Make a compass direction chart and hand sketch of parametric solutions to the S-I-R Epidemic differential equations. analyze your flow and explain whether you would be more likely to escape infection in a rubella epidemic if it began with many more or relatively few people, besides you, susceptible.

*Modify the **Flow2D.ma** NoteBook to animate the flow of an epidemic and compare the movie with your hand sketch.*

*Where does $\lim_{t \to \infty} s[t]$ show up on your phase plane and in the computer movie of the flow? How does this limit change when $s[0]$ is nearer to 1? Use the Mathematica NoteBook **AccDEsol.ma** to solve the cases of 60% of the population initially susceptible or with 90%. Start both cases with 1% infected.*

The fraction of non-immune people is s. Use your hand sketch to tell how many people in a population must be immune in order to insure that the disease decreases. In other words, which portion of your compass heading chart has i decreasing, or equivalently, $\frac{di}{dt}$ negative? Answer this in terms of parameters and s alone. Would you have an expanding epidemic in a population with s less than this?

21.6.1. S-I-R Epidemics with Births and Deaths. Suppose new babies are born without immunity to an S-I-R disease. The previous model of an epidemic would not be useful in the long term study of the disease. For simplicity we will assume that births equal deaths so that the size of our population remains constant.

VARIABLES FOR S-I-R WITH BIRTHS=DEATHS:

$$
\begin{aligned}
t &= \text{time in days with } t = 0 \text{ at outbreak} \\
s &= \text{susceptible fraction} \\
i &= \text{infected fraction} \\
r &= \text{removed fraction with } s + i + r = 1
\end{aligned}
$$

Again we have the parameters a and b where

$$
\begin{aligned}
a\,s\,i &= \text{rate of spreading the infection} \\
b\,i &= \text{rate of recovery}
\end{aligned}
$$

With births and deaths we add a rate parameter k so that there are

$$k \cdot n = \text{total number of births per day} = \text{total number of deaths per day}$$

and we assume that deaths are from causes other than the disease we study, so deaths leave the compartments at rates proportional to their sizes as indicated in the figure below.

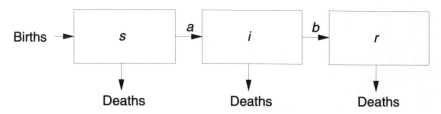

FIGURE 21.59: S-I-R with Birth and Death

The rates of deaths are

$$k \cdot s \cdot n \;=\; \text{number of dead susceptibles per day}$$
$$k \cdot i \cdot n \;=\; \text{number of dead infectives per day}$$
$$k \cdot r \cdot n \;=\; \text{number of dead removeds per day}$$

Births equals deaths because $k \cdot s \cdot n + k \cdot i \cdot n + k \cdot r \cdot n = k(s + i + r)n = k \cdot 1 \cdot n$.
The change in number of susceptibles is now

$$\text{Births - deaths of susceptibles - infections}$$

$$n\frac{ds}{dt} \;=\; kn - ksn - asin$$

so $$\frac{ds}{dt} \;=\; k - ks - asi$$

EXERCISE 21.60. *Show that the differential equation for the infectious fraction i is*

$$\frac{di}{dt} = asi - bi - ki$$

and that $r = 1 - s - i$.

The sketch of the compass headings for the equations

$$\frac{ds}{dt} = k - ks - asi$$

$$\frac{di}{dt} = asi - bi - ki$$

is a little more difficult than earlier examples. No change in i, or $\frac{di}{dt} = 0$ is easy, $asi - bi - ki = 0$ if and only if either $i = 0$ (the s-axis) or $asi - b - k = 0$, $s = (b + k)/a$ (a vertical line).

No change in s or $\frac{ds}{dt} = 0$ occurs on a hyperbola, $k - ks - asi = 0$. This might be a little confusing with all the letters. Notice that $xy = c$ is a simple hyperbola and a change of variables makes this equation look similar,

$$k - ks - asi = 0 \quad \Leftrightarrow \quad \frac{k}{a} = s\left(i + \frac{k}{a}\right) \quad \Leftrightarrow \quad c = sj$$

In any case, we can use *Mathematica* as follows:

```
k = .001;
b = 1/11.;
c = 6.8;
```

a = b c;
vertical = ContourPlot[k - k s - a s i,{s,0,1},{i,0,1},
    Contours->{0},ContourShading->False,PlotPoints->50];
horizontal = ContourPlot[a s - b - k,{s,0,1},{i,0,1},
    Contours->{0},ContourShading->False,PlotPoints->50];
Show[vertical,horizontal];

EXERCISE 21.61. *Sketch a compass heading chart for the S-I-R with births equations. Use the Mathematica computation suggested before this exercise to help, if you wish.*

*Modify the Mathematica NoteBooks **AccDEsol.ma** and **Flow2D.ma** to compute s and i for a model of S-I-R with births=deaths, $k = 1$ births per 1000 population, $k = 0.001$, and the parameters above for rubella, $c = 6.8$, $b = 1/11.0$ and $a = bc$. What does the model predict about endemic infection,*

$$\lim_{t \to \infty} i[t] \neq 0?$$

*When $\frac{a}{b+k} > 1$, use the Mathematica NoteBook **Flow2D.ma** to show that*

$$s \to \frac{b+k}{a} \qquad and \qquad i \to \frac{k}{a}\left(\frac{a}{b+k} - 1\right) \ as \ t \to \infty$$

*Verify this in several explicit examples using **AccDEsol.ma**.*

Prove that $\left(\frac{b+k}{a}, \frac{k}{a}\left(\frac{a}{b+k} - 1\right)\right)$ is an equilibrium point. Why do we need $\frac{a}{b+k} > 1$? Are other conditions needed to make $0 < i_{equil} < 1$?

Following are two explicit solutions, but you will see that interpretation of the 'phase portrait' or last frame of the flow animation reveals more about the long term dynamics of the disease.

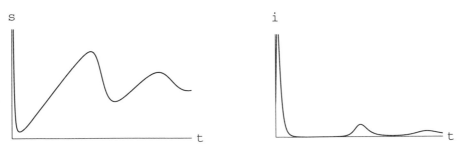

FIGURE 21.62: Two Years of Rubella Infectives with Births

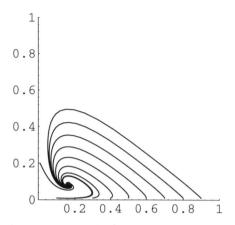

FIGURE 21.63: S-I-R with Births = Deaths and Large $k = 0.01$

21.7. Projects

The mathematical tools that we have developed in this chapter make the epidemic projects on Herd Immunity and the S-I-S Endemic Limit into simple exercises. The questions raised in those projects are important ones like, 'Why did we eradicate polio, but not measles?' The point is that calculus does give us deep insight into many scientific questions about changing quantities.

Several examples and exercises in this chapter hint at the ecological projects on competition between species. There are three extensive models in the Scientific Projects on this topic. One explores predator-prey interactions and the effect of disturbing the ecosystem. Another explores competition between species, and a third explores cooperation.

There is a model of production and exchange in the Scientific Projects on economics.

There are projects on chemical reactions and reaction stability.

The Mathematical Background has a chapter called "Differential Equations from Increment Geometry." These projects show how differential equations give several mathematically important curves. These curves are concrete, the shape of a chain hung between two poles, known as the catenary, the path of an object pulled by a tractor when the object does not start straight behind it, known as the tractrix, and an object sliding down a wire at a fixed vertical rate, known as the isochrone.

The Bungee project, the air resistance project, the project on low level bombing, the study of the real and idealized pendulum, all can be understood using the ideas of this chapter.

You are ready now to begin any of these projects and do the important first step of setting up the equations, or 'speaking the language of change.' You can solve these projects with approximate computer solutions of the associated differential equations, but the next two chapters develop additional mathematical tools that can help in that analysis.

The linear differential equations in the next chapter have important applications in their own right, such as Lanchester's combat model and the resonance and filtering Scientific Projects. Linear equations are also important mathematically. They can be solved in terms of natural exponentials as we show in Chapter 22. These solutions then give us a new kind of 'microscope.' When we 'look at a nonlinear equilibrium under a microscope,' we see an associated linear equilibrium. That local analysis is the subject of Chapter 23.

CHAPTER 22

Autonomous Linear Dynamical Systems

This chapter shows how to find symbolic solutions to two dimensional linear dynamical systems. These systems are closely related to second order one dimensional differential equations. We begin with a familiar example which is described by such an equation. First we show why the second order equation is equivalent to a two dimensional system. Then we show how to use exponentials $e^{r\,t}$ to solve the scalar equation. The solutions to linear systems give rise to a 'dynamical system microscope' which is explained in the next chapter.

22.1. Autonomous Linear Constant Coefficient Equations

This section sets up a differential equation describing the motion of an object of mass m suspended on an ideal spring with spring constant s and moving in a fluid that produces a restoring friction force proportional to speed, $-c\frac{dx}{dt}$. Newton's "$F = m\,A$" law becomes

$$m\frac{d^2x}{dt^2} = -\,c\frac{dx}{dt} - s\,x$$

where x is the displacement of the spring and t is time. Details are given below.

You may think of this equation as the one governing the reaction of your car after you go over a bump. The constants m, c and s are a fixed part of the design of your car, at least for short periods of time. As your shock absorbers wear out, c becomes smaller. How small can c become before your car oscillates after it goes over a bump? You will be able to answer this after this chapter. You know intuitively that completely worn out shocks ($c \approx 0$) make your car bounce and bounce, while very stiff shocks ($c >> 0$) jolt the inside passengers as they restore the level of the car.

22.1.1. The Phase Variable Trick.
We want to analyze this equation in several ways. We can view it as a 2 dimensional system by introducing the 'phase variable' $\frac{dx}{dt} = y$. Since $\frac{dy}{dt} = \frac{d}{dt}\left(\frac{dx}{dt}\right) = \frac{d^2x}{dt^2}$, the differential equation above becomes $m\frac{dy}{dt} = m\frac{d^2x}{dt^2} = -c\frac{dx}{dt} - s\,x = -c\,y - s\,x$. If we put these two equations together we have an equivalence between the

system of first order equations and the single equation with two derivatives:

$$\frac{dx}{dt} = y \qquad\qquad \Leftrightarrow \qquad\qquad m\frac{d^2x}{dt^2} = -c\frac{dx}{dt} - s\,x$$

$$\frac{dy}{dt} = -\frac{s}{m}x - \frac{c}{m}y$$

Physicists call the combined position and momentum the phase of a body. The use of the vector phase $(x, \frac{dx}{dt}) = (x, y)$ is why the final flow picture is often called the phase plane or phase portrait.

This system is linear, a variant of

$$\frac{dx}{dt} = a_1\,x + a_2\,y$$

$$\frac{dy}{dt} = b_1\,x + b_2\,y$$

where the constants are $a_1 = 0$, $a_2 = 1$, $b_1 = -s/m$ and $b_2 = -c/m$. Complete understanding of this system will lead to a way to 'localize' nonlinear problems and study equilibria with the basic idea of calculus that 'nonlinear phenomena locally look linear.'

22.1.2. Derivation of the Shock Absorber Equation. Now we will give the details of why the mass-spring-shock absorber system satisfies the second order differential equation above. Newton's law says, "$F = m\,A$," the total applied force equals the mass times the acceleration that force produces. We need to use the interpretation of the first and second derivatives as velocity and acceleration (see Chapter 10 for a review.)

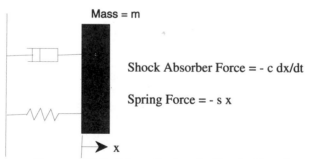

FIGURE 22.1: Mass, Spring & Shock Absorber

The position of our object of mass m is measured by x as shown in Figure 22.1.2, where the spring is relaxed when $x = 0$. The derivative $\frac{dx}{dt}$ represents the velocity of the object, or rate of change of position with respect to time, including sign, (+) to the right and (-) to the left. The second derivative $\frac{d^2x}{dt^2}$ is the acceleration of the object or rate of change of velocity (with sign, so negative acceleration is 'braking.') The $m\,A$ portion of Newton's Law becomes $m\,\frac{d^2x}{dt^2}$ and the law says,

$$m\,\frac{d^2x}{dt^2} = \text{Force of the Spring} + \text{Force of the Shock Absorber}$$

The restoring force when we stretch a spring by a distance x is $-s\,x$. It is negative because it acts opposite its displacement. This formula also 'pushes' when the spring is compressed (unlike the slack bungee cord). If the spring is compressed, $x < 0$, and $-s\,x > 0$, $[(-)(-) = (+)]$, so the spring pushes to the right. This formula makes "$F = m\,A$" look like:

$$\text{Force of the Spring} = -s\,x$$

$$m\,\frac{d^2x}{dt^2} = -s\,x + \text{Force of the Shock Absorber}$$

The force due to the shock absorber is $-c\,\frac{dx}{dt}$. This is a perfectly linear response to movement, with the force opposite the direction of movement. (This friction force is like our first model of air resistance.) The intuitive idea of this force should be familiar, if not it's linear nature. If you dip a spoon in a honey jar and move it very slowly, it resists relatively little. But if you try to move the spoon quickly, it resists a lot. We are simply making the relationship a proportion:

$$\text{Force resisting motion} \propto -\text{Speed of that motion}$$

$$\text{Force of an Ideal Shock Absorber} = -c\,\frac{dx}{dt}$$

$$m\,\frac{d^2x}{dt^2} = -s\,x + \text{Force of the Shock Absorber}$$

$$m\,\frac{d^2x}{dt^2} = -s\,x - c\,\frac{dx}{dt}$$

Hence, this is the differential equation governing the response of a 'front end' with a linear spring and shock absorber.

22.2. Symbolic Exponential Solutions

Now we proceed to try to find two 'really different' solutions to our linear differential equation. The 'method' of solution is simple. We'll guess that the solution is exponential and figure out how to make our guess work. Linear autonomous equations do have exponential solutions (in one form or another). To make the algebra simple we first consider the second order equation in the form

$$a\,\frac{d^2x}{dt^2} + b\,\frac{dx}{dt} + c\,x = 0$$

How many solutions of this can we find of the form $x(t) = e^{r\,t}$? We just plug in:

$$x = e^{r\,t} \qquad\qquad c\,x = c\,e^{t}$$

$$\frac{dx}{dt} = r\,e^{r\,t} \qquad\qquad b\,\frac{dx}{dt} = b\,r\,e^{r\,t}$$

$$\frac{d^2x}{dt} = r^2\,e^{r\,t} \qquad\qquad a\,\frac{d^2x}{dt^2} = a\,r^2\,e^{r\,t}$$

so that the sum is

$$a\,\frac{d^2x}{dt^2} + b\,\frac{dx}{dt} + c\,x = (a\,r^2 + b\,r + c)e^{r\,t}$$

We know that $e^{r\,t} \neq 0$ for any time, so the only way to make $a\frac{d^2x}{dt^2} + b\frac{dx}{dt} + c\,x = 0$ is to make

$$(a\,r^2 + b\,r + c) = 0$$

TERRIFIC! We have reduced our calculus problem, finding solutions of the differential equation, to high school algebra, solve $(a\,r^2 + b\,r + c) = 0$. This algebraic equation is called the **characteristic equation** of the differential equation and its roots are called the **characteristic roots**. The quadratic formula says the roots are

$$r_1, r_2 = \frac{-b \pm \sqrt{b^2 - 4\,a\,c}}{2\,a}$$

so two solutions to the differential equation are

$$x_1(t) = e^{r_1\,t} \quad \text{and} \quad x_2(t) = e^{r_2\,t}$$

EXERCISE 22.2. *Verify Solutions*
Show that $x_1(t) = e^{2\,t}$, $x_2(t) = e^{-3\,t}$ and $x_3(t) = k_1\,e^{2\,t} + k_2\,e^{-3\,t}$ are all solutions to the differential equation

$$\frac{d^2x}{dt^2} + \frac{dx}{dt} - 6\,x = 0$$

In each case calculate the initial values $x_1(0)$, $x_2(0)$, $x_3(0)$, $x_1'(0)$, $x_2'(0)$ and $x_3'(0)$.
    Find the roots r_1 and r_2 of the algebraic equation

$$r^2 + r - 6 = 0$$

Write the specific exponentials $e^{r_1\,t}$ and $e^{r_2\,t}$.

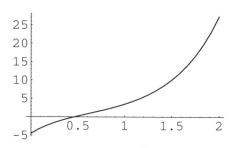

FIGURE 22.3: $x(t) = \frac{1}{2}\,e^{2\,t} - 5\,e^{-3\,t}$

Usually, we can get by with high school algebra.

EXERCISE 22.4. *Find Solutions by Algebra*
Find two distinct exponential solutions of the equations:

  (1) $\frac{d^2x}{dt^2} - \frac{dx}{dt} - 6\,x = 0$
  (2) $\frac{d^2x}{dt^2} - 5\frac{dx}{dt} + 6\,x = 0$
  (3) $\frac{d^2x}{dt^2} + \frac{dx}{dt} - x = 0$
  (4) $\frac{d^2x}{dt^2} - 6\frac{dx}{dt} + 9\,x = 0$
  (5) $\frac{d^2x}{dt^2} + \frac{dx}{dt} + x = 0$

Check and sketch your solutions with Mathematica, for example

```
x1[t_ ] := Exp[-2 t];
x2[t_ ] := Exp[3 t];
x1[t]
D[x1[t], t]
D[x1[t],{t,2}] (*The second derivative.*)
D[x1[t],{t,2}] - D[x1[t], t] - 6 x1[t]
x2[t]
D[x2[t],{t,2}] - D[x2[t], t] - 6 x2[t]
Plot[{ x1[t], x2[t] } , { t, 0, 1 } ];
```

EXERCISE 22.5. *Repeated Roots*
Show that the functions $x_1(t) = e^{rt}$ and $x_2(t) = t\,e^{rt}$ are both solutions to the differential equation $\frac{d^2x}{dt^2} + 2\frac{dx}{dt} + x = 0$, where r satisfies $r^2 + 2r + 1 = 0$. Substitute each into the left hand side. What are the characteristic roots of this equation?

Check and plot these solutions with Mathematica.

22.3. Rotation & Euler's Formula

$$e^{i\theta} = \text{Cos}[\theta] + i\ \text{Sin}[\theta]$$

Recall the following basic mathematical fact. From our peculiar point of view $z(t) = e^{rt}$ is **defined** to be the solution of

$$z(0) = 1$$
$$\frac{dz}{dt} = r\,z$$

Also recall the Existence and Uniqueness Theorem, apply its hypotheses and show that the solution to this problem is unique. We will show geometrically that the solution is $z(t) = \text{Cos}[t] + i\ \text{Sin}[t]$ when $r = i = \sqrt{-1}$. Because of the definition of the natural exponential, the initial value problem

$$z(0) = 1$$
$$\frac{dz}{dt} = i\,z$$

(1) has the solution $z(t) = e^{it}$.
(2) By geometry it also has the solution $z(t) = \text{Cos}[t] + i\ \text{Sin}[t]$.
(3) By uniqueness it has only one solution, hence, $e^{it} = \text{Cos}[t] + i\ \text{Sin}[t]$.

This is a proof of the important and useful formula of Euler. Euler's formula allows us to find real solutions to our differential equation when the characteristic roots are complex.

Complex numbers are vectors with a special multiplication peculiar to two dimensions. We may think of the complex number $z = x + iy$ as the vector

$$z = \begin{bmatrix} x \\ y \end{bmatrix}$$

We simply treat the factor of the 'imaginary' number $\sqrt{-1} = i$ as the y-axis component. The complex product can be computed algebraically: $(a+ib)(x+iy) = a\,x + i\,a\,y + i\,b\,x + i^2\,b\,y =$

$(a\,x - b\,y) + i\,(a\,y + b\,x)$. Euler's formula shows that this product also has the simple geometric meaning of adding angles and multiplying lengths, but first, here's some practice:

EXERCISE 22.6. *Complex Multiplication*

Plot the unit vectors $\begin{bmatrix} \frac{\sqrt{3}}{2} \\ \frac{1}{2} \end{bmatrix}$ *and* $\begin{bmatrix} \frac{1}{2} \\ \frac{\sqrt{3}}{2} \end{bmatrix}$. *Compute the angle each makes with the horizontal x-axis.*

Show that the product $(\frac{\sqrt{3}}{2} + i\frac{1}{2})(\frac{1}{2} + i\frac{\sqrt{3}}{2}) = i$. *(30° + 60° = 90°)*

Show that the product $(\frac{\sqrt{3}}{2} + i\frac{1}{2})(\frac{\sqrt{3}}{2} + i\frac{1}{2}) = (\frac{1}{2} + i\frac{\sqrt{3}}{2})$. *(30° + 30° = 60°)*

Plot the unit vectors $\begin{bmatrix} \frac{\sqrt{3}}{2} \\ \frac{1}{2} \end{bmatrix}$ *and* $\begin{bmatrix} \frac{1}{\sqrt{2}} \\ \frac{1}{\sqrt{2}} \end{bmatrix}$. *Compute the angle each makes with the horizontal x-axis.*

Compute the complex product $(\frac{\sqrt{3}}{2} + i\frac{1}{2})(\frac{1}{\sqrt{2}} + i\frac{1}{\sqrt{2}}) = \frac{\sqrt{6}-\sqrt{2}}{4} + i\frac{\sqrt{6}+\sqrt{2}}{4}$

Use your calculator or Mathematica to plot the unit vector $\begin{bmatrix} \frac{\sqrt{6}-\sqrt{2}}{4} \\ \frac{\sqrt{6}+\sqrt{2}}{4} \end{bmatrix}$ *and show that it makes an angle of 75° with the horizontal. (30° + 45° = 75°)*

The vector associated with the complex number $z = x + i\,y$ is perpendicular to the vector $i\,z = -y + i\,x$. To see this compute the dot product

$$\begin{bmatrix} x \\ y \end{bmatrix} \bullet \begin{bmatrix} -y \\ x \end{bmatrix} = -x\,y + x\,y = 0$$

Vectors are perpendicular when their dot product is zero.

Now consider the complex differential equation

$$\frac{dz}{dt} = iz \qquad \text{or} \qquad \frac{dx}{dt} + i\frac{dy}{dt} = -y + i\,x$$

or in vector form

$$\begin{bmatrix} \frac{dx}{dt} \\ \frac{dy}{dt} \end{bmatrix} = \begin{bmatrix} -y \\ x \end{bmatrix}$$

This vector equation simply says that the rate of change of the vector z has the same size as z and is a vector perpendicular (and pointing counterclockwise) to z. The solution is geometrically given by a point traveling around a circle in time 2π (independent of radius and starting point).

EXERCISE 22.7. *Geometric solution of* $\frac{dz}{dt} = i\,z$

Suppose we start somewhere on the circle of radius r and move counterclockwise around this circle in time. The vector pointing from the origin to our position at any time is $\begin{bmatrix} x \\ y \end{bmatrix}$.

Sketch this figure. What is the connection between x, y and the radius r?

Which way will our velocity vector point for the counterclockwise motion described?

Sketch the illegal arrow $\begin{bmatrix} -y \\ x \end{bmatrix}$ *with its tail at the tip of* $\begin{bmatrix} x \\ y \end{bmatrix}$.

If the vector $\begin{bmatrix} -y \\ x \end{bmatrix}$ *is our velocity, what is our speed? Express your answer in terms of the constant r.*

How long does it take to travel around a circle of radius r at speed r? How far do you travel?

Unit length complex numbers can now be thought of either in terms of the sine-cosine radian measure, or in terms of the complex exponential.

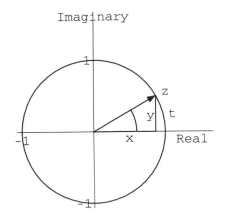

z = unit complex number

at angle t radians

$z = x + i\,y = e^{i\,t}$

$x = \text{Cos}[t]$

$y = \text{Sin}[t]$

FIGURE 22.8: $e^{i\,t} = \text{Cos}[t] + i\,\text{Sin}[t]$

We can verify the basic case solving $\frac{dz}{dt} = i\,z$ with $z(0) = 1$ by substituting the function $z(t) = \text{Cos}[t] + i\,\text{Sin}[t]$ into the equation.

$$\frac{dz}{dt}(t) = -\,\text{Sin}[t] + i\,\text{Cos}[t] = i(\text{Cos}[t] + i\,\text{Sin}[t])$$

$$\text{Cos}[0] + i\,\text{Sin}[0] = 1 + i \times 0 = 1$$

The natural exponential satisfies the important functional equation

$$e^{z+w} = e^z\,e^w$$

even when z and w are complex numbers. The idea in the proof of Section 8.2 in the chapter on natural log and exponential shows the basic step for a real constant c.

$$e^c\,e^{i\,t} = e^{c+i\,t}$$

because both the function $w_1(t) = e^c\,e^{i\,t}$ and the function $w_2(t) = e^{c+i\,t}$ satisfy the initial value problem $z(0) = e^c$ and $\frac{dz}{dt} = i\,z$. These functions must be equal by uniqueness of the solution to such initial value problems.

EXERCISE 22.9. *Addition Formulas*
We want to use Euler's formula two ways to show the connection between the functional equation of the natural exponential and the addition formulas for sine and cosine.

(1) *Use Euler's formula to write $e^{i\,\phi}$ as a sum of sine and cosine.*
(2) *Use Euler's formula to write $e^{i\,\psi}$ as a sum of sine and cosine.*
(3) *Use Euler's formula to write $e^{i(\phi+\psi)}$ as a sum of sine and cosine.*
(4) *Multiply the complex number answer to part (1) times the complex number answer to part (2) and combine the terms in the form $A + i\,B$ for real expressions A and B (involving sine and cosine).*

(5) *The answer to part (3) is $e^{i(\phi+\psi)}$ in real plus imaginary part, while the answer to part (4) is $e^{i\phi} \times e^{i\psi}$ in real plus imaginary part. Write the equality $e^{i(\phi+\psi)} = e^{i\phi} \times e^{i\psi}$ in real plus imaginary form,*

$$Ans.(3) = Ans.(4)$$

(6) *Show that this is equivalent to the pair of equations*

$$\mathrm{Cos}[\phi+\psi] = \mathrm{Cos}[\phi]\,\mathrm{Cos}[\psi] - \mathrm{Sin}[\phi]\,\mathrm{Sin}[\psi]$$
$$\mathrm{Sin}[\phi+\psi] = \mathrm{Sin}[\phi]\,\mathrm{Cos}[\psi] + \mathrm{Cos}[\phi]\,\mathrm{Sin}[\psi]$$

EXERCISE 22.10. *Use the cases shown above to prove the general case of the natural exponential equation*

$$e^z\, e^w = e^{a+ib}\, e^{x+iy} = e^{(a+x)+i(b+y)} = e^{z+w}$$

EXAMPLE 22.11. *Geometric Complex Multiplication*

Now we can use Euler's formula to give the geometric form of complex multiplication. We write two complex numbers in 'polar' form, a length (or radius) along a certain angle. The complex number $z = x + iy$ lies on a circle of radius $r = |z| = \sqrt{x^2 + y^2}$ and at some angle ϕ. The unit length vector at angle $\phi = \mathrm{ArcTan}[y/x]$ is $e^{i\phi}$ by Euler's formula. (Since we know from the radian measure definition of sine and cosine that it is the vector with components $(x_1, y_1) = (\mathrm{Cos}[\phi], \mathrm{Sin}[\phi])$.) The vector stretched to length r is therefore

$$z = r\,e^{i\phi}$$

Similarly, we may write another complex number $w = a + ib$ in the form

$$w = s\,e^{i\psi}$$

where $\psi = \mathrm{ArcTan}[b/a]$ and $s = |w| = \sqrt{a^2 + b^2}$. The product

$$z\,w = r\,s\,e^{i\,(\phi+\psi)}$$

which means that the length of the product is the product of the lengths and the angle of the product is the sum of the angles.

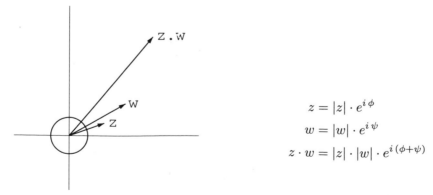

$$z = |z| \cdot e^{i\phi}$$
$$w = |w| \cdot e^{i\psi}$$
$$z \cdot w = |z| \cdot |w| \cdot e^{i\,(\phi+\psi)}$$

FIGURE 22.12: Geometric Complex Product

Now we return to our primary goal of solving linear differential equations by using Euler's formula in the form

$$e^{\alpha t + i\beta t} = e^{\alpha t} e^{i\beta t} = e^{\alpha t}(\text{Cos}[\beta t] + i\,\text{Sin}[\beta t])$$

EXAMPLE 22.13. *Real Solutions for Complex Roots*

If the characteristic roots associated with the differential equation

$$a\frac{d^2x}{dt^2} + b\frac{dx}{dt} + cx = 0$$

are the complex numbers $r_1 = \alpha + i\beta$ and $r_2 = \alpha - i\beta$, then the complex solutions $z_1(t) = e^{\alpha t + i\beta t}$ and $z_2(t) = e^{\alpha t - i\beta t}$ are solutions, but by Euler's formula, so are

$$x_1(t) = e^{\alpha t}\,\text{Cos}[\beta t] \qquad \text{and} \qquad x_2(t) = e^{\alpha t}\,\text{Sin}[\beta t]$$

EXERCISE 22.14.   (1) *For the solutions z_1, z_2, x_1 and x_2, given above, show that*

$$x_1(t) = \frac{1}{2}[z_1(t) + z_2(t)] \qquad \text{and} \qquad x_2(t) = \frac{i}{2}[z_2(t) - z_1(t)]$$

(2) *Show that*

$$\text{Cos}[\theta] = \frac{e^{i\theta} + e^{-i\theta}}{2} \qquad \text{and} \qquad \text{Sin}[\theta] = \frac{e^{i\theta} - e^{-i\theta}}{2i}$$

EXERCISE 22.15. *Verify by direct substitution that the functions*

$$x_1(t) = e^{-t}\,\text{Cos}[\sqrt{2}\,t] \qquad \text{and} \qquad x_2(t) = e^{-t}\,\text{Sin}[\sqrt{2}\,t]$$

are solutions to the differential equation

$$\frac{d^2x}{dt^2} + 2\frac{dx}{dt} + 3x = 0$$

Compute the derivatives of x_1 and x_2 carefully using the product rule on the terms e^{-t} and $\text{Cos}[\sqrt{2}\,t]$, *etc. The computation is messy, but it works.*

Also write the solutions to this equation in the complex exponential form and use Euler's formula to find a relationship between the real and complex solutions.

Check and sketch your solutions with Mathematica, for example

```
x1[t_ ] := Exp[-t] Cos[Sqrt[2] t];
x2[t_ ] := Exp[-t] Sin[Sqrt[2] t];
x1[t]
D[x1[t], t]
D[x1[t],{t,2}] (*The second derivative.*)
D[x1[t],{t,2}] + 2 D[x1[t], t] + 3 x1[t]
x2[t]
Simplify[ D[x2[t],{t,2}] + 2 D[x2[t], t] + 3 x2[t] ]
Plot[{ x1[t], x2[t] } , { t, 0, 10 } ];
```

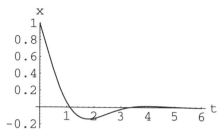

FIGURE 22.16: $x(t) = e^{-t}\,\text{Cos}[\sqrt{2}\,t]$

EXERCISE 22.17. *Find real solutions of the form*

$$x_1(t) = e^{\alpha t}\,\text{Cos}[\beta\,t] \qquad and \qquad x_2(t) = e^{\alpha t}\,\text{Sin}[\beta\,t]$$

for the equations

(1) $\frac{d^2x}{dt^2} + 4\frac{dx}{dt} + 13\,x = 0$

(2) $\frac{d^2x}{dt^2} + \frac{dx}{dt} + x = 0$

(3) $\frac{d^2x}{dt^2} + \omega^2\,x = 0$

Sketch your solution functions. Notice that they are sines or cosines attenuated or amplified by exponentials.

22.4. Superposition and The General Solution

A physical system is said to satisfy the superposition principle if the response to a sum of forces is the sum of the responses to the separate forces. We will see that the damped spring system governed by

$$m\frac{d^2x}{dt^2} + c\frac{dx}{dt} + s\,x = \text{Applied Force}$$

satisfies this principle. Mathematically, this is satisfied because x and its derivatives only appear in linear combinations on the left side of the equation. The mathematical superposition principle, or linearity of the differential equation, is the key to our complete symbolic solution of the differential equation

$$a\frac{d^2x}{dt^2} + b\frac{dx}{dt} + c\,x = 0$$

and its associated initial value problems.

So far in this section we have used real or complex exponentials to find two distinct real functions $x_1(t)$ and $x_2(t)$ satisfying the equation. Let k_1 and k_2 be constants. Mathematical superposition means that $x(t) = k_1\,x_1(t) + k_2\,x_2(t)$ is also a solution. We show this by substitution as follows.

$$c\,x_1, \qquad c\,x_2, \qquad c\,x = c(k_1\,x_x + k_2\,x_2) \qquad = \quad k_1\,c\,x_1 \quad + \quad k_2\,c\,x_2$$

$$b\,\frac{dx_1}{dt}, \qquad b\,\frac{dx_2}{dt}, \qquad b\,\frac{dx}{dt} = b(k_1\,\frac{dx_1}{dt} + k_2\,\frac{dx_2}{dt}) \qquad = \quad k_1\,b\,\frac{dx_1}{dt} \quad + \quad k_2\,b\,\frac{dx_2}{dt}$$

$$a\,\frac{dx_1^2}{dt^2}, \qquad a\,\frac{dx_2^2}{dt^2}, \qquad a\,\frac{dx^2}{dt^2} = a(k_1\,\frac{dx_1^2}{dt^2} + k_2\,\frac{dx_2^2}{dt^2}) \qquad = \quad k_1\,a\,\frac{dx_1^2}{dt^2} \quad + \quad k_2\,a\,\frac{dx_2^2}{dt^2}$$

Sum Vertically

$$0 \qquad\qquad 0 \qquad\qquad\qquad\qquad\qquad\qquad\qquad k_1 \cdot 0 \quad + \quad k_2 \cdot 0 = 0$$

We really never wanted to find solutions to the differential equation alone, but rather to the associated initial value problems. Now we have the tools to give a form for all solutions of all such problems. Two solutions $x_1(t)$ and $x_2(t)$ form a "basis" for all solutions of these initial value if they are 'really different' or linearly independent. In this case, all solutions may be written in the form

$$x(t) = k_1\,x_1(t) + k_2\,x_2(t)$$

for unknown constants k_1 and k_2. The next theorem 22.23 gives a list of basic solutions depending on the roots of the characteristic equation. First we consider a specific example.

EXAMPLE 22.18. *The Bouncing Buggy*

Suppose we have a system consisting of a unit mass, $m = 1$, a unit spring constant, $s = 1$, and unit damping friction, $c = 1$. This satisfies the second order equation $\frac{d^2x}{dt^2} = -\frac{dx}{dt} - x$. It has the characteristic equation $r^2 + r + 1 = 0$ with characteristic roots $-\frac{1}{2} + i\frac{\sqrt{3}}{2}$ and $-\frac{1}{2} - i\frac{\sqrt{3}}{2}$. This makes the basic solutions

$$e^{-\frac{t}{2}}\,\mathrm{Cos}[\frac{\sqrt{3}}{2}t] \qquad \text{and} \qquad e^{-\frac{t}{2}}\,\mathrm{Sin}[\frac{\sqrt{3}}{2}t]$$

This second order differential equation can be written as the two dimensional system

$$\frac{dx}{dt} = y$$
$$\frac{dy}{dt} = -x - y$$

Initial conditions will consist of values for both $x(0)$ and $y(0)$. This is clear from the 2-D system point of view, but why is it true physically? The velocity of the mass initially certainly effects the solution. A mass initially at $x = 0$ and at rest, $y = 0$, will remain at rest, but a fast moving mass just passing through $x = 0$ at time zero does not have a constant solution. We solve the system above with such initial conditions and show the non-trivial motion. Here are our example conditions:

$$x(0) = 0$$
$$y(0) = 1$$

We try a solution of the form

$$x(t) = k_1\, x_1(t) + k_2\, x_2(t)$$

$$= k_1\, e^{-\frac{t}{2}} \cos[\frac{\sqrt{3}}{2}t] + k_2\, e^{-\frac{t}{2}} \sin[\frac{\sqrt{3}}{2}t]$$

by simply substituting it in with unknown values of k_1 and k_2. First we will verify that this is a solution (though this step is not necessary when you know you have basic solutions.) Notice the use of the product rule for differentiation.

$$x = e^{-\frac{t}{2}}(k_1 \cos[\frac{\sqrt{3}}{2}t] + k_2 \sin[\frac{\sqrt{3}}{2}t])$$

$$\frac{dx}{dt} = -\frac{k_1}{2}e^{-\frac{t}{2}}\cos[\frac{\sqrt{3}}{2}t] - \frac{k_1\sqrt{3}}{2}e^{-\frac{t}{2}}\sin[\frac{\sqrt{3}}{2}t]$$

$$- \frac{k_2}{2}e^{-\frac{t}{2}}\sin[\frac{\sqrt{3}}{2}t] + \frac{k_2\sqrt{3}}{2}e^{-\frac{t}{2}}\cos[\frac{\sqrt{3}}{2}t]$$

$$= e^{-\frac{t}{2}}(\frac{k_2\sqrt{3} - k_1}{2}\cos[\frac{\sqrt{3}}{2}t] - \frac{k_1\sqrt{3} + k_2}{2}\sin[\frac{\sqrt{3}}{2}t])$$

$$\frac{d^2x}{dt^2} = e^{-\frac{t}{2}}(\frac{k_1 - k_2\sqrt{3}}{4}\cos[\frac{\sqrt{3}}{2}t] + \frac{k_1\sqrt{3} + k_2}{4}\sin[\frac{\sqrt{3}}{2}t]$$

$$+ \frac{k_1\sqrt{3} - k_2\,3}{4}\sin[\frac{\sqrt{3}}{2}t] - \frac{k_1\,3 + k_2\sqrt{3}}{4}\cos[\frac{\sqrt{3}}{2}t])$$

$$= e^{-\frac{t}{2}}(\frac{-k_1 - k_2\sqrt{3}}{2}\cos[\frac{\sqrt{3}}{2}t] + \frac{k_1\sqrt{3} - k_2}{2}\sin[\frac{\sqrt{3}}{2}t])$$

Adding, we obtain

$$\frac{d^2x}{dt^2} + \frac{dx}{dt} + x = e^{-\frac{t}{2}}[(k_1 + \frac{k_2\sqrt{3} - k_1}{2} + \frac{-k_1 - k_2\sqrt{3}}{2})\cos[\frac{\sqrt{3}}{2}t]$$

$$+ (k_2 - \frac{k_1\sqrt{3} + k_2}{2} + \frac{k_1\sqrt{3} - k_2}{2})\sin[\frac{\sqrt{3}}{2}t]]$$

$$= e^{-\frac{t}{2}}(0\ \cos[\frac{\sqrt{3}}{2}t] + 0\ \sin[\frac{\sqrt{3}}{2}t]) = 0$$

This verifies that this is a solution no matter which values k_1 and k_2 take.

Now we match the initial conditions by setting up equations for k_1 and k_2. The initial values are

$$x(0) = e^{-\frac{0}{2}}(k_1 \cos[\frac{\sqrt{3}}{2}0] + k_2 \sin[\frac{\sqrt{3}}{2}0])$$

$$= k_1$$

$$\frac{dx}{dt}(0) = e^{-\frac{0}{2}}(\frac{k_2\sqrt{3} - k_1}{2}\cos[\frac{\sqrt{3}}{2}0] - \frac{k_1\sqrt{3} + k_2}{2}\sin[\frac{\sqrt{3}}{2}0])$$

$$= \frac{k_2\sqrt{3} - k_1}{2}$$

Since we want

$$x(0) = 0$$
$$\frac{dx}{dt}(0) = 1$$

we take $k_1 = 0$ and $k_2 = \frac{2}{\sqrt{3}}$ making $x(t) = \frac{2}{\sqrt{3}} e^{-\frac{t}{2}} \operatorname{Sin}[\frac{\sqrt{3}}{2}t]$ the unique solution to our initial value problem.

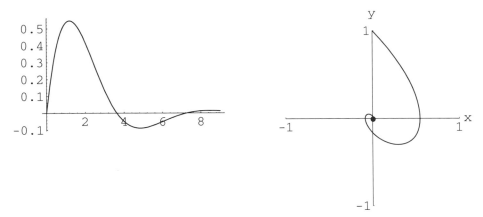

FIGURE 22.19: $x(t) = \frac{2}{\sqrt{3}} e^{-t/2} \operatorname{Sin}[\frac{\sqrt{3}t}{2}]$ & Its Phase Plot

The solutions of this initial value problem can be plotted in the phase plane as parametric curves (or in the *Mathematica* flow animation as a movie.) Notice that the velocity y starts high, $y[0] = 1$, while position x starts at $x[0] = 0$. Oscillation of the explicit solution appears as a spiral in the parametric phase plot. The solution procedure for k_1 and k_2 only selects the specific initial conditions and gives us an exact symbolic solution.

Any initial conditions for the equation $\frac{d^2 x}{dt^2} + \frac{dx}{dt} + x = 0$ can be found by setting up these equations for k_1 and k_2 in $x(0) = k_1 x_1(0) + k_2 x_2(0) = x_0$ and $x'(0) = k_1 x_1'(0) + k_2 x_2'(0) = y_0$. The linear equations in this example are:

$$\begin{bmatrix} 1 & 0 \\ -\frac{1}{2} & \frac{\sqrt{3}}{2} \end{bmatrix} \begin{bmatrix} k_1 \\ k_2 \end{bmatrix} = \begin{bmatrix} x_0 \\ y_0 \end{bmatrix}$$

If we begin with basic solutions $x_1(t)$ and $x_2(t)$ of any second order linear autonomous equation, we will always be able to solve the conditions that result from assuming that a solution has the form $x(t) = k_1 x_1(t) + k_2 x_2(t)$ and calculating k_1 and k_2 from the initial value equations $t = t_0$:

$$x_0 = x(t_0) = k_1 x_1[t_0] + k_2 x_2[t_0]$$
$$y_0 = \frac{dx}{dt}(t_0) = k_1 \frac{dx_1}{dt}(t_0) + k_2 \frac{dx_2}{dt}(t_0)$$
$$= k_1 y_1[t_0] + k_2 y_2[t_0]$$

where $y_j[t] = x_j'[t]$. Remember that x_0 and y_0 are given initial conditions and $x_1[t]$, $y_1[t]$, $x_2[t]$ and $y_2[t]$ are computed by our algebraic procedure (plus differentiation for the $y's$.) If we specify the initial time t_0, the quantities $x_1[t_0]$, $y_1[t_0]$, $x_2[t_0]$ and $y_2[t_0]$ can be computed, so the equations

$$x_0 = k_1\, x_1[t_0] + k_2\, x_2[t_0]$$
$$y_0 = k_1\, y_1[t_0] + k_2\, y_2[t_0]$$

have only k_1 and k_2 as unknowns. The computations are long and tedious, but so straight-forward that the computer could do them. They result in a system of equations that can be written in matrix form as:

$$\begin{bmatrix} x_1[t_0] & x_2[t_0] \\ y_1[t_0] & y_2[t_0] \end{bmatrix} \begin{bmatrix} k_1 \\ k_2 \end{bmatrix} = \begin{bmatrix} x_0 \\ y_0 \end{bmatrix} \qquad \Leftrightarrow \qquad \mathbf{a} \cdot \mathbf{k} = \mathbf{b}$$

where the matrices are,

$$\mathbf{a} = \begin{bmatrix} x_1[t_0] & x_2[t_0] \\ y_1[t_0] & y_2[t_0] \end{bmatrix}, \qquad \mathbf{k} = \begin{bmatrix} k_1 \\ k_2 \end{bmatrix}, \qquad \mathbf{b} = \begin{bmatrix} x_0 \\ y_0 \end{bmatrix}$$

Mathematica can solve such equations as follows.

EXAMPLE 22.20. *Mathematica Solution of Linear Equations*

Consider the linear equations

$$\begin{array}{rcl} k_1 \quad\quad -k_2 &=& 5 \\ -\tfrac{1}{2}k_1 \;+\tfrac{\sqrt{3}}{2}k_2 &=& 7 \end{array} \qquad \Leftrightarrow \qquad \mathbf{a} \cdot \mathbf{k} = \mathbf{b}$$

where the matrices are, $\mathbf{a} = \begin{bmatrix} 1 & -1 \\ -1/2 & \sqrt{3}/2 \end{bmatrix}, \mathbf{k} = \begin{bmatrix} k_1 \\ k_2 \end{bmatrix}, \mathbf{b} = \begin{bmatrix} 5 \\ 7 \end{bmatrix}$

In[1]
```
a = {{1, -1},
     {-1/2, Sqrt[3]/2}} ;
coefs = LinearSolve[ a , {5,7}];
k1 = coefs[[1]]
k2 = coefs[[2]]
```
Out[3]
$$\frac{14+5\sqrt{3}}{1+\sqrt{3}}$$
Out[3]
$$\frac{19}{1+\sqrt{3}}$$

EXERCISE 22.21. *Find the solution to $\frac{d^2x}{dt^2} + \frac{dx}{dt} + x = 0$ satisfying $x(0) = 1$ and $\frac{dx}{dt}(0) = 0$.*

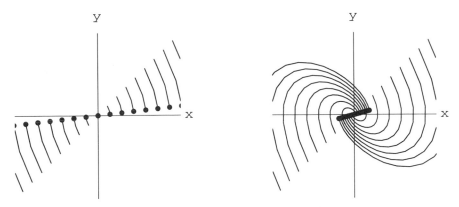

FIGURE 22.22: Flow of $x'' + x' + x = 0$

This whole section is summarized in the single result:

THEOREM 22.23. *Basic Solutions of Second Order Equations*
The following procedure gives a pair of linearly independent basic solutions to the differential equation

$$a\frac{d^2x}{dt^2} + b\frac{dx}{dt} + cx = 0$$

in the case when $a \neq 0$.
   Find the characteristic roots r_1 and r_2 of the characteristic equation $a\,r^2 + b\,r + c = 0$.

   (1) *If the root is repeated, $r_1 = r_2$, then*

$$x_1(t) = e^{r_1 t} \qquad and \qquad x_2(t) = t\,e^{r_1 t}$$

   (2) *If the roots are real and distinct, then*

$$x_1(t) = e^{r_1 t} \qquad and \qquad x_2(t) = e^{r_2 t}$$

   (3) *If the roots are a complex conjugate pair $r_1 = \alpha + i\beta$ and $r_2 = \alpha - i\beta$, then*

$$x_1(t) = e^{\alpha t}\,\mathrm{Cos}[\beta\,t] \qquad and \qquad x_2(t) = e^{\alpha t}\,\mathrm{Sin}[\beta\,t]$$

Every solution to this differential equation can be written in the form

$$x(t) = k_1\,x_1(t) + k_2\,x_2(t)$$

for suitable choices of the constants k_1 and k_2.

The next subsection turns this result into a computational procedure and symbolic *Mathematica* program with which to implement the procedure.

EXERCISE 22.24. *Transients*
The damped spring equation

$$m\frac{d^2x}{dt^2} = -c\frac{dx}{dt} - s\,x \qquad or \qquad m\frac{d^2x}{dt^2} + c\frac{dx}{dt} + s\,x = 0$$

has $c > 0$. Show that this makes every solution satisfy

$$\lim_{t \to \infty} x(t) = 0$$

by considering the 3 cases of Theorem 22.23.

Sketch the explicit graphs of $x_1(t)$ and $x_2(t)$ in the cases $m = 1$, $s = 1$, and

(1) $c = 1$

(2) $c = 2$

(3) $c = \frac{5}{2}$

*Use the Mathematica NoteBook **Flow2D.ma** to make flow animations of all the initial value problems associated with these three choices of parameters. (Note: You must use the phase variable trick to write these as equivalent first order systems when you set **Flow2D.ma** up.)*

22.5. Specific Solutions

In this subsection we combine the ingredients of the last two to solve initial value problems of the form

$$x(0) = x_0$$

$$y(0) = \frac{dx}{dt}(0) = y_0$$

$$a\frac{d^2x}{dt^2} + b\frac{dx}{dt} + cx = 0$$

This procedure is so mechanical that we will ask you to write a *Mathematica* program to implement it after you have some practice at using it.

22.6. Symbolic Solution of Second Order Autonomous I.V.P.s

1) Use Theorem 22.23 to find two basic solutions,

1.a) If $b^2 = 4\,a\,c$, the characteristic roots are repeated, so $r = -\frac{b}{2a}$ and

$$x_1(t) = e^{r\,t} \quad \text{and} \quad x_2(t) = t\,e^{r\,t}$$

1.b) If $b^2 > 4\,a\,c$, the characteristic roots are $r_1 = \frac{-b+\sqrt{b^2-4\,a\,c}}{2\,a}$ and $r_1 = \frac{-b-\sqrt{b^2-4\,a\,c}}{2\,a}$ with basic solutions

$$x_1(t) = e^{r_1\,t} \quad \text{and} \quad x_2(t) = e^{r_2\,t}$$

1.c) If $b^2 < 4\,a\,c$, the characteristic roots are complex, $\alpha \pm i\beta$ with $\alpha = -\frac{b}{2a}$ and $\beta = \frac{\sqrt{4\,a\,c-b^2}}{2\,a}$ with basic solutions

$$x_1(t) = e^{\alpha\,t}\,\text{Cos}[\beta\,t] \quad \text{and} \quad x_2(t) = e^{\alpha\,t}\,\text{Sin}[\beta\,t]$$

2) Symbolically compute the derivative of $x(t) = k_1\,x_1(t) + k_2\,x_2(t)$. (This involves the product rule in cases 1 and 3.)

3) Evaluate $x(t_0)$ and $\frac{dx}{dt}(t_0)$ numerically, except for the unknowns k_1 and k_2, and write the system of linear equations in these unknowns

$$x_0 = x(0)$$

$$y_0 = \frac{dx}{dt}(0)$$

This can be put in the matrix form $\mathbf{a} \cdot \mathbf{k} = \mathbf{b}$,

$$\begin{bmatrix} x_1[t_0] & x_2[t_0] \\ y_1[t_0] & y_2[t_0] \end{bmatrix} \begin{bmatrix} k_1 \\ k_2 \end{bmatrix} = \begin{bmatrix} x_0 \\ y_0 \end{bmatrix}$$

for constants $x_1[t_0]$, $x_2[t_0]$, $y_1[t_0]$, $y_2[t_0]$ and the given values x_0, y_0.
4) Solve this equation for k_1 and k_2.
The specific solution $x(t) = k_1\,x_1(t) + k_2\,x_2(t)$ is the answer.

EXAMPLE 22.25. *Solution when $a = 1$, $b = 0$, $c = -1$, $x_0 = y_0 = 1$*

$$\frac{d^2x}{dt^2} - x = 0 \qquad \Leftrightarrow \qquad \begin{array}{l} \frac{dx}{dt} = y \\ \frac{dy}{dt} = x \end{array}$$

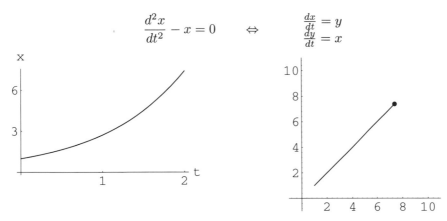

FIGURE 22.26: Explicit and Parametric Solution

The full solution flow for many initial conditions looks like the following figure.

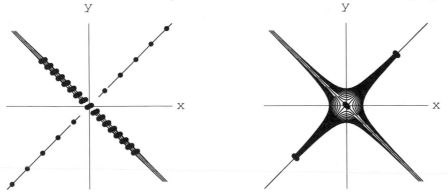

FIGURE 22.27: Flow of $x'' - x = 0$

Characteristic equation:
$$r^2 - 1 = (r - 1)(r + 1) = 0$$

Roots: $r_1 = 1$, $r_2 = -1$
Basic solutions: $x(t) = k_1\,e^t + k_2\,e^{-t}$
Symbolic derivative: $x'(t) = k_1\,e^t - k_2\,e^{-t}$
Initial values at $t = 0$:

$$x(0) = k_1\,e^0 + k_2\,e^0 = k_1 + k_2$$
$$x'(0) = k_1\,e^0 - k_2\,e^0 = k_1 - k_2$$

Matrix linear equation:
$$\begin{bmatrix} 1 & 1 \\ 1 & -1 \end{bmatrix} \begin{bmatrix} k_1 \\ k_2 \end{bmatrix} = \begin{bmatrix} x_0 \\ y_0 \end{bmatrix}$$

When $(x_0, y_0) = (1.0, 1.0)$, $k_1 = 1$ and $k_2 = 0$. This is the solution $x[t] = e^t$.

EXAMPLE 22.28. *Solution when $a = 1$, $b = 3$, $c = 1$, $x_0 = y_0 = 1$*

$$\frac{d^2 x}{dt^2} + 3 \frac{dx}{dt} + x = 0 \qquad \Leftrightarrow \qquad \begin{array}{l} \frac{dx}{dt} = y \\ \frac{dy}{dt} = -x - 3\,y \end{array}$$

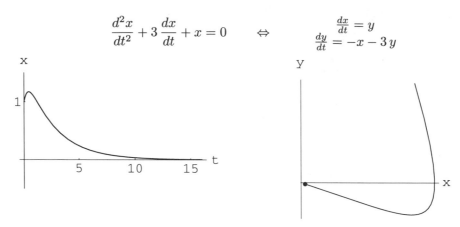

FIGURE 22.29: Explicit and Parametric Solution

A flow of many solutions with various initial conditions looks as follows:

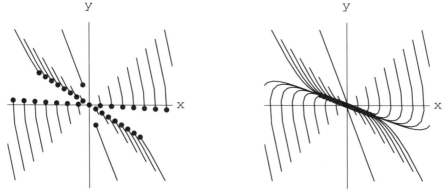

FIGURE 22.30: Flow of $x'' + 3\,x' + x = 0$

Characteristic equation: $r^2 + 3\,r + 1 = 0$

Roots: $r_1 = \frac{-3+\sqrt{5}}{2}$, $r_2 = \frac{-3+\sqrt{5}}{2}$

Basic solutions: $x(t) = k_1\, e^{\frac{-3+\sqrt{5}}{2}\, t} + k_2\, e^{\frac{-3-\sqrt{5}}{2}\, t}$

Symbolic derivative: $x'(t) = k_1\, \frac{-3+\sqrt{5}}{2} e^{\frac{-3+\sqrt{5}}{2}\, t} - k_2\, \frac{-3-\sqrt{5}}{2} e^{\frac{-3-\sqrt{5}}{2}\, t}$

Initial values at $t = 0$:

$$x(0) = k_1\, e^{\frac{-3+\sqrt{5}}{2}\, 0} + k_2\, e^{\frac{-3-\sqrt{5}}{2}\, 0} = k_1 + k_2$$

$$x'(0) = k_1\, \frac{-3+\sqrt{5}}{2} e^{\frac{-3+\sqrt{5}}{2}\, 0} + k_2\, \frac{-3-\sqrt{5}}{2} e^{\frac{-3-\sqrt{5}}{2}\, 0} = k_1\, \frac{-3+\sqrt{5}}{2} + k_2\, \frac{-3-\sqrt{5}}{2}$$

Matrix linear equation:

$$\begin{bmatrix} 1 & 1 \\ \frac{-3+\sqrt{5}}{2} & \frac{-3-\sqrt{5}}{2} \end{bmatrix} \begin{bmatrix} k_1 \\ k_2 \end{bmatrix} = \begin{bmatrix} x_0 \\ y_0 \end{bmatrix}$$

When $(x_0, y_0) = (1.0, 1.0)$, $k_1 = \frac{1+\sqrt{5}}{2}$ and $k_2 = \frac{1-\sqrt{5}}{2}$. The solution looks like the following figures.

EXERCISE 22.31. *Symbolic Solutions by Hand*
Solve the initial value problem

$$x(0) = 1$$

$$y(0) = \frac{dx}{dt}(0) = 1$$

$$a\frac{d^2x}{dt^2} + b\frac{dx}{dt} + cx = 0$$

for the cases

(1) $a = 1$, $b = 0$, $c = -4$
(2) $a = 1$, $b = 3$, $c = 2$
(3) $a = 1$, $b = 2$, $c = 2$
(4) $a = 1$, $b = 2$, $c = 1$

Use your hand calculations to verify the correctness of your *Mathematica* program in the next exercise.

EXERCISE 22.32. *Mathematica Symbolic Solutions*
Write a Mathematica program to find specific solutions to the I.V.P.

$$x(0) = x_0$$

$$y(0) = \frac{dx}{dt}(0) = y_0$$

$$a\frac{d^2x}{dt^2} + b\frac{dx}{dt} + cx = 0$$

Here is some help with the *Mathematica* commands:
Begin by assigning numbers for the coefficients and initial conditions:

$$a = 1.0;$$

$$b = 1.0;$$

$$c = 1.0;$$

The *Mathematica* "if-then" command looks like:

$$\text{If[condition, calc1, calc2]}$$

Mathematica does calculation 1 if the condition is true, otherwise it does calculation 2. Hence step 1 of the solution procedure looks like:

Clear[t,k1,k2]
If[b^ 2 == 4 a c,
    r = -b/(2 a);
    x1[t_] := Exp[r t];
    x2[t_] := t x1[t] (*Degenerate case of only one root.*);

,
If[b^ 2 - 4 a c < 0,
    r = -b/(2 a);

```
w = Sqrt[4 a c - b^ 2]/(2 a);
x1[t_ ] := Exp[r t] Cos[w t];
x2[t_ ] := Exp[r t] Sin[w t] (*Case of complex roots.*);
,
    r1 = (-b + Sqrt[b^ 2 - 4 a c])/(2 a);
    r2 = (-b - Sqrt[b^ 2 - 4 a c])/(2 a);
    x1[t_ ] := Exp[r1 t];
    x2[t_ ] := Exp[r2 t] (*Case of two real roots.*);
]]
x[t_ ] := k1 x1[t] + k2 x2[t];
x[t_ ]
```

Computation of the symbolic derivative is done as follows:

```
Clear[t,k1,k2]
    y1[t_ ] := Evaluate[ D[x1[t],t]] (*Derivative of x1.*)
    y2[t_ ] := Evaluate[ D[x2[t],t]] (*Derivative of x2.*)
```

Specific values are computed by assigning t and asking for the functions' values:

```
t = 0.0 (*Initial time.*)
    a = {{x1[t], x2[t]}, (*Matrix of initial positions *)
        { y1[t], y2[t]}} (* and initial velocities.*)
TableForm[ a ]
```

A linear equation in matrix form a.k = b = {xi,yi} is solved by

```
xi = 1.0 (*Initial position.*);
yi = 1.0 (*Initial velocity.*);
coefs = LinearSolve[ a , {xi, yi}]
k1 = coefs[[1]]
k2 = coefs[[2]]
```

Recall that your solution is defined above, $x[t] = k1x1[t] + k2x2[t]$, so display it:

```
Clear[t]
x[t]
```

Now Plot your answer.

```
Plot[ x[t], {t,0,16}];
```

Include text cells in your program that explain each part of your program.

22.7. Your Car's Worn Shocks

When a car hits a bump it sets up an initial value problem

$$x(0) = 0$$

$$\frac{dx}{dt}(0) = v_0 \neq 0 \qquad \text{an initial velocity}$$

$$m\frac{d^2x}{dt^2} + c\frac{dx}{dt} + s\,x = 0$$

The initial position is still zero, the level position of the car, but the jolt (impulse) imparts an initial velocity to the wheel.

EXERCISE 22.33. *What is the algebraic condition between the physical parameters m, c and s that determines whether or not your car oscillates after it hits a bump?*

Use your Mathematica program for solving the above initial value problem with $m = s = v_0 = 1$ and values of $c = 3$, 2.9, 2.8, $\cdots$, 2.0, $\cdots$, 0.0. Make plots of the solutions and run them together as an animation. You can use a 'Do[.]' loop to vary the c's.

22.8. Projects

22.8.1. Lanchester's Combat Models. This project studies a combat model where the size and effectiveness of the armies are compared.

22.8.2. Compartment Models for Drug Dosages. This project studies the dynamics of the blood concentration of a drug.

22.8.3. Resonance of Linear Oscillators. This project studies the response of a linear oscillator to shaking. A maximal response is called resonance.

22.8.4. A Notch Filter. This studies a two loop circuit that has a minimal response at a certain frequency. This can be used to filter out that frequency.

CHAPTER 23

Equilibria of Continuous Dynamical Systems

This chapter shows how to use calculus to "see" what is in a microscope focused at an equilibrium point of a dynamical system. The main idea of calculus is that you will see the "linear approximation," which in this case is a linear dynamical system. Our first task is to cast the results of the last chapter in terms of 2-by-2 systems, so that we know what the linear systems look like. Then we use partial derivatives to show which linear dynamical system approximates a nonlinear one near an equilibrium point.

23.1. Dynamic Equilibria in One Dimension

It is easy to find out if an equilibrium point of a one dimensional dynamical system attracts or repels. Simply plot the direction line immediately on each side of the equilibrium (before another zero) and see if both arrows point toward the equilibrium (attractor), both point away (repeller) or neither. We want to find a symbolic test for this simple condition, so we can compare the 1-D and 2-D cases. Can you conjecture what property of a smooth function $f(x)$ in the equation $dx = f(x)\,dt$ at an equilibrium will make x_e attract or repel? In two dimensions we really need help from calculus.

In the linear case

$$\frac{dx}{dt} = a\,x$$

where $f(x) = a\,x$, we know the whole story. The point $x_e = 0$ is the only equilibrium point (unless $a = 0$) and it attracts if $a < 0$ and repels if $a > 0$. You should solve this graphically by drawing direction lines in these two cases, however, the symbolic solution is

$$x(t) = x_0\,e^{a\,t}$$

and

$$\lim_{t \to \infty} x(t) = 0 \qquad \text{if} \quad a < 0$$

while

$$\lim_{t \to \infty} |x(t)| = \infty \qquad \text{if} \quad a > 0$$

where the sign of $x(t)$ remains the same as x_0. This is a symbolic proof that any x_0 makes $x(t) \to 0$ when $a < 0$ and is repelled when $a > 0$.

In the affine linear case

$$\frac{dx}{dt} = a\,x + b$$

which has its equilibrium when $dx = 0$ or $x_e = -b/a$, we can change variables to the difference between x and the equilibrium, $x = x_e + u$ or $u = x - x_e$. This makes $du = dx$ and $a\,u = a(x - \frac{-b}{a}) = a\,x + b$, so that the $a\,x + b$ equation is equivalent to

$$\frac{du}{dt} = a\,u$$

This linear equation has $u(t) \to 0$ when $a < 0$, so $x(t) = u(t) + x_e \to x_e$. Similarly, if $a > 0$, solutions are repelled from equilibrium. In either case, the coefficient a determines the type of equilibrium.

We can analyse nonlinear equilibria of the equation

$$\frac{dx}{dt} = f(x)$$

with an equilibrium at x_e by linearly approximating $f(x)$ near x_e. For smooth functions $f(x)$, we know the approximation

(Incr) $f(x) = f(u + x_e) = f(x_e) + a\,u + \varepsilon\,u$

when $x = u + x_e$, with $\varepsilon \approx 0$ for $u \approx 0$.

PROBLEM 23.1. *Increments & Equilibria*
Let x_e be an equilibrium point of the one dimensional dynamical system $\frac{dx}{dt} = f[x]$.

 (1) *Why does the approximation (Incr) above hold? What is the constant a?*
 (2) *Write the approximation (Incr) in case $f(x) = x\,(1 - x/2)$ and $x_e = 2$? What is the constant a in this case?*
 (3) *Show in general that when x_e is an equilibrium point,*

$$f(u + x_e) = a\,u + \varepsilon u \quad \text{with } \varepsilon \approx 0, \quad \text{when } u \approx 0$$

 (4) *Write this approximation for the function $f(x) = x\,(1 - x/2)$ for the two cases $x_e = 2$ and $x_e = 0$.*
 (5) *If $u = x - x_e$, show that $\frac{du}{dt} = \frac{dx}{dt}$, so the differential equation $\frac{dx}{dt} = f[x]$ written in terms of $u = (x - x_e)$ near x_e is approximately the particular linear equation*

$$\frac{dx}{dt} = f[x_e + (x - x_e)] \quad \Leftrightarrow \quad \frac{du}{dt} = f[x_e + u] \quad \approx \quad \frac{du}{dt} = a\,u$$

The derivation in the exercise above means that we have the approximation

$$\frac{dx}{dt} = f(x) \quad \approx \quad \frac{du}{dt} = a\,u \quad \text{near } x = x_e$$

EXERCISE 23.2. *Microscopic Dynamics*
Now use the idea of the previous problem with parameters.

 (1) *Sketch the direction line of*

$$\frac{dx}{dt} = f(x) = r\,x(1 - x/c)$$

assuming $r > 0$ and $c > 0$.

(2) *Compute the constant a_1 for the approximating equation at $x_{e_1} = 0$, using the method of the previous exercise.*

(3) *Sketch the direction line of $\frac{du_1}{dt} = a_1 u_1$.*

(4) *Compute the constant a_2 for the approximating equation at $x_{e_2} = c$, using the method of the previous exercise.*

(5) *Sketch the direction line of $\frac{du_2}{dt} = a_2 u_2$.*

FIGURE 23.3: Microscopic Views

23.2. Linear Equilibria in Two Dimensions

A general linear system in two dimensions may be written

$$\frac{dx}{dt} = a_1 x + a_2 y$$

$$\frac{dy}{dt} = b_1 x + b_2 y$$

for constants a_1, a_2, b_1, b_2. This may also be written in matrix form as

$$\begin{bmatrix} \frac{dx}{dt} \\ \frac{dy}{dt} \end{bmatrix} = \begin{bmatrix} a_1 & a_2 \\ b_1 & b_2 \end{bmatrix} \begin{bmatrix} x \\ y \end{bmatrix}$$

The matrix form is a little more compact and we want to use the four coefficients to compute the exponential solutions by a determinant related to the matrix.

We are already familiar with the linear system

$$\begin{bmatrix} \frac{dx}{dt} \\ \frac{dy}{dt} \end{bmatrix} = \begin{bmatrix} 0 & 1 \\ b_x & b_y \end{bmatrix} \begin{bmatrix} x \\ y \end{bmatrix} \qquad \leftrightarrow \qquad \frac{d^2 x}{dt^2} = b_y \frac{dx}{dt} + b_x x$$

as a geometric representation of the second order one dimensional equation. This is simply the 'phase variable trick' $y = \frac{dx}{dt}$ from the last chapter. Every solution of one system is also a solution of the other. In fact, if $x(t) = k_1 x_1(t) + k_2 x_2(t)$, then $y(t) = \frac{dx}{dt}$ is also a linear combination of $x_1(t)$ and $x_2(t)$. For example, if $x_1(t) = e^{r_1 t}$ and $x_2(t) = e^{r_2 t}$, then $y(t) = r_1 k_1 x_1(t) + r_2 k_2 x_2(t)$.

Because of this form of solutions, the zero equilibrium of both systems is attracting if and only if $\lim_{t \to \infty} x_1(t) = 0$ and $\lim_{t \to \infty} x_2(t) = 0$. It is repelling if and only if the real parts of the roots of the characteristic equation are positive. It is neither attracting nor repelling if one root is positive and the other is negative.

EXERCISE 23.4. *Attraction and Repulsion*
Show that the linear system

$$\begin{bmatrix} \frac{dx}{dt} \\ \frac{dy}{dt} \end{bmatrix} = \begin{bmatrix} 0 & 1 \\ b_x & b_y \end{bmatrix} \begin{bmatrix} x \\ y \end{bmatrix}$$

with $b_x \neq 0$ has zero as its only equilibrium point. Show that the equation

$$\det \begin{vmatrix} -r & 1 \\ b_x & b_y - r \end{vmatrix} = r^2 - b_y\, r - b_x = 0$$

is the same as the characteristic equation of the associated second order one dimensional equation.

The various things that can happen to second order equations $\frac{d^2x}{dt^2} = b_x\, x + b_y\, \frac{dx}{dt}$ are

(1) The roots of the characteristic equation are distinct and negative, so the basic solutions are $x_1(t) = e^{-r_1\, t}$ and $x_2(t) = e^{-r_2\, t}$, both with limit zero as $t \to \infty$.

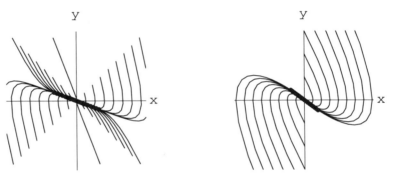

FIGURE 23.5: Distinct Negative Roots & Repeated Negative Roots

(2) The roots of the characteristic equation are negative and 'repeated,' so the basic solutions are $x_1(t) = e^{-r\, t}$ and $x_2(t) = t\, e^{-r\, t}$, both with limit zero as $t \to \infty$.
(3) The roots of the characteristic equation are complex conjugates with negative real part, so the basic solutions are $x_1(t) = e^{-\alpha\, t} \operatorname{Cos}[\beta\, t]$ and $x_2(t) = e^{-\alpha\, t} \operatorname{Sin}[\beta\, t]$, both with limit zero as $t \to \infty$.

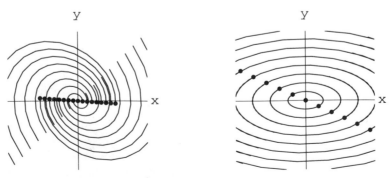

FIGURE 23.6: Complex Negative Real Part & Pure Imaginary Roots

(4) The roots of the characteristic equation are complex conjugates with zero real part, so the basic solutions are $x_1(t) = \operatorname{Cos}[\beta\, t]$ and $x_2(t) = \operatorname{Sin}[\beta\, t]$, both of absolute

value one for all t. This is 'neutral' stability; nearby solutions 'orbit' the equilibrium forever.

(5) The roots of the characteristic equation are distinct and positive, so the basic solutions are $x_1(t) = e^{+r_1 t}$ and $x_2(t) = e^{+r_2 t}$, both with limit ∞ as $t \to \infty$.

(6) The roots of the characteristic equation are positive and 'repeated,' so the basic solutions are $x_1(t) = e^{r t}$ and $x_2(t) = t\, e^{r t}$, both with limit ∞ as $t \to \infty$.

(7) The roots of the characteristic equation are complex conjugates with positive real part, so the basic solutions are $x_1(t) = e^{+\alpha t} \operatorname{Cos}[\beta\, t]$ and $x_2(t) = e^{+\alpha t} \operatorname{Sin}[\beta\, t]$, both with the limit of the absolute value ∞ as $t \to \infty$.

(8) The roots of the characteristic equation are real of opposite sign, so the basic solutions are $x_1(t) = e^{+r_1 t}$ and $x_2(t) = e^{-r_2 t}$. One tends to ∞ and the other tends to zero as $t \to \infty$. This equilibrium is neither attracting nor repelling.

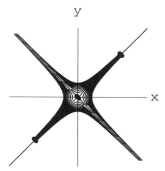

FIGURE 23.7: Roots of Opposite Sign

The *Mathematica* NoteBooks in the Linear Equilibria folder illustrate the various attracting cases. The repelling cases can be obtained by running these animations backwards (which is the solution of the equation where t is replaced by $-t$). Figures in the text are frames from these animations.

Every two dimensional linear system

$$\begin{bmatrix} \frac{dx}{dt} \\ \frac{dy}{dt} \end{bmatrix} = \begin{bmatrix} a_x & a_y \\ b_x & b_y \end{bmatrix} \begin{bmatrix} x \\ y \end{bmatrix} \quad \text{is equivalent to a second order system} \quad \frac{d^2 x}{dt^2} = \beta \frac{dx}{dt} + \gamma\, x$$

with the characteristic equation

$$\det \begin{vmatrix} a_x - r & a_y \\ b_x & b_y - r \end{vmatrix} = (a_x - r)(b_y - r) - a_y\, b_x = r^2 - (a_x + b_y)\, r + (a_x\, b_y - a_y\, b_x) = 0$$

provided

$$\det \begin{vmatrix} a_x & a_y \\ b_x & b_y \end{vmatrix} = a_x\, b_y - a_y\, b_x \neq 0 \quad \text{and either } a_y \neq 0 \text{ or } b_x \neq 0$$

In this case, the basic solutions $x_1(t)$ and $x_2(t)$ of the second order equation can be found from the roots of this equation as in Theorem 22.23 and solutions $(x(t), y(t))$ of the original system can be written as a certain specific linear combination of the basic solutions $x_1(t)$ and $x_2(t)$ of the second order equation.

The proof of the equivalence in case $a_y \neq 0$ is to solve the dx equation for y

$$\frac{1}{a_y}\frac{dx}{dt} = \frac{a_x}{a_y}x + y$$

$$y = \frac{1}{a_y}\frac{dx}{dt} - \frac{a_x}{a_y}x$$

differentiate

$$\frac{dy}{dt} = \frac{1}{a_y}\frac{d^2x}{dt^2} - \frac{a_x}{a_y}\frac{dx}{dt}$$

use the dy equation

$$b_y\,y + b_x\,x = \frac{1}{a_y}\frac{d^2x}{dt^2} - \frac{a_x}{a_y}\frac{dx}{dt}$$

$$\frac{1}{a_y}\frac{d^2x}{dt^2} - b_y\,y - b_x\,x - \frac{a_x}{a_y}\frac{dx}{dt} = 0$$

and the y equation

$$\frac{1}{a_y}\frac{d^2x}{dt^2} - b_y\left(\frac{1}{a_y}\frac{dx}{dt} - \frac{a_x}{a_y}x\right) - b_x\,x - \frac{a_x}{a_y}\frac{dx}{dt} = 0$$

$$\frac{d^2x}{dt^2} - (a_x + b_y)\frac{dx}{dt} + (a_x\,b_y - a_y\,b_x)\,x = 0$$

This has characteristic equation

$$r^2 + (-a_x - b_y)\,r + (a_x\,b_y - a_y\,b_x) = 0$$

which also can be computed from the determinant above.

Once we write a solution of the second order equation $x(t) = k_1\,x_1(t) + k_2\,x_2(t)$, we can use the equation

$$y = \frac{1}{a_y}\frac{dx}{dt} - \frac{a_x}{a_y}x$$

to find a solution to the linear system.

There are two exceptional cases where we cannot associate a second order scalar equation with a first order two dimensional linear vector equation. The linear system

$$\begin{bmatrix} \frac{dx}{dt} \\ \frac{dy}{dt} \end{bmatrix} = \begin{bmatrix} a_x & 0 \\ 0 & b_y \end{bmatrix}\begin{bmatrix} x \\ y \end{bmatrix}$$

has basic solutions $x_1(t) = e^{a_x\,t}$ and $y_2(t) = e^{b_y\,t}$. Every other solution may be written in the form

$$\begin{bmatrix} x(t) \\ y(t) \end{bmatrix} = k_1 \begin{bmatrix} x_1(t) \\ 0 \end{bmatrix} + k_2 \begin{bmatrix} 0 \\ y_2(t) \end{bmatrix}$$

Even though we do not associate this case with a second order equation, the basic solutions are still of the form $e^{r_1\,t}$ and $e^{r_2\,t}$ where r_1 and r_2 are roots of the equation

$$\det \begin{vmatrix} a_x - r & 0 \\ 0 & b_y - r \end{vmatrix} = (a_x - r)(b_y - r) = 0$$

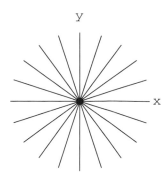 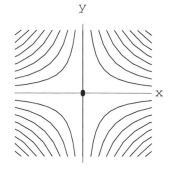

FIGURE 23.8: $a_x <= b_y < 0$, $a_y = b_x = 0$ & $a_x > 0$ and $b_y < 0$, $a_y = b_x = 0$

If the linear system

$$\begin{bmatrix} \frac{dx}{dt} \\ \frac{dy}{dt} \end{bmatrix} = \begin{bmatrix} a_x & a_y \\ b_x & b_y \end{bmatrix} \begin{bmatrix} x \\ y \end{bmatrix}$$

has

$$\det \begin{vmatrix} a_x & a_y \\ b_x & b_y \end{vmatrix} = 0$$

then the one row of the matrix of coefficients is a multiple of the other, that is, the linear system is of the form given in the next exercise. This system is degenerate from the point of view of equilibria.

EXERCISE 23.9. *Show that every point of the line $a_x \cdot x + a_y \cdot y = 0$ is an equilibrium point for the system*

$$\begin{bmatrix} \frac{dx}{dt} \\ \frac{dy}{dt} \end{bmatrix} = \begin{bmatrix} a_x & a_y \\ c\,a_x & c\,a_y \end{bmatrix} \begin{bmatrix} x \\ y \end{bmatrix}$$

Also show that the solutions travel along the lines of slope

$$\frac{dy}{dx} = c$$

so that the phase portrait consists of lines of slope c meeting a stagnant line $a_x \cdot x + a_y \cdot y = 0$.

Use one of the Linear Equilibria Mathematica NoteBooks to make an animation of the flow of a pair of linear equations where the a's are multiples of the b's.

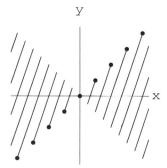

FIGURE 23.10: $b_x = 3\,a_x = 1$ and $b_y = 3\,a_y = -1$

The subsection can be summed up in this way:

THEOREM 23.11. *Linear Equilibria*
All solutions of a two dimensional linear system

$$\begin{bmatrix} \frac{dx}{dt} \\ \frac{dy}{dt} \end{bmatrix} = \begin{bmatrix} a_x & a_y \\ b_x & b_y \end{bmatrix} \begin{bmatrix} x \\ y \end{bmatrix}$$

with

$$\det \begin{vmatrix} a_x & a_y \\ b_x & b_y \end{vmatrix} = a_x\, b_y - a_y\, b_x \neq 0$$

may be written as a linear combination of (real or complex) exponentials e^{rt} (and $t\,e^{rt}$) for constants r satisfying the characteristic equation

$$\det \begin{vmatrix} a_x - r & a_y \\ b_x & b_y - r \end{vmatrix} = (a_x - r)(b_y - r) - a_y\, b_x = r^2 - (a_x + b_y)\,r + (a_x\, b_y - a_y\, b_x) = 0$$

Therefore zero is an attracting equilibrium if the real parts of the roots of the characteristic equation are negative. Zero is a repeller if the real parts are positive. Zero attracts for some initial conditions and repels for others one root is positive and the other is negative.

EXERCISE 23.12. *Checking Linear Equilibria*
The point $(0,0)$ is always an equilibrium for a linear dynamical system. Use the previous Theorem to show which of the following dynamical systems have $(0,0)$ as an attractor, repeller or neither. Modify one of the Linear Equilibria NoteBooks to animate each of the flows and verify your computations.

(1)

$$\begin{bmatrix} du \\ dv \end{bmatrix} = \begin{bmatrix} 2 & 0 \\ 0 & 1 \end{bmatrix} \begin{bmatrix} u \\ v \end{bmatrix} dt$$

(2)

$$\begin{bmatrix} du \\ dv \end{bmatrix} = \begin{bmatrix} -1 & -1 \\ -\frac{2}{3} & -\frac{1}{3} \end{bmatrix} \begin{bmatrix} u \\ v \end{bmatrix} dt$$

(3)

$$\begin{bmatrix} du \\ dv \end{bmatrix} = \begin{bmatrix} -2 & -2 \\ 0 & -\frac{1}{3} \end{bmatrix} \begin{bmatrix} u \\ v \end{bmatrix} dt$$

(4)

$$\begin{bmatrix} du \\ dv \end{bmatrix} = \begin{bmatrix} -1 & 0 \\ -2 & -1 \end{bmatrix} \begin{bmatrix} u \\ v \end{bmatrix} dt$$

23.3. Nonlinear Equilibria in Two Dimensions

Now we consider a nonlinear dynamical system

$$\begin{bmatrix} \frac{dx}{dt} \\ \frac{dy}{dt} \end{bmatrix} = \begin{bmatrix} f(x,y) \\ g(x,y) \end{bmatrix}$$

near an equilibrium point (x_e, y_e) where $f(x_e, y_e) = g(x_e, y_e) = 0$. We want to "look at the flow under a microscope focused at (x_e, y_e)." The total differential approximations for f and g say symbolically how the functions behave microscopically,

$$f(u + x_e, v + y_e) = \frac{\partial f}{\partial x}(x_e, y_e) \cdot u + \frac{\partial f}{\partial y}(x_e, y_e) \cdot v + \varepsilon_1 \sqrt{u^2 + v^2}$$

$$g(u + x_e, v + y_e) = \frac{\partial g}{\partial x}(x_e, y_e) \cdot u + \frac{\partial g}{\partial y}(x_e, y_e) \cdot v + \varepsilon_2 \sqrt{u^2 + v^2}$$

or

$$f(u + x_e, v + y_e) \approx a_x \cdot u + a_y \cdot v$$

$$g(u + x_e, v + y_e) \approx b_x \cdot u + b_y \cdot v$$

where $u = x - x_e$ and $v = y - y_e$ are the local variables, and the approximating constants are $a_x = \frac{\partial f}{\partial x}(x_e, y_e)$, $a_y = \frac{\partial f}{\partial y}(x_e, y_e)$, $b_x = \frac{\partial g}{\partial x}(x_e, y_e)$ and $b_y = \frac{\partial g}{\partial x}(x_e, y_e)$. Near the equilibrium we have

$$\begin{bmatrix} \frac{dx}{dt} \\ \frac{dy}{dt} \end{bmatrix} = \begin{bmatrix} \frac{du}{dt} \\ \frac{dv}{dt} \end{bmatrix} = \begin{bmatrix} f(u + x_e, v + y_e) \\ g(u + x_e, v + y_e) \end{bmatrix} \approx \begin{bmatrix} a_x & a_y \\ b_x & b_y \end{bmatrix} \begin{bmatrix} u \\ v \end{bmatrix}$$

So we expect that the flow of

$$\begin{bmatrix} \frac{dx}{dt} \\ \frac{dy}{dt} \end{bmatrix} = \begin{bmatrix} f(x, y) \\ g(x, y) \end{bmatrix}$$

near equilibrium microscopically looks like the flow of the linear approximation.

THEOREM 23.13. *The Equilibrum Microscope*
Let $f(x, y)$ and $g(x, y)$ be smooth functions with $f(x_e, y_e) = g(x_e, y_e) = 0$. The flow of

$$\frac{dx}{dt} = f(x, y)$$

$$\frac{dy}{dt} = g(x, y)$$

under infinite magnification at (x_e, y_e) appears the same as the flow of its linearization

$$\begin{bmatrix} \frac{du}{dt} \\ \frac{dv}{dt} \end{bmatrix} = \begin{bmatrix} a_x & a_y \\ b_x & b_y \end{bmatrix} \begin{bmatrix} u \\ v \end{bmatrix}$$

where the coefficients are given by the partial derivatives evaluated at the equilibrium

$$\begin{bmatrix} a_x & a_y \\ b_x & b_y \end{bmatrix} = \begin{bmatrix} \frac{\partial f}{\partial x} & \frac{\partial f}{\partial y} \\ \frac{\partial g}{\partial x} & \frac{\partial g}{\partial y} \end{bmatrix} (x_e, y_e)$$

Specifically, if our magnification is $1/\delta$, for $\delta \approx 0$, and our solution starts in our view,

$$(x[0] - x_e, y[0] - y_e) = \delta \cdot (a, b)$$

for finite a and b, then the solution satisfies

$$(x[t] - x_e, y[t] - y_e) = \delta \cdot (u[t], v[t]) + \delta \cdot (\varepsilon_x(t), \varepsilon_y(t))$$

where $(u[t], v[t])$ satisfies the linear equation and starts at $(u[0], v[0]) = (a, b)$ and where $(\varepsilon_x(t), \varepsilon_y(t)) \approx (0, 0)$ for all finite t.

This theorem is a formal statement of the fact that we see the linearized flow in the microscope. It is a finite time approximation result. In the limit as t tends to infinity, the nonlinear system can "look" different. Here is another way to say this. If we magnify a lot, but not by an infinite amount, then we may see a separation between the linear and nonlinear system after a very very long time. (For every screen resolution θ and every bound β on the time of observation and observed scale of initial condition, there is a magnification large enough so that if $|a| \leq \beta$ and $|b| \leq \beta$, then the error observed at that magnification is less than θ for $0 \leq t \leq \beta$.)

The point of this theorem is that it lets us compute what we would see if we looked at a nonlinear equilibrium point in a powerful microscope. Here is an example of how we "see" in this microscope:

EXAMPLE 23.14. *A Sample Equilibrium Calculation*

Consider the differential equations

$$\frac{dx}{dt} = f[x,y] = y(1 - \frac{x}{7} - \frac{y}{5})$$

$$\frac{dy}{dt} = g[x,y] = x(1 - \frac{x}{8} - \frac{y}{6})$$

This system has an equilibrium point at $(\dot{x}_e, y_e) = (0, 5)$, because

$$f[0,5] = y(1 - \frac{x}{7} - \frac{y}{5}) = 5(1 - \frac{0}{7} - \frac{5}{5}) = 0$$

$$g[0,5] = x(1 - \frac{x}{8} - \frac{y}{6}) = 0(1 - \frac{0}{8} - \frac{5}{6}) = 0$$

The matrix of symbolic partial derivatives is

$$\begin{bmatrix} \frac{\partial f}{\partial x} & \frac{\partial f}{\partial y} \\ \frac{\partial g}{\partial x} & \frac{\partial g}{\partial y} \end{bmatrix} = \begin{bmatrix} -\frac{y}{7} & 1 - \frac{x}{7} - \frac{2y}{5} \\ 1 - \frac{x}{4} - \frac{y}{6} & -\frac{x}{6} \end{bmatrix}$$

which gives the matrix for the linear system

$$\begin{bmatrix} a_x & a_y \\ b_x & b_y \end{bmatrix} = \begin{bmatrix} \frac{\partial f}{\partial x} & \frac{\partial f}{\partial y} \\ \frac{\partial g}{\partial x} & \frac{\partial g}{\partial y} \end{bmatrix}(0,5) = \begin{bmatrix} -\frac{5}{7} & -1 \\ +\frac{1}{6} & 0 \end{bmatrix}$$

The characteristic equation is

$$\det \begin{vmatrix} a_x - r & a_y \\ b_x & b_y - r \end{vmatrix} = \det \begin{vmatrix} -\frac{5}{7} - r & -1 \\ \frac{1}{6} & -r \end{vmatrix} = (-\frac{5}{7} - r)(-r) + \frac{1}{6} = r^2 + \frac{5}{7}r + \frac{1}{6} = 0$$

with roots

$$r_1, r_2 = \frac{-\frac{5}{7} \pm \sqrt{\left(\frac{5}{7}\right)^2 - 4 \cdot \frac{1}{6}}}{2} = -\frac{5}{14} \pm \frac{i}{14}\sqrt{\frac{23}{3}} \approx -0.357143 \pm i\,0.197777$$

These complex roots mean that a microscopic view of the equilibrium point at $(0,5)$ of the system spirals in as follows:

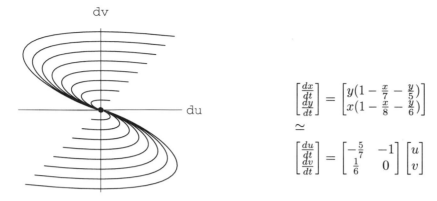

$$\begin{bmatrix} \frac{dx}{dt} \\ \frac{dy}{dt} \end{bmatrix} = \begin{bmatrix} y(1 - \frac{x}{7} - \frac{y}{5}) \\ x(1 - \frac{x}{8} - \frac{y}{6}) \end{bmatrix}$$

$$\simeq$$

$$\begin{bmatrix} \frac{du}{dt} \\ \frac{dv}{dt} \end{bmatrix} = \begin{bmatrix} -\frac{5}{7} & -1 \\ \frac{1}{6} & 0 \end{bmatrix} \begin{bmatrix} u \\ v \end{bmatrix}$$

FIGURE 23.15: A Microscopic View near $(0,5)$

We shall not prove the Equilibrium Microscope Theorem in the main text, but ask you to use it in the following exercise. It simply means that the flow of the nonlinear system 'looks like' the flow of the associated linear system under a microscope focused at the equilibrium. Nonlinear equilibria can look quite different in larger neighborhoods; examples are given in the Mathematical Background Chapter on Theory of Initial Value Problems.

EXERCISE 23.16. *Nonlinear Equilibria*
Show that the nonlinear system

$$\begin{bmatrix} \frac{dx}{dt} \\ \frac{dy}{dt} \end{bmatrix} = \begin{bmatrix} 2\,x(1 - \frac{x}{4} - \frac{y}{4}) \\ y(1 - \frac{x}{3} - \frac{y}{6}) \end{bmatrix}$$

has equilibria at $(x,y) = (0,0)$, $(2,2)$, $(4,0)$ *and* $(0,6)$. *Also show that the linear approximation at these equilibria are as follows:*
At $(0,0)$:

$$\begin{bmatrix} du \\ dv \end{bmatrix} = \begin{bmatrix} 2 & 0 \\ 0 & 1 \end{bmatrix} \begin{bmatrix} u \\ v \end{bmatrix} dt$$

a local repeller.
At $(2,2)$:

$$\begin{bmatrix} du \\ dv \end{bmatrix} = \begin{bmatrix} -1 & -1 \\ -\frac{2}{3} & -\frac{1}{3} \end{bmatrix} \begin{bmatrix} u \\ v \end{bmatrix} dt$$

a 'saddle point,' neither attracting nor repelling.
At $(4,0)$:

$$\begin{bmatrix} du \\ dv \end{bmatrix} = \begin{bmatrix} -2 & -2 \\ 0 & -\frac{1}{3} \end{bmatrix} \begin{bmatrix} u \\ v \end{bmatrix} dt$$

a local attractor.
At $(0,6)$:

$$\begin{bmatrix} du \\ dv \end{bmatrix} = \begin{bmatrix} -1 & 0 \\ -2 & -1 \end{bmatrix} \begin{bmatrix} u \\ v \end{bmatrix} dt$$

a local attractor.
    Check your 4 local results by using the Mathematica NoteBook **Flow2D.ma** *to make a flow animation that includes initial points around all these equilibria.*

The local stability of an equilibrium point for a dynamical system is formulated as the next result. Notice that stability is an "infinite time" result, whereas the localization of the previous theorem is a finite time result after magnification. In effect, the stability result says that when the real parts of the characteristic roots are negative, we can look in the microscope infinitely long and never tell the difference between the linear and nonlinear flows.

THEOREM 23.17. *Local Stability*
Let $f(x,y)$ and $g(x,y)$ be smooth functions with $f(x_e, y_e) = g(x_e, y_e) = 0$. The coefficients given by the partial derivatives evaluated at the equilibrium

$$\begin{bmatrix} a_x & a_y \\ b_x & b_y \end{bmatrix} = \begin{bmatrix} \frac{\partial f}{\partial x} & \frac{\partial f}{\partial y} \\ \frac{\partial g}{\partial x} & \frac{\partial g}{\partial y} \end{bmatrix} (x_e, y_e)$$

define the characteristic equation of the equilibrium,

$$\det \begin{vmatrix} a_x - r & a_y \\ b_x & b_y - r \end{vmatrix} = (a_x - r)(b_y - r) - a_y\, b_x = r^2 - (a_x + b_y)\, r + (a_x\, b_y - a_y\, b_x) = 0$$

Suppose that the real parts of both of the roots of this equation are negative. Then there is a real neighborhood of (x_e, y_e) or a non-infinitesimal $\varepsilon > 0$ such that when a solution satisfies

$$\frac{dx}{dt} = f(x, y)$$
$$\frac{dy}{dt} = g(x, y)$$

with initial condition in the neighborhood, $|x(0) - x_e| < \varepsilon$ and $|y(0) - y_e| < \varepsilon$, then

$$\lim_{t \to \infty} x(t) = x_e \qquad and \qquad \lim_{t \to \infty} y(t) = y_e$$

The proof of this result is given in the Mathematical Background Chapter on Theory of Initial Value Problems. A local repeller result is also given. You may find it interesting to look at the phase diagrams of the various 'exceptional' cases given there.

PROBLEM 23.18. *S-I-R with Births = Deaths*
Show that the equilibrium point of the dynamical system

$$\frac{ds}{dt} = k - k\, s - a\, s\, i$$
$$\frac{di}{dt} = a\, s\, i - b\, i - k\, i$$

is an attractor when the parameters produce a positive equilibrium. In particular, show that the equilibrium is at

$$s_e = \frac{b + k}{a} \qquad and \qquad i_e = k\, \frac{a - b - k}{a\,(b + k)}$$

with derivative matrix at equilibrium,

$$\begin{bmatrix} -\frac{a\,k}{b+k} & -(b + k) \\ \frac{k(a-b-k)}{b+k} & 0 \end{bmatrix}$$

and the characteristic equation at equilibrium,

$$r^2 + \frac{ak}{b+k} r + k[a - (b+k)] = 0$$

In general, $a > 0$, $b > 0$, $k > 0$ and k is small compared with a and b. Mathematically, we need $a > b + k$ in order that $s_e > 0$ and $i_e > 0$. Why are the real parts of the roots of this characteristic equation negative? Why does this make the equilibrium point an attractor?

For rubella with $a = 6.8/11.0$, $b = 1/11.0$ and $k = 0.001$, show that this is a spiral attractor (with complex roots).

Mathematica can help with the computations needed in using the local stability theorem, as in the general NoteBook **LocalStability.ma**.

EXERCISE 23.19. *Mathematica and Stability*
Suppose we wish to investigate the equilibria of the particular case of the system of equations:

$$\frac{dx}{dt} = f[x, y] = x(1 - x/7 - y/7)$$
$$\frac{dy}{dt} = g[x, y] = y(1 - x/10 - y/5)$$

We find the equilibria by Entering the commands
f = x(1 - x/7 - y/7);
g = y(1 - x/10 - y/5);
Solve[{f == 0, g == 0},{x,y}]
We can have Mathematica compute the partial derivatives for us as follows:
aX = D[f,x];
aY = D[f,y];
bX = D[g,x];
bY = D[g,y];
The characteristic equation

$$\det \begin{vmatrix} a_x - r & a_y \\ b_x & b_y - r \end{vmatrix} = (a_x - r)(b_y - r) - a_y b_x = r^2 - (a_x + b_y) r + (a_x b_y - a_y b_x) = 0$$

can be solved at the equilibria by Entering the additional commands
x = 4;
y = 3;
charPoly = (aX - r)(bY - r) - aY bX
Solve[charPoly == 0, r]
Yielding the two roots.
1) Use Mathematica to show that the equilibrium $(4, 3)$ of the above system of differential equations is a local attractor near $(4, 3)$ with characteristic roots -1 and $-6/35$.
Use Mathematica to classify the other equilibria of this system at $(0, 0)$, $(7, 0)$ and $(0, 5)$. Compare your analytical results with the geometric analysis of the first competition example in section 5 of this chapter.

2) Use Mathematica to show that the positive equilibrium of the system of differential equations below is neither attracting nor repelling, because the characteristic roots are -1 and $+6/35$.

$$\frac{dx}{dt} = f[x, y] = x(1 - x/10 - y/5)$$

$$\frac{dy}{dt} = g[x, y] = y(1 - x/7 - y/7)$$

Use Mathematica to classify the other equilibria of this system at $(0, 0)$, $(10, 0)$ and $(0, 7)$. Compare your analytical results to the geometric results in the example on fierce competition in section 5 of this chapter.

23.4. Explicit Solutions, Phase Portraits & Invariants

The energy in an undamped spring

$$m \frac{d^2 x}{dt^2} + s\,x = 0$$

is constant. If it is moving, it has kinetic energy dependent on the mass, while if the spring is compressed or extended, it has potential energy captured in the spring. As the system oscillates, it trades potential and kinetic energy forth and back between these two things.

Energy is a very important example of a general mathematical idea associated with dynamical systems, an invariant quantity.

The energy in the damped spring

$$m \frac{d^2 x}{dt^2} + c \frac{dx}{dt} + s\,x = 0$$

dissipates due to the shock absorber friction term $c\frac{dx}{dt}$. In this case we can show that the analytical expression for energy is a decreasing quantity. Since the level sets of the energy function are ellipses, we know that the flow of the dynamical system must steadily move inside smaller and smaller ellipses and thus approach a stable equilibrium at zero. We already know stability in this case from explicit solutions, but the new approach via energy or a decreasing quantity is new and can apply to nonlinear systems when we do not know explicit solutions.

23.4.1. Energy in the Spring is Invariant. The undamped spring

$$m \frac{d^2 x}{dt^2} + s\,x = 0$$

can be written as the two dimensional system

$$\frac{dx}{dt} = y$$

$$\frac{dy}{dt} = -\frac{s}{m} x$$

Without worrying about why this might work, divide the two equations and cancel dt

$$\frac{dy}{dx} = -\frac{s\,x}{m\,y}$$

separate variables and integrate

$$m\,y\,dy = -s\,x\,dx$$

$$\int m\,y\,dy = -\int s\,x\,dx$$

$$\frac{1}{2}m\,y^2 + \frac{1}{2}s\,x^2 = k \quad \text{a constant of indefinite integration}$$

Perhaps these steps are illegal, so let's build an interpretation of this calculation. We will show that any solution of the dynamical system $(x(t), y(t))$ makes

$$\frac{1}{2}m\,y^2(t) + \frac{1}{2}s\,x^2(t) = k \qquad \text{a constant for all } t$$

To prove that the quantity $E(x(t), y(t)) = \frac{1}{2}m\,y^2(t) + \frac{1}{2}s\,x^2(t)$ is constant, it suffices to prove that its derivative is zero. Differentiate and use the differential equations:

$$\frac{dE}{dt} = m\,y(t)\frac{dy}{dt}(t) + s\,x(t)\frac{dx}{dt}$$

$$= m\,y(t)[-\frac{s}{m}\,x] + s\,x(t)[y]$$

$$= -s\,y(t)\,x(t) + s\,x(t)\,y(t)$$

$$= 0$$

The expression $\frac{1}{2}m\,y^2$ is the kinetic energy and $\frac{1}{2}s\,x^2$ is the potential energy. We have just proved mathematically that the model conserves energy or that E is constant on solutions. Another way to say this is that energy is an invariant of the dynamical system.

What if we add damping $-c\frac{dx}{dt}$ and ask what happens to energy? Our differential equations are now

$$\frac{dx}{dt} = y$$

$$\frac{dy}{dt} = -\frac{s}{m}\,x - \frac{c}{m}\,y$$

We will show that energy is decreasing by showing that $\frac{dE}{dt} < 0$ on solutions.

$$\frac{dE}{dt} = m\,y(t)\frac{dy}{dt}(t) + s\,x(t)\frac{dx}{dt}$$

$$= m\,y(t)[-\frac{s}{m}\,x - \frac{c}{m}\,y] + s\,x(t)[y]$$

$$= -s\,y(t)\,x(t) + s\,x(t)\,y(t) - \frac{c}{m}\,y^2(t)$$

$$= -\frac{c}{m}\,y^2(t)$$

$$< 0$$

The level sets of E are shown in the figure with lower levels corresponding to smaller ellipses, hence solutions move into smaller and smaller ellipses.

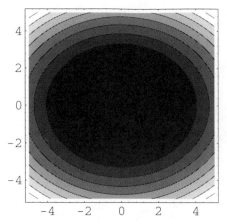

FIGURE 23.20: Constant Energy

23.4.2. S-I-R Revisited. Compute the invariant of the S-I-R differential equations forming

$$\frac{di}{ds} = \frac{b}{a}\frac{1}{s} - 1 = \frac{1}{c\,s} - 1$$

and integrating both sides to obtain

$$i + s - \frac{1}{c}\operatorname{Log}[s] = k \qquad \text{for some constant } k.$$

EXERCISE 23.21. *S-I-R Invariant*
Use the S-I-R differential equations to prove that

$$Q(s(t), i(t)) = i(t) + s(t) - \frac{1}{c}\operatorname{Log}[s(t)]$$

is constant for any solution $(s(t), i(t))$.

Make a ContourPlot of Q using Mathematica and compare your figure with the phase portrait or flow of the S-I-R system.

In Chapter 2 we used this equation to measure the contact number c from data. This required finding the smallest root of the equation $Q = k$. Why do we need the smallest root?

23.4.3. Lotka-Volterra Predator-Prey. The Lotka-Volterra predator-prey equations are

$$dx = (p - m\,y)x\,dt$$
$$dy = (-q + n\,x)y\,dt$$

for positive constants. The sketch of the phase plane you did in Exercise 21.57 or the **FoxRabbit.ma** NoteBook suggests closed loops in the flow. The computer animation shows solutions that at least are very nearly closed loops. Closed loop solutions have the ecological interpretation of persistent oscillations in the populations - years of boom and bust - for a population disturbed from equilibrium.

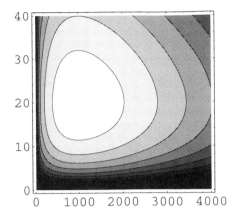

EXERCISE 23.22. *Closed Orbits*
Calculate an invariant for the Lotka-Volterra model and plot the contours of that invariant using the Mathematica ContourPlot function. Why are the orbits of the model closed loops as shown in the figure?

23.5. Review: Looking at Equilibria

Calculus lets us "see" in a powerful microscope without actually first graphing the non-linear object. The way we "see" is by doing computations: first we use rules for derivatives to find the linear equation that we will see, then we simply graph the linear thing. The continuous equilibrium microscope is the most sophisticated one we have learned. First, we need to use partial derivatives for two functions, but even after we calculate the associated linear dynamical system, we have substantial work left in sketching its flow.

It is helpful to remember the five basic cases, which could either be flowing in or out depending on the signs of the characteristic roots. These cases are: (1) Real roots of the same sign. (2) Real roots of opposite sign. (3) Purely imaginary roots. (4) Complex roots. These spiral in if the real part of the roots is negative. (5) A 'repeated' real root, or non-distinct real root. (6) There is also the possibility of a degeneracy in the linear problem, for example, if the equations are just multiples, so that $y(t) = k\,x(t)$ for some constant k. Here are the flows for these cases. Except for the exact directions, these are the views in a microscope focused at an equilibrium point of a smooth dynamical system.

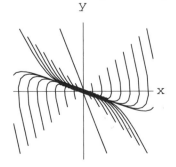

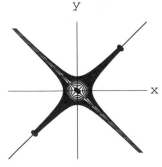

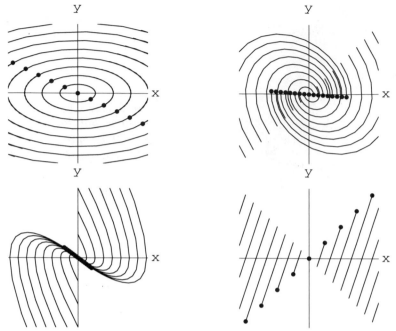

LOOKING AT EQUILIBRIA

To "see" the equilibria of the dynamical system

$$\frac{dx}{dt} = f(x, y)$$

$$\frac{dy}{dt} = g(x, y)$$

compute the following things in the following order: (You can even make a *Mathematica* program to do this for you.)

(1) Find the solutions (x_{e1}, y_{e1}), (x_{e2}, y_{e2}), (x_{e3}, y_{e3}), $\cdots$ to the equilibrium equations

$$0 = f(x, y)$$
$$0 = g(x, y)$$

(2) Use rules of differentiation (the symbolic part of the 'microscope') to find the four symbolic partial derivatives:

$$\frac{\partial f}{\partial x}(x, y) \qquad\qquad \frac{\partial f}{\partial y}(x, y)$$

$$\frac{\partial g}{\partial x}(x, y) \qquad\qquad \frac{\partial g}{\partial y}(x, y)$$

(3) For each equilibrium point (x_{e1}, y_{e1}), (x_{e2}, y_{e2}), (x_{e3}, y_{e3}), $\cdots$
　　(a) Evaluate the symbolic partial derivatives at the specific equilibrium to obtain a matrix of constants

$$\begin{bmatrix} \frac{\partial f}{\partial x}(x_e, y_e) & \frac{\partial f}{\partial y}(x_e, y_e) \\ \frac{\partial g}{\partial x}(x_e, y_e) & \frac{\partial g}{\partial y}(x_e, y_e) \end{bmatrix} = \begin{bmatrix} a_x & a_y \\ b_x & b_y \end{bmatrix}$$

(b) Calculate the roots of the characteristic polynomial

$$\det \begin{vmatrix} a_x - r & a_y \\ b_x & b_y - r \end{vmatrix} = r^2 - (a_x + b_y)\,r + (a_x\,b_y - a_y\,b_x) = 0$$

(c) Characterize the equilibrium according to the basic cases given above.
As long as the characteristic values are not zero, what you "see" in the calculus microscope is one of the five basic pictures either flowing in or out depending on the sign of the characteristic roots. (Zero characteristic values produce degenerate pictures like the sixth.)

EXERCISE 23.23. *A Look in the Microscope*

1) The differential equation

$$x'' + 2x' + x + 15x^3 = 0$$

is "autonomous." What does that mean? Why can't we study flows of non-autonomous equations, but rather only their explicit solutions?
2) This differential equation is equivalent to the system

$$dx = y\ dt$$
$$dy = (-x - 15x^3 - 2y)\ dt$$

Why?
3) If we write this system of differential equations in the form

$$\frac{dx}{dt} = f(x,y)$$
$$\frac{dy}{dt} = g(x,y)$$

then $f(x,y) = ?$ and $g(x,y) = ??$ and the system satisfies the hypotheses of the Existence and Uniqueness Theorem. Why?
4) Show that the only equilibrium point of this system is $(x_e, y_e) = (0,0)$.
5) At the equilibrium, show that

$$\frac{\partial f}{\partial x}(x_e, y_e) = 0 \qquad\qquad \frac{\partial f}{\partial y}(x_e, y_e) = 1$$

$$\frac{\partial g}{\partial x}(x_e, y_e) = -1 \qquad\qquad \frac{\partial g}{\partial y}(x_e, y_e) = -2$$

6) The characteristic roots of the system

$$\frac{du}{dt} = v$$
$$\frac{dv}{dt} = -u - 2v$$

are repeated, that is, there really is only one, $r = -1$.
7) Which of the figures above is similar to the flow of the (u,v) system?
8) What does the flow of the (x,y) system look like under a powerful microscope focused at the point $(0,0)$?

EXERCISE 23.24. *The system of differential equations*

$$dx = y(1 - \frac{x}{10} - \frac{y}{5}) \, dt$$
$$dy = x(1 - \frac{x}{7} - \frac{y}{7}) \, dt$$

has equilibria at the points $(0,0)$, $(0,5)$, $(7,0)$ *and* $(4,3)$. *What you would see in an infinitesimal microscope focused at each of these points?*

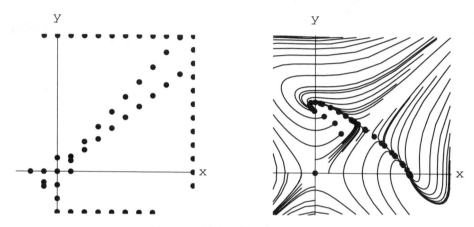

FIGURE 23.25: Nonlinear Equilibria

EXERCISE 23.26. *What do you see if you look inside a microscope at a smooth flow at a non-equilibrium point?*

EXERCISE 23.27. *Uniqueness*
A 2-D flow given by smooth differential equations has a large closed orbit, in fact, points move around the circle of radius 1,000 centered at (2,5). For example, the initial point (1002,5) travels around the circle and returns at t = 100. Can any solution that starts inside this circle tend to infinity? In other words, could a solution starting inside the circle cross the circle? (HINT: See the title of this exercise.)

23.6. Projects

23.6.1. The Second Derivative Test for Max-Min in 2-D as Stability. This Chapter of the Mathematical Background shows how to test a critical point in two variables as a max or min. The idea is to imagine being on a mountain, moving down in the steepest direction. If a place with a horizontal tangent plane is a minimum, then it is a stable equilibrium for this motion.

23.6.2. The Perfecto Skier. The gradient "motion" in the mathematical project on max-min in 2 variables is quite interesting mathematically, but it does not correspond to sliding down a mountain. If a skier slides down a slippery slope, the component of force acting tangent to the mountain produces an acceleration or second derivative. This produces different motion than the gradient flow.

23.6.3. Vectors, Bombing, The Pendulum and Close Encounters with Jupiter.
Vector geometry underlies the derivation of many differential equations in science. There
are projects on low level bombing, the pendulum and shooting a satellite near Jupiter to
'sling shot' it out of the solar system.

23.6.4. Differential Equations from Increment Geometry. Many important curves
in mathematics are given by their differential equation. The tractrix, isochrone and catenary
are three such examples explored in the Mathematical Background Chapter on Differential
Equations from Increment Geometry.

23.6.5. Applications to Ecology. We have seen competition and predator-prey equations in the last three chapters. The Scientific Projects on Predator-Prey interactions and
on Competition and Cooperation study these and related examples in more detail.

23.6.6. Chemical Reactions. Two projects on chemical reactions ar given. One studies stability of a chemical reaction using the mathematical stability theory. The other studies
approximations to the nonlinear chemical dynamics of enzyme reactions.

Part 6

Infinite Series

CHAPTER 24

Geometric Series

Series are simply discrete sums

$$u_1 + u_2 + u_3 + \cdots$$

for example

$$\frac{1}{2} + \frac{1}{4} + \frac{1}{8} + \frac{1}{16} + \cdots$$

or

$$1 + x + \frac{x^2}{2!} + \frac{x^3}{3!} + \cdots$$

or

$$\mathrm{Cos}[x] + \frac{1}{2}\,\mathrm{Cos}[3x] + \frac{1}{2^2}\,\mathrm{Cos}[3^2 x] + \cdots$$

Finite series end at a specified point such as

$$1 + r + r^2 + r^3 + \cdots + r^{356}$$

whereas infinite series 'keep going to infinity.'

Various kinds of infinite series are important as approximations to functions. Power series and Fourier series are the most important kinds of series approximations (for different reasons.) In this case, "approximation" means that you can get 'close enough' to the 'infinite sum' by taking 'sufficiently many' terms. The expression

$$e^x = 1 + x + \frac{x^2}{2!} + \frac{x^3}{3!} + \cdots$$

means that for 'sufficiently large' n,

$$e^x \approx 1 + x + \frac{x^2}{2!} + \frac{x^3}{3!} + \cdots + \frac{x^n}{n!}$$

However, there are some difficulties with adding sums of functions. If we find "sufficiently many" terms to approximate at one value of x, we may not necessarily assume a good approximation at another value of x.

The *Mathematica* NoteBooks **ConvergSeries.ma** and **FourierSeries.ma** visually illustrate the way various approximations depend on the value of x. You can look at the animations now and return to the electronic exercises when they arise in the text. Some of the figures associated with the power series

$$\text{Cos}[x] = 1 - \frac{x^2}{2} + \frac{x^4}{24} + \cdots + \frac{(-1)^n x^{2n}}{(2n)!} + \cdots$$

are shown below. Notice that the worst errors always occur at x values of largest absolute value. This is a general property of power series which helps to make their theory simple and powerful.

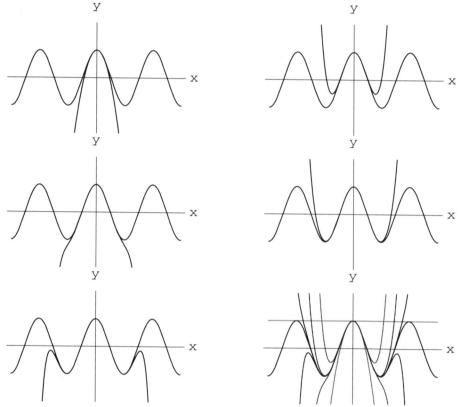

FIGURE 24.1: Power Series Approximations to $y = \text{Cos}[x]$

Approximation or convergence of Fourier series can be more complicated than for power series. For example, the approximation

$$\frac{x}{2} = \text{Sin}[x] + \frac{-1}{2}\text{Sin}[2x] + \frac{1}{3}\text{Sin}[3x] + \cdots + \frac{(-1)^{n+1}}{n}\text{Sin}[nx] + \cdots$$

is valid for $-\pi < x < \pi$, but since all the sine terms are 2π periodic, the limit of the series must also be periodic. When we repeat a piece of the curve $y = x$, for $-\pi < x < \pi$, periodically, we get a 'sawtooth' wave with jump disconuities at multiples of π. Convergence of the infinite Fouries series is necessarily complicated near the jumps, in fact, you can see that every partial sum is a bad approximation at some x near π. However, if we stay away from the trouble spot, this remarkable combination of sines approximates a linear function.

Notice in particular that, if $x = \pi/2$,

$$\frac{\pi}{4} = 1 + 0 - \frac{1}{3} + 0 + \frac{1}{5} + 0 - \frac{1}{7} + \cdots$$

This curious numerical fact falls out of the symbolic expression once we know which values of x produce a valid approximation.

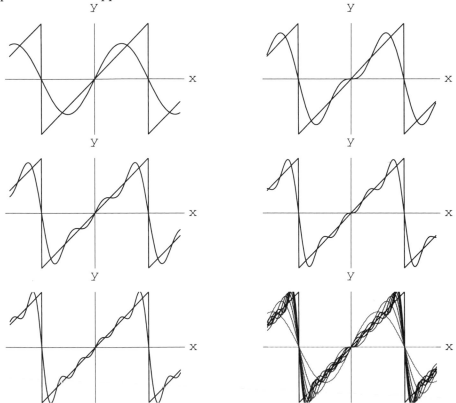

FIGURE 24.2: Fourier Series Approximations to $y = x$

The goal of this chapter is to learn how to calculate series approximations of functions and to learn when the approximations are valid. Approximation is not just a question of estimating errors in one series, but also includes operations on series. We will see that we have a lot of freedom to treat power series like very long polynomials and this means that we can study transcentental functions with polynomial computations plus approximations. Fourier series do not allow us as much computational freedom as power series, but they are important for other reasons.

24.1. Geometric Series

The geometric series

$$1 + r + r^2 + \cdots + r^n + \cdots$$

is the most important series in mathematics. It is a fundamental building block for the theory of series and arises in many aplications. We have the convergence formula

$$\frac{1}{1-r} = 1 + r + r^2 + \cdots + r^n + \cdots$$

$$= \mathrm{Sum}[r^k, \{k, 0, \mathrm{Infinity}\}], \qquad \text{for } -1 < r < 1$$

Moreover, if $|x| \le r < 1$, then the error between $\frac{1}{1-x}$ and $1 + x + \cdots + x^n$ is less than or equal to the error between $\frac{1}{1-r}$ and $1 + r + \cdots + r^n$, that is, the worst error occurs at the largest x. You can see this on the next figure and can see it even better in the *Mathematica* animation of the NoteBook **ConvergSeries.ma**.

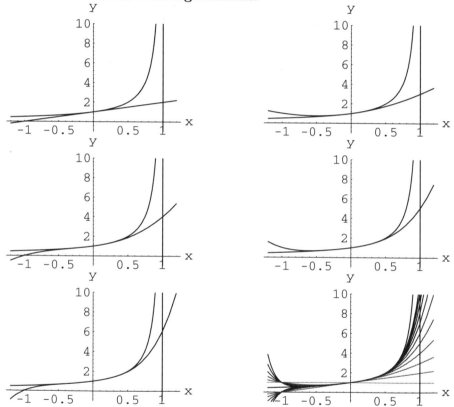

FIGURE 24.3: More and more terms of $1 + x + x^2 + x^3 + \cdots$

We want you to understand this convergence and its limitations. In other words, we want you to be able to estimate errors in finite approximations and to know why the formula does not work for all values of x. Once you understand this simple example throughly, you will be able to extend your knowledge to many rather complicated series.

First, the good news. How can we see the formula

$$\lim_{n \to \infty} [1 + r + r^2 + \cdots + r^n] = \frac{1}{1-r}, \qquad \text{for } |r| < 1$$

Here is one case that is intuitively clear:

EXERCISE 24.4. *Approaching the Cliff*
You visit Yosemite National Park and are thrilled to see the magnificant view from the top of the cliff at the south overlook. However, after spending time in Iowa you are a little afraid of heights. (Midwesterners are known for being sensible, if a little boring...) Beginning from a distance of 1 meter from the edge of a 900 m vertical cliff, you step ahead half way toward the edge.

You have traveled $\frac{1}{2}$ m and have $\frac{1}{2}$ m to go.

Cautiously, you step ahead again half of the way from where you are to the edge, a distance of $\frac{1}{4}$ meter. How far have you traveled total? How far remains to the edge?

Again you step half way, this time $\frac{1}{8}$ m. How far have you traveled? How far remains to the edge?

Write a formula for the total distance travled after n steps. Write a formula for the distance that remains after n steps.

How much is the limit

$$\frac{1}{2} + \frac{1}{4} + \frac{1}{8} + \frac{1}{16} + \cdots$$

How much is the error between a finite approximation to this limit,

$$\frac{1}{2} + \frac{1}{4} + \frac{1}{8} + \frac{1}{16} + \cdots + \frac{1}{2^n}$$

and the total limit? Write this error two ways, first as an infinite sum, and second, in terms of the distance remaining.

Your cousin from California is visiting the Park with you. She is much bolder and, starting from the same place, steps ahead 2/3 if the way to the edge. Each successive step, she steps 2/3 of the way to the cliff. Show that her distance traveled after n steps is

$$\frac{2}{3} + \frac{2}{3}\left(\frac{1}{3}\right) + \frac{2}{3}\left(\frac{1}{3} \cdot \frac{1}{3}\right) + \cdots + \frac{2}{3}\left(\frac{1}{3}\right)^{n-1}$$

How much is the infinite sum

$$\frac{2}{3} + \frac{2}{3}\left(\frac{1}{3}\right) + \frac{2}{3}\left(\frac{1}{3} \cdot \frac{1}{3}\right) + \cdots + \frac{2}{3}\left(\frac{1}{3}\right)^{n-1} + \cdots$$

How much of the distance to the edge remains after your cousin takes n steps?

PROOF OF THE GEOMETRIC SERIES FORMULA

$$1 + r + r^2 + \cdots + r^n + \cdots = \frac{1}{1-r}$$

1: We need to use the result of an Exercise you worked last semester.

EXERCISE 24.5. $\lim_{n\to\infty} r^n$

$$\lim_{n\to\infty} r^n = 0, \qquad \text{for } |r| < 1$$
$$= 1, \qquad \text{for } r = 1$$
$$= \infty, \qquad \text{for } r > 1$$
$$= \text{diverges by oscillation}, \qquad \text{for } r \le -1$$

Y

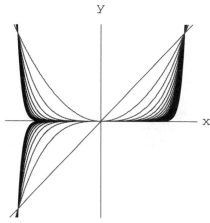

FIGURE 24.6: Limit of r^n

2: Multiply a finite partial sum by $(1 - r)$,

$$(1 + r + r^2 + \cdots + r^{n-1} + r^n)(1 - r) = 1 + r + \cdots + r^n$$
$$- r - \cdots - r^n - r^{n+1}$$
$$= 1 - r^{n+1}$$

so

$$1 + r + r^2 + \cdots + r^{n-1} + r^n = \frac{1 - r^{n+1}}{1 - r}$$
$$= \frac{1}{1 - r} - \frac{r^{n+1}}{1 - r}$$

The limit is calculated using the exercise as follows:

$$\lim_{n \to \infty} [1 + r + \cdots + r^n] = \lim_{n \to \infty} [\frac{1}{1 - r} - \frac{r^{n+1}}{1 - r}]$$
$$= \frac{1}{1 - r} - \frac{1}{1 - r} \lim_{n \to \infty} r^{n+1}$$
$$= \frac{1}{1 - r}$$

when $|r| < 1$.

Let's practice using the formula.

EXERCISE 24.7. *Start at Zero - Somehow*
Use the formula $1 + r + r^2 + \cdots + r^n + \cdots = \frac{1}{1-r}$, *for* $|r| < 1$ *to calculate the following:*

(1) $\frac{1}{2} + \frac{1}{4} + \frac{1}{8} + \cdots$ *(HINT: Use* $-1 + [1 + \frac{1}{2} + \frac{1}{4} + \frac{1}{8} + \cdots]$.)
(2) $\frac{1}{2} + \frac{1}{4} + \frac{1}{8} + \cdots$ *(HINT: Use* $\frac{1}{2}[1 + \frac{1}{2} + \frac{1}{4} + \frac{1}{8} + \cdots]$.)
(3) $\frac{1}{3} + \frac{1}{9} + \frac{1}{27} + \cdots$
(4) $1 - \frac{3}{4} + \frac{9}{16} - \frac{27}{64} + \cdots$
(5) $3 + \frac{3}{2} + \frac{3}{4} + \frac{3}{8} + \cdots = 3[1 + \frac{1}{2} + \frac{1}{4} + \frac{1}{8} + \cdots]$
(6) $a\,x + a\,x^2 + a\,x^3 + \cdots$ *for a constant* a *and variable* x. *State restrictions on* a *or* x *if needed.*

The biggest pitfall in estimating the accuracy of a partial sum of a series

$$\text{Sum}[u_k, \{k, 1, \infty\}] = u_1 + u_2 + u_3 + \cdots$$

is to sum the terms until u_n is small and then assume that the error

$$\text{Error}_n = \text{Sum}[u_k, \{k, 1, \infty\}] - \text{Sum}[u_k, \{k, 1, n\}]$$

is small. This estimate is almost always false. It can be infinite! The error itself is an infinite series,

$$\text{Error}_n = \text{Sum}[u_k, \{k, n+1, \infty\}] = u_{n+1} + u_{n+2} + \cdots$$

In other words, the error is the potential amount that remains to be added from n on to infinity. The formula in the geometric series is very simple, while, on the other hand, it tells exactly how much more is left to be added on from n to infinity.

The proof of the formula for the geometric series gives the error as a closed formula. The error term of the proof is analoglus to the question, 'How far is the remaining distance to the cliff?' in the cliffhanger exercise above.

When $|r| < 1$, we have

$$\text{Sum}[r^k, \{k, 0, n\}] =$$

$$1 + r + r^2 + \cdots + r^n = \frac{1}{1-r} - \frac{r^{n+1}}{1-r}$$

and

$$\text{Sum}[r^k, \{k, 0, \infty\}] =$$

$$1 + r + r^2 + \cdots + r^n + \cdots = \frac{1}{1-r}$$

The error in summing only n terms of the infinite series is

$$[1 + r + r^2 + \cdots + r^n + \cdots] - [1 + r + r^2 + \cdots + r^n] = [r^{n+1} + r^{n+2} + \cdots]$$

$$\text{Sum}[r^k, \{k, 0, \infty\}] - \text{Sum}[r^k, \{k, 0, n\}] = \text{Sum}[r^k, \{k, n+1, \infty\}]$$

$$= [\frac{1}{1-r}] - [\frac{1}{1-r} - \frac{r^{n+1}}{1-r}]$$

$$= \frac{r^{n+1}}{1-r}$$

Rephrasing this observation, the error for summing n terms instead of infinitely many is

$$\text{Error}_n(r) = \frac{r^{n+1}}{1-r}$$

However, for the material ahead on more general series, it is important to realize that the error is also an infinite series,

$$\text{Error}_n(r) = r^{n+1} + r^{n+2} + \cdots$$

The error formula in this case tells us how much more there is to add from n to infinity. In more general series we will use this basic formula to estimate the error caused by stopping at n.

EXAMPLE 24.8. *Distance to the Edge*

In the cliff example above where you step 1/2 way to the edge of the cliff each step, the total distance is the series,

$$\frac{1}{2} + \frac{1}{4} + \frac{1}{8} + \cdots = \frac{1}{2}\left(1 + \frac{1}{2} + \frac{1}{4} + \frac{1}{8} + \cdots\right)$$

$$= \frac{1}{2}\frac{1}{1-r}, \qquad \text{with } r = \frac{1}{2}$$

$$= \frac{1}{2}\frac{1}{1-1/2} = \frac{1}{2}\frac{1}{\frac{1}{2}} = 1$$

The distance remaining after n steps is the error formula computation

$$\left(\frac{1}{2} + \frac{1}{4} + \frac{1}{8} + \cdots\right) - \left(\frac{1}{2} + \frac{1}{4} + \frac{1}{8} + \cdots + \frac{1}{2^n}\right)$$

$$= \frac{1}{2}\left[\left(1 + \frac{1}{2} + \frac{1}{4} + \frac{1}{8} + \cdots\right) - \left(1 + \frac{1}{2} + \frac{1}{4} + \frac{1}{8} + \cdots + \frac{1}{2^{n-1}}\right)\right]$$

$$= \frac{1}{2}\left(\frac{1}{2^n} + \frac{1}{2^{n+1}} + \cdots\right)$$

$$= \frac{1}{2} \cdot \frac{r^{n-1+1}}{1-r}, \qquad \text{with } r = \frac{1}{2}$$

$$= \frac{1}{2} \cdot \frac{1/2^n}{1-1/2} = \frac{1}{2^n}$$

as you showed by direct reasoning in the exercise on approaching the cliff above.

Similarily, your bold cousin travels the distance,

$$\frac{2}{3} + \frac{2}{3}\left(\frac{1}{3}\right) + \frac{2}{3}\left(\frac{1}{3}\frac{1}{3}\right) + \cdots = \frac{2}{3}\left(1 + \frac{1}{3} + \frac{1}{9} + \frac{1}{27} + \cdots\right)$$

$$= \frac{2}{3} \cdot \frac{1}{1-r}, \qquad \text{with } r = \frac{1}{3}$$

$$= \frac{2}{3} \cdot \frac{1}{1-1/3} = \frac{2}{3}\frac{1}{\frac{2}{3}} = 1$$

Her distance remaining after n steps is the error formula computation

$$\left(\frac{2}{3} + \frac{2}{3}\left(\frac{1}{3}\right) + \frac{2}{3}\left(\frac{1}{9}\right) + \cdots\right) - \left(\frac{2}{3} + \frac{2}{3}\left(\frac{1}{3}\right) + \frac{2}{3}\left(\frac{1}{9}\right) + \cdots + \frac{2}{3}\frac{1}{3^{n-1}}\right)$$

$$= \frac{2}{3}\left[\left(1 + \frac{1}{3} + \frac{1}{9} + \frac{1}{27} + \cdots\right) - \left(1 + \frac{1}{3} + \frac{1}{9} + \frac{1}{27} + \cdots + \frac{1}{3^{n-1}}\right)\right]$$

$$= \frac{2}{3}\left(\frac{1}{3^n} + \frac{1}{3^{n+1}} + \cdots\right)$$

$$= \frac{2}{3} \cdot \frac{r^{n-1+1}}{1-r}, \qquad \text{with } r = \frac{1}{3}$$

$$= \frac{2}{3} \cdot \frac{1/3^n}{1-1/3} = \frac{1}{3^n}$$

Again, this computation should only confirm your direct calculation from the exercise above on approaching the cliff.

EXERCISE 24.9. *Geometric Errors*
In each of the specific numerical series of the Exercise 24.7, compute the value of n so that the sum of n terms of that series is within 6 significant digits (machine precision) of the final limit. (You need a calculator or Mathematica to take logarithms.)

Compute r^n and $\frac{r^{n+1}}{1-r}$ in each of these cases. When is the error equal to the last term added?

24.1.1. The Bouncing Ball.

A nice new superball dropped vertically from a height h onto a hard surface, rebounds a large fraction of the way back to its starting position, say to height hr for a value of r like $r = 0.8$. An old mushy tennis ball also rebounds a fraction of its height, but with a value like $r = 0.3$ If the bouncing is perfectly elastic and vertical (with no other energy loss or transfer to spin), the top of the second bounce is a fraction r times the hight at the top of the first.

EXERCISE 24.10. *Bouncing*
Calculate the total vertical distance traveled by a perfectly elastic ball with rebound coefficient r when dropped from an initial height h. Test some numerical cases such as a superball with r = 0.8 dropped from a second floor window, h = 7m.

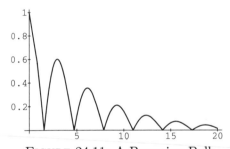

FIGURE 24.11: A Bouncing Ball

24.1.2. Compound Interest.

You find a snazzy red roadster, a 1984 FireBall XL, for sale at a local used car dealer with a window price of $5,000. The dealer offers financing with no down payment. The salesman explains that the car is only $200 per month with payments for 3 years. It sounds like a great start to the summer, since you expect to have a job that will more than cover the payment. But wait, that's $ 200 times 36 months, $200 \times 36 = 7,200$. Whoa!

You tell the salesman, "That's seven thousand two hundred dollars!" I thought the car was only five thousand. He pulls out his calculator and computes

$$\frac{7200 - 5000}{5000} \frac{1}{3} = .14 \cdots$$

saying, that's only fourteen percent and you have no down payment. (Actually, his calculator says nearly fifteen precent.) You really like the car, but know the calculation smells fishy. You would like to know the actual rate of interest r.

Monthly computation of interest at a (nominal) annual rate r is done as follows

$$5000 \times [1 + \frac{r}{12}]$$

This is the amount you would owe at rate r (as a decimal) after one month. At the end of the month you pay your two hundred dollars,

$$\text{Balance after one payment} = 5000 \times [1 + \frac{r}{12}] - 200$$

At the end of the second month you owe interest on this balance,

$$[5000 \times [1 + \frac{r}{12}] - 200][1 + \frac{r}{12}] = 5000[1 + \frac{r}{12}]^2 - 200[1 + \frac{r}{12}]$$

Then you make your payment, leaving the outstanding balance

$$\text{Balance after two payments} = 5000[1 + \frac{r}{12}]^2 - 200[1 + \frac{r}{12}] - 200$$

At the end of the third month you owe $[1 + \frac{r}{12}]$ times this, but make another payment,

$$\text{Balance after three payments} = 5000[1 + \frac{r}{12}]^3 - 200[1 + \frac{r}{12}]^2 - 200[1 + \frac{r}{12}] - 200$$

EXERCISE 24.12. *Installment Interest*
Write a formula for the balance owed on an initial debt of five thousand dollars borrowed at annual rate r after n monthly payments of two hundred dollars. Express your answer as a series in terms of the combined quantity $m = [1 + \frac{r}{12}]$.
Use the formula for a finite geometric series to re-write your series as the formula

$$\text{Balance after } n \text{ payments} = 5000\, m^n - 200\frac{1 - m^n}{1 - m}$$

In the car payment problem, after 36 months you owe nothing. This makes the computation of the true interest rate a root finding problem,

$$\text{Balance after 36 payments} = 0$$

or

$$5000\, m^{36} = 200\frac{1 - m^{36}}{1 - m}$$

EXERCISE 24.13. *The True Rate*
Use the FindRoot or Solve Mathematica command with an initial guess of 16 percent to find the true interest rate in a five thousand dollar loan with 36 monthly payments of two hundred dollars.
Also use Mathematica to print a table of the remaining balance on a five thousand dollar loan with two hundred dollar monthly payments that really is only charging fourteen percent (annual) interest. How many months does it take to pay this loan off?

24.1.3. The Bad and Strange News. The geometric series diverges for $x \geq 1$, clearly

$$1 + x + x^2 + x^3 + \cdots + x^n \geq 1 + 1 + 1 + \cdots + 1 = n + 1 \to \infty$$

If $x < -1$, successive partial sums change by increasing amounts that grow in magnitude, for example,

$$1$$
$$1 - 2 = -1$$
$$1 - 2 + 4 = 3$$
$$1 - 2 + 4 - 8 = -5$$
$$1 - 2 + 4 - 8 + 16 = +11$$

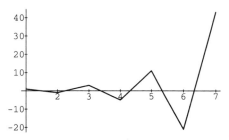

FIGURE 24.14: Zigzag Geometric Growth

There is one strange case of divergence in our formula

$$1 + x + x^2 + x^3 + \cdots = \frac{1}{1 - x}$$

Suppose that we take the limit as x tends to -1 on both sides of our expression. There is no problem with

$$\lim_{x \to -1} \frac{1}{1 - x} = \frac{1}{2}$$

EXERCISE 24.15.

(1) *What is the limit of the expression*

$$\lim_{x \to -1} [1 + x + x^2 + x^3 + \cdots]$$

(2) *What are the partial sums of* $1 - 1 + 1 - 1 + \cdots$ *?*

$$1$$
$$1 - 1$$
$$1 - 1 + 1$$
$$1 - 1 + 1 - 1$$
$$\cdots$$

(3) *What sense can we make of the formula*

$$1 - 1 + 1 - 1 + \cdots = \frac{1}{2}$$

if any?

24.2. Convergence by Comparison

Series of functions

$$u_0(x) + u_1(x) + \cdots$$

can often be compared to known series of numbers

$$M_0 + M_1 + \cdots$$

The best comparison, when the M_n series converges is

$$|u_n(x)| \leq M_n, \qquad \text{for all } n \text{ and all } a \leq x \leq b$$

This makes the maximum error in the function series no worse than the (single) error in the numerical series,

$$\text{Error}_n(x) =$$
$$|u_{n+1}(x) + u_{n+2}(x) + \cdots| \leq |u_{n+1}(x)| + |u_{n+2}| + \cdots$$
$$\leq M_{n+1} + M_{n+2} + \cdots$$
$$\leq \text{M-series Error}_n, \qquad \text{for all } a \leq x \leq b$$

This error estimate is called the Weierstrass majorization estimate. It is very simple, provided we can find the majorizing series and its error. The simplicity is that then we get a single maximum error for all x (perhaps in an interval).

24.2.1. Uniform Geometric Estimation. We want to view the geometric series as a sum of the functions, 1, x, x^2, ... and make a single estimate of error over a whole interval of x values. This is illustrated graphically in the *Mathematica* NoteBook **ConvergSeries.ma**. We really have treated the geometric series as a function series with the variable r, but now we just want the maximum error for an interval. We are being a little pedantic, becuase we are trying to illustrate estimation in a very simple case.

Suppose $r < 1$ is a fixed value (parameter). We want to consider the maximum error between a partial sum of $1 + x + x^2 + \cdots$ and its limit $1/(1-x)$ for $-r \leq x \leq r$. We want an estimate for

$$\text{Max Error}_n(x) =$$
$$\text{Max}|[1 + x + x^2 + \cdots + x^n + \cdots] - [1 + x + x^2 + \cdots + x^n]| =$$
$$= \text{Max}|[x^{n+1} + x^{n+2} + \cdots]|$$
$$\text{Max}|\text{Sum}[x^k, \{k, 0, \infty\}] - \text{Sum}[x^k, \{k, 0, n\}]| = \text{Max}|\text{Sum}[x^k, \{k, n+1, \infty\}]|$$

where the maximum is for $|x| \leq r$ and $0 < r < 1$ is fixed. Each term of the error series satisfies $|x^k| \leq r^k$ and we know that finite sums satisfy

$$|x^{n+1} + x^{n+2} + \cdots + x^{n+N}| \leq |x^{n+1}| + |x^{n+2}| + \cdots + |x^{n+N}|$$
$$\leq r^{n+1} + r^{n+2} + \cdots + r^{n+N}$$

and so, when $|x| \leq r$,

$$\text{Error}_n(x) = \lim_{N \to \infty} |x^{n+1} + x^{n+2} + \cdots + x^{n+N}|$$

$$\leq \lim_{N \to \infty} r^{n+1} + r^{n+2} + \cdots + r^{n+N} = \frac{r^{n+1}}{1-r}$$

This estimate does not depend on the particular x, so

$$\text{Max Error}_n \leq \frac{r^{n+1}}{1-r}$$

DEFINITION 24.16. *Uniform Absolute Convergence*
A series of functions

$$u_0(x) + u_1(x) + \cdots + u_n(x) + \cdots$$

is said to be uniformly absolutely convergent on $[a,b]$, if the maximum error in computing a partial sum of $|u_0(x)| + |u_1(x)| + \cdots + |u_n(x)| + \cdots$ tends to zero as n tends to infinity.

The estimates just before this definition show that the geometric series $1 + x + x^2 + \cdots + x^n + \cdots$ is uniformly absolutely convergent on any closed interval strictly inside $-1 < x < 1$. The Weierstrass majorization estimate always produces an absolutely uniformly convergent series.

THEOREM 24.17. *Convergence of The Geometric Series*
A geometric series $a + ax + ax^2 + \cdots + ax^n + \cdots$ converges uniformly absolutely to the limit $\frac{a}{1-x}$ for $|x| \leq r$ provided the constant $r < 1$. The maximum error between the sum of n terms and the limit $\frac{a}{1-x}$ is

$$\frac{ar^{n+1}}{1-r}$$

24.2.2. Weierstrass' Wild Wiggles. In chapter 4 when we first studied local approximation by a linear function, we mentioned Weierstrass' famous example of a function which is continuous, but has a kink (or corner) at every point on its graph. Specifically, Weierstrass' function is nowhere differentiable. This means that if we look at the graph with microscopes of arbitrary power, we will always see wiggles - it never straightens out to look like its tangent, because it never has a tangent...

Weierstrass function is given by the series

$$W(x) = \text{Cos}[x] + \frac{\text{Cos}[3x]}{2} + \frac{\text{Cos}[9x]}{4} + \cdots + \frac{\text{Cos}[3^n x]}{2^n} + \cdots$$

This series is easy to majorize since

$$\left| \frac{\text{Cos}[3^n x]}{2^n} \right| \leq \frac{1}{2^n}, \qquad \text{for all } x$$

EXERCISE 24.18. *Weierstrass' Wild Wiggles*
Write a Mathematica program to plot a piece of Weierstrass' function $W(x)$ over an arbitrary interval $a \leq x \leq b$ which is specified at the beginning of your program. Use the same PlotRange in the y-coordinate, for example by including the graphics option $PlotRange ->$ $\{W(c) - d, W(c) + d\}$ in the Plot command where $c = (a+b)/2$ and $d = (b-a)/2$.
You can define

$$W(x\_) = Sum[Cos[3^k \, x]/2^k, \{k, 0, n\}]$$

but you must find an estimate of the error caused by using n instead of ∞. You can use a geometric series estimate to find the n so that the error is less than the screen resolution of about 1 in 250 (for example, there are a million total pixels on a NeXT monitor, 1,000 by 1,000, but we use about 1/4th of the screen for a plot.) In the case that you plot in a y-range of width 2 d you want the error to be less than d/250. Estimate with the geometric majorant and use logs to solve for n.

What is a formula for the n such that the sum of n terms of Weierstrass' series has an error less than d/250?

Experiment with your program using small intervals (in other words, large magnification.) For example plot the graph for .499 < x < .501.

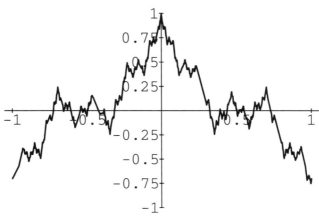

FIGURE 24.19: Weierstrass' Wild Wiggles

CHAPTER 25

Power Series

Power series are series built from power functions in the following way,

$$\text{Sum}[a_k\, x^k, \{k, 0, \infty\}] = a_0 + a_1\, x + a_2\, x^2 + \cdots + a_n\, x^n + \cdots$$

where the sequence a_0, a_1, a_2, $\cdots$, may depend on the index n, but not on x. For example, the geometric series is a power series,

$$\frac{1}{1-x} = 1 + x + x^2 + x^3 + \cdots + x^n + \cdots$$

where $a_n = 1$ for all n. The exponential series,

$$e^x = 1 + x + \frac{1}{2}\, x^2 + \frac{1}{6}\, x^3 + \cdots + \frac{1}{n!}\, x^n + \cdots$$

is also a power series with $a_n = \frac{1}{n!}$. (Note: The exclamation sign ! denotes the factorial function, $n! = n \cdot (n-1) \cdot (n-2) \cdots 3 \cdot 2$. This function is also built into *Mathematica* as !)
  The series

$$|t| = 1 + \frac{1}{2}[t^2 - 1] + \frac{\frac{1}{2}(\frac{1}{2} - 1)}{2}[t^2 - 1]^2 + \frac{\frac{1}{2}(\frac{1}{2} - 1)(\frac{1}{2} - 2)}{3 \cdot 2}[t^2 - 1]^3 + \cdots$$
$$\cdots + \binom{\frac{1}{2}}{n}[t^2 - 1]^n + \cdots$$

is a convergent series for $|t| < 1$ and it is 'built up from power functions,' but it is not considered a power series. Notice that each new term of this series would produce lower powers of t after simplification, $[t^2 - 1]^{n+1} = (-1)^{n+1} + (-1)^n(n+1)t^2 + \cdots + t^{2n+2}$. Power series have the property that the partial sum $\text{Sum}[a_k\, x^k, \{k, 0, n\}]$ contains all the powers lower than the nth that ever occur in the series.

25.1. Computation of Power Series

One importance of power series is that we have some freedom to calculate with them as if they were polynomials. For example, the geometric series

$$\frac{1}{1-u} = 1 + u + u^2 + \cdots + u^n + \cdots$$

converges for $|u| < 1$. If we let $u = -x$, the resulting series will still converge for $|x| < 1$,

$$\frac{1}{1 - (-x)} = 1 + (-x) + (-x)^2 + \cdots + (-x)^n + \cdots$$

$$\frac{1}{1 + x} = 1 - x + x^2 - x^3 + \cdots + (-1)^n x^n + \cdots$$

EXERCISE 25.1. *Use a substitution to find a power series expansion for*

$$\frac{1}{1 + x^2} = 1 + a_1 x + a_2 x^2 + \cdots + a_n x^n + \cdots$$

We can also differentiate and integrate convergent power series term-by-term. At first it probably seems strange to suggest that this needs proof, but it does. Before we worry about that, let's see why we would want to do these operations. Integrate both sides of the expression for $1/(1 + x)$ from 0 to u,

$$\frac{1}{1 + x} = 1 - x + x^2 + \cdots + (-1)^n x^n + \cdots$$

$$\int_0^u \frac{1}{1 + x} dx = \int_0^u 1 \, dx - \int_0^u x \, dx + \int_0^u x^2 \, dx + \cdots + (-1)^n \int_0^u x^n \, dx + \cdots$$

$$\text{Log}[1 + u] = u - \frac{1}{2}u^2 + \cdots + (-1)^n \frac{1}{n + 1} u^{n+1} + \cdots$$

or written in terms of x,

$$\text{Log}[1 + x] = x - \frac{1}{2}x^2 + \cdots + (-1)^n \frac{1}{n + 1} x^{n+1} + \cdots$$

We expect this series to converge for $|x| < 1$ (and know that there is trouble with the logarithm when $x = -1$.)

Why do we want a theorem to justify term-by-term integration of power series? Such a theorem would tell us the above expression is a valid expansion for the logarithm. Notice that this gives the curious special case

$$\text{Log}[2] = 1 - \frac{1}{2} + \frac{1}{3} - \frac{1}{4} + \cdots$$

when we take $x = 1$. We could use this to approximate the natural log of 2 with pencil and paper calculations.

EXERCISE 25.2. ArcTan$[x]$ *Series*
The integral of $\frac{1}{1+x^2}$ from 0 to u is ArcTan$[u]$. Use your series from the exercise above to show that

$$\text{ArcTan}[x] = \text{Sum}[(-1)^k \frac{x^{2k+1}}{2k + 1}, \{k, 0, \infty\}]$$

and express this in the more familiar form

$$\text{ArcTan}[x] = x - ? + ? \cdots$$

The tangent of $\pi/4$ is 1, so the arc tangent of 1 is $\pi/4$. Substitute $x = 1$ in your series to obtain the special case

$$\frac{\pi}{4} = 1 - \frac{1}{3} + \frac{1}{5} - \frac{1}{7} + \cdots$$

EXAMPLE 25.3. *The Exponential Series,*

$$e^x = 1 + x + \frac{1}{2}x^2 + \frac{1}{3 \cdot 2}x^3 + \frac{1}{4 \cdot 3 \cdot 2}x^4 + \cdots + \frac{1}{n!}x^n + \cdots$$

Term-by-term differentiation is also allowed for convergent power series. We can use this to find a power series expansion for e^x. Recall our official definition that $y = e^x$ is the unique solution to

$$y(0) = 1$$
$$\frac{dy}{dx} = y$$

Suppose we have a series

$$y = a_0 + a_1 x + a_2 x^2 + \cdots$$

We want to find coefficients a_0, a_1, $\cdots$, so that y is also a solution to this initial value problem. If the series converges and differentiation is valid, this y must equal e^x, by the uniqueness of the solution to the initial value problem.

Let $x = 0$ to obtain

$$y(0) = a_0 = 1$$

Differentiate y term-by-term to obtain

$$\frac{dy}{dx} = a_1 + 2a_2 x + 3a_3 x^2 + 4a_4 x^3 + \cdots$$

In order for the series to satisfy the differential equation we must have $y = \frac{dy}{dx}$ or

$$1 + a_1 x + a_2 x^2 + a_3 x^3 + \cdots = a_1 + 2a_2 x + 3a_3 x^2 + 4a_4 x^3 + \cdots$$

We can recursively solve for the coefficients. If $x = 0$, we get $a_1 = 1$,

$$1 + x + a_2 x^2 + a_3 x^3 + \cdots = 1 + 2a_2 x + 3a_3 x^2 + 4a_4 x^3 + \cdots$$

Subtract 1 from both sides, divide by x and we see

$$1 + a_2 x + a_3 x^2 + \cdots = 2a_2 + 3a_3 x + 4a_4 x^2 + \cdots$$

Let $x = 0$ to obtain $2a_2 = 1$ or $a_2 = \frac{1}{2}$.

The preceding algebra simply says we can equate coefficients of like powers, obtaining

$$a_0 = 1 \qquad \text{or} \qquad a_0 = 1$$
$$a_n = (n+1)a_{n+1} \qquad a_{n+1} = \frac{1}{n+1}a_n$$

so

$$a_0 = 1$$

$$a_1 = a_0 = 1$$

$$a_2 = \frac{1}{2}a_1 = \frac{1}{2}$$

$$a_3 = \frac{1}{3}a_2 = \frac{1}{3} \cdot \frac{1}{2}$$

$$a_4 = \frac{1}{4}a_3 = \frac{1}{4} \cdot \frac{1}{3} \cdot \frac{1}{2}$$

In general

$$a_n = \frac{1}{n!}$$

So term-by-term differentiation and proof of convergence would justify the important formula

$$e^x = 1 + x + \frac{x^2}{2} + \frac{x^3}{3 \cdot 2} + \cdots + \frac{x^n}{n!} + \cdots$$

which does converge and gives an expansion for the natural exponential at all values of x.

EXERCISE 25.4. $Sin[x]$ and $Cos[x]$ *Power Series*

(1) *Substitute the complex variable* $x = i\theta$ *into the series above for* e^x. *Note that* $(i\theta)^2 = -\theta$, $(i\theta)^3 = -i\theta^3$, $(i\theta)^4 = \theta^4$, $\cdots$. *Re-write the series in the form*

$$e^{i\theta} = \left[1 - \frac{\theta^2}{2} + \cdots + (-1)^n \frac{\theta^{2n}}{(2n)!} + \cdots\right] + i\left[\theta - \frac{\theta^3}{6} + \cdots + (-1)^n \frac{\theta^{2n+1}}{(2n+1)!} + \cdots\right]$$

and write out the first five terms of both the real and complex parts.

(2) *Use Euler's formula* $e^{i\theta} = Cos[\theta] + i\, Sin[\theta]$ *and the previous part of the exercise to find series for sine and cosine.*

(3) *Check your series for cosine by showing that the series satisfies*

$$y(0) = 1$$
$$y'(0) = 0$$
$$\frac{d^2y}{dx^2} = -y$$

(4) *Check your series for sine by showing that the series satisfies*

$$y(0) = 0$$
$$y'(0) = 1$$
$$\frac{d^2y}{dx^2} = -y$$

Between the text discussion and the exercises, we now have all the basic power series:

$$e^x = 1 + x + \frac{x^2}{2} + \frac{x^3}{3 \cdot 2} + \cdots + \frac{x^n}{n!} + \cdots$$

$$= \text{Sum}[\frac{x^k}{k!}, \{k, 0, \infty\}], \qquad \text{for all } x$$

$$\text{Cos}[x] = 1 - \frac{x^2}{2} + \frac{x^4}{4 \cdot 3 \cdot 2} + \cdots + (-1)^n \frac{x^{2n}}{(2n)!} + \cdots$$

$$= \text{Sum}[(-1)^k \frac{x^{2k}}{(2k)!}, \{k, 0, \infty\}], \qquad \text{for all } x$$

$$\text{Sin}[x] = x - \frac{x^3}{3 \cdot 2} + \frac{x^5}{5 \cdot 4 \cdot 3 \cdot 2} + \cdots + (-1)^n \frac{x^{2n+1}}{(2n + 1)!} + \cdots$$

$$= \text{Sum}[(-1)^k \frac{x^{2k+1}}{(2k + 1)!}, \{k, 0, \infty\}], \qquad \text{for all } x$$

$$\frac{1}{1 - x} = 1 + x + x^2 + \cdots + x^n + \cdots$$

$$= \text{Sum}[x^k, \{k, 0, \infty\}], \qquad \text{for } |x| < 1$$

$$\text{Log}[1 + x] = x - \frac{x^2}{2} + \frac{x^3}{3} + \cdots + (-1)^n \frac{x^{n+1}}{n + 1} + \cdots$$

$$= \text{Sum}[(-1)^k \frac{x^{k+1}}{k + 1}, \{k, 0, \infty\}], \qquad \text{for } |x| < 1$$

$$\text{ArcTan}[x] = x - \frac{x^3}{3} + \frac{x^5}{5} + \cdots + (-1)^n \frac{x^{2n+1}}{(2n + 1)} + \cdots$$

$$= \text{Sum}[(-1)^k \frac{x^{2k+1}}{2k + 1}, \{k, 0, \infty\}], \qquad \text{for } |x| < 1$$

EXERCISE 25.5. *ConvergSeries.ma NoteBook*
*Run the Mathematica **ConvergSeries.ma** NoteBook and complete the electronic exercises of making animations of the convergence of the series for sine, log and arc tangent.*

The animations of cosine and the geometric series are already complete in that NoteBook, but you should check the intervals of convergence and compare the animations with the theory.

The purpose of the animations is to show you which parts of the graphs converge to the limit functions. Convergence that is "uniform" in x for an interval simply means that the graphs converge over that interval.

The binomial theorem from algebra says

$$(1+x)^n = 1 + n\,x + \cdots + \binom{n}{k}\,x^k + \cdots + x^n$$
$$= \mathrm{Sum}[\binom{n}{k}\,x^k, \{k, 0, n\}]$$

where $\binom{n}{k} = \frac{n\cdot(n-1)\cdots(n-k+1)}{k\cdot(k-1)\cdot(k-2)\cdots 1}$ are the binomial coefficients. For example, $(1+x)^3 = 1 + \frac{3}{1}x + \frac{3\cdot 2}{2\cdot 1}x^2 + \frac{3\cdot 2\cdot 1}{3\cdot 2\cdot 1}x^3 = 1 + 3\,x + 3\,x^2 + x^3$. We can generalize this to THE BINOMIAL SERIES:

$$(1+x)^p = \mathrm{Sum}[\binom{p}{k}\,x^k, \{k, 0, \infty\}] \qquad \text{for } |x| < 1$$

where the generalized binomial coefficients are given by

$$\binom{p}{k} = \frac{p\cdot(p-1)\cdots(p-k+1)}{k\cdot(k-1)\cdot(k-2)\cdots 1}$$

and p is any real power, not necessarily an integer or even positive. Here is the way to do this:

EXERCISE 25.6. *Finding the Binomial Series*

(1) *Show that* $y = (1+x)^p$ *satisfies the initial value problem*

$$y(0) = 1$$
$$(1+x)\frac{dy}{dx} = p\,y$$

(2) *Show that the generalized binomial coefficients defined above have the property*

$$(k+1)\binom{p}{k+1} = (p-k)\binom{p}{k}$$

for any real p and any positive integer k. Use this to show

$$(k+1)\binom{p}{k+1} + k\binom{p}{k} = p\binom{p}{k}$$

(3) *Differentiate the series*

$$y = \mathrm{Sum}[\binom{p}{k}\,x^k, \{k, 0, \infty\}]$$

term-by-term and show that it satisfies

$$y' = \mathrm{Sum}[k\binom{p}{k}\,x^{k-1}, \{k, 1, \infty\}] = \mathrm{Sum}[(n+1)\binom{p}{n+1}\,x^n, \{n, 0, \infty\}]$$

(4) *Multiply $(1+x)$ times y' to obtain*

$$(1+x)y' = \mathrm{Sum}[\{(k+1)\binom{p}{k+1} + k\binom{p}{k}\}x^k, \{k, 0, \infty\}]$$
$$= p\,\mathrm{Sum}[\binom{p}{k}\,x^k, \{k, 0, \infty\}]$$
$$= p\,y$$

(5) *Use the uniqueness theorem for initial value problems to prove the binomial series, given that the series converges for $|x| < 1$ and that term-by-term differentiation is valid.*

(6) *Write the binomial series for the case $p = -\frac{1}{2}$ and check the first six terms of the series for $1/\sqrt{1-x} = 1 + \frac{1}{2}x + \frac{3}{8}x^2 + \frac{5}{16}x^3 + \frac{35}{128}x^4 + \frac{63}{256}x^5 + \cdots$.*

Mathematica can help with binomial series. Type the following to see the series for $(1 + x)^p$:

In[1]
  p = 1/2
  Sum[Binomial[p,k] x∧k ,{k,0,10}]
Out[2]

$$1+\frac{1}{2}x-\frac{1}{8}x^2+\frac{1}{16}x^3-\frac{5}{128}x^4+\frac{7}{256}x^5-\frac{21}{1024}x^6+\frac{33}{2048}x^7-\frac{429}{32768}x^8+\frac{715}{65536}x^9-\frac{2431}{262144}x^{10}$$

Try several values of the power p.

PROBLEM 25.7. *Substitute the expression* $x = t^2 - 1$ *into the series for* $\sqrt{1+x}$ *(given just before this exercise), obtaining a series for* $\sqrt{1 + [t^2 - 1]} = \sqrt{t^2} = |t| = ?$. *Is this a power series?*

25.2. The Ratio Test for Convergence of Power Series

THEOREM 25.8. *The Ratio Test*
Consider a power series

$$a_0 + a_1\,x + a_2\,x^2 + \cdots + a_n\,x^n + \cdots$$

Suppose that $\lim_{n\to\infty} \frac{|a_n|}{|a_{n+1}|}$ *exists or tends to* $+\infty$. *Then the series converges uniformly and absolutely for all* $|x| \le \rho$, *for any constant* $\rho < \lim_{n\to\infty} \frac{|a_n|}{|a_{n+1}|}$.

PROOF: We will use the Weierstrass majorization estimate with a geometric series.
  There is an integer N such that for all $|x| < \rho$ and all $n \ge N$, the ratio of successive terms, $|a_{n+1}\,x^{n+1}|$ over $|a_n\,x^n|$ satisfies

$$|x|\frac{|a_{n+1}|}{|a_n|} \le \rho\frac{|a_{n+1}|}{|a_n|} \le r < 1$$

for some constant $r < 1$. This is because $\rho \lim_{n\to\infty}\left|\frac{a_{n+1}}{a_n}\right| < 1$, so eventually $\rho\frac{|a_{n+1}|}{|a_n|} \le r < 1$, for a constant $r < 1$.
  This makes the error in the absolute series,

$$|a_{n+1}\,x^{n+1}| + |a_{n+2}\,x^{n+2}| + \cdots \le$$
$$\le |a_{n+1}\,\rho^{n+1}| + |a_{n+2}\,\rho^{n+2}| + \cdots$$
$$\le |a_n\,\rho^n|[r^1 + r^2 + \cdots] = |a_n\,\rho^n|\frac{r}{1-r}$$

because

$$\frac{|a_{n+1}\,\rho^{n+1}|}{|a_n\,\rho^n|} \le r \quad \text{or} \quad |a_{n+1}\,\rho^{n+1}| \le |a_n\,\rho^n|\,r$$

$$\frac{|a_{n+2}\,\rho^{n+2}|}{|a_{n+1}\,\rho^{n+1}|} \le r \quad \text{or} \quad |a_{n+2}\,\rho^{n+2}| \le |a_{n+1}\,\rho^{n+1}|\,r \le |a_n\,\rho^n|\,r^2$$

$$\frac{|a_{n+3}\,\rho^{n+3}|}{|a_{n+2}\,\rho^{n+2}|} \le r \quad \text{or} \quad |a_{n+3}\,\rho^{n+3}| \le |a_{n+2}\,\rho^{n+2}|\,r \le |a_n\,\rho^n|\,r^3$$

and this error tends to zero as n tends to infinity - uniformly in x, since x does not appear in the expression.

The proof that $\lim_{n \to \infty} a_n \, \rho^n = 0$ is as follows. For all $n \geq N$,

$$a_n \, \rho^n \leq a_N \, \rho^N r^{n-N} = \frac{a_N \rho^N}{r^N} r^n \to 0 \qquad \text{as } n \to \infty \qquad \text{(for } N \text{ fixed.)}$$

for the constant $r < 1$ above.

That's all there is to the proof - just remember that it amounts to comparison to a geometric series by estimating ratios.

The moral of the story is clear:

(1) Compute $\lim_{n \to \infty} \frac{|a_n|}{|a_{n+1}|}$

(2) Let $|x| \leq \rho < \lim_{n \to \infty} \left| \frac{a_n}{a_{n+1}} \right|$, for any constant ρ.

This is your interval of safe convergence. Be careful of values of x which make

$$|x| \lim_{n \to \infty} \frac{|a_{n+1}|}{|a_n|} = 1$$

The theorem makes no guarantees in that case. In other words, you cannot take $\rho = \lim_{n \to \infty} \frac{|a_n|}{|a_{n+1}|}$ and apply the theorem. We'll see why below; sometimes the series still converges, sometimes not.

EXAMPLE 25.9. *Radius of Convergence*

The coefficients in the exponential series are $a_n = \frac{1}{n!}$, so

$$\lim_{n \to \infty} \frac{a_n}{a_{n+1}} = \lim_{n \to \infty} \frac{(n+1)!}{n!} = \lim_{n \to \infty} n + 1 = \infty$$

This means that for any positive ρ, the series

$$1 + x + \frac{x^2}{2} + \frac{x^3}{3 \cdot 2} + \cdots + \frac{x^n}{n!} + \cdots$$

converges uniformly absolutely for $|x| \leq \rho < \infty$.

The coefficients of the logarithm series are $\frac{(-1)^n}{n+1}$, so

$$\lim_{n \to \infty} \left| \frac{a_n}{a_{n+1}} \right| = \lim_{n \to \infty} \frac{n+1}{n} = \lim_{n \to \infty} 1 + \frac{1}{n} = 1$$

This means that the log series converges uniformly absolutely for $|x| \leq \rho < 1$. The theorem does not apply when $x = 1$ and we know there is trouble when $x = -1$, because we have $\text{Log}[1 - 1] = \text{Log}[0] = -1 - \frac{1}{2} - \frac{1}{3} - \cdots$.

EXERCISE 25.10. *Write each of the following series in the old-fashioned style like*

$$e^x = 1 + x + \frac{x^2}{2} + \frac{x^3}{3 \cdot 2} + \cdots$$

For which values of the constant ρ will the following series converge absolutely and uniformly for $|x| \leq \rho$? What is the exact symbolic value of the series?

(1) $\text{Sum}[k \, x^k, \{k, 0, \infty\}]$
(2) $\text{Sum}[(-1)^k \, k \, x^k, \{k, 0, \infty\}]$

(3) $\mathrm{Sum}[\frac{x^k}{3^k}, \{k, 0, \infty\}]$
(4) $\mathrm{Sum}[\frac{x^k}{3^{k+1}}, \{k, 0, \infty\}]$
(5) $\mathrm{Sum}[(-2)^k \frac{k+2}{k+1} x^k, \{k, 0, \infty\}]$
(6) $\mathrm{Sum}[\frac{x^k}{(k+3)!}, \{k, 0, \infty\}]$

The important series for sine and cosine skip terms, so $a_n = 0$ for every other coefficient in the formula of the ratio test theorem. The limit of $\frac{a_n}{a_{n+1}}$ does not exist, because it is alternately zero and undefined. However, the same basic idea of comparison with a geometric series still works.

EXERCISE 25.11. *Ratios for Series that Skip Terms*
Suppose a power series skips every other term as in the sine or cosine series,

$$\mathrm{Cos}[x] = 1 + 0 \cdot x - \frac{x^2}{2} + 0 \cdot x^3 + \frac{x^4}{4 \cdot 3 \cdot 2} + \cdots$$

Successive nonvanishing terms have ratios $x^2 \frac{a_{n+2}}{a_n}$. Suppose that the $\lim_{n \to \infty} \left| \frac{a_n}{a_{n+2}} \right|$ exists or tends to $+\infty$. Show that the series is absolutely uniformly convergent for $|x| \le \sigma$ for constants σ related to the limit.

Apply your test to the series, and if possible give the exact symbolic sum of the series.

(1) *The sine series.*
(2) *The cosine series.*
(3) *The arc tangent series.*
(4) $\mathrm{Sum}[\frac{x^{3k}}{k!}, \{k, 0, \infty\}]$
(5) $\mathrm{Sum}[\frac{(-1)^k}{2k+1} \left[\frac{x}{2}\right]^{2k}, \{k, 0, \infty\}]$

25.3. Integration of Series

Integration of uniformly convergent series is always possible term-by-term. You might wonder why we don't always have

$$\lim_{n \to \infty} \int_a^b f_n(x) \, dx = \int_a^b \lim_{n \to \infty} f_n(x) \, dx$$

where $f_n(x) = u_0(x) + u_1(x) + \cdots + u_n(x)$. Here is a simple reason. The functions

$$f_n(x) = (n+1) \, 2x \left[1 - x^2\right]^n$$

all have area 1 under their graphs between zero and one,

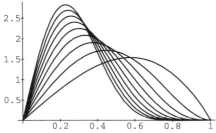

FIGURE 25.12: Equal Areas Under $f_n(x) = (n+1) \, 2x \left[1 - x^2\right]^n$

EXERCISE 25.13. *Show that*

$$\int_0^1 f_n(x) = \int_0^1 (n+1)\, 2x \left[1 - x^2\right]^n \, dx = 1$$

The limit of these functions is still zero, though not uniformly.

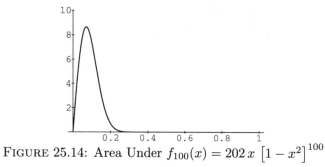

FIGURE 25.14: Area Under $f_{100}(x) = 202\, x \left[1 - x^2\right]^{100}$

EXERCISE 25.15.

(1) *Show that* $\lim_{n\to\infty} f_n(0) = \lim_{n\to\infty} f_n(1) = 0$
(2) *If* $0 < x < 1$, *show that* $r = [1 - x^2] < 1$, *so* $\lim_{n\to\infty} f_n(x) = 0$
(3) *Find* $Max[f_n(x) : 0 \le x \le 1] = M_n$ *and show that* $M_n \to \infty$. *The maximum error between* $f_n(x)$ *and its zero limit does not tend to zero.*

In this example we have $\int_0^1 f_n(x)\, dx = 1$ for all n, so

$$\lim_{n\to\infty} \int_0^1 f_n(x)\, dx = 1$$

On the other hand, $\lim_{n\to\infty} f_n(x) = 0$ for every x, so

$$\int_0^1 \lim_{n\to\infty} f_n(x)\, dx = 0$$

This is because we do not have the whole graph close to its limit. When the whole graph is close, we can interchange limits.

THEOREM 25.16. *Term-by-Term Integration of Series*
Suppose that the series of continuous functions

$$u_o(x) + u_1(x) + \cdots$$

converges uniformly to a 'sum'

$$S(x) = \lim_{n\to\infty} u_0(x) + \cdots + u_n(x)$$

on the interval $[a, b]$. *That is,* *suppose*

$$M_n = Max[|S(x) - [u_0(x) + \cdots + u_n(x)]| : a \le x \le b] \to 0$$

Then the limit $S(x)$ is continuous and

$$\int_a^b \lim_{n\to\infty} u_0(x) + \cdots + u_n(x) \ dx = \lim_{n\to\infty} \int_a^b u_0(x) + \cdots + u_n(x) \ dx$$

PROOF We shall omit the proof of continuity, it simply means that if $x_1 \approx x_2$, then $S(x_1) \approx S(x_2)$.

The integral part is easy,

$$\left| \int_a^b S(x) \ dx - \left[\int_a^b u_0(x) \ dx + \cdots + \int_a^b u_n(x) \ dx \right] \right| =$$

$$\left| \int_a^b S(x) - [u_0(x) + \cdots + u_n(x)] \ dx \right| =$$

$$\leq \int_a^b |S(x) - [u_0(x) + \cdots + u_n(x)]| \ dx$$

$$\leq \int_a^b M_n \ dx = (b-a) M_n \to 0$$

That's the proof.

EXAMPLE 25.17.

Let's justify the computation of the log series. We know

$$\frac{1}{1+x} = 1 - x + x^2 - x^3 + \cdots$$

converges uniformly absolutely on any interval $-r \leq x \leq r$, for $r < 1$. This means that

$$\int_0^u \frac{1}{1+x} \ dx = \int_0^u 1 \ dx - \int_0^u x \ dx + \int_0^u x^2 \ dx - \int_0^u x^3 \ dx + \cdots$$

$$\mathrm{Log}[u] = u - \frac{1}{2}u^2 + \frac{1}{3}u^3 - \frac{1}{4}u^4 + \cdots$$

for $|u| \leq r < 1$.

Notice that this still does not justify the formula

$$\mathrm{Log}[2] = 1 - \frac{1}{2} + \frac{1}{3} - \frac{1}{4} + \cdots$$

because we would need to integrate all the way to $u = 1$. We take this up later.

EXERCISE 25.18. *The series for arc tangent can be obtained by integrating*

$$\frac{1}{1+x^2} = 1 - x^2 + x^4 - x^6 + \cdots$$

from zero to u and using the fact that $\mathrm{ArcTan}[0] = 0$, as you showed in an exercise above. For which values of u is this computation valid?

We can use term-by-term integration to derive more power series formulas.

EXERCISE 25.19. *ArcSin[x]*
Substitute $-x^2$ *in the binomial series to obtain a series for* $1/\sqrt{1-x^2}$, *then integrate to obtain*

$$\text{ArcSin}[x] = x + \text{Sum}[\frac{1 \cdot 3 \cdot 5 \cdots (2k-1)}{2 \cdot 4 \cdot 6 \cdots (2k)} \frac{x^{2k+1}}{2k+1}, \{k, 1, \infty\}]$$

For which values of x *is this series uniformly absolutely convergent? Hint: Make the substitution* $x = \text{Sin}[u]$, *so*

$$\int \frac{dx}{\sqrt{1-x^2}} = \int \frac{\text{Cos}[u] \, du}{\sqrt{\text{Cos}^2[u]}} = \int du$$

Here is a non-power series to integrate term-by-term.

EXERCISE 25.20. *The Dirichlet convergence theorem from Fourier series shows that for all* $0 \le x \le 2\pi$,

$$\frac{x^2}{4} - \pi \frac{x}{2} + \frac{\pi^2}{6} = \text{Sum}[\frac{1}{k^2} \text{Cos}[k\,x], \{k, 1, \infty\}] = \text{Cos}[x] + \frac{1}{4}\text{Cos}[2\,x] + \frac{1}{9}\text{Cos}[3\,x] + \cdots$$

Use this fact to deduce the following special cases:

$$\frac{\pi^2}{6} = 1 + \frac{1}{4} + \frac{1}{9} + \cdots = \text{Sum}[\frac{1}{k^2}, \{k, 1, \infty\}]$$

and

$$\frac{\pi^3}{32} = \text{Sum}[\frac{(-1)^{n+1}}{(2n-1)^3}, \{k, 1, \infty\}]$$

Mathematics is not magic. If the hypotheses are not satisfied, the conclusions need not hold.

EXERCISE 25.21. *Integration*
Is the following proof that $\text{Log}[3] = \frac{8}{3} - \frac{16}{4} + \frac{32}{5} - \frac{64}{6} + \frac{128}{7} + \cdots$ *correct? If so, explain the steps. If not, explain the error(s). We have* $\frac{1}{1-x} = 1 + x + x^2 + x^3 + \cdots$ *so, letting* $x = -u$, *we get*

$$\frac{1}{1+u} = 1 - u + u^2 - u^3 + \cdots$$

Integrating both sides from 0 *to* x *gives*

$$\int_0^x \frac{1}{1+u} \, du = \int_0^x 1 \, du - \int_0^x u \, du + \int_0^x u^2 \, du + \cdots$$

$$\text{Log}[1+x] = x - \frac{x^2}{2} + \frac{x^3}{3} + \cdots + (-1)^{n+1}\frac{x^n}{n} + \cdots \quad \text{and letting } x = 2, \text{ we obtain,}$$

$$\text{Log}[1+2] = 2 - 2 + \frac{8}{3} + \cdots + (-1)^{n+1}\frac{2^n}{n} + \cdots$$

$$\text{Log}[3] = \frac{8}{3} - \frac{16}{4} + \frac{32}{5} - \frac{64}{6} + \frac{128}{7} + \cdots$$

Sum several terms of this with Mathematica or a calculator and show that it is nonsense.

The next Example gives a different formula for $\text{Log}[3]$.

EXAMPLE 25.22. $\text{Log}[3] = 2\left(\frac{1}{1\cdot 2} + \frac{1}{3\cdot 2^3} + \frac{1}{5\cdot 2^5} + \cdots\right)$

We know

$$\text{Log}[1 + x] = x + \frac{x^2}{2} + \frac{x^3}{3} + \cdots$$

so substituting $-x$ gives

$$-\text{Log}[1 - x] = x - \frac{x^2}{2} + \frac{x^3}{3} + \cdots$$

We also know $\text{Log}[a] - \text{Log}[b] = \text{Log}[a/b]$ from high school and the Mathematical Background Chapter on Functional Identities. Adding the series above and using this identity, we have

$$\text{Log}\left[\frac{1 + x}{1 - x}\right] = 2\left(x + \frac{x^3}{3} + \frac{x^5}{5} + \cdots\right)$$

Some algebra gives us

$$\text{Log}[3] = \text{Log}\left[\frac{1 + \frac{1}{2}}{1 - \frac{1}{2}}\right]$$

Together, these computations yield

$$\text{Log}[3] = 2\left(\frac{1}{2} + \frac{1}{2^3 \cdot 3} + \frac{1}{2^5 \cdot 5} + \cdots\right)$$

These computations are all valid, because the log series is uniformly absolutely convergent for $|x| \leq 1/2$.

25.4. Differentiation of Power Series

THEOREM 25.23. *Differentiation of Power Series*
Suppose that a power series converges uniformly absolutely for $|x| \leq \rho$ to $S(x)$,

$$S(x) = a_0 + a_1\, x + a_2\, x^2 + a_3\, x^3 + \cdots$$

Then the derivative of $S(x)$ exists and the series obtained from term-by-term differentiation converges uniformly absolutely to it on $|x| \leq \rho$,

$$\frac{dS(x)}{dx} = a_1 + 2\, a_2\, x + 3\, a_3\, x^2 + \cdots + n\, a_n\, x^{n-1} + \cdots$$

We omit the proof, but warn the reader that this theorem only applies to power series. Here are some familiar examples to remind you of this limited applicability.

Weierstrass' nowhere differentiable function

$$W(x) = \text{Cos}[x] + \frac{1}{2}\text{Cos}[3x] + \frac{1}{2^2}\text{Cos}[3^2 x] + \cdots$$

does not have a derivative at any value of x. Differentiate the series term-by-term and see what happens.

The identity

$$|t| = 1 + \frac{1}{2}[t^2 - 1] + \frac{\frac{1}{2}(\frac{1}{2} - 1)}{2}[t^2 - 1]^2 + \frac{\frac{1}{2}(\frac{1}{2} - 1)(\frac{1}{2} - 2)}{3 \cdot 2}[t^2 - 1]^3 + \cdots$$

$$\cdots + \left(\begin{smallmatrix}\frac{1}{2}\\n\end{smallmatrix}\right)[t^2 - 1]^n + \cdots$$

is perfectly valid for $|t| \leq 1$. However, the absolute value does not have a derivative at $t = 0$. Term-by-term differentiation of this series is still possible, but what does the series converge to? This series might be fun to explore with *Mathematica* . You can define it by

f[n _] := Sum[Binomial[1/2,k] x∧k , { k,0,n}]
f[10]
x = t∧2 -1
f[10]
Expand[f[10]]

Some partial sums look as follows:

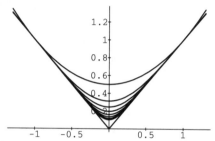

FIGURE 25.24: Non-Power Series for $|t|$

If we have a convergent power series, its derivative is also a convergent power series. Hence we can differentiate that and obtain another convergent power series. In other words, convergent power series have infinitely many derivatives on intervals of convergence.

EXERCISE 25.25. *Show that the series* $y = \mathrm{Sum}[\frac{x^{4k}}{(4k)!}, \{k, 0, \infty\}]$ *converges for all* x *(uniformly and absolutely on* $|x| \leq \rho$, *for any* ρ*) and satisfies*

$$\frac{d^4 y}{dx^4} = y$$

with certain initial conditions.

What initial conditions are needed to make the solution unique? (Hint: Apply the phase variable trick enough times to make it a first order system.)

EXERCISE 25.26. *Cosh[x] and Sinh[x]*
The hyperbolic sine and cosine arise in some applications such as the catenary project. They can be defined as solutions to initial value problems, as series, or by identities with the natural exponential function. All these descriptions are useful.

(1) *Find power series solutions to the initial value problems*

$$
\begin{array}{lll}
y = \mathrm{Cosh}[x] : & & y = \mathrm{Sinh}[x] : \\
y(0) = 1 & \textit{and} & y(0) = 0 \\
y'(0) = 0 & & y'(0) = 1 \\
\dfrac{d^2 y}{dx^2} = y & & \dfrac{d^2 y}{dx^2} = y
\end{array}
$$

(2) *Where do your series converge?*

(3) *Prove the connections with the natural exponential*

$$\text{Cosh}[x] = \frac{e^x + e^{-x}}{2} \qquad \text{Sinh}[x] = \frac{e^x - e^{-x}}{2}$$

(Note: You can use either series or differential equations.) Notice the analogy to Euler's formula, which gives

$$\text{Cos}[x] = \frac{e^{i\,x} + e^{-i\,x}}{2} \qquad \text{Sin}[x] = \frac{e^{i\,x} - e^{-i\,x}}{2\,i}$$

(4) *Show that* $\text{Cosh}[x] = \text{Cos}[i\,x]$ *and* $\text{Sinh}[x] = \frac{\text{Sin}[i\,x]}{i}$.
(5) *Show that* $\frac{d\,\text{Cosh}[x]}{dx} = \text{Sinh}[x]$ *and* $\frac{d\,\text{Sinh}[x]}{dx} = \text{Cosh}[x]$.
(6) *Show that* $\text{Cosh}^2[t] - \text{Sinh}^2[t] = 1$. *This implies that these functions parametrize a hyperbola. (Sine and cosine parametrize the circle* $x^2 + y^2 = 1$.) *If you let* $x = \text{Cosh}[t]$ *and* $y = \text{Sinh}[t]$, *Mathematica's ParametricPlot[.] command will let you draw this hyperbola.*

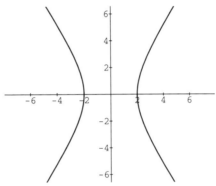

FIGURE *25.27: Hyperbola*

(7) *Show that these hyperbolic functions satisfy the nonlinear first order differential equations*

$$\frac{dy}{dx} = \sqrt{y^2 - 1} \qquad and \qquad \frac{dy}{dx} = \sqrt{y^2 + 1}$$

(Hint: You can just solve it. Separate variables and integrate with the change of variables $\text{Sinh}[u] = y$, *so* $dy = \text{Cosh}[u]\,du$ *and* $1 + y^2 = 1 + \text{Sinh}^2[x] = \text{Cosh}^2[u]$, *making*

$$\int \frac{dy}{\sqrt{1 + y^2}} = \int \frac{\text{Cosh}[u]\,du}{\sqrt{\text{Cosh}^2[u]}} = \int du$$

Can you think of an easier way?)

EXERCISE 25.28. *Bessel Functions*
The Bessel functions $J_0[x]$ *and* $J_1[x]$ *are defined as solutions of the differential equations*

$$x^2 \frac{d^2 y}{dx^2} + x \frac{dy}{dx} + (x^2 - n^2)y = 0$$

for $n = 0$ and $n = 1$ respectively. Bessel functions arise in the study of wave propagation or heat flow when there is cylindrical symmetry. (In this case the partial differential equations describing those phenomena specialize to the differential equation.)

Notice that there is a difficulty with these equations when we re-write them as two dimensional first order systems,

$$\frac{dy}{dx} = z = f(x, y, z)$$

$$\frac{dz}{dx} = -\frac{1}{x} z - \frac{x^2 - n^2}{x^2} y = g(x, y, z)$$

We can not apply the uniqueness theorem for initial conditions at zero, because $g(x, y, z)$ is discontinuous at $x = 0$. However,

(1) *Show that the series*

$$J_0[x] = \mathrm{Sum}[(-1)^k \frac{x^{2k}}{(k!)^2 \, 2^{2k}}, \{k, 0, \infty\}]$$

$$J_1[x] = \mathrm{Sum}[(-1)^k \frac{x^{2k+1}}{k! \, (k+1)! \, 2^{2k+1}}, \{k, 0, \infty\}]$$

converge for all x.

(2) *Show that these series satisfy the differential equations with the respective values of $n = 0$ and 1.*

(3) *Show the identities $J_0'[x] = -J_1[x]$ and $x \, J_0[x] = \frac{d(x \, J_1[x])}{dx}$.*

EXERCISE 25.29. *Taylor's Formula*

Suppose that a function is represented by a uniformly absolutely convergent power series,

$$f(x) = a_0 + a_1 \, x + a_2 \, x^2 + \cdots + a_n \, x^n + \cdots$$

for $|x| \le \rho$. We know that the derivative exists and is given by

$$f'(x) = a_1 + 2 \, a_2 \, x + \cdots + n \, a_n \, x^{n-1} + \cdots$$

Let $x = 0$ in this series to show that $f'(0) = a_1$.

Differentiate again, obtaining the convergent series

$$f''(x) = 2 \, a_2 + 3 \cdot 2 a_3 \, x + \cdots + n(n-1) \, a_n \, x^{n-2} + \cdots$$

Set $x = 0$ in this series to see that $f''(0) = 2 \, a_2$.

Differentiate again, set $x = 0$ and show that $f^{(3)}(0) = 3 \cdot 2 \, a_3$, or that $a_3 = \frac{1}{3 \cdot 2} f^{(3)}(0)$.

Generalize and show Taylor's formula

$$a_n = \frac{1}{n!} \, f^{(n)}(0)$$

CHAPTER 26

The Edge of Convergence

This chapter investigates some weakly convergent series. Inside their interval of convergence, power series are strongly convergent. For example, the ratio test says that convergence is stronger than a geometric series. At the radius of convergence, a power series may either converge or diverge. Also, Fourier series often can not be estimated uniformly. In these cases, the series are converging by cancellations. We begin the chapter with the simplest kind of convergence by cancellation.

26.1. Alternating Series

The alternating series

$$1 - \frac{1}{2} + \frac{1}{3} - \frac{1}{4} + \cdots = \text{Sum}[(-1)^{k+1}\frac{1}{k}, \{k, 1, \infty\}]$$

is called a conditionally convergent series. This is convergence, but a very sensitive kind. The harmonic series

$$1 + \frac{1}{2} + \frac{1}{3} + \frac{1}{4} + \cdots = \text{Sum}[\frac{1}{k}, \{k, 1, \infty\}]$$

diverges to infinity, in fact, we will see that

$$1 + \frac{1}{2} + \frac{1}{3} + \frac{1}{4} + \cdots + \frac{1}{n} > \text{Log}[n] \to \infty$$

The reason that the alternating harmonic series converges is simply decreasing cancellation. First we add 1. Then we subtract a half. Next, we add a third. We never go back above one, nor back below a half, because the size of the oscillations decreases. At each point in this process, we never move more that the next term in the direction of the sign of that term. This is an error estimate, but a very dangerous one. It simply says that the error in summing n terms is no more than the next term. This is dangerous because alternating series are about the only general case where this estimate works. Usually, the next term is not a good estimate of the error in an infinite series. Moreover, numerically in floating point computer arithmetic, this kind of cancellation produces errors that are often large compared with the answer. In other words, computer arithmetic renders the estimate useless. (This is explored in "The Big Bite of the Subtraction Bug" example in the Mathematical Background Chapter on Series. Conditionally convergent improper integrals are explored in the Mathematical Background Chapter on Integrals.)

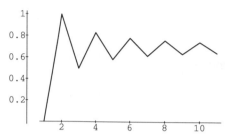

FIGURE 26.1: Alternating Decreasing Moves

What we want to conclude from the observation is simply that if we take the log series

$$\text{Log}[1 + x] = \text{Sum}[(-1)^{k+1}\frac{x^k}{k}, \{k, 1, \infty\}] = x - \frac{x^2}{2} + \frac{x^3}{3} + \cdots$$

the limit of the terms on the right hand side is a convergent series. On the left side of this identity (which we know for $|x| < 1$) we know

$$\lim_{x \to 1} \text{Log}[1 + x] = \text{Log}[2]$$

For a partial sum, we know

$$\lim_{x \to 1}[1 - \frac{x^2}{2} + \cdots + (-1)^{n+1}\frac{x^n}{n}] = 1 - \frac{1}{2} + \frac{1}{3} - \frac{1}{4} + \cdots + (-1)^n\frac{1}{n}$$

Our estimate, described above, for alternating series shows that

$$|\text{Error}_n(x)| \leq \frac{1}{n+1}$$

for all x. Hence, we do have the formula

$$\text{Log}[2] = 1 - \frac{1}{2} + \frac{1}{3} + \cdots$$

EXERCISE 26.2. *Verify that*

$$\frac{\pi}{4} = 1 - \frac{1}{3} + \frac{1}{5} - \frac{1}{7} + \cdots$$

A general fact about power series is that if we can find a point of even conditional convergence, then we can use geometric comparison to prove convergence at smaller values.

THEOREM 26.3. *If the power series*

$$a_0 + a_1 x + a_2 x^2 + \cdots + a_n x^n + \cdots$$

converges for a particular $x = x_1$, then the series converges uniformly and absolutely for $|x| \leq \rho < |x_1|$, for any constant ρ.

PROOF Since the series converges at x_1, we must have $a_n x_1^n \to 0$. If $|x| \le \rho < |x_1|$, then

$$|a_n x^n| = |a_n x_1^n| \left| \frac{x}{x_1} \right|^n \le \left| \frac{\rho}{x_1} \right|^n = r^n$$

is a geometric majorant for the tail of the series.

26.2. Telescoping Series

Recall the idea of the Fundamental Theorem of Integral Calculus: In order to find

$$\int_a^b f(x) \, dx$$

We first find an antiderivative, $F(x)$ such that its differential satisfies $dF(x) = f(x) \, dx$. The increment equation for this new function says

$$F(x + \delta x) - F(x) = f(x) \, \delta x + \varepsilon \delta x$$

so the defining sum for the integral collapses or 'telescopes,'

$$\int_a^b f(x) \, dx = \text{Sum}[f(x)\delta x, \{x, a, b, step\delta x\}]$$
$$\approx \text{Sum}[F(x + \delta x) - F(x), \{x, a, b, step\delta x\}] + \text{Max}|\varepsilon| \cdot (b - a)$$
$$\approx [F(a + \delta x) - F(a)] + [F(a + 2\delta x) - F(a + \delta x)]$$
$$+ \cdots + [F(b) - F(b - \delta x)]$$
$$= F(b) - F(a)$$

Sometimes we can use this idea to find the sum of an infinite series. We can not antidifferentiate, but we can occasionally find a difference,

$$\text{Sum}[\frac{1}{k^2 - 1}, \{k, 2, n\}] = \frac{1}{2} \text{Sum}[(\frac{1}{k - 1} - \frac{1}{k + 1}), \{k, 2, n\}]$$

because

$$\frac{1}{k^2 - 1} = \frac{1}{2}\left[\frac{1}{k - 1} - \frac{1}{k + 1}\right]$$
$$= \frac{1}{2}\left[\frac{k + 1}{(k - 1)(k + 1)} - \frac{k - 1}{(k - 1)(k + 1)}\right]$$
$$= \frac{1}{2}\frac{k + 1 - (k - 1)}{k^2 - 1} = \frac{1}{2}\frac{2}{k^2 - 1}$$

The partial sum

$$\text{Sum}[\frac{1}{k^2 - 1}, \{k, 2, n\}] = \frac{1}{2}\left[(1 - \frac{1}{3}) + (\frac{1}{2} - \frac{1}{4}) + (\frac{1}{3} - \frac{1}{5}) + \cdots + (\frac{1}{n - 1} - \frac{1}{n + 1})\right]$$
$$= \frac{1}{2}\left[1 + \frac{1}{2} - \frac{1}{n} - \frac{1}{n + 1}\right] \to \frac{3}{4}$$

EXERCISE 26.4. *Telescoping Series*
Find the sum of the series

(1) $\text{Sum}[\frac{1}{(2k-1)(2k+1)}, \{k, 1, \infty\}]$

(2) $\text{Sum}[\frac{2k+1}{[k(k+1)]^2}, \{k, 1, \infty\}]$

Series we know can be used to estimate the error in series we don't know.

EXERCISE 26.5. *The Next Term is a Bad Estimate*
How fast does the series

$$\text{Sum}[\frac{1}{k^2}, \{k, 1, \infty\}]$$

converge? Use the estimates

$$\frac{1}{k} - \frac{1}{k+1} = \frac{1}{k(k+1)} \leq \frac{1}{k^2} \leq \frac{1}{k-1} - \frac{1}{k} = \frac{1}{k^2-1}$$

and sum the telescoping terms. The error series satisfies

$$\frac{1}{n} \leq \text{Sum}[\frac{1}{k^2}, \{k, n, \infty\}] \leq \frac{1}{n-1}$$

How much is the difference between these estimates of error? This difference gives us

$$Error_n = \text{Sum}[\frac{1}{k^2}, \{k, n, \infty\}] = \frac{1}{n} + \varepsilon$$

with ε no more than $\frac{1}{n-1} - \frac{1}{n} = ???$

If we sum 100 terms of the series of terms $\frac{1}{k^2}$, how much is the error? How does the error compare with the next term, $\frac{1}{101^2}$?

26.3. Comparison of Integrals and Series

Another way to estimate series above or below is to compare them with integrals. For example, the continuous function $\frac{1}{x}$ decreases, so it stays below $\frac{1}{n}$ for $n \leq x \leq n+1$. This means that one term of the series $1 + \frac{1}{2} + \frac{1}{3} + \cdots$ satisfies

$$\frac{1}{n} \geq \int_n^{n+1} \frac{1}{x}\, dx$$

and

$$1 + \frac{1}{2} + \frac{1}{3} + \cdots + \frac{1}{n-1} \geq \int_1^2 \frac{1}{x}\, dx + \int_2^3 \frac{1}{x}\, dx + \int_3^4 \frac{1}{x}\, dx + \cdots + \int_{n-1}^n \frac{1}{x}\, dx$$

$$\geq \int_1^n \frac{1}{x}\, dx = \text{Log}[n] \to \infty$$

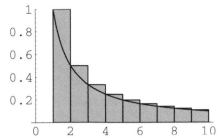

FIGURE 26.6: An Integral Below $1 + \frac{1}{2} + \frac{1}{3} + \cdots$

This shows that the harmonic series diverges.

EXERCISE 26.7. *Log goes to infinity reluctantly*
Run the Mathematica NoteBook **LogGth.ma** *and see how slowly the harmonic series grows. Though it does tend to infinity, it certainly takes its time...*

EXERCISE 26.8. *Lower Integral Comparison*
Test the following series for divergence by squeezing an integral below them:

(1) $\mathrm{Sum}[\dfrac{1}{k^{2/3}}, \{k, 2, \infty\}]$

(2) $\mathrm{Sum}[\dfrac{1}{k(\mathrm{Log}[k])^{2/3}}, \{k, 2, \infty\}]$

An estimate of series with integrals can be used to prove convergence. We know $\dfrac{1}{x^2}$ is decreasing, so we have

$$\frac{1}{n^2} \le \frac{1}{x^2} \qquad \text{for } n - 1 \le x \le n$$

This makes

$$\frac{1}{2^2} + \cdots + \frac{1}{n^2} \le \int_1^2 \frac{1}{x^2}\, dx + \int_2^3 \frac{1}{x^2}\, dx + \cdots + \int_{n-1}^n \frac{1}{x^2}\, dx$$

$$= \int_1^n x^{-2}\, dx$$

$$= \frac{1}{(1-2)} x^{1-2} \,|_1^n$$

$$= \frac{n^{-1} - 1}{-1} = 1 - \frac{1}{n} \to 1$$

In fact, the error for the series satisfies

$$\frac{1}{(n+1)^2} + \frac{1}{(n+2)^2} + \cdots \le \lim_{N \to \infty} \int_n^N x^{-2}\, dx = \frac{1}{n}$$

Compare this with Exercise 26.5.

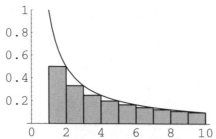

FIGURE 26.9: An Integral Above $\frac{1}{2^2} + \frac{1}{3^2} + \cdots$

EXERCISE 26.10. *Upper Integral Comparison*
Test the following series for convergence by squashing them below an integral:

(1) $\text{Sum}[\dfrac{1}{k^{3/2}}, \{k, 2, \infty\}]$

(2) $\text{Sum}[\dfrac{1}{k(\text{Log}[k])^{3/2}}, \{k, 2, \infty\}]$

The idea of this section can be summarized as follows.

THEOREM 26.11. *Integral Comparison*
Suppose that $f(x)$ is a positive continuous function, defined and decreasing for $x \geq n$. The series of positive decreasing terms, $a_k = f(k)$,

$$a_n + a_{n+1} + a_{n+2} + \cdots \quad \textit{converges, if} \quad \lim_{N \to \infty} \int_n^N f(x) \, dx \; \textit{converges}$$

and

$$a_n + a_{n+1} + a_{n+2} + \cdots \quad \textit{diverges to } \infty, \textit{ if} \quad \lim_{N \to \infty} \int_n^N f(x) \, dx = \infty$$

The two previous simple integral comparison exercises generalize as follows:

EXERCISE 26.12. *Series with Powers*
Show that the following series diverge if $p \leq 1$ and converge if $p > 1$:

(1) $\text{Sum}[\dfrac{1}{k^p}, \{k, 1, \infty\}]$

(2) $\text{Sum}[\dfrac{1}{k(\text{Log}[k])^p}, \{k, 2, \infty\}]$

(3) $\text{Sum}[\dfrac{1}{k \, \text{Log}[k](\text{Log}[\text{Log}[k]])^p}, \{k, 4, \infty\}]$

The series $1 - \frac{1}{3} + \frac{1}{5} - \frac{1}{7} + \cdots$ converges.

EXERCISE 26.13. *Does $1 + \frac{1}{3} + \frac{1}{5} + \frac{1}{7} + \cdots$ converge?*

26.4. Limit Comparisons

Each time we learn a new convergent or divergent series, we can use it to compare to many other series. For example, we know from the exercise above that $1 + 1/2^p + \cdots + 1/n^p + \cdots$ converges for any $p > 1$. We also know from the rate of growth of log that

$$\lim_{n \to \infty} (\text{Log}[n]/n^q) = 0 \qquad \text{for any } q > 0$$

The series

$$\text{Sum}[\frac{\text{Log}[k]}{k\sqrt{k}}, \{k, 2, \infty\}]$$

must therefore converge, since $\lim_{n \to \infty} (\text{Log}[n]/n^{1/6}) = 0$ so that eventually

$$1/n^{\frac{4}{3}} > \text{Log}[n]/(n^{1/6} \, n^{4/3}) = \text{Log}[n]/n^{3/2}$$

and from that point on,

$$\frac{\text{Log}[n]}{n^{3/2}} + \frac{\text{Log}[(n+1)]}{(n+1)^{3/2}} + \cdots < \frac{1}{n^{4/3}} + \frac{1}{(n+1)^{4/3}} + \cdots < \infty$$

EXERCISE 26.14. *General Comparison*
Test the following series for convergence or divergence.

(1) $\text{Sum}[\frac{\text{Log}[k]}{k\sqrt{k+1}}, \{k, 2, \infty\}]$
(2) $\text{Sum}[\frac{1}{\text{Log}[k]^p}, \{k, 2, \infty\}]$

We can also use each new numerical series in function estimates. For example, the Fourier series

$$\frac{\pi}{2} - \frac{\pi}{4}\left(\frac{\text{Cos}[x]}{1} + \frac{\text{Cos}[3\,x]}{3^2} + \frac{\text{Cos}[5\,x]}{5^2} + \cdots + \frac{\text{Cos}[(2n+1)\,x]}{(2n+1)^2} + \cdots\right)$$

is convergent absolutely and uniformly, because

$$\left|\frac{\text{Cos}[(2n+1)\,x]}{(2n+1)^2}\right| \leq \frac{1}{(2n+1)^2} \qquad \text{for all } x$$

and $1 + 1/9 + 1/25 + \cdots$ converges.

26.5. Fourier Series

Fourier series arise in many mathematical and physical problems. The one above actually converges to the function that equals $|x|$ for $-\pi < x \leq \pi$ and is then repeated periodically. Some approximating graphs are shown in the next figure.

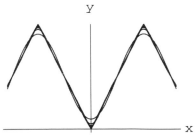

FIGURE 26.15: Fourier Series for $f(x) = |x|$

The Mathematical Background Chapter on Series shows you the simple formula for Fourier coefficients and gives the interesting convergence theorem for functions. Fourier series can converge delicately. For example, the identity

$$x = 2\left(\text{Sin}[x] - \frac{\text{Sin}[2x]}{2} + \frac{\text{Sin}[3x]}{3} + \cdots + (-1)^{n+1}\frac{\text{Sin}[nx]}{n} + \cdots\right)$$

is a valid convergent series for $-\pi < x < \pi$. However, the Weierstrass majorization does not yield a simple convergence estimate, because

$$\left|(-1)^{n+1}\frac{\text{Sin}[nx]}{n}\right| \leq \frac{1}{n}$$

is a useless upper estimate by a divergent series. This series converges, but not uniformly and its limit function is discontinuous, since repeating x periodically produces a jump at π.

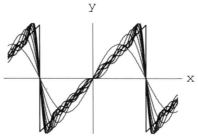

FIGURE 26.16: Fourier Series for $f(x) = x$

This more complicated kind of convergence is left to the Mathematical Background.

Index

acceleration, 161
addition formulas for sine and cosine, 453
additivity of integrals, 269
angle between vectors, 295
animation, Mathematica, 47
antiderivative, 250
arc tangent derivative, 116
arc tangent series, 508, 511, 515, 517
area between curves, 228, 279
attracting equilibria, 419, 434
attractor, discrete, 213, 214, 217
autonomous D E, 418, 432, 434

Bessel functions, 521
binomial series, 512
Bugs Bunny's Law of Gravity, 131, 167

canary, cool, 58
canary, dead, 81
canary's postmortem, 130
cartesian coordinates, 285
chain rule, 97
change of variables in integrals, 253, 257
chaos, 219
closed and bounded interval, 173
compact, 391, 392, 402
compact, 1-D, 173
comparison of series, 504, 514, 515, 526–529
comparison to geometric series, 524
compass directions, 436
composition of functions, 40
compound interest, 501
constrained max-min, 406
continuity, 57, 80
continuity of derivative, 80
cosine derivative, 74, 87
cosine series, 510, 511, 515
critical points, 1-D, 170, 171
critical points, 2-D, 390, 406
cross product, vector, 298

delta and epsilon, 84
derivative of powers, 86
derivative, total, 362, 363, 371
derivatives, partial, 362, 364
differentiable, definition, 68, 70
differentiation of power series, 519
direction field, 434, 436
direction line, 418
distance to a line, 327
distance to a plane, 338

dot product, 294
Duhamel's principle, 269
dynamical systems, continuous, 413
dynamical systems, discrete, 201

epidemic, 15
epidemic DEs, 24, 442, 480, 484
epsilon and delta, 84
equilibrium, continuous, 419, 434
esponential growth, 414
Euler's method of solution, 82, 125, 126
Euler's solution method, 423, 424
exponential decay, 158
exponential derivative, 79, 88
exponential derivatives, other bases, 99
exponential function, 38
exponential growth, 129
exponential series, 509, 511
exponential, natural, 123
exponential, official definition, 125, 126
extreme value theorem, 1-D, 173
extreme value theorem, 2-D, 391

flow of a DE, 437, 440, 441
Fourier series, 493–495, 518, 523, 529
Fundamantal Theorem of Integral Calculus, 242, 525
Fundamental Theorem, Part 2, 245

geometric series, 493, 495, 497, 499, 503–505, 507, 511, 513, 514, 523, 525
gradient, 363, 377, 379, 382
graphing, 144

harmonic series, 523, 527
Hubble's Law, 83
hyperbolic sine and cosine, 520

illegal vectors, 291
implicit curve, 344, 355
implicit differentiation, 109, 187, 400
implicit plane, 323, 336
implicit tangents, 383
increment principle, 69, 71, 362
indefinite integral, 250
infinitesimals, computation rules, 66
initial value problem, 413, 414, 417, 423, 427
integral, definite, 223, 234
integration by parts, 261
integration of series, 515, 516
interval notation, 173